BOILER OPERATOR'S GUIDE

Boiler Construction, Operation, Inspection, Maintenance, and Repair, with Typical Boiler Questions and Answers for Plant, Operating, and Maintenance Engineers

Harry M. Spring, Jr.

Late Boiler Inspector Commissioned by the National Board of Boiler and Pressure Vessel Inspectors, and by Code States; Licensed Chief Engineer; Member of the American Society of Mechanical Engineers

Anthony Lawrence Kohan

Manager, Boiler and Machinery Technical Specialists, Boiler and Machinery Department, Royal Insurance Company; Member of the American Society of Mechanical Engineers; National Board Commissioned Inspector (various state boiler inspector commissions); Certified Safety Professional; Member of the National Society of Professional Engineers; Member of the American Welding Society

Second Edition

McGraw-Hill Book Company

New York St. Louis San Francisco Auckland
Bogotá Hamburg Johannesburg London Madrid Mexico
Montreal New Delhi Panama Paris São Paulo
Singapore Sydney Tokyo Toronto

Library of Congress Cataloging in Publication Data

Spring, Harry Mortimer
Boiler operator's guide.

Bibliography: p.
Includes index.
1. Steam-boilers—Handbooks, manuals, etc.
I. Kohan, Anthony Lawrence, joint author.
II. Title.
TJ289.S6 1981 621.1'94 80–25368

Copyright © 1981 by McGraw-Hill, Inc. All rights reserved.

Copyright 1940 by McGraw-Hill, Inc. All rights reserved.

Printed in the United States of America. Except as permitted under the Copyright Act of 1976, no part of this publication may be reproduced or distributed in any form or by any means, or stored in a data base or retrieval system, without the prior written permission of the publisher.

1234567890 KPKP 8987654321

ISBN 0-07-060511-4

The editors for this book were Robert L. Davidson and James T. Halston, the designer was Blaise Zito Associates, and the production supervisor was Teresa F. Leaden. It was set in Gael by The Kingsport Press.

Printed and bound by The Kingsport Press

CONTENTS

Preface to the Second Edition

The purpose of this book is to provide useful information to the many owners, operators, plant engineers, inspectors, fabricators, repairers, and installers of all types of boilers and boiler auxiliary equipment to help them maintain productive output in their business.

Progress in the boiler and pressure-vessel field has necessitated numerous changes from the first edition. Particular attention has been given to the metallurgy of metals with the increasing use of high-strength steel in boiler and pressure-vessel construction. Also emphasized are welding, nondestructive testing, stress analysis, failure prediction through fracture mechanics, quality control, automatic combustion controls on industrial boilers, and the advent of central control-room operation on large units with the need for adequate instrumentation, alarms, and safety trips.

The American Society of Mechanical Engineers (ASME) and National Board (NB) requirements on construction, installation, and repairs are stressed, as are operating problems involving equipment failures, water treatment, and plant management problems such as qualified personnel, log maintenance, and shift schedules.

Boiler-strength calculations are provided to show how components are designed to provide safe, continuous service under normal operating conditions and how repairs aim to restore the original strength of the damaged part.

A chapter on nuclear reactors has been included in order to demonstrate that certain steam-plant principles also apply to nuclear power plant components. These include ASME Code construction requirements and the maintenance of the highest quality control standards in the fabrication and erection of the pressure-vessel components.

Finally, the questions and answers at the end of each chapter have been retained, changed, and enlarged as needed so that readers can test their knowledge as well as prepare themselves for tests that may be needed for a license or other purpose, such as better employment opportunities.

Many corporations and individuals have provided numerous pictures, illustrations, and information, and this assistance is most gratefully acknowledged. The editors of *Power* magazine have been very helpful, as have the professional societies who publish diverse material on boilers and pressure vessels. These include, among others, the American Society of Mechanical Engineers, American Boiler Manufacturers Association, American Welding Society, National Fire Protection Association, Industrial Risk Insurers, Factory Mutual Engineering, British Engine, National Board of Boiler and Pressure Vessel Inspectors, and numerous others whose assistance is again gratefully acknowledged.

Mention should be made of another text, *Standard Boiler Operators' Ques-*

tions and Answers (by Elonka and Kohan, McGraw-Hill Book Company, 1969), from which some of the material in this text was taken, especially in the questions-and-answers sections. The glossary of boiler terms and definitions given in App. 1 should also be mentioned. Some of these definitions were obtained from a publication entitled "Lexicon—Steam Generating Equipment," published by the American Boiler Manufacturers Association. This lexicon should be referred to for the latest accurate definitions of steam-generating apparatus.

Anthony Lawrence Kohan

Preface to the First Edition

The purpose of this book is to supply a modern handbook specializing on boilers and boiler equipment.

The chapter on construction should be of especial interest to the cadet boiler inspector, to the operating engineer interested in license examinations, and to those wishing to know more of standard shop practices. Methods are shown for calculating safe pressures for many types of boiler.

Succeeding chapters describe installation, design, and distinctive features of a wide variety of boiler from the common fire-tube types to the high-pressure steam generators in this country and abroad.

A chapter on appliances and auxiliaries attempts to give the reader a picture of equipment necessary for safe, economical boiler operation. This is followed by a chapter on plant management, which includes personnel problems, wage incentives, and hourly shift schedules used in many plants. The concluding chapter covers many operating problems of a wide variety of equipment and their practical solutions.

An effort has been made to make the text sufficiently nontechnical to benefit the man starting at the foot of the ladder and yet contain information that will interest and be of value to the man at the top.

The questions and answers at the end of many chapters are based on each common type of boiler and the plant equipment described in the chapter. They are typical of those asked in examinations for firemen's, engineers', and boiler inspectors' licenses by various states. The author disapproves of so-called "trick" questions, but some, representative of those occasionally asked, are included.

The author gratefully acknowledges the generous cooperation of the many manufacturers and individuals who have aided in the preparation of this book.

In addition to the photographs whose sources are acknowledged throughout the pages, greatly appreciated data have been received from the American Iron and Steel Institute; the Babcock and Wilcox Co.; Besler Systems; Brown, Boveri & Co., Ltd.; Cleaver-Brooks Co.; Combustion Engineering Company, Inc.; Crane Co.; The Dow Chemical Company; Elliott Company, Foster Wheeler Corporation; E. Keeler Company; Lukens Steel Company; Morehead Manufacturing Company; Riley Stoker Corporation; Siemens, Incorporated; and from Vitkovice Mines, Steel and Iron Works, Corp.

Special mention is made of the courtesies extended by *Power,* F. A. Annett and S. A. Tucker, associate editors, and L. N. Rowley, assistant editor; and by Frederick G. Straub, Research Associate Professor of Chemical Engineering, University of Illinois; Paul R. Sidler, Electrical Engineer; Arthur J.

Herschmann, agent in United States for Vitkovice Mines, Steel and Iron Works, Corp.; and E. R. Doherty, an authority on steam-boiler design and inspection.

Canton, Mass. Harry M. Spring, Jr.
March, 1940.

Abbreviations and Symbols

A or *a*	area
ASME	American Society of Mechanical Engineers
ASTM	American Society for Testing and Materials
AWS	American Welding Society
Bhn	Brinell hardness number
Btu	British thermal unit
C	carbon
C	coulomb
C	a constant
Ca	calcium
°C	degree Celsius (centigrade)
cm	centimeter
cm^3	cubic centimeter
CO	carbon monoxide
CO_2	carbon dioxide
Code, the	ASME Boiler and Pressure Vessel Codes
Cu	copper
D	diameter of a shell or drum
E	Young's modulus of elasticity = unit stress divided by unit strain (29 million for steel)
°F	degree Fahrenheit
Fe	iron
FS	factor of safety
ft	foot
gal	gallon
gr/gal	grain per gallon (concentration)
g/min	gallon per minute flow
H	hydrogen
H_2O	water
HAZ	heat-affected zone
HTHW	high-temperature hot-water system
HP	horsepower
hr	hour
HRT boiler	horizontal-return-tubular boiler
HS	water heating surface
ID	inside diameter
in.	inch

J	joule
k	a constant
kg	kilogram
kW	kilowatt
L	liter
l or L	length, in inches, unless otherwise specified
lb	pound
max	maximum
Mg	magnesium
$MgSO_4$	magnesium sulphate
min	minimum
min	minute
mm	millimeter
Mn	manganese
NB	National Board of Boiler and Pressure Vessel Inspectors
N	nitrogen
NaOH	sodium hydroxide
$NaSiO_2$	sodium silicate
NDE	nondestructive examination
NDT	nondestructive testing
Ni	nickel
O	oxygen
OD	outside diameter
OH	hydroxide
oz	ounce
p	pitch, in inches, usually of a series of holes
P	maximum allowable working pressure
Si	silicon
SM boiler	scotch marine boiler
SiO_2	silica
SMAW	shielded metal arc welding
SO_4	sulfate
std	standard
t	thickness, in inches, unless otherwise stated
T	thickness, in sixteenths of an inch
temp	temperature
TS	tensile strength
V	vanadium
VT boiler	vertical tubular boiler
W	watt
yd	yard
YP	yield point
%	percent
μm	micrometer (formerly micron)

1

Boiler Systems, Classifications by Use, and Regulations

The need for better-maintained and efficient heat utilization equipment, in order to preserve the earth's limited fuel resources, has become more apparent with developing fuel shortages. In addition to considering the factor of safe, continuous, and reliable service, designers, manufacturers, owners, operators, and maintenance and inspection personnel must now strive for more efficient utilization of boiler equipment. In today's industrial operations, increased fuel costs can erode the profitability of an enterprise. Without foreseeable relief in the near future, it is becoming necessary for industrial managers to more tightly control the use of energy, both technically and economically, through improved operation and maintenance practices. The demands on industrial energy systems, which include boilers, are now even more critical for the following reasons:

1. Energy costs have skyrocketed and often may represent the greatest single component of any plant's operating budget.
2. The impact of government regulations has required tighter control of energy systems.
3. More modern equipment design criteria require cleaner heat-transfer surfaces, and energy costs require the optimization of system cleanliness for all existing as well as new equipment.
4. Costs of unit downtime are very high.

These factors have combined to create a quite different situation from that of the past when energy costs, as well as other costs, were lower and heat equipment operations were not monitored closely.

Before we review boiler types, construction details, stress, operation, and boiler problems, because of the variety of boiler systems, it is appropriate to review briefly such terms as *boilers; steam generators; critical-pressure boilers; low-pressure, high-pressure, steam,* and *hot-water-heating boilers;* and *hot-water-supply boilers;* as well as some jurisdictional requirements.

Many of these terms are referred to in legal terminology drawn up by state and city laws. They are of importance not only to operators, engineers, and maintenance and service people, but also to those in management. Included are people in charge of plant safety; fiscal and legal policy affected by plant insurance, costs, and hazards; and the city or state governments having jurisdiction over plant equipment.

The following definitions of boilers usually are found in state laws and codes on boilers in reference to installation or reinspection requirements as well as engineer-licensing laws for operating this type of equipment.

A *boiler* is a closed pressure vessel in which a fluid is heated for use external to itself by the direct application of heat resulting from the combustion of fuel (solid, liquid, or gaseous) or by the use of electricity or nuclear energy.

A *high-pressure steam boiler* is one which generates steam or vapor at a pressure of more than 15 pounds per square inch gauge (psig). Below this pressure it is classified as a *low-pressure steam boiler.* Small high-pressure boilers are classified as *miniature boilers.*

According to Section I of the Boiler and Pressure Vessel Code of the American Society of Mechanical Engineers (ASME), a *miniature high-pressure boiler* is a high-pressure boiler which does not exceed the following limits: 16-inch (in.) inside diameter of shell, 5-cubic-feet (ft^3) gross volume exclusive of casing and insulation, and 100-psig pressure. If it exceeds any of these limits, it is a *power boiler.* Most states follow this definition. The welding requirements for these small boilers are not as severe as for the larger boilers.

A *power boiler* is a steam or vapor boiler operating above 15 psig and exceeding the miniature boiler size. This also includes hot-water-heating or hot-water-supply boilers operating above 160 psi or 250 degrees Fahrenheit (°F). Power boilers are also called *high-pressure boilers.*

A *low-pressure boiler* is defined as a steam boiler that operates below 15-psig pressure or a hot-water boiler that operates below 160 psig or 250°F.

A *hot-water-heating boiler* is a boiler in which no steam is generated, but from which hot water is circulated for heating purposes and then returned to the boiler, and which operates at a pressure not exceeding 160 psig or a water temperature not over 250°F at or near the boiler outlet. These types of boilers are considered low-pressure heating boilers, built under Section IV of the Heating Boiler Code part of the ASME Boiler Codes. If the pressure or temperature conditions are exceeded, the boilers must be designed as high-pressure boilers under Section I of the Code.

A *hot-water-supply boiler* is completely filled with water and furnishes hot water to be used externally to itself (not returned) at a pressure not exceeding 160 psig or a water temperature not exceeding 250°F. These types of boilers are also considered low-pressure boilers, built to Section IV (Heating Boiler) requirements of the ASME Code. If the pressure or temperature is exceeded, these must be designed as high-pressure boilers.

A *waste-heat boiler* uses by-product heat such as from a blast furnace in a steel mill or exhaust from a gas turbine or by-products from a manufacturing process. The waste heat is passed over heat-exchanger surfaces to produce steam or hot water for conventional use.

The same basic ASME Code construction rules apply to waste-heat boilers as are applied to fired units, and the usual auxiliaries and safety features normally required on a boiler are also required for a waste-heat unit.

Engineers prefer to use the term *steam generator* instead of *steam boiler* because *boiler* refers to the physical change of the contained fluid whereas *steam generator* covers the whole apparatus in which this physical change is taking place. But in ordinary use, both are essentially the same. Most state laws are still written under the old, basic boiler nomenclature.

A *packaged boiler* is a completely factory-assembled boiler, watertube, firetube, or cast-iron, and it includes boiler firing apparatus, controls, and boiler safety appurtenances. A shop-assembled boiler is less costly than a field-erected unit of equal steaming capacity. While a shop-assembled boiler is not an off-the-shelf item, generally it can be put together and delivered much faster than a field-erected boiler; installation and start-up times are substantially shorter. Shop-assembled work usually can be better supervised and done at lower cost.

A *supercritical boiler* operates above the supercritical pressure of 3206.2 pounds per square inch absolute (psia) and 705.4°F saturation temperature. Steam and water have a critical pressure at 3206.2 psia. At this pressure, steam and water are at the same density, which means that the steam is compressed as tightly as the water. When this mixture is heated above the corresponding saturation temperature of 705.4°F for this pressure, dry, superheated steam is produced to do useful high-pressure work. This dry steam is especially well suited for driving turbine generators.

Supercritical pressure boilers are of two types: once-through and recirculation. Both types operate in the supercritical range above 3206.2 psia and 705.4°F. In this range the properties of the saturated liquid and saturated vapor are identical; there is no change in the liquid-vapor phase, and therefore no water level exists, thus requiring no steam drum as such.

Boilers are also classified by the nature of services intended. The traditional classifications are stationary, portable, locomotive, and marine, defined as follows. A *stationary boiler* is installed permanently on a land installation. A *portable boiler* is mounted on a truck, barge, small riverboat, or any other such mobile-type apparatus. A *locomotive boiler* is a specially designed boiler, specifically meant for self-propelled traction vehicles on rails (it is also used for stationary service). A *marine boiler* is usually a low-head-type special-design boiler meant for ocean cargo and passenger ships with an inherent fast-steaming capacity.

The type of construction also distinguishes boilers as follows:

Cast-iron boilers are low-pressure heating units manufactured by casting the pressure components in sections from iron, bronze, or brass. The usual types manufactured are further classified by the manner in which the cast sections are arranged or assembled—by means of push nipples, external headers, and screwed nipples. Three types of cast-iron boilers are:

1. Vertical sectional cast-iron boilers have their sections stacked or assembled vertically one above the other, similar to pancakes, with push nipples interconnecting the sections.
2. Horizontal sectional cast-iron boilers have their sections stacked or assembled horizontally so that the sections stand together like slices in a loaf of bread.
3. Small cast-iron boilers are also built in one-piece, or single, casting. These are generally smaller boilers used primarily in the past for hot-water-supply service.

See Chap. 7 on cast-iron boilers for further details on construction.

Steel boilers can be of the high-pressure or low-pressure type and today are usually of welded construction. They are subdivided into two classes:

1. In *fire-tube boilers*, the products of combustion pass through the inside of tubes with the water surrounding the tubes. Fire-tube boilers are described in detail in later chapters.
2. In *watertube boilers*, the water passes through the tubes, and the products of combustion pass around the tubes.

Fire-tube boilers generally are used for capacity up to 35,000 pounds per hour (lb/hr) and 300-psi pressure; above this capacity and pressure, watertube boilers are used. Fire-tube boilers are classified as shell boilers. Water and steam are confined to a shell. This arrangement limits the volume of steam that can be generated without making the shells prohibitively large, and with respect to pressure, the thickness required would become too expensive to fabricate.

PROPERTIES OF STEAM AND BOILER SYSTEMS

A brief review of some properties of steam will also assist in differentiating boiler systems. A book of steam tables is necessary for computing boiler efficiency. The standard in the United States is *Thermodynamic Properties of Steam* by Keenan and Keyes, published by John Wiley & Sons Inc., New York. For data based on temperature, use Table 1 in Fig. 1-1. Use Table 2 if you know the pressure. All pressures in these tables are absolute. To get absolute pressure, just add 14.7 psi to the gauge pressure (15 psi is close enough).

For properties of superheated steam, use Table 3 in Fig. 1-1. This table of superheated steam must be used with the absolute pressure (gauge pressure plus 15) and with the *total* steam temperature, not the degrees of superheat. This total temperature is the saturation temperature (also given in the table) plus the degrees of superheat.

Enthalpy means the heat content of the fluid. In dealing with water and steam, three enthalpies are to be noted:

1. Enthalpy of saturated liquid [in British thermal units (Btu)], which is the heat content of the water at a certain pressure and temperature under consideration
2. Enthalpy of evaporation (Btu), which is the heat required to evaporate 1 lb of water to steam at that pressure and temperature
3. Enthalpy of saturated vapor (Btu), which is the heat content of the saturated steam at the pressure and temperature being considered

The enthalpy of saturated steam is thus a sum of the enthalpy of saturated liquid and the enthalpy of evaporation, or the *total* heat content of the saturated steam in Btu per pound.

Tables 1 and 2 in Fig. 1-1 give the properties of water and of saturated steam. The only difference is that in Table 1 we enter with the boiler temperature, while in Table 2 we enter with the boiler pressure (psia). For example, Table 1 shows that for water to boil at 100°F, the absolute pressure must be 0.95 psi. Table 2 shows that at 40 psia, water boils at 267°F. It is not necessary to use all the digits given in the table. Most practical work does not require it. Engineers rarely need to figure water temperatures to closer than the nearest degree, or heats or enthalpies to closer than the nearest Btu.

Sat liquid means liquid water at the saturation or boiling temperature; *sat vapor* means steam at the boiling temperature. When water is boiling in a closed container, both the water and the steam over it are in a saturated condition. Steam is saturated when generated by a boiler without a superheater. For steam, *saturated* means steam that contains no liquid water yet is *not* superheated (still at boiling temperature). Note that the absolute pressure is gauge pressure plus about 15 lb. Now, in Table 2, try reading across the line for 50 psia (35 psig).

Boiling temperature is 281°F. At this temperature 1 lb of water fills 0.0713 ft^3 and 1 lb of saturated steam fills 8.51 ft^3. Specific volume is in cubic feet per pound of water or steam. Thus it takes 250 Btu to heat the pound of water from 32°F to the boiling point and another 924 Btu to evaporate it, making a total of 1174 Btu. As mentioned, enthalpy used to be called *heat* in the old steam tables, and it is given in Btu per pound. The last three columns of the old tables were labeled *heat of the liquid, heat of vaporization,* and *total heat.*

Example: A boiler generates saturated steam at 135 psig (150 psia). The enthalpy, or heat of the final steam, is 1194 Btu/lb. The amount of heat required to produce this steam in an actual boiler will depend on the temperature of the feedwater. Suppose the feedwater temperature is 180°F. Table 1 in Fig. 1-1 shows that the heat in the water is 148 Btu. Then the heat supplied to turn this water into steam is merely the difference, or 1194 − 148 = 1046 Btu.

It is easy from this to figure the boiler efficiency. Let us say the boiler

Table 1. Saturation, Temperatures

Temp, °F	Abs press, psi	Specific vol		Enthalphy (heat)		
		Sat liquid	Sat vapor	Sat liquid	Evap	Sat vapor
32	0.08859	0.01602	3304.7	0.01	1075.5	1075.5
40	0.12170	0.01602	2444	8.05	1071.3	1079.3
50	0.17811	0.01603	1703.2	18.07	1065.6	1083.7
60	0.2563	0.01604	1206.7	28.06	1059.9	1088.0
70	0.3631	0.01606	867.9	38.04	1054.3	1092.3
80	0.5069	0.01608	633.1	43.02	1048.6	1096.6
90	0.6982	0.01610	468.0	57.99	1042.9	1100.9
100	0.9492	0.01613	350.4	67.97	1037.2	1105.2
110	1.2748	0.01617	265.4	77.94	1031.6	1109.5
120	1.6924	0.01620	203.27	87.92	1025.8	1113.7
130	2.2225	0.01625	157.34	97.90	1020.0	1117.9
140	2.8886	0.01629	123.01	107.9	1014.1	1122.0
150	3.718	0.01634	97.07	117.9	1008.2	1126.1
160	4.741	0.01639	77.29	127.9	1002.3	1130.2
170	5.992	0.01645	62.06	137.9	996.3	1134.2
180	7.510	0.01651	50.23	147.9	990.2	1138.1
190	9.339	0.01657	40.96	157.9	984.1	1142.0
200	11.526	0.01663	33.64	168.0	977.9	1145.9
212	14.696	0.01672	26.80	180.0	970.4	1150.4
220	17.186	0.01677	23.15	188.1	965.2	1153.4
240	24.969	0.01692	16.323	208.3	952.2	1160.5
280	49.203	0.01726	8.645	249.1	924.7	1173.8
300	67.013	0.01745	6.466	269.6	910.1	1179.7
340	118.01	0.01787	3.788	311.1	879.0	1190.1
380	195.77	0.01836	2.335	353.5	844.6	1198.1
400	247.31	0.01864	1.8633	375.0	826.0	1201.0

Table 2. Saturation, Pressures

Abs press, psi	Temp, °F	Specific vol		Enthalpy (heat)		
		Sat liquid	Sat vapor	Sat liquid	Evap	Sat vapor
0.50	79.58	0.01608	641.4	47.6	1048.8	1096.4
1.0	101.74	0.01614	333.6	69.7	1036.3	1106.0
5.0	162.24	0.01640	73.52	130.1	1001.0	1131.1
10	193.21	0.01659	38.42	161.2	982.1	1143.3
14.7	212.00	0.01672	26.80	180.0	970.4	1150.4
15	213.03	0.01672	26.29	181.1	969.7	1150.8
20	227.96	0.01683	20.089	196.2	960.1	1156.3
25	240.07	0.01692	16.303	208.5	952.1	1160.6
30	250.33	0.01701	13.746	218.8	945.3	1164.1
40	267.25	0.01715	10.498	236.0	933.7	1169.7
50	281.01	0.01727	8.515	250.1	924.0	1174.1
60	292.71	0.01738	7.175	262.1	915.5	1177.6
70	302.92	0.01748	6.206	272.6	907.9	1180.6
80	312.03	0.01757	5.472	282.0	901.1	1183.1
90	320.27	0.01766	4.896	290.6	894.7	1185.3
100	327.81	0.01774	4.432	298.4	888.8	1187.2
110	334.77	0.01782	4.049	305.7	883.2	1188.9
120	341.25	0.01789	3.728	312.4	877.9	1190.4
130	347.32	0.01796	3.455	318.8	872.9	1191.7
140	353.02	0.01802	3.220	324.8	868.2	1193.0
150	358.42	0.01809	3.015	330.5	863.6	1194.1
200	381.79	0.01839	2.288	355.4	843.0	1198.4
250	400.95	0.01865	1.8438	376.0	825.1	1201.1
300	417.33	0.01890	1.5433	393.8	809.0	1202.8
350	431.72	0.01913	1.3260	409.7	794.2	1203.9
400	444.59	0.0193	1.1613	424.0	780.5	1204.5

Table 3. Superheated Steam

Abs pressure, psi (sat temp)	*	Sat liquid	Sat vapor	Temperature, °F							
				300	400	500	600	700	800	900	1000
15	v	0.016	26.29	29.91	33.97	37.99	41.99	45.98	49.97	53.95	57.93
(213.03)	h	181.1	1150.8	1192.8	1239.9	1287.1	1334.8	1383.1	1432.3	1482.3	1533.1
20	v	0.016	20.09	22.36	25.43	28.46	31.47	34.47	37.46	40.45	43.44
(227.96)	h	196.2	1156.3	1191.6	1239.2	1286.6	1334.4	1382.9	1432.1	1482.1	1533.0
40	v	0.017	10.498	11.040	12.628	14.168	15.688	17.198	18.702	20.20	21.70
(267.25)	h	236.0	1169.7	1186.8	1236.5	1284.8	1333.1	1381.9	1431.3	1481.4	1532.4
60	v	0.017	7.175	7.259	8.357	9.403	10.427	11.441	12.449	13.452	14.454
(292.71)	h	262.1	1177.6	1181.6	1233.6	1283.0	1331.8	1380.9	1430.5	1480.8	1531.9
80	v	0.018	5.472	.....	6.220	7.020	7.797	8.562	9.322	10.077	10.830
(312.03)	h	.....	282.10	1183.1	1230.7	1281.1	1330.5	1379.9	1429.7	1480.1	1531.3
100	v	0.018	4.432	.....	4.937	5.589	6.218	6.835	7.446	8.052	8.656
(327.81)	h	298.4	1187.2	.....	1227.6	1279.1	1329.1	1378.9	1428.9	1479.5	1530.8
150	v	0.018	3.015	.....	3.223	3.681	4.113	4.532	4.944	5.352	5.758
(358.42)	h	330.5	1194.1	.....	1219.4	1274.1	1325.7	1376.3	1426.9	1477.8	1529.4
200	v	0.018	2.288	.....	2.361	2.726	3.060	3.380	3.693	4.002	4.309
(381.79)	h	355.4	1198.4	.....	1210.3	1268.9	1322.1	1373.6	1424.8	1476.2	1528.0
300	v	0.0189	1.5433	.....	.....	1.7675	2.005	2.227	2.442	2.652	2.859
(417.33)	h	393.8	1202.8	.....	.....	1257.6	1314.7	1368.3	1420.6	1472.8	1525.2
400	v	0.0193	1.1613	.....	.....	1.2851	1.4770	1.6508	1.8161	1.9767	2.134
(444.59)	h	424.0	1204.5	.....	.....	1245.1	1306.9	1362.7	1416.4	1469.4	1522.4
500	v	0.0197	0.9278	.....	.....	0.9927	1.1591	1.3044	1.4405	1.5715	1.6996
(467.01)	h	449.4	1204.4	.....	.....	1231.3	1298.6	1357.0	1412.1	1466.0	1519.6
600	v	0.0201	0.7698	.....	.....	0.7947	0.9463	1.0732	1.1899	1.3013	1.4096
(486.21)	h	471.6	1203.2	.....	.....	1215.7	1289.9	1351.1	1407.7	1462.5	1516.7
800	v	0.0209	0.5687	.....	.....	.....	0.6779	0.7833	0.8763	0.9633	1.0470
(518.23)	h	509.7	1198.6	.....	.....	.....	1270.7	1338.6	1398.6	1455.4	1511.0
1000	v	0.0216	0.4456	.....	.....	.....	0.5140	0.6084	0.6878	0.7604	0.8294
(544.61)	h	542.4	1191.8	.....	.....	.....	1248.8	1325.3	1389.2	1448.2	1505.1
1200	v	0.0223	0.3619	.....	.....	.....	0.4016	0.4909	0.5617	0.6250	0.6843
(567.22)	h	571.7	1183.4	.....	.....	.....	1223.5	1311.0	1379.3	1440.7	1499.2
1400	v	0.0231	0.3012	.....	.....	.....	0.3174	0.4062	0.4714	0.5281	0.5805
(587.10)	h	598.7	1173.4	.....	.....	.....	1193.0	1295.5	1369.1	1433.1	1493.2

Fig. 1-1 Pressure and temperature relationships of water and steam. Use Table 1 for data based on temperature, Table 2 for data based on pressure, and Table 3 for data on superheated steam.

generates 10 lb steam per pound of coal burned and the coal contains 13,000 Btu/lb. Then, for every 13,000 Btu put in as fuel, there is delivered in steam $10 \times 1046 = 10{,}460$ Btu.

The efficiency of any power unit is its output divided by its input, so here $10{,}460/13{,}000 = 0.805$, or 80.5 percent efficiency.

For most purposes Table 1 in Fig. 1-1 is not needed to get a close value of the heat of the liquid. Just subtract 32 from the water temperature. For example, the enthalpy of water at 180°F is the heat required to raise it from 32 to 180°F, or a difference of 148°F. This takes about 148 Btu. But it will not work out so closely for very high temperatures. Take water at 300°F. Table 1 in Fig. 1-1 gives 269.7 Btu, while our simple method gives $300 - 32 = 268$ Btu, close enough for most purposes.

To use the steam tables for superheated steam, the first column of Table 3 in Fig. 1-1 gives the absolute pressure and (directly below it in parentheses) the corresponding saturation temperature, or boiling temperature. In the next column, v and h stand for volume of 1 lb and its heat content. For example, at 150 psia the volume of 1 lb is 0.018 ft³ for liquid water and 3.015 ft³ for saturated steam. The corresponding heat contents of 1 lb are 330.5 and 1194.1 Btu.

The temperature columns give the volume and heat content per pound for superheated steam at the indicated temperature. Take steam at 150 psi, superheated to a total temperature of 600°F. Look in the 600°F column opposite 150 psi. The volume is 4.113 ft³, as against 3.015 ft³ for saturated steam at the same pressure. This is natural because steam expands as a gas when superheated. Also, the heat content is naturally higher, 1325.7 instead of 1194.1 Btu. Note that this table gives the actual temperature of the superheated steam rather than the degrees of superheat, which is a different thing. If the steam has been superheated from a saturation temperature of 358 to 600°F, the superheat is

$$600 - 358 = 242°\text{F}$$

These superheat tables are used similarly to the saturation tables. Let us take a problem. How much heat does it take to convert 1 lb of feedwater at 205°F into superheated steam at 150 psia and 600°F? The heat in the steam is 1325.7 (1326) Btu. The heat in the water is $205 - 32 = 173$ Btu. Then the heat required to convert 1 lb of steam is $1326 - 173 = 1153$ Btu.

To calculate boiler efficiency, the method is the same as that for finding the efficiency of practically any other piece of power equipment; namely, efficiency is the useful energy output divided by the energy input. For example, if we get out three-quarters of what we put in, the efficiency is ¾, or 0.75 percent. In the case of a boiler unit, we feed in Btu in the form of coal, oil, or gas, and we get out useful Btu in the form of steam. Thus, the first method states that boiler efficiency can be figured directly from the total fuel burned in a given period and the total water evaporated into steam in the same period. It is more common to figure first the evaporation per pound of fuel fired and then, from this, the efficiency. Another method uses data on heat lost up the stack. Figuring this way, we get

$$\text{Boiler efficiency} = \frac{\text{fuel energy input} - \text{energy lost up stack}}{\text{fuel energy input}}$$

STEAM BOILERS FOR POWER GENERATION SYSTEMS

Most utility boilers used for electric power generation are of the supercritical or subcritical type. The steam generator is an important element in power generation. Figure 1-2 is a simplified flow diagram of a basic power plant. Its three most important components are the steam generator (boiler), shown at the left; the turbine generator set, shown coupled together at the right; and the condenser, located beneath the turbine. The principal element that ties together the three pieces of equipment is steam, often called the working medium produced by a high-pressure boiler. The steam travels in succession from the steam generator to the turbine to the condenser. The feedwater cycle, also shown in the diagram, completes this path by making the flow continuous from the condenser back to the boiler. Thus, at the high-temperature end of the cycle, the steam generator transfers heat energy from the fuel to heat energy in the form of superheated steam. The turbine then transfers the heat in the steam to do mechanical work and then to drive the generator which is coupled to the turbine. The generator, in turn, transforms this mechanical energy to electric energy.

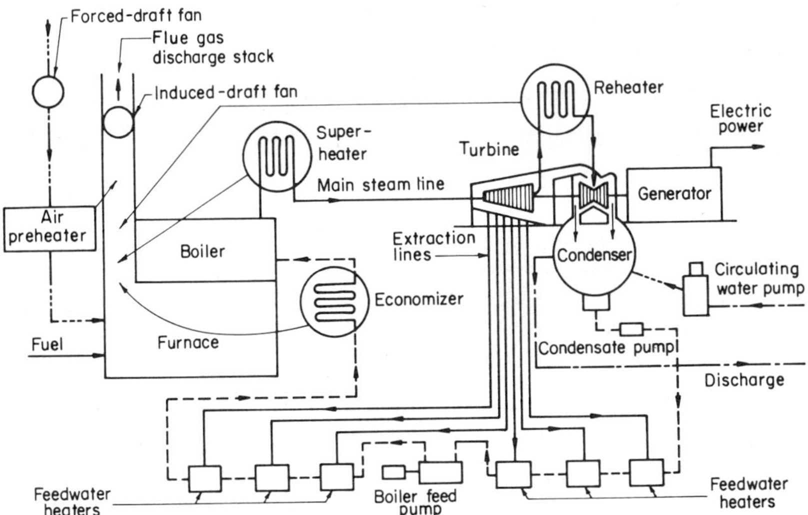

Fig. 1-2 Basic power-plant steam loop encompasses boiler, steam turbogenerator, condenser, feedwater heaters, and boiler feedwater pumps.

By adding auxiliaries and other components such as the heaters, superheaters, reheaters, and preheaters shown in Fig. 1-2, greater efficiencies for modern utility plants can be attained. Steam for electric generation is also produced by heat from a nuclear-powered reactor. In the boiling-water reactor (BWR) system shown in Fig. 1-3*a*, the reactor vessel supports and contains the reactor core and supplies the necessary flow paths for fluid entering the core and steam leaving it. Water passing over the hot core generates steam, which travels through steam-water separators inside the reactor vessel and then through dryers, where the steam's moisture content is reduced. The steam then passes through the steam line directly into the turbine generator, as shown.

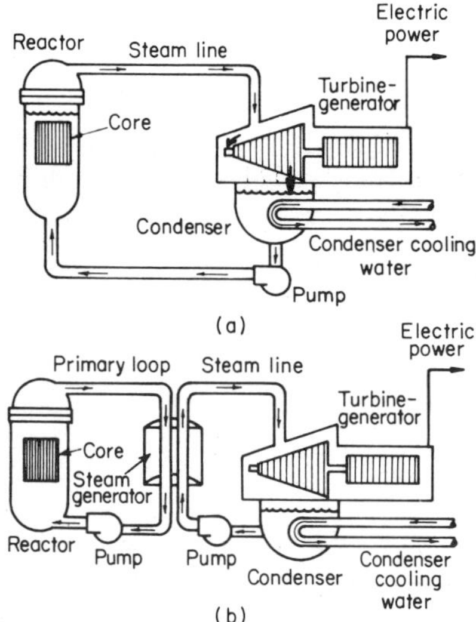

Fig. 1-3 Two types of nuclear-power reactor systems: *(a)* boiling-water reactor (BWR) and *(b)* pressurized-water reactor (PWR).

The pressurized water reactor system shown in Fig. 1-3*b* has a reactor vessel and core somewhat similar to the BWR type, but the fluid passes through the reactor (primary loop) and does not mix with that passing through the steam line on the turbine side. The heat is transferred from the reactor system to the turbine system in the steam generator. Actually, the water in the reactor and primary loop does not boil, even at 600°F, for example, because it is kept under very high pressure. In the steam generator, however, this water passes through tubes that are surrounded by water from the turbine loop, which is at a much lower pressure. To use the same example, a transfer of 500°F to the lower-pressure turbine loop is adequate to boil the water and produce the steam necessary to operate the turbine generator. Having given up most of the heat, the water in the primary loop is pumped back to the reactor to be reheated for use again.

STEAM-HEATING BOILER SYSTEMS

Steam-heating boilers are usually low-pressure units of cast-iron or steel construction, although high-pressure steel boilers may also be used for large buildings or for large, complex areas. Usually if this is done, pressure-reducing valves in the steam lines lower the pressure to the radiators, convectors, or steam coils. The term *steam heating* also generally implies that all condensate is returned to the boiler in a closed-loop system. The maximum pressure allowed on a low-pressure steam-heating boiler is 15 psig.

Cast-iron boilers for steam use are limited to a maximum working pressure (MWP) of 15 psig by the ASME Heating and Boiler Code. Cast-iron boilers are specifically restricted by the ASME Code, Section IV, to be used exclusively for low-pressure steam heating. If they were used for process work, this usually would mean heavy-duty service of continuous steaming and heavy makeup of fresh cold water. This will cause rapid temperature changes in a cast-iron boiler, resulting in cracking of the cast-iron parts. Thus the Code restricts their use to steam-heating service only.

Low-pressure heating boilers usually are operated automatically by on-off or modulating burner controls. Quite often the mistake is made of thinking that low-pressure boilers thus automated are perfectly reliable because they operate as a robot. However, as Fig. 1-4 shows, serious explosions can occur from faulty operation or controls that malfunction. For example, if a limit control fails or is blocked so that it cannot operate when needed, overfiring can result. Overfiring may be caused in numerous ways:

1. Failure of a limit control to stop the burner because of a relay or mechanical defect
2. Mechanical failure of a fuel valve or dirt lodged in a valve so it cannot close
3. Burner on manual operation with no one watching the temperature
4. Residual heat with coal firing, with no one watching the temperature
5. Burner considerably oversized in relation to the boiler and the system (also if demand is mild on a day of use with pump not operating)
6. Wiring short, causing controls to be bypassed
7. Fusing of contacts on a stop-go switch into the go position
8. Solenoid or air-operated valves isolating the boiler from the load because of mechanical or electrical defect of the controls on the solenoid or on the air stop-go device

Steam-heating systems use gravity or mechanical condensate-return systems. Their differences are as follows. When all the heating elements (such as radiators, convectors, and steam coils) are located above the boiler and no pumps are used, it is called a *gravity return,* for all the condensate returns to the boiler by gravity. If traps or pumps are installed to aid the return of condensate, the system is called a *mechanical return system.* In addition to traps, this system usually includes a condensate tank, a condensate pump, or a vacuum tank or vacuum pump.

Fig. 1-4 A 72-in. scotch marine, low-pressure steam-heating boiler explosion caused property damage.

Minimum protective devices required on steam-heating boiler systems are also outlined in the ASME Heating Boiler Code. Among the most prominent are the following:

1. Each steam-heating boiler must have a steam pressure gauge with a scale in the dial graduated to not less than 30 psi nor more than 60 psi. Connections to the boiler must be not less than ¼-in. standard pipe size; but if steel or wrought-iron pipe is used, it should be not less than ½ in.

2. Each steam-heating boiler must have a water gauge glass attached to the boiler by valve fittings not less than ½ in. and with a drain on the gauge glass not less than ¼ in. The lowest visible part of the gauge glass must be at least 1 in. above the lowest permissible water level as stipulated by the boiler manufacturer.

3. Two pressure controls are required on automatically fired steam-heating boilers:

   (a) An operating-pressure cutout control that cuts off the fuel supply when the desired operating pressure is reached.

   (b) An upper-limit control set no greater than 15 psi which backs the

operating-pressure limit control so that the fuel is shut off when the operating-pressure control does not function.

4. An automatically fired steam-heating boiler must have a low-water fuel cutoff located so that the device will cut off the fuel supply when the water level drops to the lowest visible part of the water gauge glass. Low-water fuel cutoffs must be connected to the boiler with nonferrous tees on Y's not less than ½-in. pipe size and must also have ¾-in. drains if embodying a chamber for the low-water fuel-cutoff device, so that the chamber and connected piping can be flushed of sludge periodically. This drain also permits testing of the low-water fuel cutoff as the level in the chamber drops during blowdown.

5. Each steam-heating boiler must have at least one safety valve of the spring-loaded pop type, adjusted and sealed to discharge at a pressure not greater than the maximum allowable pressure of the boiler. No safety valve can be smaller than ½ in. or greater than 4½ in. The capacity of the safety valves must exceed the output rating in pounds per hour of the boiler, but in no case should the capacity be less, so that with the fuel-burning equipment firing at maximum capacity, the pressure cannot rise 5 psi above the stamped maximum allowable pressure of the boiler.

6. All electric control circuitry on automatically fired steam-heating boilers must be positively grounded and operate at 150 volts (V) or less. The wiring system must include a grounded neutral as well as equipment grounding.

7. Automatically fired steam-heating boilers must be equipped with flame safeguard safety controls as mentioned in the controls for hot-water-heating boilers.

Stop valves on the steam supply line are required for a single-boiler installation that is used for low-pressure heating, if there are not other restrictions in the steam and condensate line and all condensate is returned to the boiler. But if a stop valve (or trap) is placed in the condensate-return line, a valve is required on the steam supply line. A stop valve is required on the steam supply line where more than one heating boiler is used on the same steam supply system and also on the condensate-return line to each boiler.

HOT-WATER SYSTEMS

There are three general classes of hot-water systems: hot-water supply systems for washing and similar uses, space-heating systems of the low-pressure type, often referred to as building heating systems, and high-temperature high-pressure water systems, also referred to as supertherm systems, operating at temperatures of over 250°F and pressures of over 160 psi.

Both the hot-water-heating system and the high-temperature hot-water systems require some form of expansion tank in order to permit the water to expand as heat is supplied, without a corresponding increase in pressure. A common problem of hot-water-heating systems is that expansion tanks lose their air cushion, so that the water system can no longer expand without raising the pressure of the system. If this problem is neglected, pressure can build up to the point where the relief valve may open and dump water in the property. Thus periodic draining of the expansion tank is necessary to re-establish the air cushion.

The ASME Heating Boiler Code requires some minimum protective devices on hot-water-heating boiler systems. Among these are the following:

1. A pressure or altitude gauge is required on the hot-water boiler with a scale on the dial graduated to not less than 1½ times nor more than 3 times the pressure at which the relief valve is set.

2. A thermometer gauge is needed on the hot-water boiler that is located and connected so that it can be read when the pressure or altitude on the boiler is noted. Graduation of the thermometer must be in degrees Fahrenheit, and the thermometer must be located so that the water temperature in the boiler is measured at or near the outlet of the heated hot water.

3. Two temperature controls are required in automatically fired hot-water boilers:

   (a) An operating limit control that cuts off the fuel supply when the water temperature reaches the desired operating limit.

   (b) An upper-limit control that backs up the operating-limit control and cuts off the fuel supply. This upper-limit control is set at a temperature above the desired operating temperature, but must be set so that the water temperature cannot exceed 250°F at the boiler outlet.

4. A low-water fuel cutoff is required on automatically fired hot-water boilers with heat inputs greater than 400,000 Btu per hour (Btu/hr). It must be installed so that it cuts off the fuel when the water level drops below the safe, permissible water level established by the boiler manufacturer.

5. All electric control circuitry on automatically fired hot-water boilers as well as one steam-heating boilers must be positively grounded and operate at 150 V or less. The wiring system must include a grounded neutral as well as equipment grounding.

6. A hot-water-heating boiler must be equipped with spring-loaded ASME-approved relief valves set at or below the maximum stamped allowable pressure of the boiler. The minimum size of valve is ¾ in., and the maximum permitted size is 4½ in. Capacity must be greater than the stamped output of the boiler, but in no case should the pressure rise more than 10 percent above the maximum allowable pressure if the fuel-burning equipment operates at maximum capacity.

7. Automatically fired hot-water-heating boilers and steam-heating boilers must also be equipped with flame safeguard safety controls that cut off the fuel when an improper flame (or combustion) exists by the burner. The ASME Code makes reference to other nationally recognized standards for further requirements. These usually include pilot and main-flame proving, as well as prefiring and postfiring purging cycles.

Officially rated ASME pressure-relief valves must be used in hot-water boilers. An officially rated ASME pressure-relief valve is stamped for its pressure setting and its Btu-per-hour relieving capacity. Also, it must be equipped with a manual test lever, must be spring-loaded, and must not be of the adjustable screw-down type.

Low water can occur in a hot-water-heating type of boiler for numerous reasons, such as the following: (1) Loss of water due to carelessness in *(a)* draining the boiler for repair or summer lay-up without eliminating the possibility of firing, *(b)* drawing hot water from the boiler; (2) loss of water in the distribution system because of *(a)* leaks in the piping caused by expansion breakage or corrosion, *(b)* leaks in the boiler, *(c)* leaks through the pump or other operating equipment; (3) relief-valve discharge caused by overfiring; (4) closed or stuck city makeup line.

In addition to an ASME pressure safety relief valve, a low-water fuel cutoff for an automatic-fired boiler should be installed.

Boiler-Output Rating Terminology Boiler output can be expressed in horsepower, pounds per hour, Btu per hour, and, for utility boilers, the capability of generating so many megawatts of electricity. Heating boilers can also be rated in horsepower, pounds per hour, and Btu per hour, but their output is also described in terms related to heat-transfer area needed for a space. For example, *equivalent square feet of steam radiation surface* is a measure of the heat-transfer area needed in a room that will use steam as a heat source.

A *boiler horsepower* (boiler HP) is defined as the evaporation into dry saturated steam of 34.5 lb/hr of water at a temperature of 212°F. Thus 1 boiler HP by this method is equivalent to an output of 33,475 Btu/hr, and was commonly taken as 10 square feet (ft^2) of boiler heating surface. But 10 ft^2 of boiler heating surface in a modern boiler will generate anywhere from 50 to 500 lb/hr of steam. Today the capacity of larger boilers is stated as so many pounds per hour of steam, or Btu per hour, or megawatts of power produced.

The term *heating surface* is also used to define or relate to the output of a boiler. The heating surface of a boiler is the area, expressed in square feet, that is exposed to the products of combustion. The following surface parts of boilers must be considered in determining the amount of heating surface that may be available for producing steam or hot water: tubes, fireboxes, shell surfaces, tube sheets, headers, and furnaces.

A comparison of output ratings based on horsepower, heating surface, and pounds per hour can be made by assuming a boiler has a nominal horsepower rating of 500 HP.

1. The heating-surface rating would be 5000 ft^2 under the old rule of 10 ft^2/HP.
2. The steam output in pounds per hour would be

$$500 \times 34.5 = 17{,}250 \text{ lb/hr}$$

3. For a hot-water-heating boiler, the output would be

$$500 \times 33{,}475 = 16{,}737{,}500 \text{ Btu/hr}$$

The pounds-per-hour rating often guaranteed by the manufacturer system of rating boilers is a measure of the capacity at which a boiler can be operated continuously. The peak output of a boiler for a 2-hr period is usually set 10 to 20 percent above the maximum continuous output. The pounds-per-hour rating usually is expressed in pounds of steam at the design temperature and pressure for the boiler. Low-pressure boilers are also rated by heating contractor code requirements as well as pounds per hour or Btu per hour.

In heating-load calculations, the terms IBR-rated, SBI-rated, and EDR are often used. These terms affect the output rating of a boiler. Thus they are important in sizing a boiler for heating a certain size space. They also affect the safety valve required on a boiler. They are defined as follows:

The acronym IBR stands for the Institute of Boiler and Radiator Manufacturers, which rates cast-iron boilers. Usually IBR-rated boilers have a nameplate indicating net and gross output in Btu per hour. Gross output is further defined as the net output plus an allowance for starting, or pickup load, and a piping heat loss. The net output will show the actual useful heating effect produced. The ASME Code states that it is the gross heat output of the equipment that should be matched in specifying relief-valve capacity.

The acronym SBI stands for the Steel Boiler Institute. The nameplate data shown on SBI-rated boilers are not uniform, but the style or product number may be shown. The manufacturer's catalog will often show an SBI rating and an SBI net rating. The SBI rating tends to show the sum of SBI net ratings and 20 percent extra for piping loss, not including the pickup allowances noted under IBR ratings. Thus, it is difficult to obtain the true gross output to determine safety relief capacity from these data. But the SBI does require the number of square feet of heating surface to be stamped on the boiler. With this, the ASME rule of minimum steam safety-valve capacity in pounds per hour per square foot of heating surface is used.

And, EDR stands for equivalent direct radiation. Specifically it refers to equivalent square feet of steam radiation surface. It is further defined as a surface which emits 240 Btu/hr with a steam temperature of 215°F at a room temperature of 70°F. With hot-water heating, the value of 150 Btu/hr is used with a 20°F drop between inlet and outlet water. This term is used by architects and heating engineers in determining the area of heat-transfer equipment required to heat a space. Thus boiler capacity is obtained indirectly from a summation of the EDRs.

The following ratings are also often noted on heating boiler specifications:

American Gas Association Rating This rating method is used by the American Gas Association (AGA) and is applied to boilers designed for gas firing. The rating is expressed as maximum boiler output in Btu per hour, and it reflects 80 percent of the AGA-approved input rating as determined by performance tests described in the "American Standard Approval Requirements for Central Heating Appliances." For all practical purposes, AGA output ratings are equivalent to gross SBI and gross IBR ratings.

Mechanical Contractors Association Rating The Mechanical Contractors Association (MCA) of America (formerly the Heating, Piping, and Air Conditioning Contractors National Association) has adopted methods for rating boilers that are expressed on a net-load basis in square feet of EDR of steam.

The MCA has also adopted a Testing and Rating Code for Boiler-Burner Units which they apply to the rating of commercial sizes of steel heating boiler units fired with oil or gas fuel. This code allows a higher rating than is permissible under the SBI Code. A gross output is established with certain limiting factors applying to flue-gas temperature, carbon dioxide, efficiency, and quality of steam. This output is divided by 1.5 to determine the net MCA rating.

American Boiler Manufacturers Association Rating This rating method, developed by the Packaged Firetube Branch of the American Boiler Manufacturers Association (ABMA), is generally subscribed to by manufacturers of packaged boilers and by a few manufacturers of steel firebox and cast-iron boilers. The ratings are established by performance tests in accordance with the ASME Power Test Code for Steam Generating Units and are usually expressed as maximum guaranteed Btu output at the outlet nozzle or similar output rating.

Codes and Regulations The ASME Boiler Code and the National Board of Boiler and Pressure Vessel Inspectors Inspection Codes are important source documents for legal requirements in the various states and municipalities that have adopted boiler safety laws. In addition to maintaining active boiler and pressure-vessel committees in order to keep the published Codes up to date with developing technology, the ASME issues to qualified manufacturers, assemblers, material suppliers, and nuclear power plant owners Code symbol stamps indicating that the manufacturer has received authorization from the ASME to build boilers and pressure vessels to the ASME Code.

A fundamental principle of the ASME Boiler and Pressure Vessel Code is that a boiler or pressure vessel, to be stamped ASME Code-designed, must receive third-party authorized inspection during construction for compliance with the prevailing Code requirements. Most third-party inspections are performed by authorized boiler and pressure-vessel inspectors who have appropriate experience and have passed a written examination in a jurisdiction. They must be employed either by the state or by an insurance company licensed to write boiler and pressure-vessel insurance in the jurisdiction where the boiler or pressure vessel is to be built, and in some cases the installation's location also must be considered. With uniform requirements for inspectors that have been promoted and implemented by the National Board of Boiler and Pressure Vessel Inspectors, a boiler or pressure vessel inspected by a properly credited National Board inspector will generally be accepted in all jurisdictions.

The manufacturer or contractor who wishes to build or assemble boilers or pressure vessels under an ASME certificate of authorization must first agree with an authorized inspection agency that Code inspections will be performed by the agency. This is usually arranged by both parties signing a contract with the inspection work done on a fee basis.

The National Board of Boiler and Pressure Vessel Inspectors is composed of chief inspectors of states and municipalities in the United States and Canadian provinces. This organization has established criteria for boiler inspectors' experience requirements, the promotion and conductance of uniform examinations, and testing that are used by the jurisdictions. The National Board issues commissions to inspectors passing an NB examination, which are accepted on a reciprocal basis by most jurisdictions, thus providing a "portability" feature to a credential.

Most areas of the United States and all jurisdictions in Canada require that high-pressure boilers be subjected to periodic inspection by an authorized inspector. In most jurisdictions, this consists of annual internal inspection of power boilers and biennial inspection of heating boilers and usually of pressure vessels for those states that have adopted laws on low-pressure boilers or unfired pressure vessels. If the results prove satisfactory, the jurisdiction issues an inspection certificate, authorizing use of the vessel for a specific period.

There are three types of inspectors who make the legal inspections and reports to a jurisdiction that a boiler is safe or unsafe to operate or that it requires repairs before it can be operated:

1. State, province, or city inspectors see that all provisions of the boiler and pressure-vessel law, and all the rules and regulations of the jurisdiction, are observed. Any order of these inspectors must be complied with, unless the owner or operator petitions (and is granted) relief or exception.
2. Insurance company inspectors are qualified to make ASME Code inspections. If commissioned under the law of the jurisdiction where the unit is located, they can also make the required periodic reinspection. As commissioned inspectors, they require compliance with all the provisions of the law and rules and regulations of the authorities. In addition, they may recommend changes that will prolong the life of the boiler or pressure vessel.
3. Owner-user inspectors are employed by a company to inspect unfired pressure vessels for direct use and not for resale by such a company. They also must be qualified under the rules of any state or municipality which has adopted the Code. Most states do not permit this group of inspectors to serve in lieu of state or insurance company inspectors.

Figures 1-5 and 1-6 list the states, cities, and counties in the United States that have some form of installation and periodic reinspection requirements on boilers and some unfired pressure vessels. These laws vary a great deal. For example, on low-pressure boilers, reinspection requirements may be limited to installations located in places of public assembly. Others include all heating boilers, except those located in private residences or in apartment houses with six families or less. Therefore local or state laws should be checked for more specific requirements.

The same principle applies to operating-engineer licensing laws listed in

Fig. 1-5 States having boiler and pressure-vessel reinspection laws.

State	Accept insurance company reports (X = yes)	Require inspection for: High-pressure boilers	Low-pressure boilers	Unfired pressure vessels
Alaska	X	X	X	X
Alabama	No law	—	—	—
Arizona	X	X	X	—
Arkansas	X	X	X	X
California	X	X	—	X
Colorado	X	X	X	—
Connecticut	X	X	X	—
Delaware	X	X	X	—
District of Columbia	X	X	X	—
Florida	No law	—	—	—
Georgia	No law	—	—	—
Hawaii	X	X	X	X
Idaho	X	X	X	X
Illinois	X	X	X	—
Indiana	X	X	X	X
Iowa	X	X	X	X
Kansas	X	X	—	—
Kentucky	X	X	X	—
Louisiana	X	X	X	X
Maine	X	X	X	—
Maryland	X	X	X	X
Massachusetts	X	X	X	X
Michigan	X	X	X	—
Minnesota	X	X	X	X
Mississippi	X	X	X	X
Missouri	No law	—	—	—
Montana	Do not	X	X	—
Nebraska	X	X	X	X
New Mexico	No law	—	—	—
Nevada	X	X	X	X
New Hampshire	X	X	X	X
New Jersey	X	X	X	X
New York	X	X	X	—
North Carolina	X	X	X	X
North Dakota	X	X	X	—
Ohio	X	X	X	—
Oklahoma	X	X	—	—
Oregon	X	X	X	X
Pennsylvania	X	X	X	X
Rhode Island	X	X	X	—
South Carolina	No law	—	—	—
South Dakota	X	X	X	—
Tennessee	X	X	X	X
Texas	X	X	X	—
Utah	X	X	X	—
Vermont	X	X	X	X
Virginia	X	X	X	X
Washington	X	X	X	X
West Virginia	X	X	—	—
Wisconsin	X	X	X	X
Wyoming	No law	—	—	—

Fig. 1-6 Cities and counties having boiler and pressure-vessel reinspection laws.

City or county	Accept insurance company reports (X = yes)	High-pressure boilers	Low-pressure boilers	Unfired pressure vessels (UPV)
Albuquerque, N.Mex.	X	X	X	—
Buffalo, N.Y.	X	X	X	—
Chicago, Ill.	No	X	X	—
Dearborn, Mich.	X	X	X	X
Denver, Colo.	No	X	X	X
Des Moines, Iowa	X	X	X	—
Detroit, Mich.	UPV only	X	X	X
E. St. Louis, Mich.	No	X	X	X
Greensboro, N.C.	X	X	X	X
Kansas City, Mo.	X	X	X	X
Los Angeles, Calif.	X	X	X	X
Memphis, Tenn.	X	X	X	X
Miami, Fla.	X	X	X	X
Milwaukee, Wisc.	X	X	X	X
New Orleans, La.	X	X	X	X
New York City, N.Y.	X	X	X	—
Oklahoma City, Okla.	X	X	X	—
Omaha, Neb.	X	X	X	—
Phoenix, Ariz.	X	X	X	X
St. Louis, Mo.	X	X	X	X
San Francisco, Calif.	X	X	X	X
San Jose, Calif.	X	X	X	—
Seattle, Wash.	X	X	X	X
Spokane, Wash.	X	X	X	X
Tacoma, Wash.	X	X	X	X
Tampa, Fla.	X	X	X	X
Tucson, Ariz.	X	X	X	X
Tulsa, Okla.	No	X	X	X
University City, Mo.	No	X	X	—
White Plains, N.Y.	X	X	X	—
Arlington County, Va.	X	X	X	—
Dade County, Fla.	X	X	X	X
Fairfax County, Va.	X	X	X	X
Jefferson Parish, La.	X	X	X	X
St. Louis County, Mo.	X	X	X	X

Fig. 1-7. Local requirements vary quite a bit on experience needed, grades, and degree of responsibility. It is suggested that the appropriate jurisdiction be contacted for further details.

Environmental Regulations Technology plays a central role in maintaining the standard of living to which society has become accustomed and which can affect the daily life of the average citizen. Some of these impacts have been negative—for example, major power blackouts, noise near jetports, pollution of air and water resources, etc. In response to some of these negative aspects, the public attitudes toward the social value of technology have been

Fig. 1-7 Jurisdictions having operating-engineers' licensing laws for boilers.

Jurisdiction	High-pressure boilers	Low-pressure boilers
U.S. cities and counties		
Buffalo, N.Y.	X	—
Chicago, Ill.	X	—
Dearborn, Mich.	X	X
Denver, Colo.	X	X
Des Moines, Iowa	X	X
Detroit, Mich.	X	X
E. St. Louis, Ill.	X	X
Kansas City, Mo.	X	X
Los Angeles, Calif.	X	X
Memphis, Tenn.	X	X
Miami, Fla.	X	X
Milwaukee, Wis.	X	X
New Orleans, La.	X	X
New York City, N.Y.	X	—
Oklahoma City, Okla.	X	X
Omaha, Neb.	X	X
St. Joseph, Mo.	X	X
St. Louis, Mo.	X	X
San Jose, Calif.	X	—
Spokane, Wash.	X	X
Tacoma, Wash.	X	X
Tampa, Fla.	X	X
Tulsa, Okla.	X	X
University City, Mo.	X	X
White Plains, N.Y.	X	—
Jefferson Parish, La.	X	X
St. Louis Co., Mo.	X	X
States		
Alaska	X	X
Arkansas	X	X
District of Columbia	X	X
Massachusetts	X	—
Minnesota	X	X
Montana	X	X
Nebraska	—	X
New Jersey	X	X
New York	X	X
Ohio	X	X
Pennsylvania	X	X
Canadian provinces		
Alberta	X	X
British Columbia	X	X
Manitoba	X	X
New Brunswick	X	X
Newfoundland and Labrador	X	X
N.W. Territory	X	X
Nova Scotia	X	—
Ontario	X	X
Quebec	X	X
Saskatchewan	X	X
Yukon Territory	X	X

Note: Due to variations in the laws, it is necessary to check the jurisdiction for specific requirements on licensed operators.

changing, and government and private groups have become more actively involved in questioning and even suggesting restraining the advancement of technology. As a result, some new criteria for acceptance have been added to technology. For example, safety from chronic effects, such as the long-term effects of radiation or the long-term exposure to potential carcinogen-type products or toxic materials, must be evaluated; and the disposal of both waste and used products that may be potentially harmful to the environment requires attention.

Thus, fuel-burning systems for boilers, and nuclear-energy systems, must be designed and then operated and maintained so that air pollution and waste disposal will have minimal effects on the environment. As a minimum they also must comply with legal requirements on established threshold limits for air and water pollution, as well as radiation levels in nuclear plant applications.

Legal requirements on boilers and nuclear power plant equipment no longer are limited to establishing safe construction codes. They have been expanded into requirements on controls, on devices to prevent furnace explosions, and on measurements to limit air pollution and radioactive contamination. Owners and operators must periodically review their operation and maintenance practices in order to make sure they comply with these additional legal requirements of the jurisdiction in which the equipment is located.

Questions and Answers

1-1 How would you define a boiler?

Ans. A boiler is a closed pressure vessel in which a fluid is heated for use external to itself by the direct application of heat resulting from the combustion of fuel (solid, liquid, or gaseous) or by the use of electricity or nuclear energy.

1-2 What is a steam boiler?

Ans. A steam boiler is a closed vessel in which steam or other vapor is generated for use external to itself by the direct application of heat resulting from the combustion of fuel (solid, liquid, or gaseous) or by the use of electricity or nuclear energy.

1-3 What is a high-pressure steam boiler?

Ans. A high-pressure steam boiler generates steam vapor at a pressure of more than 15 psig. Below this pressure it is classified as a low-pressure steam boiler.

1-4 Define a miniature high-pressure boiler.

Ans. According to Section I of the ASME Boiler and Pressure Vessel Code, a miniature boiler is a high-pressure boiler which does not exceed the following limits: 16-in. inside diameter of shell, 5-ft^3 gross volume exclusive of casing and insulation, and 100-psig pressure. If it exceeds any of these limits, it is a power boiler. Most states follow this definition.

1-5 What is a power boiler?

Ans. A power boiler is a steam or vapor boiler operating above 15 psig and exceeding the miniature-boiler size. This also includes hot-water-heating or hot-water-supply boilers operating above 160 psi or 250°F.

1-6 Define a hot-water-heating boiler.

Ans. A hot-water-heating boiler is a boiler used for space hot-water heating, with the water returned to the boiler. It is further classified as low-pressure if it does not exceed 160 psi or 250°F. But if it exceeds any of these, it becomes a high-pressure boiler.

1-7 What is a hot-water-supply boiler?

Ans. A hot-water-supply boiler furnishes hot water to be used externally to itself for washing, cleaning, etc. If it exceeds 160 psi or 250°F, it becomes a high-pressure power boiler.

1-8 What is meant by a boiler horsepower?

Ans. A boiler horsepower (boiler HP) is defined as the evaporation into dry saturated steam of 34.5 lb/hr of water at a temperature of 212°F. Thus one boiler HP by this method is equivalent to an output of 33,475 Btu/hr. In the past it was commonly taken as 10 ft^2 of boiler heating surface.

1-9 The symbol NB is often noted on boilers, with a number following it. What does this stand for?

Ans. The acronym NB stands for National Board of Boiler and Pressure Vessel Inspectors. It means that the boiler's design and fabrication were followed in the shop by an NB-commissioned inspector, including the witnessing of the hydrostatic test and signing of data sheets required by the ASME.

1-10 What is meant by heating surface in a boiler?

Ans. This is the (fireside) area in a boiler exposed to the products of combustion. This area is usually calculated on the basis of areas on the following boiler-element surfaces: tubes, fireboxes, shells, tube sheets, and projected area of headers. See later chapters on safety-valve calculations.

1-11 Define the terms IBR rate, SBI-rated, and EDR.

Ans. The acronym IBR stands for the Institute of Boiler and Radiator Manufacturers, which rates the output of cast-iron boilers in net and gross output in Btu per hour. Gross output is further defined as the net output plus an allowance for starting, or pickup load, and a piping heat loss.

The acronym SBI stands for the Steel Boiler Institute. The SBI boiler-output rating tends to show the sum of SBI net ratings in Btu per hour or pounds per hour, plus 20 percent extra for piping loss, not including the pickup allowances noted under IBR ratings. The SBI requires the number of square feet of heating surface to be stamped on the boiler.

The acronym EDR stands for equivalent direct radiation. It specifically refers to equivalent square feet of steam radiation surface. It is further defined as a surface which emits 240 Btu/hr with a steam temperature of 215°F at a room temperature of 70°F. With hot-water heating, the value of 150 Btu/hr is used with a 20°F drop between inlet and outlet water. This term is used by architects and heating engineers in determining the area of heat-transfer equipment required to heat a space.

1-12 Name three terms used to indicate boiler output.

Ans. These three terms are often used with pressure and temperature listings:

1. For steam boilers, the actual evaporation is in *pounds per hour.* For hot-water boilers, the *Btu-per-hour* outputs for the given pressures and temperatures are stamped on the boiler. Today this is the preferred method.
2. Square feet of heating surface.
3. Boiler horsepower.

1-13 What is a supercritical once-through boiler?

Ans. This is a boiler which operates above the supercritical pressure of 3206.2 psia and 705.4°F saturation temperature and which has no fluid recirculation when operating at full pressure and temperature. The fluid is brought up to pressure and temperature in series-connected fluid passes; thus the term *once-through* is applied.

1-14 Above what pressure and temperature does a hot-water boiler system become a high-temperature hot-water boiler system?

Ans. A hot-water boiler becomes a high-temperature hot-water (HTHW) system when the water temperature exceeds 250°F and the pressure is over 160 psi.

1-15 What is the main reason for using thermal fluids such as dowtherm and glycol in processes requiring heat?

Ans. The thermal fluids are used to get high temperatures at low pressures which may be difficult to obtain with ordinary steam boiler equipment. Note a pressure of 3206.2 psia is needed to obtain a saturation steam temperature of 705.4°F. This temperature can be obtained with some thermal fluids at a pressure below 50 psi.

1-16 Name three pressure-limiting devices needed on a steam-heating boiler per ASME Code requirements.

Ans. The three pressure-limiting devices are:

1. An operating-pressure cutoff switch that automatically cuts off the fuel supply when the desired pressure is reached
2. An upper-limit pressure control switch, set no greater than 15 psi, that automatically cuts off the fuel supply when the upper pressure is reached
3. At least one spring-loaded pop-type safety valve, set and sealed to discharge at a pressure not greater than the maximum allowable working pressure of the boiler and with a capacity sufficient that the pressure on the boiler cannot rise 5 psi above the stamped maximum allowable pressure on the boiler

1-17 What is the output in pounds per hour and Btu per hour of a boiler rated at 750 HP?

Ans. Output = 750(34.5) = 25,875 lb/hr
Output = 750(33,475) = 25,116,250 Btu/hr

1-18 What causes water hammer and how can it be corrected?

Ans. Water hammer is the passage of high-velocity slugs of water through a steam pipe. The impact of the slugs on pipe bends or elbows can cause pipe failures. Slugs or droplets of water are caused by pockets in the steam lines that are not adequately drained and by opening steam valves too rapidly. Providing adequate drains at low points in the steam lines will prevent pockets of water. Valves on cold steam lines should be opened slowly.

1-19 When are double stop valves required on the steam lines connecting a boiler to another boiler's line, and what are the requirements for these valves?

Ans. Boilers connected to common headers must have two stop valves with a free-blow drain valve between the valves. The valve next to the boiler should be a nonreturn type, and the second stop valve should be of the outside-screw-and-yoke type.

1-20 What is the purpose of the free-blow drain valve when double stop valves are used to connect boilers in battery to a common steam header?

Ans. The free-blow drain will tell an operator if a valve is leaking before the operator enters a drum for maintenance and inspection work. This avoids scalding of the operator.

1-21 What is the minimum size of safety valve allowed on a cast-iron boiler to be used for steam-heating or hot-water-heating service?

Ans. ASME rules require a minimum ¾-in. safety valve for both steam-heating and hot-water-heating cast-iron boilers.

1-22 Name two undesirable conditions which should be evaluated when steam-pipe systems are supported.

Ans. Vibration and expansion.

1-23 Calculate boiler efficiency, using the steam generated versus the fuel consumed. You are given that for one calendar month of regular operation, the coal consumed is 682,000 lb and the steam generated is 6.4 million lb at 179 psig and superheated to a total temperature of 520°F.

Ans. First, the actual evaporation per pound of coal fired is

$$\frac{6{,}400{,}000}{682{,}000} = 9.40 \text{ lb steam/lb coal}$$

Assume that the heat content is 13,260 Btu/lb of coal as fired. Remember that this pound of coal produces 9.4 lb of steam.

The absolute steam pressure is

$$170 + 15 = 194 \text{ lb abs}$$

Then, the steam tables show that the total heat of 1 lb of steam at 194-lb absolute pressure and 520°F is 1280.4 Btu. Assuming that the feedwater temperature is 208°F, its heat content above water at 32°F is merely 208 −

32 = 176 Btu. Thus the heat put into each pound of steam produced by the boiler is 1280.4 − 176.0 = 1104.4 Btu. The heat put into 9.4 lb of steam will be 10,381 Btu. Then, boiler efficiency equals heat put into 9.4-lb steam divided by the heat in 1 lb of coal, or

$$\frac{10{,}381}{13{,}260} \times 100\% = 78.3\% \text{ efficiency}$$

1-24 Does a cast-iron boiler require a bottom blowoff pipe and valve?

Ans. Yes. The ASME Code requires each boiler to have a blowoff pipe connection fitted with a valve or cock, of not less than ¾-in. pipe size. It must be connected with the lowest water space practicable.

1-25 What is the minimum size of pipe required for connecting a water column to a steam-heating boiler?

Ans The minimum size of ferrous or nonferrous pipe must be of 1-in. diameter.

1-26 What *hydrostatic test* is required on hot-water or hot-water-supply boilers built of cast iron and operating over a pressure of 30 psi?

Ans. Each section of a cast-iron boiler must be subjected to a hydrostatic test of 2½ times the maximum allowable pressure *at the shop* where it is built. Cast-iron boilers marked for working pressures over 40 psi must be subjected to a hydrostatic test of 1½ times the maximum allowable pressure in the field (when erected and ready for service). After the boiler is in service and a hydrostatic test is required, the test shall be at 1½ times the maximum allowable pressure.

1-27 What *stamping* does the ASME require on cast-iron boilers?

Ans. The marking must consist of the following: manufacturer's name, maximum allowable pressure in pounds per square inch, and capacity in pounds per hour for steam or Btu per hour for water service.

2
The Metallurgy of Steel and Material Selection

Most pressure parts of boilers are made from ferrous alloys, which signifies that they are manufactured from iron and iron-base alloys. Except for cast-iron boilers and some special heating boilers, all other boilers usually have steel parts that are joined by fusion welding. There is an increased demand for higher service pressures which requires better material and fabrication technology. This is being achieved by sophisticated alloying, vacuum degassing in steel-making processes, multiple heat treatment, extensive nondestructive testing, and high quality-control standards, to name a few of the methods being used to obtain better material. Quality control of the welding procedure used and control of the welding operation have been materially improved by greater emphasis being placed on more stringent specifications, inspection, and documentation requirements.

MATERIAL STRUCTURE

If a piece of metal is carefully polished, immersed for a short time in an acid or other appropriate reagent, and then examined under a microscope, it will be found to be composed of small particles or crystals. The metal, instead of being perfectly uniform, is built up of these small units of matter.

Materials which appear to be perfectly homogeneous in reality are composed of an aggregate of grains or crystals of distinctly different materials. A piece of the material will therefore have different properties at different points within the piece. Even pure metals are made up of an aggregate of crystals having different properties in different directions.

One convenient method for explaining some of the similarities and differences in the characteristics of metals is to study the arrangement of atoms in the material.

Space Lattice A crystal of a given metal is composed of atoms arranged in a definite and regular geometric pattern. This pattern is known as a space lattice and is determined by x-ray studies. The atoms in iron at *room temperature,* for example, are arranged in a body-centered cubic lattice. That is, the atoms are located as at the corners of a cube, with one atom in the center, as indicated in Fig. 2-1*a*. This pattern repeats itself throughout a single crystal,

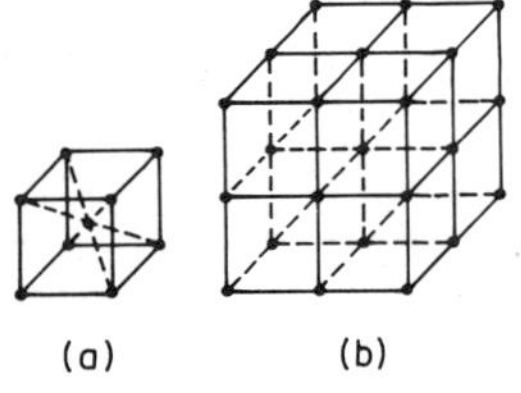

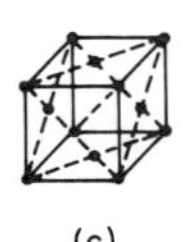

Fig. 2-1 Atoms in metals are arranged in space lattices as follows: *(a)* body-centered cubic lattice, *(b)* unit cubes in crystal, *(c)* face-centered cubic lattice.

as indicated in Fig. 2-1*b*. Two adjacent crystals in a bar of iron will have the same lattice formation, but their orientation, or the directions of the axes of the space lattice, will be different. Chromium, vanadium, molybdenum, and tungsten also have the body-centered cubic lattice.

Iron at high temperature crystallizes in the face-centered cubic lattice, or with an atom at each corner of a cube and an atom in the center of each face, as illustrated in Fig. 2-1*c*. There are a total of 14 lattice systems.

A pure metal crystallizes in cooling through its freezing point; if a nucleus or seed is present, crystallization consists of growth of crystals around a nucleus at the freezing point of the metal. Once a nucleus is established on a pure liquid metal at or below the freezing point, other atoms from the molten liquid start to attach to the nucleus. The atoms deposit along defined directions of the crystal, or space lattices.

A study of space-lattice or crystal formations that are created when metals are cooled to freezing or solidification temperatures assists in determining the effects of alloying, rate of cooling, and similar metallurgical considerations. Iron, for example, is transformed to various crystal types, and the corresponding space-lattice arrangements for these types are as follows:

Type of iron	*Type of crystal structure*
Alpha, delta, and ferrite iron	Body-centered cubic (BCC) lattice
Gamma and austenite iron	Face-centered cubic (FCC) lattice

Alpha iron reaches the BCC structure below 1670°F.
Delta and ferrite iron with the BCC structure exist above 2534°F.
Gamma and austenite iron with the FCC structure exist between 1670 and 2534°F.

At room temperature, iron is composed of a body-centered cubic lattice (see Fig. 2-2). In this form it is known as alpha iron or alpha ferrite, and it is soft, ductile, and magnetic. Upon heating above about 1415°F, alpha iron loses its magnetism, but retains its body-centered crystalline structure. This structure changes to face-centered cubic at about 1670°F, at which temperature alpha iron is transformed to gamma iron and remains nonmagnetic. In continuing upward in temperature, another phase change occurs at 2570°F, when delta iron is formed. The latter is identical in crystal structure (body-centered) to that of the low-temperature alpha iron. It is magnetic and is stable to the melting point. There are no known phase changes in the liquid form (above about 2800°F). On cooling very slowly from the liquid state, the atomic rearrangements described above occur in reverse order.

A difference of 2 or 3 percent in the carbon content, a difference in heat treatment, and a difference in the amount of mechanical working during the shaping process may cause the ultimate tensile strength to vary from 20,000 to 300,000 psi. The elongation at failure may be varied from 50 to 0.1 percent, and the elastic strength may be varied from 10,000 to 270,000 psi.

The variations in strength, ductility, and other properties of an iron-carbon alloy may be explained by referring to a typical equilibrium diagram for a portion of the iron-carbon system, as shown in Fig. 2-3. The temperature

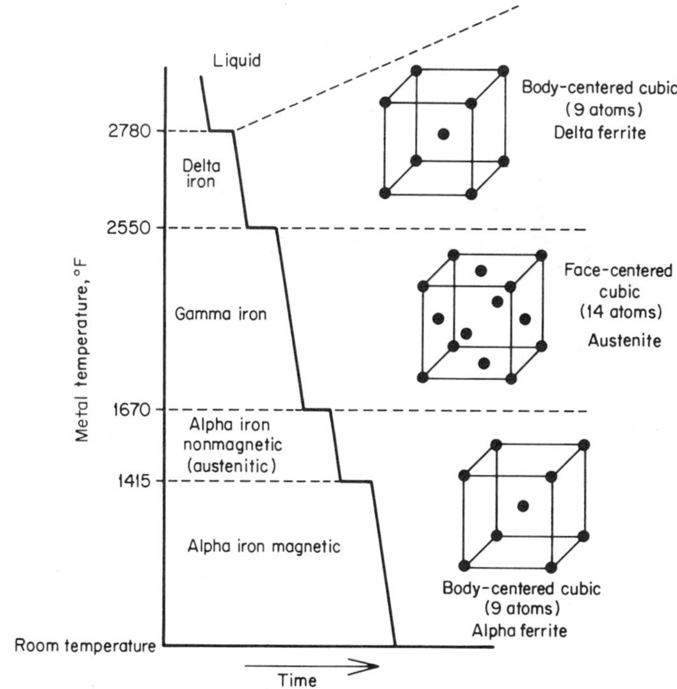

Fig. 2-2 Crystal forms of iron at different temperatures.

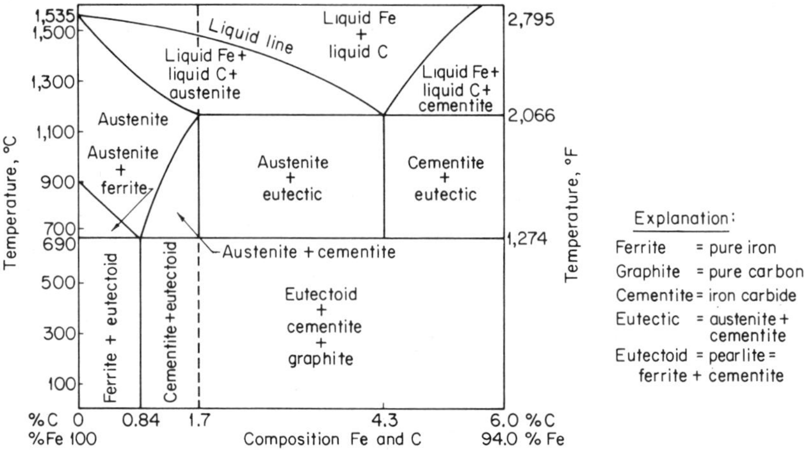

Fig. 2-3 Iron-carbon equilibrium diagram.

range is illustrated from 0 to 1535°C (2795°F), the melting point of pure iron. The range of carbon contents is from 0 to 6 percent, representing the range of materials used. Alloys containing more than 6 percent carbon are too weak and brittle. Wrought iron contains between 0 and 0.12 percent carbon; steel, between 0.12 and 1.7 percent; and cast iron, over 1.7 percent.

Figure 2-3 is a combination of the solid-solution, eutectic, and eutectoid types. It presents only the condition of slow cooling that would occur normally in air and is only for combinations of iron and carbon without other alloying elements. An increase in the rate of cooling of the material by immersing it in oil or water, or the addition of an alloying element such as nickel, silicon, or chromium, will change the diagram.

The properties of the iron-carbon alloys at room temperature are dependent not only on the percentage of carbon but also on the size of the crystals and on the degree of dispersion of the carbon. Small crystals and highly dispersed carbon tend to increase the strength and hardness. The size of the crystals also may be controlled by the shaping operations used in fabricating the member, and both the size of the crystals and the degree of dispersion of the carbon may be controlled by modifying the rate of cooling or by subjecting the material to additional cycles of heating and cooling after the initial cooling to room temperature (heat treatment).

Transformation Temperature In carbon steels, only three phases can be present in solid state under equilibrium conditions: *Austenite, ferrite,* and *cementite.* The temperature interval within which austenite forms on heating, and also the temperature interval within which austenite disappears on cooling, is called the transformation-temperature range. The transformations on heating do not occur at the same temperatures as on cooling, except at almost infinitely slow rates of temperature change. The temperatures of transformation are dependent on carbon content in accordance with the equilibrium diagrams.

Effects of Reheating In general, the effects of reheating a metal are the reverse of those obtained in slow cooling. For example, a steel containing 0.20 percent carbon and consisting of pearlite and ferrite is reconverted to austenite as the alpha iron is transformed to gamma iron in the critical range. However, for this steel, the temperatures of transformation are about 30°C or 86°F higher for heating than for slow cooling. Changes in the carbon content and the presence of other alloying elements affect the temperature differential.

Because they indicate the effects of different cooling rates, isothermal and continuous transformation diagrams are sometimes used by metallurgists in predicting the results of various welding processes. In the arc welding of steel, the structure may range from ferrite to martensite, depending on welding conditions. In spot welding, large quantities of martensite are obtained, and a postweld treatment is essential if the carbon content is higher than approximately 0.25 percent. A transformation diagram will indicate, approximately, the type of microstructure in the zone adjacent to the weld. The type of microstructure desired will indicate also the most suitable welding process to use.

MANUFACTURE OF CAST IRON AND STEEL

Blast Furnace The first step in the production of cast iron and steel is the extraction of the iron from the ore. This is accomplished in the blast furnace. The blast furnace is a vertical, tubular steel chamber lined with refractory. The furnace may be 5 to 25 ft in diameter and up to 100 ft high. See Fig. 2-4. It is charged from the top with iron ore, flux, and coke. The usual flux is some form of limestone. When the "charge" is fired up, the coke burns at extremely high temperatures—up to 3600°F. The iron is melted out of the ore and flows to the bottom of the furnace. The molten iron is poured into molds, where it solidifies into "pigs." Pig iron contains a high percentage of carbon, which causes low ductility. Consequently, further refinement is necessary before the physical and chemical properties of pig iron are satisfactory for its use in boiler construction.

Cast iron is produced by remelting pig iron and pouring it into molds of the desired shape. The purpose of the melting is to reduce the amount of impurities and to secure a more uniform product than would be obtained by casting the pig iron directly as it came from the blast furnace. There are two types of furnaces in general use for the remelting of pig iron. In the production of most of the ordinary gray cast iron, a cupola is used, while an air furnace, known as a reverberatory furnace, is more commonly used for the better grades of gray cast iron and the cast iron which is to receive heat treatment.

Cast-Iron Types and Properties Compared with steel, cast iron is decidedly

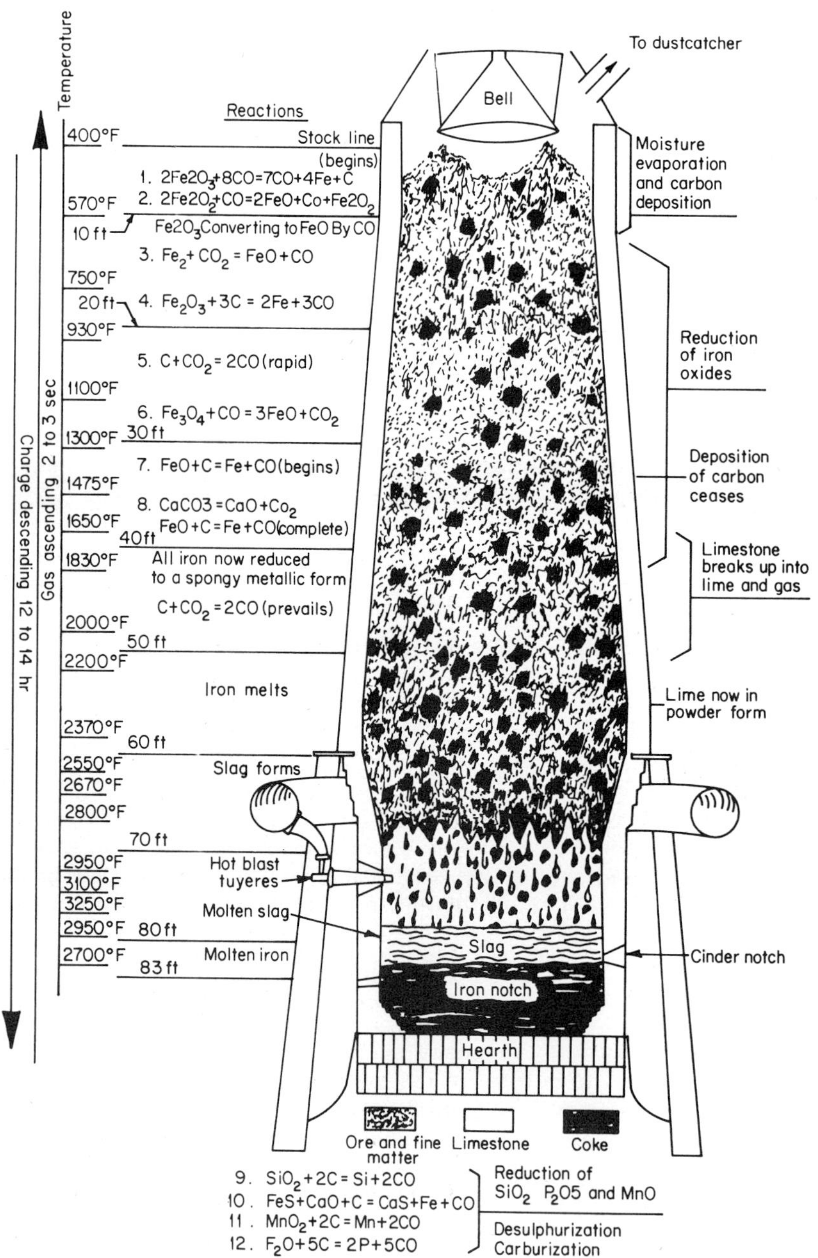

Fig. 2-4 Blast furnace. *(Courtesy CF&I Steel Corp.)*

inferior in malleability, strength, toughness, and ductility. The most important types of cast iron are the white and the gray cast irons.

White Cast Iron White cast iron is so known because of the silvery luster of its fracture. In this alloy, the carbon is present in combined form as iron carbide (Fe_3C), known metallographically as cementite.

Gray Cast Iron Gray iron is the most widely used of cast metals. In this iron, the carbon is in the form of graphite flakes that form a multitude of notches and discontinuities in the iron matrix. The appearance of the fracture of this iron is gray because the graphite flakes are exposed. The strength of the iron increases as the graphite-crystal size decreases and the amount of cementite increases. Gray cast iron is easily machinable because the graphite carbon acts as a lubricant for the cutting tool and also provides discontinuities which break the chips as they are cast. Gray iron, having a wide range of tensile strength, from 20,000–30,000 to 90,000 psi, can be made by alloying with nickel, chromium, molybdenum, vanadium, and copper. Permissible stresses per the ASME Code, Section IV, are as follows:

Class	*Ultimate tensile strength, kilopounds per square inch (kips/in.^{2})*	*Allowable stress in tensile strength, kips/in.^{2}*
20	20.0	4.0
25	25.0	5.0
30	30.0	6.0
35	35.0	7.0
40	40.0	8.0

Physical Defects Some of the most common physical defects which may be present in cast iron and which will weaken it are blowholes, cracks, segregation of the impurities, and coarse-grain structure.

Sulfur, which combines with the manganese or the iron to form a sulfide, makes the cast iron brittle (hot short) at high temperatures. It also increases shrinkage. Hence, its amount is usually limited to less than 0.1 percent in specifications for cast iron.

Phosphorus, in amounts of more than 2 percent, makes the iron brittle and weakens it. However, it also has the effect of increasing the fluidity and decreasing the shrinkage. So it is desirable in making sharp castings for ornamental parts where strength is unimportant.

MANUFACTURE OF STEEL

The production of steel from pig iron involves the removal of as much of the impurities as is practicable, the adjusting of the carbon content to the desired value, and the addition of such alloying as may be required to alter the properties.

Acid and Basic Processes The steel-making processes may be classified as acid or basic, according to the nature of the charge. Pig iron normally contains phosphorus, which tends to give the charge an acid reaction. If no basic materials are added and if the lining of the furnace is acid or neutral, the charge remains acid throughout the process, and the phosphorus will be retained in the steel, making it brittle at ordinary temperatures. However, if limestone is added to the charge, it will react with most of the phosphorus in the iron to form a material which is retained in the slag. The addition of limestone results in a basic charge requiring a basic lining for the furnace, and this produces a less brittle metal.

The manufacture of steel starts with pig iron. Pig iron is transformed into steel by the oxidation of the impurities with air, oxygen, or iron oxide, since the impurities combine more readily with oxygen than does iron. Pig iron may be refined by oxidation alone in the acid process or by oxidation in conjunction with a strong basic slag in the basic process. Carbon, silicone, and manganese are removed by both processes, but phosphorus and some of the sulfur in the pig iron can be removed only in the basic process. Phosphorus stays persistently with the iron if the slag is strongly acidic or when it is high in silica and therefore is not removed. The principal methods of manufacturing steel for subsequent rolling or forging are the basic-oxygen process, the basic open-hearth process, and the basic electric-furnace process. Cast steel can be made by the above methods and also by the acid electric and acid open-hearth methods. The furnace charge is steel scrap and pig iron in both the basic-oxygen and basic open-hearth processes and selected steel scrap in the electric-furnace process. Prereduced iron pellets may also be used in these processes as part of the charge.

Basic-Oxygen Process The basic-oxygen process uses a basic-lined cylindrical furnace with a solid bottom and an open top. The furnace is tilted to receive the charge which usually consists of 30% iron scrap and the balance hot pig iron. The furnace is returned to the upright position, and an oxygen lance is lowered into the opening at the top to start the melting cycle.

Open-Hearth Process Another process of steel manufacture is the open-hearth method, shown in Fig. 2-5. The chemical properties of steel made by this process are more accurately proportioned. Open-hearth steel is required for steam-boiler parts under pressure. Briefly, the open hearth is a bowl-shaped container which holds the molten pig iron. A mixture of gas and air burning at high temperature is blown onto its surface continuously until the carbon and other impurities have burned out. Then the desired elements are added in correct proportion and amount. The advantages of the open-hearth method are that less metal is lost, there is less oxidation due to surface combustion, and a somewhat more accurate control of chemical properties is possible.

Basic Process The basic open-hearth process is carried out in a large regenerative furnace lined with basic refractories of magnesite and/or dolomite.

Acid Process The acid open-hearth process is essentially a melting process, and, as in all acid processes, the removal of phosphorus and sulfur is impossible. It is necessary to use a charge low in these impurities to produce satisfactory

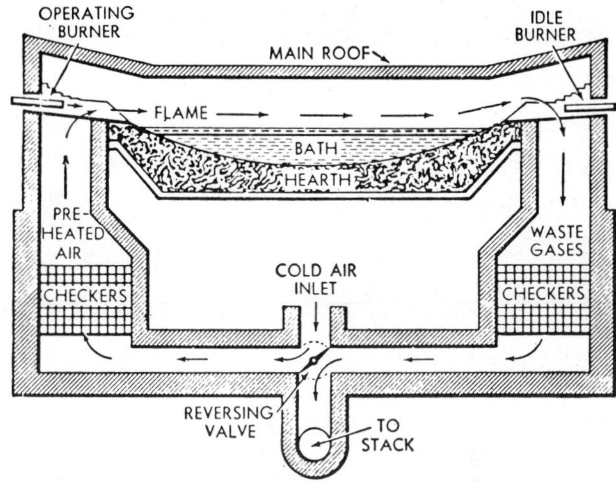

Fig. 2-5 Cross-sectional view of an open-hearth furnace.

steel. The smaller heats usually made in the acid process and the higher cost restrict the use of acid open-hearth steel to castings and special products. Most of the production is killed carbon and alloy steels.

Electric-Furnace Process High-grade alloy steels are usually manufactured in the electric furnace (see Fig. 2-6) because the electric-furnace process offers the following advantages over the other processes:

1. Desired temperatures can be obtained because of good control of temperature within close limits.
2. Oxidizing, reducing, or neutral conditions can be maintained in the furnace. This allows the addition of alloying elements to the melt with better quality.
3. The melt is free of contamination from burning fuel.
4. Sulfur and solid nonmetallic impurities, as well as occluded gases, are significantly eliminated.

The control of quality and composition in the electric furnace is better than in the open-hearth process, but the cost is somewhat higher.

There are two general types of electric furnaces: the arc type and the induction type. The arc type consists of a well-insulated chamber in which the charge is placed and into which the carbon electrodes extend, inserted in the charge. Either single-phase or three-phase current may be used. The induction furnace consists of an insulated chamber around which is wound a coil. The charge is placed in the chamber, and usually high-frequency current is passed through the surrounding coil, which functions as the primary winding of a transformer, with the charge itself serving as the secondary. The eddy currents induced in the charge develop sufficient heat to melt it.

Some steel is melted under vacuum. This procedure holds dissolved gases

Fig. 2-6 Electric-arc furnace discharging a 150-ton melt. *(Courtesy Lukens Steel Co.)*

such as oxygen, nitrogen, and hydrogen to a minimum and significantly reduces the number of nonmetallic inclusions.

Shaping of Steel As the steel comes from the basic-oxygen furnace, open-hearth furnace, or electric furnace, it may be cast directly into the desired shape or into ingots weighing from 3 to 10 tons. After solidification, the ingots are given a preliminary shaping by being rolled or forged into billets, which may then be reduced to final dimensions by rolling, forging, drawing, or other operations. Each of the operations will affect the crystalline structure, thereby changing the properties of the final product. Figure 2-7 shows typical stress-

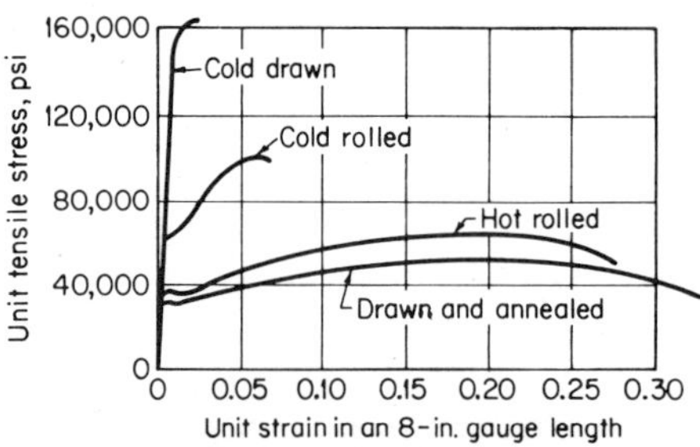

Fig. 2-7 Effect of shaping operations influences ultimate strength of carbon steel.

Fig. 2-8 A 140-in. rolling mill processing ingots into plate steel. *(Courtesy Lukens Steel Co.)*

strain diagrams for specimens of low-carbon steel shaped by some of the more common manufacturing methods.

Casting is widely used for the production of complex shapes which would involve large amounts of waste if other shaping techniques were employed. Rolling is used for the production of plates, sheets, rods, and the common structural shapes such as angles, channels, and I beams. Figure 2-8 is a photograph of a rolling mill in action. Most of the steel shapes formed by rolling are hot-rolled; but where dimensions must be held to relatively close tolerances, cold rolling is employed. This process gives the steel a shiny surface, at a higher strength than does the hot-rolled process. Forging is commonly used for the production of thick plate or special formed shapes such as rotors of steam turbines.

The temperature at which steel is mechanically worked has a profound effect on its properties. Cold work increases the hardness, tensile strength, and yield point, but its indices of ductility-elongation and reduction of area are decreased.

Seamless tubes for boilers are made by piercing hot, solid ingot pieces in special piercing machines especially designed for the purpose. Some cold drawing is performed where tubes require finer finish or close tolerance. Some seamless tubes are now made by the hot extrusion press method.

Electric-resistance-welded steel tubing is made by forming flat strips into a tubular shape called a skelp and then welding the edges together by an electric-resistance welding machine. Compared with seamless tubes, electric-

Fig. 2-9 Hot spinning of steel plate to form a concave head. *(Courtesy Lukens Steel Co.)*

resistance-welded boiler tubes have a smoother surface, a more uniform wall thickness, and less eccentricity on the outside diameter of the tubes.

Figure 2-9 shows a hot steel plate being formed into a head on a hot spinning machine at a steel company. Heads are used as end closures for boiler drums, unfired pressure vessels, nuclear reactors, and all types of petroleum and chemical process vessels.

Impurities from Steel Manufacture The strength, ductility, and related properties of the iron-carbon alloys can be affected by the presence of harmful elements such as sulfur, phosphorus, oxygen, hydrogen, and nitrogen.

Sulfur Sulfur has the same effect on steel as on cast iron, making the metal hot-short, or brittle at high temperatures. As a result, it may be harmful in steel which is to be used at elevated temperatures or, more particularly, may cause difficulty during hot-rolling or other shaping operations. Most specifications for steel limit its amount to less than 0.05 percent.

Phosphorus Phosphorus makes steel cold-short, or brittle at low temperatures; so it is undesirable in parts which are subjected to impact loading when cold. However, it has the beneficial effect of increasing both fluidity, which tends to make hot rolling easier, and the sharpness of castings. Since cast iron is brittle anyway, phosphorus is sometimes added to make the castings clean-cut. Most specifications for structural steel limit the phosphorus content to less than 0.05 percent.

Oxygen When iron is in the liquid state, any free oxygen combines readily with iron to form iron oxide. In the finished steel or iron, the iron oxide usually appears in the form of tiny inclusions distributed throughout the metal. These inclusions introduce points of weakness and increased brittleness, concentration from voids, and stress, which are undesirable, for they promote the formation of cracks that may result in progressive fracture.

Hydrogen Hydrogen, such as is produced when steel is immersed in sulfuric acid to remove mill scale before cold drawing, makes steel brittle. The hydrogen may be removed by heating the steel for a few hours, or hydrogen will gradually work out of the steel at ordinary temperatures. When present, hydrogen increases the hardenability of the steel.

Nitrogen Nitrogen has a hardening and embrittling effect on steel. This may be objectionable, or it may be desirable in producing a hard surface on the steel, under controlled manufacturing conditions. For example, in the nitriding process, the steel is exposed to ammonia gas at about 600°C (1112°F) to produce a hard, wear-resisting surface without lowering the ductility of the center of the piece.

Nonmetallic Inclusions Nonmetallic inclusions occur during the plate-rolling operation in steel making. Quite often, these inclusions cause laminations, or planes in the plate where metal separation or voids exist.

The quality of steel plate can be significantly affected by nonmetallic inclusions. The presence of inclusions, such as sulfides and oxides, primarily affects the ductile behavior of the steel. The quality of a particular grade of steel can be improved by eliminating or minimizing inclusions. Nondestructive testing is used extensively for critical-pressure applications, such as nuclear reactors, in order to find inclusions.

Alloying Elements on Steel The purpose of adding alloying elements to carbon steel is to impart to the finished product desirable physical or chemical properties which are not available in carbon-steel parts fabricated by standard procedures. These properties could involve desirable electrical, magnetic, or thermal characteristics, as well as engineering considerations such as (1) high tensile strength or hardness without brittleness; (2) resistance to corrosion; (3) high tensile strength or high creep limit at elevated or subzero temperatures; (4) other desirable physical features required to resist special loading.

Carbon Up to 1.2 percent carbon in iron increases the strength and ductility of steel. When the carbon content is above 2 percent, graphite formation is promoted, which lowers both the strength and the ductility of steel. Carbon content above 5 or 6 percent causes the metal to be brittle with very low strength for any load-resistance application.

Manganese By combining with sulfur, this element prevents the formation of iron sulfide at the grain boundaries. This minimizes surface ruptures at steel-rolling temperatures (red shortness) which results in significant improvement in surface quality after rolling.

Nickel This element increases toughness or resistance to impact. In this respect it is the most effective of all the common alloying elements in improving low-temperature toughness.

Chromium This element contributes to corrosion resistance and heat resistance in alloy steels. A strong carbide-former, chromium is frequently used in carburizing grades and in high-carbon-bearing steels for superior wear resistance. An 18 percent chromium and 8 percent nickel (18-8 nickel-chrome) alloy is widely used as a high-strength stainless steel in pressure equipment requiring strengths and corrosion resistance.

Molybdenum Like nickel, molybdenum does not oxidize in the steel-making process, a feature which facilitates precise control of hardenability. It markedly improves high-temperature tensile and creep strength and reduces a steel's susceptibility to temper brittleness.

Boron Since boron does not form a carbide nor strengthen ferrite, a particular level of hardenability can be achieved without an adverse effect on machinability and cold formability that may occur with the other common alloying elements.

Aluminum In amounts of 0.95 to 1.30 percent, aluminum is used in nitriding steels because of its strong tendency to form aluminum nitride, which contributes to high surface hardness and superior wear resistance.

Silicon Silicon combines with carbon to form hard carbides which, when properly distributed throughout the alloy, have the effect of increasing the elastic strength without loss of ductility.

Tungsten Tungsten forms hard stable carbides when it is added to steel. It raises the critical temperature, thus increasing the strength of the alloy at high temperatures.

Vanadium This element acts as a deoxidizing agent as aluminum does on molten steel. It forms very hard carbides, thus increasing the elastic and tensile strength of low-carbon and medium-carbon steels.

Other elements may be added to improve the rolled strength and toughness of steels in the high-strength low-alloy category.

Heat Treatment of Steels When steel is heated to certain temperatures and then rapidly or slowly cooled, its physical properties such as elastic limit, ultimate strength, and hardness can be changed.

Heat treatments fall into two general categories: those which increase strength, hardness, and toughness by quenching and tempering and those which decrease hardness and promote uniformity by slow cooling from above the transformation range or by prolonged heating within or below the transformation range, followed by slow cooling.

Annealing consists of heating steel to a certain temperature and cooling it by a relatively slow process. Annealing may be used to remove stresses such as are produced in forgings and castings, to refine the crystalline structure of steel, or to alter the ductility or toughness of steel.

Stress-relief annealing involves heating to a temperature approaching the transformation range, holding for a sufficient time to achieve temperature uniformity throughout the part, and then cooling to atmospheric temperature. The purpose of this treatment is to relieve residual stresses induced by normalizing, machining, straightening, or cold deformation of any kind. Some softening and improvement of ductility may be experienced, depending on the temperature and time involved.

Normalizing involves heating to a uniform temperature about 100 to 150°F above the transformation range, followed by cooling in still air.

Hardening consists of heating steel to above its transformation temperature range and cooling it suddenly by quenching in water, oil, or some other cooling medium that absorbs heat rapidly.

Quenching is defined as a process of rapid cooling from an elevated temperature, by contact with liquids, gases, or solids. Quenching increases the hardness of steel if its carbon content is 0.20 percent or higher. It also raises the elastic limit and ultimate strength and reduces the ductility; however, it induces internal stresses, and the metal is apt to become brittle.

Properties of materials for engineering structures are generally described under the following headings:

1. *Physical.* These would describe the material's composition, structure, homogeneity, specific weight, thermal conductivity and ability to expand and contract, and resistance to corrosion.
2. *Formability.* These would relate to the manufacture of the material, such as fusibility (welding), forgeability, malleability, and ability to shape the material by bending or machining.
3. *Mechanical.* Mechanical properties describe the ability of a material to resist applied loads and are usually obtained from tests.

The basic mechanical properties are elastic limits, moduli of elasticity, ultimate strengths, endurance limits, and hardness. Secondary mechanical characteristics determined from the basic ones or simultaneously with them are resilience, toughness, ductility, and brittleness.

Strength depends on the type and nature of loading. The static strength of a material is expressed by the corresponding elastic limit stress. The impact strength is measured by the corresponding modulus of resilience. The endurance strength is expressed by the corresponding endurance limit. See later chapters on stress, pressures, and forces.

Qualities other than strength are also important.

Hardness is a relative characteristic. There are several methods of measuring it, all of an arbitrary nature. The Brinell hardness number (Bhn) is obtained as follows: A hardened steel ball 10 millimeters (mm) in diameter is pressed under a certain load, F kilograms (kg), into the smooth surface of the material to be tested; the diameter D of the indentation is measured in millimeters, and the depth h is calculated from it. The hardness number, Bhn, is then expressed as

$$\text{Bhn} = \frac{F}{10\pi h}$$

In using the Shore scleroscope, a small cylinder of steel with a hardened point is allowed to fall on the smooth surface of the material, and the height of the rebound of the cylinder is taken as the measure of hardness.

The hardness number obtained with the Rockwell instrument is based on the additional depth to which a test point is driven by a heavy load beyond the depth to which the same penetrator has been driven by a definite, lighter load.

Fig. 2-10 Carbon, low-alloy, and high-alloy steel plate, allowable stresses per Section I of ASME Code.

Spec no.	Designation	Ultimate tensile strength, kips/in.²	Representative allowable stresses not exceeding metal temperature, °F	
			−20 to 650	800
A Carbon steels—plate				
SA 285A	Carbon, C	45.0	11.3	8.3
SA 285B	Carbon, C	50.0	12.5	9.0
SA 285C	Carbon, C	55.0	13.8	10.2
SA 442 Gr55	C-MN-Si	55.0	13.8	10.2
SA 515 Gr55	C-Si	55.0	13.8	10.2
SA 516 Gr55	C-Si	55.0	13.8	10.2
SA 442 Gr60	C-MN-Si	60.0	15.0	10.8
SA 516 Gr60	C-Si	60.0	15.0	10.8
SA 515 Gr60	C-Si	60.0	15.0	10.8
SA 515 Gr65	C-Si	65.0	16.3	11.4
SA 516 Gr65	C-MN-Si	65.0	16.3	11.4
SA 515 Gr70	C-Si	70.0	17.5	12.0
SA 516 Gr70	C-MN-Si	70.0	17.5	12.0
SA 299	C-MN-Si	75.0	18.8	12.0
B Low-alloy steels—plate				
SA 204A	C-½ Mo	65.0	16.3	16.2
SA 204B	C-½ Mo	70.0	17.5	17.5
SA 204C	C-½ Mo	75.0	18.8	18.8
SA 302A	MN-½ Mo	75.0	18.8	17.7
SA 302B	MN-½ Mo	80.0	20.0	18.8
SA 302C	MN-½ Mo-½ Ni	80.0	20.0	18.8
SA 302D	MN-½ Mo-¾ Ni	80.0	20.0	18.8
SA 225A	MN-V	70.0	17.5	14.8
SA 225B	MN-V	75.0	18.8	12.0

SA 202A	½ Cr-1¼ MN-Si	75.0	18.8	12.0		
SA 202B	½ Cr-1¼ MN-Si	85.0	21.3	12.0		
SA 203A & D	2½ Ni & 3½ Ni	65.0	16.3	11.4		
SA 203B & E	2½ Ni & 3½ Ni	70.0	17.5	12.0		
SA 387 2C1.1	½ Cr-½ Mo	55.0	13.8	13.5		
SA 387 12C1.1	1 Cr-½ Mo	55.0	13.8	13.8		
SA 387 11C1.1	1¼Cr-½ Mo-Si	60.0	15.0	14.8		
SA 387 22C1.1	2¼Cr-1 Mo	60.0	15.0	15.0		
SA 387 21C1.1	3 Cr-1 Mo	60.0	15.0	13.9		
SA 387–5	5 Cr-½ Mo	60.0	15.0	12.8		

C High-alloy steels—plate

			−20 to 100	300	500	700
SA-240–405	12 Cr-1A1	60.0	15.0	13.3	12.9	12.1
SA-240–304	18 Cr-8Ni	75.0	18.8	16.6	15.9	15.9
SA-240–316	16 Cr-12Ni-2 Mo	75.0	18.8	18.4	18.0	16.3
SA-240–321	18 Cr-10Ni-Ti	75.0	18.8	17.3	17.1	15.8
SA-240–347	18 Cr-10Ni-Cb	75.0	18.8	15.5	14.9	14.7

A material is *ductile* if it is capable of undergoing a large, permanent deformation and yet offers great resistance to rupture. The measure of ductility is the percentage of elongation or the percentage of reduction of area during a tensile test carried to rupture, and it is used as a relative measure. Ductility helps to relieve localized stress concentration through local yielding. It is a necessary characteristic of a material used to take live loads, especially where concentrated stresses may occur.

Brittleness is a characteristic opposite to ductility and toughness. A material may be considered brittle if its elongation at rupture through tension is less than 5 percent in a specimen 2 in. long.

Toughness is a term used to denote the capacity of a material to resist failure under dynamic loading. The *modulus of toughness* is defined as the amount of energy per unit volume that a material can withstand or absorb without fracture occurring. The modulus of toughness is useful as an index for comparing the resistance of materials to dynamic loads, and it is especially applicable in the design of moving parts of machinery.

Figure 2-10 lists some of the steels that are permitted to be used in high-pressure boilers as stipulated by Section I of the ASME Boiler and Pressure Vessel Code. Listed are carbon and low-alloy and high-alloy steel plates and tubes. Each specification number shows the principal alloying elements in the steel.

ASME Code Material Requirements Certain procedures must be followed by a fabricator or repairer in order to make sure only Code-specified material is used in boiler construction. It also is part of the responsibility of the authorized inspector to help implement a quality control procedure in order to make sure Code material is used. These controls may include the following:

1. The material to be used for a boiler or pressure vessel must be specified in the section of the Code under which the boiler or pressure vessel is built. For example, if a high-pressure boiler is involved, it must be listed as a permissible material in Section I (Power Boilers), or data must be presented to show that it has the same chemical and physical characteristics as a Code-listed material.
2. The fabricator of the boiler or pressure vessel generally orders permissible Code material from the steel mills. The steel mill is responsible for making the necessary tests to the specifications given in Section II of the ASME Boiler and Pressure Vessel Code.
3. Section II test requirements that the steel manufacturer may have to perform include the following:
   - *(a)* Chemical analysis of the steel to determine if it is within Code limits for the specification.
   - *(b)* Tests to determine if the metallurgical grain structure is within Code-specified limits.
   - *(c)* Inspection of plate or tube to note if defects such as blowholes, slag, laminations, and any other imperfections may be present and whether these are within Code-permissible tolerance.

(*d*) Tension and bend tests as stipulated in the Code to note if these are within Code specifications.
(*e*) Notch toughness tests to check on fatigue-failure strength.
(*f*) Mill test report showing that the material complies to Code specifications; this must be certified by a responsible person of the material testing laboratory of the steel manufacturer.

Code Stress Tables

1. When allowable pressures are calculated, the allowable stress of the material, as shown in the stress table of the Code section, must be used.
2. Some stress values from Section I are shown in Figs. 2-10 and 2-11.
3. The stress value under expected mean operating temperature is used in the calculations.

Plate steel for any part of a boiler subject to pressure and exposed to the fire, or products of combustion, must be of firebox quality. If not exposed to fire or the products of combustion, the plate can be of flange quality. Some firebox-quality steels are specification SA-201 carbon-silicon steel, specification SA-202 chromium-manganese-silicon steel, and specification SA-204 molybdenum steel. Check the Code for other firebox-quality steels and refer to the ASME Material Specification, Section II, for their physical and chemical characteristics. Seamless steel drum forgings made in accordance with specifications SA-266 and SA-336 for alloy steel can be used for any part of a boiler for which either firebox or flange quality is permitted.

Pipes and tubes may be made of open-hearth, electric-furnace, basic-oxygen, or acid deoxidized bessemer steel pipe or tubing, according to Code specifications.

Superheater pressure parts, whether the integral type or separately fired, must be of wrought steel, puddled or knobbled charcoal wrought iron, or carbon or alloy steel, according to Code specifications.

Rivets must be of steel or iron of the quality designated under specification SA-31 or SA-84 for wrought iron.

Stays and stay bolts fabricated by forge welding must be of SA-84 stay-bolt wrought iron. Threaded stay bolts must be of wrought iron of SA-84 grade, of steel complying with specification SA-31, or of annealed nickel-copper alloy of specification SB-164.

If the pressure does not exceed 250 psi and the temperature does not exceed 450°F, specification SA-278 gray-iron castings may be used for power boiler parts such as pipe fittings, water columns, and valves and their bonnets. Cast iron cannot be used for nozzles or flanges for any pressure or temperature. But this does not apply to low-pressure boilers. The same pressure parts as enumerated for cast iron can be made of malleable iron, except that the pressure is limited to a maximum of 350 psi and the temperature to 450°F.

The minimum thickness is ¼ in. for plate subjected to pressure. An exception is for miniature boilers of seamless construction, where the minimum plate thickness may be $\frac{3}{16}$ in. The minimum thickness of tube sheets is ⅜ in., except on miniature boilers where it is $\frac{5}{16}$ in. The plate material must be not more

Fig. 2-11 Steel tube material, allowable stresses per Section I of ASME Code.

Spec no.	Composition	Form	Ultimate	Allowable stress				
				−20 to 650°F	700°F	800°F		
A Carbon steel—tubes								
SA 192	C-Si	Seamless	47.0	18.8	11.5	9.0		
SA 178A	C	Welded	47.0	11.8	11.5	7.7		
SA 226	C-Si	Welded	47.0	11.8	11.5	7.7		
SA 210 A-1	C	Seamless	60.0	15.0	14.4	10.8		
SA 178 C	C	Welded	60.0	15.0	14.4	9.2		
SA 210 C	C-Mn	Seamless	70.0	17.5	16.6	12.0		
B Low-alloy steel—tubes								
SA 209 T1b	C-½ Mo	Seamless	53.0	13.3	13.2	13.1		
SA 250 T1b	C-½ Mo	Welded	53.0	11.3	11.2	11.1		
SA 250 T1	C-½ Mo	Welded	55.0	11.7	11.7	11.7		
SA 209 T1	C-½ Mo	Seamless	55.0	13.8	13.8	13.7		
SA 213 T2	½ Cr-½ Mo	Seamless	60.0	15.0	15.0	14.4		
SA 423-1	¾Cr-½ Ni-Cu	Seamless	60.0	15.0	15.0	—		
SA-213-T12	1 Cr-Mo	Seamless	60.0	15.0	15.0	14.8		
SA-213-T11	1¼ Cr-½ Mo-Si	Seamless	60.0	15.0	15.0	15.0		
SA-213-T3b	2 Cr-½ Mo	Seamless	60.0	15.0	15.0	14.7		
SA-213-T22	2¼ Cr-1 Mo	Seamless	60.0	15.0	15.0	15.0		
				−20 to 100	300	500	700	800
SA 213-T21	3 Cr-1 Mo	Seamless	60.0	15.0	15.0	15.0	14.8	14.5
SA 213-T5	5 Cr-½ Mo	Seamless	60.0	15.0	15.0	15.0	13.4	12.8
SA 213-T7	7 Cr-½ Mo	Seamless	60.0	15.0	15.0	14.5	13.4	12.5
SA 213-T9	9 Cr-Mo	Seamless	60.0	15.0	15.0	14.5	13.4	12.8
C High-alloy steel—tubes								
SA 268-TP405	12 Cr-1A1	Seamless	60.0	15.0	13.3	12.9	12.1	—
SA 268-TP446	27 Cr	Seamless	70.0	17.5	15.6	14.5	14.1	—
SA 213-TP304	18 Cr-8 Ni	Seamless	75.0	18.8	16.6	15.9	15.9	15.2
SA 213-TP316	16 Cr-12 Ni & 2MO	Seamless	75.0	18.8	18.4	18.0	16.3	15.9
SA 213-TP321	18 Cr-10 Ni & Ti	Seamless	75.0	18.8	17.3	17.1	15.8	15.5
SA 213-TP347	18 Cr-10 Ni & Cb	Seamless	75.0	18.8	15.5	14.9	14.7	14.7

than 0.01 in. thinner than that required for the plate by the formula used to calculate its strength, provided the tolerance in fabrication (or when the plate is ordered) also has this tolerance of not less than 0.01 in.

Questions and Answers

2-1 Name some typical mechanical properties that describe the ability of a steel material to resist loads.

Ans. Mechanical properties that are important to describe a steel material are ultimate strength, elastic limit, endurance limit, hardness, and modulus of elasticity.

2-2 What are the three most prominent names given to the method used in measuring the hardness of steel material?

Ans. Brinell, Rockwell, and the Shore sclerescope are the most prominent methods used in describing the hardness quality of a steel material.

2-3 How is a space lattice defined?

Ans. A space lattice is the geometric arrangement of atoms in the crystal formation of a given metal, such as face-centered cubic lattice or body-centered cubic lattice, describing the arrangement of atoms in an iron crystal.

2-4 What is the space lattice of alpha iron?

Ans. Alpha iron is composed of body-centered cubic lattices.

2-5 What is meant by a eutectic?

Ans. A eutectic is the alloy of a solution that has the lowest melting point in the solution.

2-6 What is the transformation temperature in carbon steels?

Ans. This is the temperature within which austenite (nonmagnetic steel) forms on heating and also the temperature interval within which austenite disappears in cooling in an iron-carbon solution. The temperatures of transformation are dependent on the carbon content, as noted in iron-carbon equilibrium diagrams for the solution.

2-7 Name some typical physical defects that may be found in cast iron.

Ans. Blowholes, cracks, segregation of impurities, and coarse grain or lack of uniform grain structure are the most common defects.

2-8 What are the two major chemical steps in the manufacture of steel from ore?

Ans. (1) To remove oxygen from the ore in producing pig iron. (2) To remove excess carbon from the pig or cast iron to produce steel.

2-9 How does steel differ from cast iron?

Ans. The carbon content is much lower, the tensile strength is much higher, and the ductility and resistance to shock load are higher.

2-10 Briefly, how is excess carbon removed from iron to form steel?

Ans. By blowing air through the molten iron. The oxygen in the air unites with the carbon by combustion.

2-11 How is the highest-quality steel produced?

Ans. By the electric-furnace method, owing somewhat to more accurate control and to a minimum of oxidizing factors.

2-12 Why is the electric-furnace method not always used?

Ans. The expense of manufacture by this method is greater.

2-13 What are the ASME Code requirements for chemical properties of open-hearth firebox and flange steel?

Ans. For open-hearth firebox steel, the maximum carbon content for plates not over ¾ in. thick is 0.25 percent, and for plates over ¾ in. thick, 0.30 percent; the maximum phosphorus (P) content for steel made by the acid method is 0.04 percent, and for steel made by the basic method 0.035 percent; the maximum sulfur (S) content is 0.04 percent.

For open-hearth flange steel, the maximum manganese (Mn) content is 0.80 percent; the maximum phosphorus content if the steel is made by the acid method is 0.05 percent; if the steel is made by the basic method, it is 0.04 percent; and the maximum sulfur content is 0.05 percent.

2-14 What is the minimum yield point required for steel of either flange or firebox grade?

Ans. One-half the tensile strength.

2-15 How much below specified thickness may a boiler plate be rolled and still be acceptable?

Ans. 0.010 in.

2-16 Where and how shall a boiler plate be stamped?

Ans. In at least two places not less than 1 ft from the edges should be the manufacturer's name or label, the test number, the type of steel, and the lowest tensile strength of the 10,000-lb range.

2-17 What determines the hardness of steel?

Ans. The carbon content, the alloy content, and the grain structure often due to heat treatment.

2-18 How may grain structure and hardness be changed?

Ans. By heat treatment at temperatures above the lower critical point.

2-19 What are the major differences between cast iron and malleable iron?

Ans. Malleable iron is more ductile. It will withstand safely a considerably greater shock load than cast iron. It possesses a higher tensile strength.

2-20 What are ductility and malleability?

Ans. These terms are practically synonymous. They describe the ability of a material to withstand a comparative degree of deformation without failure or impairment of its strength or other physical properties.

2-21 What is the common measure of ductility?

Ans. The percentage elongation as found in making a test of tensile strength.

2-22 What is hardness?

Ans. It is the property of a material to resist surface deformation on application of an external force.

2-23 What is brittleness?

Ans. It is the tendency of a material to fracture on application of shock load.

2-24 Is there any definite relation between hardness and brittleness?

Ans. Usually not. For example, some of the so-called white metals are quite brittle but soft.

2-25 Is there any definite relation between hardness and toughness?

Ans. Usually not. Toughness is a physical property enabling a material to withstand deformation as well as abrasion, such as is required for the teeth of a power dredge.

2-26 What are fatigue and fatigue failure?

Ans. Fatigue is "tiring" of the metal. It is the breaking through of the crystalline grain or fibrous structure of a material after a number of stress applications or stress reversals. Fatigue endurance is the ability of a material to withstand a great number of stress applications or reversals without failure.

2-27 In what direction does the fiber run in plates of a boiler, and does it make any difference?

Ans. The fiber runs in the direction in which the plate is rolled—usually parallel to its longest dimension. In fabrication the fiber runs circumferentially, because the longest dimension of the plate is usually required in this direction, and to keep the greatest tensile load on the plate in the same direction as the so-called fibrous structure.

2-28 How much does a piece of boiler plate 12 in. by 12 in. by 1 in. weigh?

Ans. The average weight of steel boiler plate is 0.28 lb/in.^{3}; therefore, $12 \times 12 \times 1 \times 0.28 = 40.3$ lb.

2-29 How would you tell wrought iron from steel?

Ans. Wrought iron is of much greater fibrous structure than steel. If a spot is polished on the surface of the metal with a file or emery cloth, the fibrous structure usually can be seen. Sparking the metal on an emery wheel will show a reddish spark from wrought iron and an exploding yellowish spark from steel.

2-30 What is a laminated boiler plate?

Ans. One that contains a stratum of slag, rolled in by accident.

2-31 What harm does a lamination do?

Ans. It is an insulator that prevents free conduction of heat. If the plate is in a high-temperature zone, a blister may result because of overheating of the plate on the fire side of the lamination.

2-32 What type and grade of steel are required for boiler plates under pressure (nonalloy, non-high-tensile-strength steel)?

Ans. Open-hearth firebox steel if the plate is to be exposed to the products of combustion; open-hearth flange steel if not.

2-33 Is the actual tensile strength stamped on steel plate or is there a range for stamping per the ASME Code?

Ans. For a carbon steel, the actual tensile strength may be anywhere between 55,000 and 65,000 lb/in.^{2}, but the plate is stamped 55,000 lb, the minimum of the 10,000-lb range. If the plate fell slightly below 55,000-lb/in.^{2} tensile strength, it would be stamped 45,000.

2-34 What is mill scale, and what harm does it do?

Ans. It is an iron oxide scale formed on the surfaces of a plate when, after rolling at high temperatures, it is exposed to the air. It may be a cause of plate deterioration in boiler service, or it may promote and localize corrosion.

2-35 Where can the exact tensile strength be found?

Ans. From the mill test report.

2-36 What else does the mill test report show?

Ans. The thickness; the chemical properties; other required physical properties, namely, elongation and yield point; the manufacturer's name; and the heat and slab numbers.

2-37 What is the effect of sulfur and phosphorus when they are melted into steel?

Ans. Sulfur makes the steel brittle at high temperature, often referred to as hot-short. Phosphorus makes steel brittle at low temperatures, so it is undesirable in parts which are subjected to impact loading when cold. This brittleness when cold is sometimes called cold-short.

2-38 In making a shop inspection of plate material for ASME-constructed boilers, what should be especially checked by a Code inspector?

Ans. The plate should be checked for the following:

1. Examine mill test reports and note if the plate complies with Code specification for that material.

2. Check plate stamping and note if stamping identifies the plate with the mill test report, including tensile strength, thickness, and grade and quality of finish.

3. Make a visual examination of the plate for defects such as cracks, laminations, inclusions, pits, and similar questionable conditions.

2-39 When may firebox or flange steel be used per the ASME boiler code? What process of manufacture is stipulated?

Ans. Firebox-quality steel must be used when parts of the boiler are exposed to fire or products of combustion. For other service, either firebox- or flange-quality steel may be used. Steel should be made by an open-hearth or electric-furnace process by either the acid or basic method.

2-40 What is the maximum allowable working pressure and temperature on a water column made of cast iron and malleable iron?

Ans. Per the ASME Code, for cast iron, 250-psi pressure and 450°F; for malleable iron, 350 psi and 450°F.

2-41 Who is responsible for stamping the steel plates used in high-pressure boilers?

Ans. The steel mill in which the plates are made.

2-42 Name some typical types of cast iron.

Ans. Cast-iron types are white cast iron, gray cast iron, malleable iron, and ductile cast iron. Cast irons are also classified by their ultimate strengths—20,000, 30,000, 40,000, etc.

2-43 What process is used in fabricating tubes?

Ans. Tubes are made by either the seamless method or the butt welding process.

3 Welding and Nondestructive Testing

WELDING

Welding is the chief method employed in the manufacture of boilers and pressure vessels. There are still some riveted boilers in service that will require repairs; some of the requirements for riveted boilers are detailed in App. 3. Threaded connections as well as expanded connections are also used in the joining or assembling of boiler components. The Code states that threaded connections larger than 3-in. pipe size shall *not* be used when the maximum allowable pressure exceeds 100 psi. But this 3-in. pipe size restriction does not apply to plug closures used for inspection openings or end closures used for similar purposes. Expanded connections for high-pressure boilers are permitted provided the pipe, tube, or forging does not exceed a 6-in. outside diameter (OD) so that the opening meets all reinforcement requirements. Also, the expanded connections must meet all requirements on expanded tubes for fire-tube and watertube boilers.

Welding is a localized coalescence (fusing together) or consolidation of metal where joining is produced by heating to fusion temperatures, with or without the application of pressure and with or without the use of a filler metal. The filler metal (when used) has a melting point of approximately that of the pieces (base metal) joined together. The weld is that portion which has been melted during welding. And the welded joint is the union of two or more members produced by the welding process.

The most common method of welding pressure parts is by *fusion* (melting) of the metal, the heat being supplied in one of several different ways. In fusion welding, no pressure is applied between the pieces being welded. Arc welding, gas welding, and Thermit welding are classified as fusion welding, but arc welding is the most common.

Arc welding is a localized progressive melting and flowing together of adjacent edges of the base-metal parts, caused by heat produced by an electric arc between a metal electrode, or rod, and the base metal. Both the welding material (welding rod or electrode) and the adjacent base metal are melted. On cooling they solidify, thus joining the two pieces with continuous material.

In *metal arc welding,* welding rods can be of two types: *bare* electrodes and *coated* electrodes. Plain arc welding is used with bare electrodes. Shielded arc welding is used with coated electrodes. The reason is that the flux or coating (Fig. 3-1) on the electrode protects the deposited metal from oxidation. Then as the welding rod melts, a more reliable weld is obtained than with bare electrodes. Shielded metal arc welding is abbreviated as SMAW.

In submerged arc welding, coalescence (fusing together) is produced by heating with an electric arc, or arcs, between a bare metal electrode, or electrodes, and the work. The welding is shielded by flux, which is a blanket of granular, fusible material on the work. Pressure is not used, and filler metal is obtained from the electrode or sometimes from a supplementary welding rod.

In gas tungsten arc welding, coalescence is produced by heating with an electric arc between a single tungsten (nonconsumable) electrode and the work. Shielding is obtained from a gas or gas mixture (which may contain an inert gas). Filler metal is usually added separate from the electrode. This process is also called tungsten inert gas (TIG) welding.

Gas welding is a group of welding processes in which coalescence is produced by heating with a gas flame, or flames, with or without the application of pressure. Filler metal is added (usually) to the heated base metals to be welded.

Oxyacetylene welding is a gas welding process in which coalescence is produced by heating with a gas flame, or flames, of about 6000°F obtained from the combustion of acetylene with oxygen as filler metal is added.

Thermit welding is a group of welding processes in which coalescence is produced by heating with superheated liquid metal and slag resulting from a chemical reaction between a metal oxide and aluminum. Usually no pressure

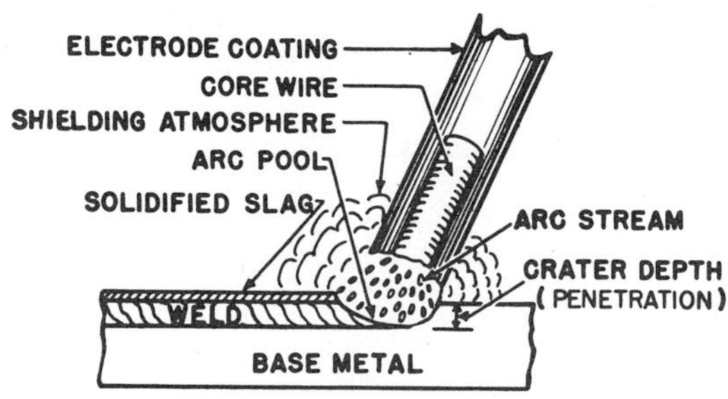

Fig. 3-1 Shielded metal arc welding.

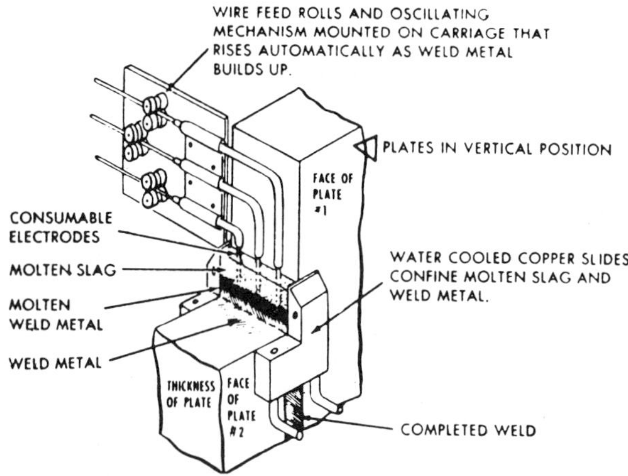

Fig. 3-2 Basic elements of electroslag welding.

is applied. Filler metal, when used, is obtained from the liquid metal. The Thermit process is useful for very heavy welds, such as parts with thicknesses of 3 in. and over.

In electroslag welding (see Fig. 3-2), parts are joined in one pass with the welding groove in a vertical plane. The slag on top conducts the electric arc that generates the heat required to melt the slag or flux and the weld filler metal. A water-cooled dam is used to confine the molten pool of flux, filler metal, and parent metals within the welding groove until the weld is completed and solidified. Deposit rates are high with this type of welding. The weld is considered coarse-grain-structured and requires further heat treatment in order to refine the grain structure of the weld and the heat-affected zones.

Some other welding processes are *electron beam welding,* where a concentrated beam of high-velocity electrons generates heat of fusion on the surface to be joined; *plasma arc welding,* an inert-gas fusion welding process using a constricted arc, similar to the gas tungsten arc process; and *laser welding and cutting,* which is accomplished by the energy of light concentrating a beam through an optical device called the optical maser or laser. The radiation source in a laser is concentrated with only small beam divergence, which results in high power intensities of up to 10^9 watts per square centimeter (w/cm^2).

The ASME Section I Code permits the following welding processes to be used on power boilers: shielded metal arc, submerged metal arc (SAW), gas metal arc (GMAW), gas tungsten arc (GTAW), plasma arc, atomic hydrogen metal arc, oxyhydrogen, and oxyacetelene. Pressure welding processes allowed are flash, induction, resistance, pressure Thermit, and pressure gas.

Fillet welds are made with certain tolerances to be followed, and it is necessary to define the terms *throat, toe, face, legs,* and *root.* (See Fig. 3-3, showing a full fillet weld.) The terms are defined as follows. The *throat* of a weld is

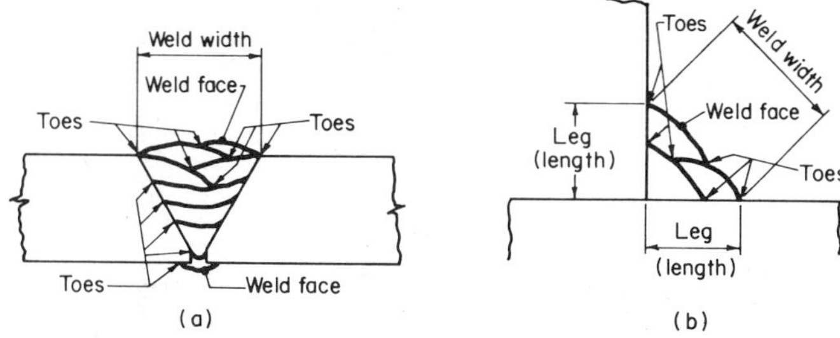

Fig. 3-3 Examples of weld terminology: *(a)* butt-weld details and *(b)* fillet-weld details.

the shortest distance from the root of a fillet weld to the face. The *toe* of a weld is the juncture between the face of the weld and the base metal. The *face* of a weld is the exposed surface on the side from which welding was done. The *root* of a weld is the deepest beginning of the weld which intersects the base-metal surfaces. The *leg* of a weld is the distance from the root of the joint to the toes of a fillet weld. Groove welds are used extensively for butt joints (Fig. 3-3*a*).

In double-butt welding, the metal is deposited in the welding groove from both sides of the plate, whereas in single-butt welding the metal is deposited from only one side. Single-butt welding on longitudinal joints of drums is permitted, provided a backing strip is used to ensure full penetration welding of the joint. But single-butt welding can be used only where the inside of the weld is inaccessible for welding. On longitudinal joints, the backing must be removed.

After the first side of a double-welded butt joint has been welded, the second side should be chipped, ground, or melted out to secure a clean surface of proper shape, to ensure fusion of the weld metal on that side without porosity, slag, or voids. The base metals must be clean and free from grease, dirt, rust, paint, or other foreign substances. Sometimes, prior to welding, a linseed oil coating is specified on the base metals. A common welding problem is slag inclusion. This is a nonmetallic solid material entrapped in a welded joint or between weld metal and the base metal. The slag is objectionable because it prevents metal-joint strength based on metal characteristics. Porosity of a weld is the voids or gas pockets left in a weld as a result of incorrect welding or welding difficulties. When fusing takes place, the base metals welded are affected in a zone called the *heat-affected zone* (HAZ). The heat-affected zone is that portion of the base metal which has not been melted but where the structural properties of the base metal have been altered by the heat of welding (or cutting).

Locked-up stresses are those internal stresses remaining in the weld metal and adjoining base material when a weld has been made. They are caused by the high concentrated heat in the weld compared to the adjoining cooler metal. This sets up a thermal gradient, leading to nonuniform expansion and contraction, which sets up internal stresses in the weld. Peening and heat treatment will reduce locked-up stresses. This is called *stress relieving.*

Preheating consists of heating the base metals to a temperature of around 150 to 400°F, depending on the materials welded, before welding. This reduces the chances of locked-up stresses when the weld is made. Stress relieving is done by heating a welded joint from 1100 to 1200°F after the joint is welded and keeping it at this temperature *one hour for each inch of thickness.* The entire vessel or only the joint may be stress-relieved. It is common practice to complete all welding and then stress-relieve the entire vessel. Stress relieving affects the metal structure in its crystalline form and thus reduces the concentration of locked-up stresses in the weld. After the required temperature is held for the specified time, cooling to 600°F must be gradual, after which the metal can be cooled to room temperature. Peening is a mechanical working of metal by means of hammer blows so as to reduce the residual stress in a welded joint caused by the heat of welding.

Defects in welding usually can be repaired by chipping or grinding out the defective section (voids, heat-affected cracks, lack of penetration, etc.) to sound metal and then rewelding. Generally, nondestructive testing must be used to make sure that the defective section has been completely eliminated; the repair also has to be retested by nondestructive testing. Openings or holes are permitted in boiler welded joints. However, the weld must have been stress-relieved and radiographed, and the weld at the opening must be examined for cracks on both sides by the magnetic-particle method. Figure 3-4

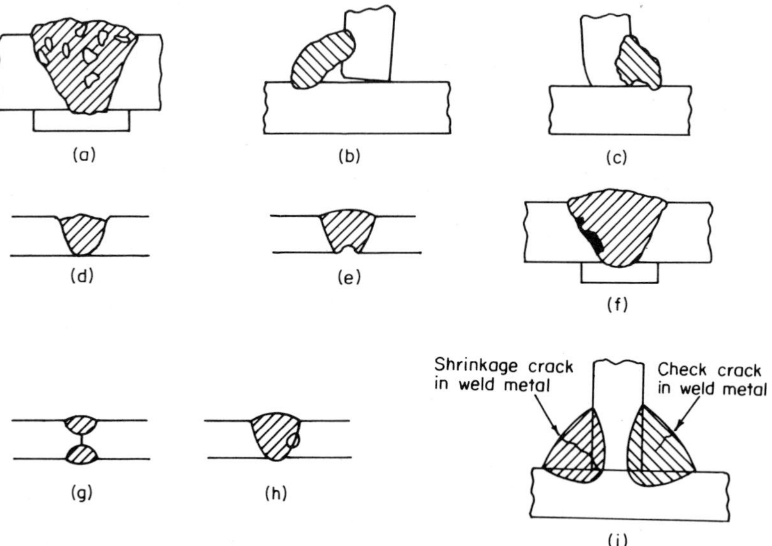

Fig. 3-4 Common welding problems: *(a)* porosity, *(b)* inadequate penetration, *(c)* incomplete fusion, *(d)* incomplete groove, *(e)* root cavity, *(f)* slag inclusion, *(g)* incomplete root penetration, *(h)* incomplete side fusion, *(i)* shrinkage cracks.

lists some common welding problems and the steps to be taken to avoid these problems.

Metallurgical considerations are many when one contemplates a welding process. The following must be considered:

1. Weldability of the parts to be joined. The Code may have severe restrictions.

2. Surface conditions needed and the compatibility of the chemical and mechanical properties of the base metals to the welding process being used. Moisture on the surface of ferrous material as well as on the electrode can result in porosity and hydrogen under bead cracking.

3. The matching of filler metals or electrodes to the base metals in order to avoid possible embrittlement of the weld.

4. Suitability of the welding process for the type, thickness, and alloys of the metals to be joined.

5. Evaluation of the effects of preheat and postwelding heat treatment. For metals with high hardenability rates, slow cooling will prevent heat-affected zone cracking. Face-centered cubic-lattice material may develop coarse-grain crystal structures when cooled slowly.

6. Fluxes. They float when melted on top of the liquid metals, protecting or shielding the metals from the atmosphere. Chemical reactions between the metal and the flux are possible at the interface. If the flux is high in silicon dioxide, a chemical reaction is possible which may produce higher silicon in the weld metal than anticipated.

Section IX of the ASME Boiler and Pressure Vessel Code contains details on procedure specifications that must be prepared by a manufacturer and the tests that must be made by a welder in order to be considered a qualified welder.

A qualified welder is one who is capable of performing manual or semiautomatic welding. A welding operator is one who operates a machine or automatic welding equipment. Boiler Code work requires the use of a qualified welder on most of the work where Code welding is to be done.

Qualification tests may be given by responsible manufacturers or contractors. On pressure-vessel work, the welding procedure of the fabricator or contractor must also be qualified *before* the welder can be qualified. Under other codes this is not necessary. To become qualified, the welder must make specified welds, using the required welding process, type of metal, thickness, electrode type, position, and joint design. Test specimens must be made to standard sizes and under the observation of a qualified person. In most government specifications, a government inspector must witness the making of welding specimens. Specimens must also be properly identified and prepared for testing. The common test is the guided-bend test. However, x-ray examinations, fracture tests, or other tests are also used. Satisfactory completion of test specimens (providing that they meet acceptability standards) will qualify the welder for specific types of welding. Code certification, in general, is based on the range of thicknesses to be welded, the positions to be used, and the materials to be welded.

Each manufacturer or contractor who is to do boiler Code welding is required to record in detail the procedure to be used. Each procedure requires testing of the welds to be made by reduced-section test specimens and guided-bend specimens. The variables requiring a new procedure and new test plates are very numerous. Among these are changes in base materials, grouped in ASME Welding Qualifications (Section IX) into "P" numbers. For example, P-1 includes mostly carbon steels, P-2 consists of wrought iron, P-3 consists of chrome-moly steels with chromium content below 0.75 percent, and with a total alloy content not exceeding 2 percent. The P numbers range to P-10, so the base-material variable on procedure qualification is large.

The next variable is the electrode and welding rod selection, which ranges from F-1 to F-7. Any change in electrode or welding rod selection requires a new set of test plates or procedure qualification. Weld metal is classified by weld-metal analysis numbers A-1 to A-8. These are related to equivalent P numbers of the base material. Again, changes of weld metal from equivalent base-metal classification require a new set of test plates or procedure qualification.

The thickness of the plate, or pipe, to be welded is another variable. Classification is from $\frac{1}{16}$ to $\frac{3}{8}$ in., from $\frac{3}{8}$ to $\frac{3}{4}$ in., and over $\frac{3}{4}$ in. Each classification requires new test plates as shown in the Code. The ASME Welding Code specifies other variables to consider in requiring a new procedure qualification test, and these should be consulted for specific variables.

Remember: Welding boiler parts is under close control of the many variables involved in welding, so check the ASME Welding Code.

On groove welds, one face-bend test and one root-bend test are required for operator qualification for each position to be welded. For fillet welds, a test plate is required according to the Code, but passing the groove-weld test will also qualify a welder for fillet welding. Procedure qualification requires two face-bend tests, two root-bend tests, and two reduced-section tension tests, as illustrated in the Welding Qualification Code of the ASME.

It might be appropriate to define the terms *reduced-section tension test, free-bend test, root-bend test, face-bend test,* and *side-bend test* since they are used extensively in Code qualifications of welding. See ASME Code, Section IX, for typical weld test specimens. The reduced-section tension test is used for qualifying the procedure that the shop, or contractor, is to use in welding. When broken in tension, it must have an ultimate tensile strength at least that of the minimum range of the plate which is welded (base material), and the elongation of stretch must be a minimum of 20 percent.

The side-bend test is used for qualifying operators. The specimen is subjected to bending against the side of the weld. In the face-bend test, the specimen is subjected to bending against the surface, or face, of the weld. In the root-bend test, the specimen is subjected to bending against the bottom, or root, of the weld. The free-bend test is a shop- or contractor-qualifying procedure test. The test consists of bending the specimen cold, and the outside fibers of the weld must elongate at least 30 percent before failure occurs.

In order to pass each test, guided-bend specimens must have no cracks or other open defects exceeding ⅛ in., measured in any direction on the convex surface of the specimen *after bending,* except that cracks occurring on the corners of the specimen during testing are not considered unless these occur from slag inclusions or other welding technique defects.

Three questions arise quite often from welders working on boilers:

1. Who is responsible for conducting tests of welding procedures and for qualifying welding operators?
2. How long does a qualified welder's approval test remain in effect?
3. Who keeps the records for procedure and operator's approval tests?

1. The manufacturer or contractor who is to do Code welding on boilers and pressure vessels (or nuclear vessels) is responsible for conducting procedure qualification tests and welding operator's qualification tests for work to be done by his or her organization.

2. The operator's qualification remains in effect as long as the operator is employed by the same manufacturer or organization and does welding on a continuous basis. But if the operator changes employment, she or he is no longer considered qualified and thus must take the test again. If the operator has not done any welding for a period of over 6 months in the position, material, etc., for which she or he is qualified, the operator must be requalified.

3. The manufacturer or contractor is responsible for keeping all records of procedure and operator's qualification tests. These are needed as evidence of the shop's, or welder's, ability to do acceptable Code work. An inspector has the right, however, to ask for retests if there is reason to believe the welding is not acceptable by Code requirements.

Unqualified welders may be used on Code-constructed boilers to weld parts only where the strength of the boiler does not depend on the weld in any way and, in addition, the effect of welding heat will in no way affect the boiler components' strength or setup stresses on adjoining parts to the area being welded.

It is not permissible to weld any steel for boiler pressure parts. It must be material permitted by the Code and also must be carbon or alloy steel having a carbon content of not more than 0.35 percent. This rule also applies to oxygen cutting or other thermal cutting processes.

On longitudinal joints of welded power boilers, weld reinforcements should be removed substantially flush with the surface of the plate. If they are not, the Code provides a penalty on the permissible longitudinal joint efficiency. But removal of weld reinforcements is not mandatory on other joints.

The *efficiency* of a welded joint is defined the same way as it is on a riveted joint, namely, the strength of the joint divided by the strength of the solid plate. However, since a welded joint cannot be calculated in the same way as a riveted joint, the Code allows the following if all Code requirements are met:

1. Ninety percent longitudinal joint efficiency if the joint is stress-relieved and radiographed but the weld reinforcement is not removed substantially flush with the surface of the plate.

2. One hundred percent longitudinal joint efficiency if it is stress-relieved, radiographed, and the weld reinforcement is removed substantially flush with the surface of the plate.

Welded boiler drums can have a limited distortion. The drum must be circular at any section within a limit of 1 percent of the mean diameter. If necessary to meet this requirement, plates may be reheated, rerolled, or re-formed. Furnaces must be rolled (scotch marine boilers), with a maximum permissible deviation from the true circle of not more than ¼ in.

Plates of unequal thickness can be welded provided a tapered transition section, having a length not less than 3 times the offset between the adjacent plate surfaces, is provided at joints between plates that differ in thickness by more than one-fourth of the thickness of the thinner plate, or by more than ⅛ in.

There are many Code requirements detailed in Section I as well as requirements in the Heating Boiler Code, Section IV. In addition, general welding requirements are covered quite extensively in Section IX of the Code, and these should be reviewed for details on requirements.

NONDESTRUCTIVE TESTING

The term *nondestructive testing* (NDT) is used to describe a method of testing or inspecting material for soundness (or lack of defects) without affecting the material physically or chemically. Nondestructive testing can involve the following methods: visual examination, hydrostatic or leak testing, radiography, magnetic particle, dye penetrant, ultrasonic, and eddy current. Also being developed are acoustic emission and holographic testing with both requiring computer application to track signals and to note changes in the signals from previous readings.

The purpose of nondestructive testing (NDT) is to detect incipient-type faults such as cracks, inclusions, voids, porosity, lack of fusion in welds, laminations, lack of penetrations, undercuts, shrinkage, and similar defects so that repairs can be made before the defects may cause a serious failure in service. Mechanical properties of welds are also checked. These include bend tests and tensile tests, but these are considered destructive because permanent deformation has taken place. Hardness tests are also used to check the metallurgical changes that may have occurred in the heat-affected zone (HAZ) of a weld, and this test can be considered nondestructive.

A brief review of some of the NDT methods employed in boiler fabrication and repair might be appropriate, but exact requirements are detailed in the ASME Codes.

Radiographic Examination Radiographic inspection includes x-ray and gamma rays obtained from isotopes, such as cobalt 60 and iridium 192 where the resultant radiant energy can be safely controlled. The radiographic method of testing basically involves passing rays through materials to be tested. The rays impinge on a film or screen, and by noting the contrast of the film, it is possible for an experienced radiographer to detect and detail the internal structure of the object under test. The focal spot is a small area in the x-ray tube from which radiation emanates. In gamma radiography, an isotope like cobalt 60 is the radiation source. When radioactive isotopes are used, strength in curies is important, as is the physical size of the source. The smaller the radiation source, the closer to the material it can be placed; at the same time, the smaller the size, the weaker the source in curies and the longer the exposure time needed. See Fig. 3-5.

Fig. 3-5 ***(a)*** **Typical x-ray machines used in radiography to check for metal flaws.**

Maximum voltage (kV peak)	Screens	Applications and approximate practical thickness limits
50	None	Extremely thin metallic sections. Wood, plastics, biological specimens, etc.
150	None or lead foil	Light alloys, 5-in. aluminum or equivalent; 1-in. steel, or equivalent
	Fluorescent	1½-in. steel, or equivalent
250	Lead foil	2-in. steel, or equivalent
	Fluorescent	3-in. steel, or equivalent
400	Lead foil	3-in. steel, or equivalent
	Fluorescent	4-in. steel, or equivalent
1000	Lead foil	5-in. steel, or equivalent
	Fluorescent	8-in. steel, or equivalent
2000	Lead foil	8-in. steel, or equivalent
15–24	Lead foil	16-in. steel, or equivalent
MeV*	Fluorescent	20-in. steel, or equivalent

* Million electronvolts.

Fig. 3-5 ***(b)*** **Radioactive gamma-ray sources used in radiography to check for metal flaws.**

Material	Half-life	Million electronvolts of rays	Steel thickness application, in.
Cobalt 60	5.3 yr	1.17, 1.33	1.5–5
Cesium 137	33 yr	0.66	1.0–4
Iridium 192	70 days	0.137–0.651	0.5–2.5

Intensity of radiation is inversely proportional to the square of the distance from its source. However, it is not possible to bring the source and the material as close together as possible. It will be found that the shorter the source-to-film distance, the greater the parallax, or halo, effect on the radiographic film. This can blur or dissipate the image of any inclusion that might be in the test material and thus mask a defect. The minimum source-to-film distance, according to type of source, size of focal spot, and the thickness involved, is determined by a standard formula for geometrical unsharpness:

$$Ug = \frac{Ft}{D}$$

where

Ug = geometric unsharpness
F = source size, in.
D = distance from source to object being radiographed, in.
t = thickness of weld or object being radiographed, in.

Generally, geometric unsharpness shall be no more than 0.020 in. for material with a thickness under 2 in., no more than 0.030 in. for material with a thickness from 2 to 4 in., and no more than 0.050 in. for thickness greater than 4 in. A qualified radiographer is required to demonstrate the ability to meet these conditions.

The exposure time for any radiograph must be sufficient to meet film-density standards, which is a measure of the darkening effect on exposed film. Radiographs that are too light may not reveal defects such as inclusions, and those that are too dark may make it impossible to separate the defects from the background. Density requirements per Section V of the ASME Code stipulate that "transmitted film density through the radiographic image of the body of the appropriate penetrameter and the area of interest" shall be

1.8 for single film viewing of x-ray film
2.0 for radiographs of gamma-ray source
2.6 for double film exposures
4.0 for maximum density permitted for single or composite viewing

On welded areas that are to be radiographed, the surfaces must be ground so as to eliminate any surface irregularities that might interfere with proper interpretation of the film.

In all radiographic work, a health hazard exists for all personnel in the working area unless safety rules are followed. These include safety regulations of government bodies as well as those of plant radiation safety officers. Safety regulations generally specify the use of proper monitoring devices for the detection of radiation; the need to cordon off working area and the posting of warning signs; the establishment of dosage limits and the recording of the dosage received by workers; and the action to be taken when the limits are exceeded or violated.

Radiographic inspection requires an interpretation of pictures on a film.

The Code provides guides or standards in order to judge whether an indication is acceptable. Some of the standards or requirements are as follows:

1. The first requirement is visual inspection of the weld. Joints must have complete joint penetration, and be free of undercutting, overlaps, or abrupt ridges and valleys. Weld reinforcement is specified not to exceed the following:

Plate thickness, in.	*Maximum thickness of reinforcement, in.*
Up to ½-in. inclusive	1/16
Over ½- to 1-in. inclusive	3/32
Over 1- to 2-in. inclusive	⅛
Over 2-in. inclusive	5/16

2. Welded joints to be radiographed must be free of ripples or weld surface irregularities to such a degree that the resulting radiographic contrast due to any irregularities cannot mask or be confused with the image of any objectionable defect.

3. Gauges called *penetrameters* must be used for every exposure of a film. The penetrameter serves as a comparison gauge on the film to compare faults in the weld. This is done by making a strip of metal for each exposure and drilling holes in the strip prior to exposure; these strips serve as guides to detect flaws within 2 percent of the plate thickness being welded. These holes are usually drilled with a minimum hole diameter of 1/16 in. specified. The proper method of penetrameter location is shown in Fig. 3-6.

When an x-ray picture is taken, the penetrameter will be included, and the holes will serve as a guide for detecting faults right on the film. By comparing the holes on a penetrameter to the holes noted on a weld when the film is developed, an immediate comparison gauge is included on each film. The Code stipulates tolerances for weld rejects based on penetrameter data.

The Code also provides maximum porosity indications and porosity charts which serve as guides for the permissible number of pores (voids) and sizes of voids permitted in each 6-in. length of weld. These should be referred to in determining unacceptable voids.

Defects such as cracks, slag inclusions, lack of penetration, voids, and others appear as darkened areas on the film because they have a lower density than solid metal. Typical faults are noted as follows. *Cracks* appear as dark irregular lines. *Slag inclusions* show up as small, dark spots with irregular outlines. *Gas pockets* appear as small, dark spots with smooth outlines with occasional teardrop fails. *Lack of penetration* is evidenced by a smooth, dark line most often located in the middle of a weld.

Magnetic-Particle Testing This test is used to detect surface faults by means of setting up a magnetic field, or magnetic lines of force, between two electrodes. Powdered magnetic material is sprinkled over the work to be tested. The magnetic field will affect the magnetic powder, and these particles will

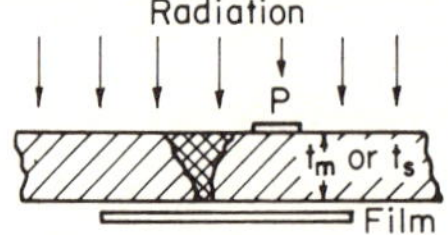

Single wall, no reinforcement,
no backup strip

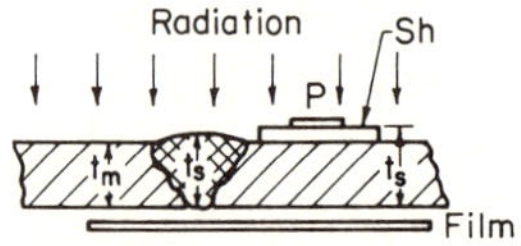

Single wall, weld reinforcements,
no backup strip

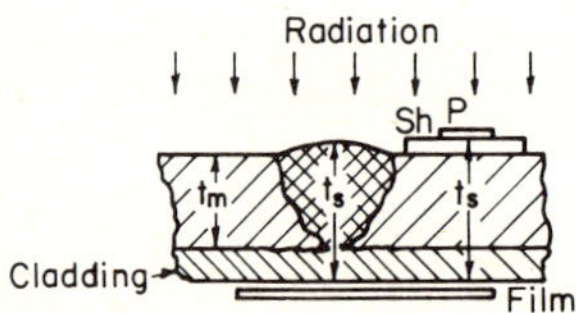

Single wall, weld reinforcement, stainless
cladding redeposited over weld in base metal

Symbols

P = penetrameter
Sh = shim
t_m = design material thickness upon which the penetrameter is based
t_s = specimen thickness

(a)

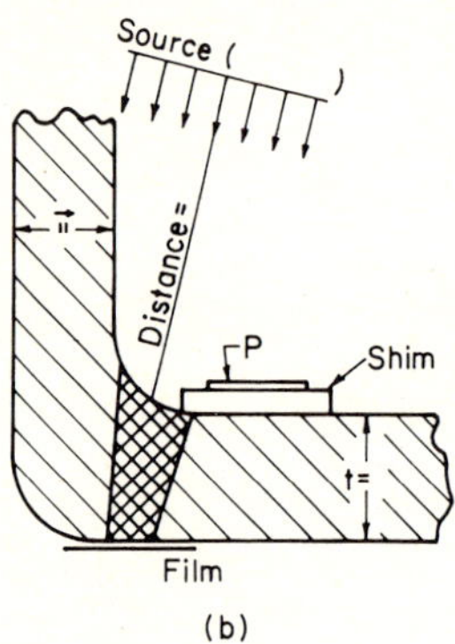

(b)

Fig. 3-6 Penetrameter and shim locations: *(a)* butt welds, *(b)* nozzle welds.

align themselves in a fault, as shown in Fig. 3-7. But the correct interpretation of the gatherings of the magnetic powder requires experience and practice. Magnetic-particle inspection is a practical means for spotting close-lipped discontinuities at or near the surface of a part. Both the wet and dry methods are presently available.

Discontinuities in a magnetized material give rise to localized leakage fields. And these fields attract finely divided magnetic particles. The latter point a finger at the defect and mark its extent on the surface of the part under inspection.

Both direct and alternating currents are used for magnetizing. Direct current (dc) is useful in finding subsurface discontinuities and is commonly used for inspecting welds and castings. Alternating current (ac) is usually employed when highly finished machined parts are checked.

Generally *dc magnetization* is considered where subsurface and surface cracklike defects must be found. With dc, the magnetic field extends within the part itself, and magnetic leakage fields are produced on the surface by interruption in the magnetic path below the surface.

Half-wave dc is a practical supply for subsurface defects, especially where the outer surface is rough as on casting or weldment. Current is rectified half-wave from a single-phase ac source at high amperage and low voltage levels. Most fabricators of heavy industrial equipment use such tests to check

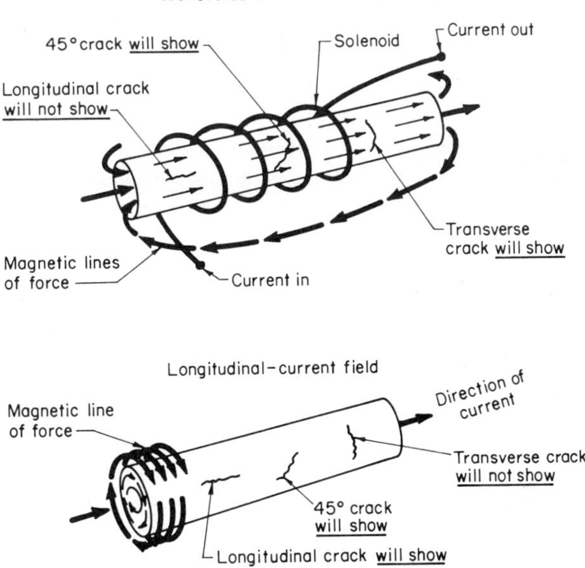

Fig. 3-7 Magnetic-particle inspection—discontinuities in magnetized material develop localized magnetic leakage fields. *(Courtesy Power magazine.)*

for cracks on welds and castings. Heavy, thick welds can be inspected several times during the building up of the weld, after each welding pass if desired.

Dry magnetic powders are used for most portable types of inspection. Red, gray, or black powders give needed contrast with the color of the part being tested. Dry powders permit maximum sensitivity for deep, subsurface cracklike defects.

The *wet-method technique* uses a paste mixture of magnetic particles combined with oil or water. Either the part to be inspected is dipped into the mixture, or the mixture is flowed over the piece. In the dry method, which is more sensitive to subsurface defects, magnetic powder is dusted or blown over the part to be inspected.

For many applications, magnetic particles are coated with a fluorescent material and used in conjunction with black light. Thus, the resultant indication glows and attracts the eye. In effect, this increases the sensitivity of the inspection.

The principal limitation of the magnetic-particle method is that it applies to magnetic materials only and is not suited for very small, deep-seated defects. The deeper the defect is below the surface, the larger it must be to show up. Subsurface defects are easier to find when they have a cracklike shape, such as lack of fusion in the weld. On large, heavy objects, when extremely sensitive inspection is desired, the operation takes more time. This holds true on medium-sized critical parts such as aircraft propellers. With magnetic-particle testing, the surface to be inspected must be available to the operator. This means shafts or other equipment cannot be inspected without removing pressed wheels, pulleys, or bearing housings.

The advantages of magnetic-particle testing are many. Magnetic-particle testing can be used on magnetic-type material, which is very common in industrial products. It is a positive method of finding all cracks at the surface. And because cracks at the surface are most serious and may lead to failure, this is important. The method is flexible and permits effective use of portable equipment. The cost of both equipment and worker hours required for the test compares favorably with that of a simple visual-inspection method.

Liquid-Penetrant (Dye) Inspection This method of testing is used somewhat as magnetic-particle testing, except that it is used primarily on nonmagnetic material. But it can be used on magnetic material. The dye penetrant contains a visible dye, usually red. Indications of defects appear as red lines or dots against the white developer background. It is primarily a surface-defect indicator, and it is applied as follows: A dye penetrant is applied to the part by dip, brush, or spray and is allowed to sit for a specified time. After suitable penetration time, the excess penetrant is removed from the surface, and a developer is applied. The penetrant becomes entrapped in a defect and is brought to the surface by the action of a developer. See Fig. 3-8. Cracks are detected by noting the contrast between the white color of the developer and the red penetrant.

Another penetrant method used is the fluorescent penetrant method, which contains a material that fluoresces brilliantly under black lights. Indications

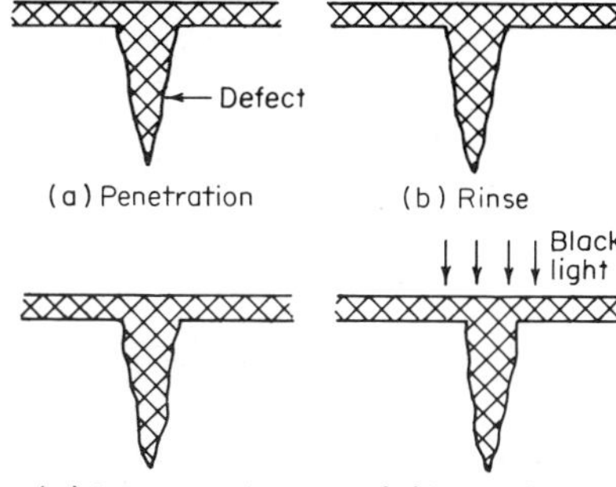

Fig. 3-8 Dye-penetrant inspection involves four steps: *(a)* dye penetration *(red dye)*, *(b)* rinse, *(c)* development *(white color)*, *(d)* inspection (rate of bleeding indicates depth).

of defects appear as fluorescent lines or dots against a nonfluorescent background.

The advantages of the dye-penetrant method are as follows: It provides fast, on-the-spot inspection during overhaul or shutdown periods; the initial cost of the test is relatively low. A perfectly white or blank surface indicates freedom from cracks or other defects that are open to the surface. The disadvantages are that it is not practical on very rough surfaces and color contrast is limited on some surfaces. Also it detects only defects open to the surface.

Ultrasonic Inspections Ultrasonic testing makes use of high-frequency sound waves of 0.5 to 10.0 megahertz (MHz) for inspection of material for flaws and for measuring plate or wall thickness. The basic principle used in an ultrasonic system is the transformation of an electric impulse into mechanical vibrations and then the retransformation of the mechanical vibrations into electric pulses that can be measured or displayed on a screen called a cathode-ray tube (CRT) screen. The transfer of mechanical energy to electric energy is done by means of a transducer. A transducer is a device which transforms energy from one form to another. Ultrasonic transducers transform high-frequency alternating electric energy into high-frequency mechanical energy through magneto restrictive elements up to 100 kilohertz (kHz); and above this range they are usually of the piezoelectric type. In the ultrasonic NDT field, piezoelectric transducers are chiefly used. Piezoelectric transducers depend on the expansion and contraction of certain crystalline and ceramic materials under the influence of an applied fluctuating electric potential. If mechanical pressure is applied to certain faces of specific crystals, then an electric potential is generated across them which is proportional to the applied pressure; conversely, an electric potential of appropriate frequency will cause a crystal to vibrate and so generate pressure waves in any medium with which it is in intimate contact. Figure 3-9 shows the arrangement of a probe in which the crystal or transducers are mounted to generate longitudinal or angle mechanical energy (sound) into a piece to be tested.

The basic component of a typical pulse-echo type of flaw detector is a short electric pulse generated and applied to the electrodes of the probe, which converts the pulse from electric to mechanical vibration energy. This ultrasonic

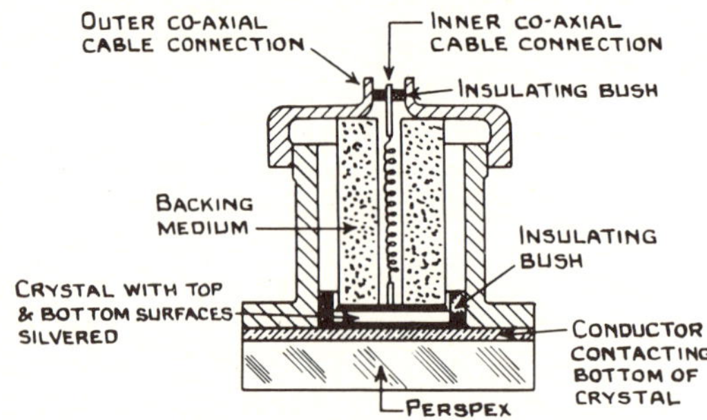

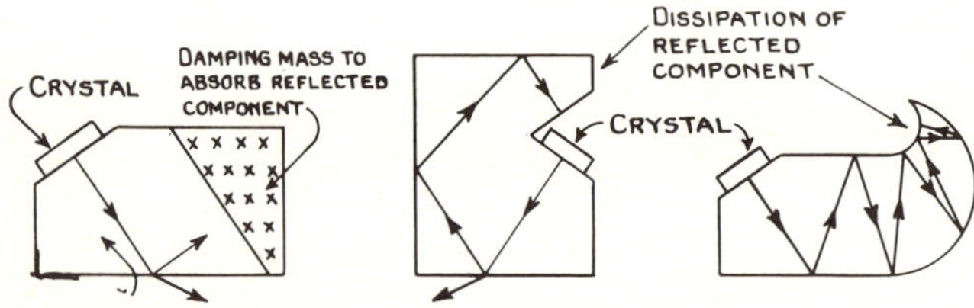

Fig. 3-9 Details of ultrasonic probe detector crystals for longitudinal and angle detection.

pulse travels through the material, and part is reflected to the receiver probe by a reflector which could be the end of the material under test or an internal defect in the material. The returning pulse in the receiver generates an electric pulse which is amplified and displayed by a cathode-ray tube.

The CRT is a familiar component of a television set. It is employed in many applications to render visible any wave that can be converted to an electron beam. In order to evaluate the information displayed on the screen of the CRT, calibration techniques must be followed so that defects can be compared or picked up in relation to standard reference blocks. The main purposes of these blocks are to:

1. Calibrate the instrument for a time base for longitudinal or transverse waves.
2. To measure certain characteristics of probes and the apparatus.
3. To adjust sensitivity and periodically check to make certain it remains constant throughout a test.
4. To have a written procedure so that the settings and calibration can be duplicated at a later date.

Ultrasonic tests are grouped into three basic categories: pulse-echo testing, through transmission, and resonance testing. The pulse-echo method involves transmitting a short burst of high-frequency sound through the piece being tested and then detecting the echoes that are received from either a construction detail, such as a shoulder or hole, or a defect in the material with a separation or void in the material sufficiently large that sound cannot be transferred across the interface. In operation, a pulse-echo unit will produce, through an electronic pulser, a short burst of high-frequency electric signal. This is transmitted to the transducer which is forced to vibrate, usually at its resonant frequency. The probe must be coupled to the test piece with oil, water, or some other liquid or grease. The sound wave train then travels through the test piece until some form of discontinuity or boundary is encountered. This interruption in the medium then causes the sound wave to be reflected to the receiver transducer. The vibrational energy of the sound wave sets the transducer in motion to produce an electric impulse which is fed into an amplifier. The output of the amplifier is displayed on a cathode-ray tube which shows the signals on a linear time baseline. If a linear amplifier is used, the amplitude of the returned echoes can be used as a measure of the area producing the reflected signal.

Resonance testing makes use of a tunable, continuous wave system. This method is usually employed for measuring small or thin walls to 2 or 3 in. The resonance of the crystal is tuned to the piece under test. In practice, a loud pip is heard and can be also seen because the electronic circuit is also in resonance electrically. By proper calibration, direct thickness readings can be made.

Crystals are made for longitudinal waves, shear waves, and surface waves. High-frequency sound waves obey the same physical laws as light waves and can be reflected, refracted, absorbed, and polarized. Using an ultrasonic beam to scan materials such as acoustically transparent steel is analogous to using a light beam to scan optically transparent materials such as glass or water. Since most sonic testing equipment is portable, ultrasonic testing is well suited for field application. There are many battery-operated instruments on the market capable of performing many qualitative tests.

As with all other methods of nondestructive testing, there are distinct advantages and disadvantages associated with ultrasonic tests. The advantages of ultrasonic tests are:

1. Highly sensitive, permitting detection of minute defects
2. Penetrating power, allowing inspection of thick sections
3. Accuracy in locating and measuring flaw size
4. Fast response, permitting rapid inspection

The disadvantages of ultrasonic tests are that the test can be limited by unfavorable part geometry, such as size, contour, complexity, and relative defect orientation; and material structure, such as grain size, porosity, inclusion content, or fine dispersed precipitates.

Eddy-Current Testing The underlying principle of eddy-current testing is the measurement of impedance of electron flow in the part being tested. Weak electric currents or eddy currents are induced by a probe containing inducing and sensing coils. Any changes in the geometry of the part such as a pit, crack, or thinning will affect the flow of the eddy currents. This change in the flow of the eddy currents will be detected by the sensing coil and displayed on a strip chart, a CRT display, or both. The accuracy of characterization of the detected flaw depends on the quality of the standard used in calibrating the eddy-current instrument. A piece of tubing of material identical to the tubes being inspected should be used for calibration. The standard should have a range of defects similar to those expected to be found in the tubes being tested.

Eddy-current testing can indicate tubes that are not leaking, but which may be deteriorated as a result of thinning, erosion, or rubbing. By periodically checking tubes and keeping accurate records, tube replacements can be conveniently scheduled in advance during turnaround plant-maintenance periods, and not on a break-down basis during production runs. Eddy-current inspections from the inner diameter of exchanger tubing allows detection of external and/or internal pits, thickness reduction, cracks, voids, inclusions, dezincification, and carbonization. These conditions can be identified by experienced testers and located along the tube wall. The severity of the test defect can be judged from the results by comparing with the defects in a standard specimen.

Acoustic Emission When a material stretches or "gives," some sound is emitted, and by placing sensors on the material it is possible to pick up these sounds. It has been found that the acoustic sounds emitted by most structural material are above the hearing range of human beings and are generally above 100 kHz. This makes it possible to instrument for sound levels above most spurious noise sources, which will be at lower frequencies. For example, if we assume a small subcritical crack growth exists in a thick steel section, as the crack propagates forward, the atoms are displaced by a small distance in a very short time. This displacement leads to a transient elastic stress wave which then travels outward from the crack-tip area. This stress wave in the material is called the *acoustic emission.* Very large crack extensions that travel to the surface would be heard. However, by placing transducers on the surface, crack propagation can be detected as the crack develops and progresses. An acoustic emission monitoring system provides a means to continuously scan a loaded structural member for changes in acoustic emissions in order to detect any developing flaw. When the stress wave reaches a transducer, an electric signal is generated which can then be processed to provide information about the acoustic-emission event. By detecting and processing the information contained in the acoustic-emission stress wave, the growth of flaws can be tracked and studied in real time as they occur. Present systems include data acquisition for evaluating the severity of flaws. These data are stored in a computer for present analysis and for future reference. Acoustic emission has great potential in continuously monitoring critical pressure vessels for flaw formations such as exist in nuclear power plants, where, because of radiation hazards, access for inspection may be restrictive. This method is still in the developmental stage for such applications.

Nondestructive testing examination and inspection require extensive experience in the method of NDT to be applied for flaw detection. This specialized experience is detailed in SNT-TC-lA, "Recommended Practice for Nondestructive Testing Personnel Qualification and Certification," published by the American Society for Nondestructive Testing (ASNT). Three levels or grades of qualifications are possible in each of the NDT methods previously described. The level 3 person is the most qualified, which generally requires not only knowledge of operating an instrument or applying a method but also theoretical knowledge of the NDT methods, their advantages and shortcomings, and the interpretation of test results. The ASME Boiler and Pressure Vessel Codes refer to the ASNT Society as the source for details on qualifying and certifying NDT "examiners." The ASME Code inspector must still make sure that the written procedures can detect the discontinuities by the NDT method to be used which are not acceptable in the section of the Code to which a boiler or pressure vessel is being built. For example, there are radiographic standards in Section I that may differ somewhat from those involving a nuclear pressure vessel. The "authorized" inspector making the Code inspections will review the NDT results in order to make sure discontinuities found by NDT are within Code-permissible limits. Close coordination is thus essential between the NDT examiner and the authorized inspector.

Questions and Answers

WELDING

3-1 Name some typical dimensional defects that can occur in a welded joint.

Ans. Dimensional defects possible in a welded joint are:

1. Warpage of plate
2. Incorrect joint preparation or misalignment of the joint
3. Incorrect weld size for the thickness of plate to be welded
4. Incorrect weld profile such as overlap, excess convexity, excess weld reinforcement
5. Incorrect final dimension

3-2 Name some typical structural discontinuities that can be found in a welded joint.

Ans. Structural discontinuities that may be found in welded joints are: (1) porosity; (2) slag inclusion; (3) tungsten inclusion; (4) incomplete fusion; (5) undercut; (6) cracks; and (7) surface irregularities.

3-3 *(a)* What is electric arc welding? *(b)* What are the base or parent metals?

Ans. *(a)* Welding two pieces of steel together, with the fusion brought about by heating the metal by the electric arc. *(b)* The metals which are to be welded.

3-4 What should be the condition of the base metals at the time of welding?

Ans. Metals should be clean and free from grease, dirt, dust, paint, or other foreign substances except for a thin coating of linseed oil if this has been specified.

3-5 What precautions must be taken by the operator while welding or the inspector of welding while watching the welding process?

Ans. The face and neck of the operator or inspector should be covered by a helmet. The glass in the helmet should be darkened to protect the eyes. The welder should also wear gloves to protect the hands.

3-6 Which is better, the long or short arc?

Ans. The short arc because it makes a purer weld, gives better fusion and penetration, and causes less puttering of the molten metal.

3-7 How should a condemned weld be removed in order to permit the joint to be rewelded?

Ans. The weld metal should be chipped or ground out to good metal before the weld is redone.

3-8 What is meant by the terms *forehand* and *backhand* as applied to oxyacetylene welding?

Ans. In forehand welding, the welding flames are pointed in the direction in which the weld is progressing or is being made. In backhand welding, the welding flame is pointed in the direction opposite to that in which the weld is being made.

3-9 What is meant by double-butt welding and single-butt welding?

Ans. In double-butt welding, the metal is deposited in the welding groove from both sides of the plate, whereas in single-butt welding the metal is deposited from only one side with the plates to be welded butted together instead of being lapped one over the other.

3-10 After the first side of double-welded butt joint has been welded, what should be done to the joint before the second side is welded?

Ans. The second side should be chipped, ground, or melted out to secure a clean surface of proper shape to ensure fusion of the weld metal on that side.

3-11 What is meant by reinforcement of a weld?

Ans. The amount of weld above the surface of plate on the face side. This ensures full penetration for full thickness of plates to be joined.

3-12 Why is it common practice to use heavily coated electrodes in metallic arc welding?

Ans. The coating shields the arc from the atmosphere and produces a slag covering on the weld, both of which give the weld metal good tensile strength and ductility.

3-13 Is single-butt welding permitted in the fabrication of fusion-welded boiler drums?

Ans. No, unless a backing strip is used. If a backing strip is used to ensure complete penetration of the weld metal, the weld may be made from one side only.

3-14 *(a)* What is the least elongation acceptable in applying a free-bend test?

(b) How many sets of test plates are required in a test of welder operators?

(c) How often should welders be required to take the qualification test?

Ans. *(a)* The specimen shall be bent cold under free-bending conditions until the least elongation measured within or across approximately the entire weld on the outside fibers of the bend-test specimen is *30 percent* or 700,000 divided by *U* plus 20 percent, whichever is less. Here *U* represents the minimum ultimate specified tensile strength of the material to be welded, in pounds per square inch, from Code stress tables.

(b) Two sets, one face-bend and one root-bend test per the thickness of the material, position, and welding process to be used. See Section IX of the ASME Code.

(c) When a welder or welding operator has not used the specific welding process such as metal arc, submerged arc, etc. to weld ferrous on nonferrous materials for a period of 6 months or more, or when there is a specific reason to question the welder's ability to make welds that meet specifications.

3-15 What welds does the ASME Power Boiler Code require to be subjected to radiographic examinations?

Ans. The following welds must be radiographed on power boilers:

1. If a quality factor of 100 percent is desired for steel castings, radiographic examination is required of all critical areas on castings 4½ in. or less.

2. Complete radiographic examination is required of all castings over 4½ in. thick.

3. All weld repairs on steel castings which exceed 1 in. in depth or 20 percent of the section thickness, whichever is less, must be radiographically inspected.

4. Longitudinal and circumferential joints of power boiler drums, except those made of pipe material not exceeding 10-in. nominal pipe size or 1⅛-in. thickness, must be radiographed in their entire length.

5. Circumferential welded joints of pipes, tubes, and headers must be radiographed if: *(a)* the pipe containing steam is 16-in. nominal pipe size and over or 1⅝-in. thickness and over; *(b)* the pipe containing water is 10-in. nominal pipe size and over or 1⅛-in. thickness and over; *(c)* the pipes or tubes with welds in contact with furnace gases are 6-in. nominal pipe size or ¾-in. thickness and over; and *(d)* the pipes or tubes with welds in contact with furnace gases and subject to radiation are 4-in. nominal pipe size and ½-in. thickness and over.

3-16 What are locked-up stresses in welding? What causes them?

Ans. Locked-up stresses are those internal stresses remaining in the weld metal and adjoining base material when a weld has been made. They are caused because the weld metal and the base material are, of necessity, at considerably different temperatures when a weld is made and therefore do not expand and contract uniformly, thus setting up internal stresses from HAZ effects.

3-17 What is meant by stress relieving and how is it accomplished?

Ans. Stress relieving or postweld heat treatment is the uniform heating of a structure or parts thereof (welded joint) to a temperature just below the critical metal range. The purpose of stress relieving is to relieve the major residual stress that may exist in a welded joint from the heat effects caused by the welding process.

The Code specifies preheating temperatures for the different types of P-number metals that may be welded. Generally, the preheat temperature is below 450°F. Postweld heat-treatment temperatures to be applied to the different types of metals are also specified in the Code by P-number groupings. For plain carbon steels, this is generally 1100°F to be held one hour per inch of plate thickness.

3-18 What is an "approved" or "qualified" process of welding?

Ans. One that has been written out in detail and followed in the making of test plates as required by the ASME Code, such test plates having been subjected to the tests specified in the Code and found to meet the requirements of the Code for the class of work to be done. This includes considering material (P numbers), thickness, welding process, electrodes, welding position, and heat treatment to be applied.

3-19 What is meant by the terms *lack of fusion, lack of penetration,* and *slag inclusions?* How are they caused and how should such conditions be remedied?

Ans. Lack of fusion is the lack of fusion or melting together of the base metals or weld metal being deposited. It is caused by improper welding techniques, among which could be the heat applied being too low because of low amperage, too fast welding and travel, improper surface preparation, and similar causes. When lack of fusion is discovered or noted, it is necessary to grind or chip out to sound weld metal and reapply weld metal to get fusion of the base metal.

Lack of penetration means that the welding or fusion of plates was not made for the entire thickness of the plate, leaving parts of the plates not joined. It is caused by improper joint preparation or fit, poor welding techniques, or improper surface preparation. It is corrected by chipping and grinding back to sound metal and filling up the area with a good penetrating weld per Code requirements.

Slag inclusions mean there is foreign matter in the weld, which can weaken the joint. This is generally caused by poor joint penetration or cleaning. To correct, the affected area is ground or chipped out and rewelded per Code requirements.

3-20 What is a backing strip as used in metallic welding? When a joint is welded from one side only, can it be considered the equivalent to a double-welded butt joint?

Ans. It is a strip of metal usually of the same material as that being welded, from 1 to 3 in. wide and ⅛ to ¼ in. thick, placed on the underside of the joint to be welded, to enable the welder to obtain penetration of the weld throughout the entire thickness of the plate or pipe and avoid the dripping through of molten metal. Welding from one side only can be considered a double-butt joint when complete penetration and reenforcement on both sides are achieved, as by the use of a backing strip or chill ring.

3-21 What is the maximum permissible offset of the abutting edges of welded joints at any one point in longitudinal joints and in girth joints?

Ans.

	Maximum offset permitted	
Thickness of plate	*Longitudinal*	*Circumferential*
To ½ in. inclusive	¼T	¼T
Over ½ to ¾ in. inclusive	⅛ in.	¼T
Over ¾ to 1½ in. inclusive	⅛ in.	3/16 in.
Over 1½ to 2 in. inclusive	⅛ in.	⅛T
Over 2 in.	Lesser of 1/16T or ⅜ in.	Lesser of ⅛T or ¾ in.

3-22 What preparations are made before joining together by a butt weld a ⅞-in. head flange with a ½-in. shell plate? Why are those preparations necessary?

Ans. A 3-to-1 taper is required in the thicker plate for the purpose of avoiding stress concentration at the transition section between the plates of unequal thickness.

3-23 A boiler drum 5 in. thick has a fusion-welding longitudinal joint. What is the minimum temperature under which it must be stress-relieved and for how long a period?

Ans. At least 1100°F for a period of 5 hr if we assume a low P number for the material.

3-24 Has an inspector any right to disagree with the manufacturer's records and call for a requalification of a welding operator or to have physical tests made?

Ans. If the inspector has any doubts regarding the ability of a welder or of the process, he or she can require a requalification test of the process or of the welder.

3-25 In what position must the test plates be when a welder makes test welds for the purposes of qualifications?

Ans. In whatever position the welder will encounter in actual work. If the welder makes the test plate in a flat position, she or he is not permitted to do work in other than that position, such as "vertical" or "overhead" welding.

3-26 *(a)* Is it permissible to locate a hole in a welded joint? *(b)* Is it permissible to roll and expand tubes into holes located in a welded joint?

Ans. *(a)* Yes, provided that magnetic-particle examination indicates the weld is sound after the hole is made. *(b)* Yes, as in part *a*.

3-27 When an unqualified welder is available for making a fusion-weld boiler repair, what two principles would determine the type of repair that may be made?

Ans. (1) The strength of the boiler must not be dependent on the weld. (2) The effect of welding heat and welding stresses must not be detrimental.

3-28 What is the maximum permitted length of a course in a horizontal boiler of either riveted or welded construction?

Ans. 12 ft.

3-29 *(a)* May the longitudinal riveted seams of power boilers be seal-welded?

(b) May circumferential seams be seal-welded?

(c) When repairs are made to the caulking edges of girth seams by fusion welding, what precautions should be taken regarding the rivets?

Ans. *(a)* No.

(b) Yes.

(c) The rivets should be removed and the holes reamed after welding, and new rivets should be driven. This is to prevent the rivets being distorted from the heat of welding and possibly cracking.

3-30 Describe briefly the necessary hydrostatic and other physical tests to be applied to a welded boiler drum or vessel before it can be accepted.

Ans. Welded joints on boiler drums must be radiographed in sections of the drum specified by the Code as well as stress-relieved. A hydrostatic test is required after the boiler has been completed with water at a temperature not less than 70°F. The pressure shall be applied gradually up to 1½ times the maximum allowable pressure to be stamped on the boiler. Control of the pressure must be such that the test pressure is never exceeded by more than 6 percent. The boiler is carefully examined for leakage when the pressure is dropped back to the maximum allowable pressure to be stamped on the boiler.

3-31 *(a)* Who is responsible for conducting tests of welding procedures and for qualifying welding operators?

(b) How long does a qualified welder's approval test remain in affect?

(c) Who keeps the qualification records of procedures and welder's approval and acceptance?

Ans. *(a)* The manufacturer is responsible for the tests, subject to an inspector's approval.

(b) Indefinitely if regularly employed and employed in the same type of welding, thickness, and material.

(c) The manufacturer.

3-32 What is meant by preheating and stress relieving?

Ans. Preheating is the application of heat to the base metals prior to a welding or cutting operation. Stress relieving or postweld heat treatment is the uniform heating of a structure or parts thereof (welded joint) to a temperature just below the critical metal range. The purpose of stress relieving is to relieve the major residual stress that may exist in a welded joint from the heat effects caused by the welding process.

3-33 What temperature should be used if it is found necessary to preheat and stress-relieve the welding of low-carbon steel?

Ans. The Code specifies preheating temperatures for the different types of P-number metals that may be welded. Generally, the preheat temperature is below 450°F. Postweld heat-treatment temperatures to be applied to the different types of metals are also specified in the Code by P-number groupings. For plain carbon steels, this is generally 1100°F to be held one hour per inch of thickness of plate.

3-34 *(a)* What is meant by a nondestructive test of a welded joint? Give some typical examples.

(b) What would cause you to reject a welder's test piece?

Ans. *(a)* A test which does not destroy the boiler, such as radiograph, hydrostatic, ultrasonic, dye-penetrant, and magnetic-particle tests.

(b) Porosity, lack of penetration, slag inclusions, and cracks on the surface of the weld, or any other such defects not meeting Code requirements.

3-35 State the essential requirements that must be met to permit the welding of pipe or tube connections which do not require radiographic examination or the hammer test.

Ans. If the procedure and operator have been qualified per Code requirements, circumferential welded joints on pipes or tubes do not require radiographic examination provided:

1. The steam pipe does not exceed 16-in. nominal pipe size or 1⅝-in. thickness.
2. The water pipe does not exceed 10-in. nominal pipe size or 1⅛-in. thickness.
3. For pipe or tubes that are in contact with furnace gases, but there is no radiation from the furnace, the pipes or tubes do not exceed 6-in. nominal pipe size or ¾-in. thickness.
4. The pipe or tubes will be in contact with furnace gas and subject to radiation from the furnace pipes or tubes.

NONDESTRUCTIVE TESTING

3-36 Define the half-life of an isotope.

Ans. The term *half-life* refers to the decrease in strength of radiation of an isotope; the half-life is the number of years it takes for the isotope to decrease one-half its present strength (time of reference).

3-37 If an isotope has a strength of 80 curies (Ci) now with a half-life of 10 years (yr), what will its strength be in 30 yr?

Ans. At the end of the first 10 yr, the strength will be 40 Ci. At the end of 20 yr, the strength will be 20 Ci. At the end of 30 yr, the strength will be 10 Ci.

3-38 Name two survey meters that are used to check on the radiation dosage to which an individual may be exposed.

Ans. Pocket dosimeters and film badges are used to check on the cumulative dosage of an individual.

3-39 *(a)* How is the energy level of an x-ray machine rated?

(b) What penetrates material better, short or long radiation waves?

(c) Does high or low voltage produce short waves?

Ans. *(a)* The energy level of an x-ray machine is rated by the peak voltage at which it can drive electrons into the tube target, and it is expressed as follows: kVP = peak kilovolts (1000 V) and MV = megavolt (1 million V)

(b) and *(c)* Short waves have greater penetrating power, and high voltages of the radiation source produce shorter wavelengths; therefore high voltage has greater penetrating power.

3-40 *(a)* Name two factors that control contrast in radiography.

(b) Name the two principal types of screens used in radiography.

Ans. *(a)* The voltage or energy level of the source of radiation that the radiography will be subjected to and the type of film used.

(b) Fluorescent and lead foil.

3-41 *(a)* Name two radiation sources permitted for radiographic examination in the Power Boiler Code.

(b) What class of film shall be used for radiographic examination of power boilers?

Ans. *(a)* X-rays and radioactive isotopes.

(b) Type 4 film may be permitted for power boilers as well as type 2, as stipulated in Section V.

3-42 What must appear on a radiographic image in order to prove the technique and sensitivity of the radiographic examination?

Ans. The radiographic image must display the penetrameter and the specified hole in the penetrameter in order to prove that the technique used was of the proper sensitivity. In addition, the radiographs should display the radiograph identifying number and letters so the area radiographed can be traced.

3-43 *(a)* What is the purpose of a penetrameter? What is penetrameter sensitivity?

(b) From what type of material should penetrameters be fabricated when they are to be used to radiograph welds?

Ans. *(a)* A penetrameter is used to obtain evidence on a radiograph that the technique used in exposing a joint to radiography will show certain minimum defects considered acceptable, but will show defects considered unacceptable by Code requirements. Penetrameter sensitivity refers to the image quality and smallest hole that can be "sensed" or seen; therefore, it is a measure of the lowest or minimum discontinuity that may be measured or observed by radiography on a particular thickness of metal.

(b) Penetrameters should be fabricated of material radiographically similar to the object being inspected.

3-44 *(a)* Where are penetrameters normally placed in relation to the weld and the source of radiation?

(b) If the penetrameter is placed on the film side of the joint, how is this identified?

Ans. *(a)* Penetrameters are normally placed adjacent to the weld seam and on the source side of the seam to be radiographed.

(b) A lead letter *"F"* at least as high as the identification number must be placed adjacent to the penetrameter for those cases where the penetrameter is placed on the film side, and not on the source side.

3-45 How large must the diameter of the hole be in a penetrameter in relation to penetrameter thickness *t*?

Ans. $2t$ or twice the thickness of the penetrameter, except for penetrameters under 0.010-in. thickness, where the hole diameter must be a minimum of 0.020 in.

3-46 *(a)* Why is it necessary to remove weld ripples when a weld is to be radiographed?

(b) List some typical blemishes that must be avoided on radiographic film.

Ans. *(a)* It is necessary to remove irregularities on weld surfaces so that any radiographic image taken of the weld does not mask or hide, or in other ways make it difficult to identify, any unacceptable weld discontinuity that may appear on the radiographic image.

(b) Radiographs must be free from the following blemishes so that discontinuities can be properly identified: fogging; streaks, water marks, stains; scratches, finger marks, crimps, dirt, smudges, tears; loss of detail on the image.

3-47 *(a)* May backing strips on single-welded circumferential butt joints for power boilers be left in place even when the joint is to be radiographed?

(b) What procedure must be followed in using penetrameters for weld radiographing for those cases where the weld reinforcement or backing strip is not removed?

Ans. *(a)* Yes, backing strips on single-welded circumferential butt joints may be left in place, provided that they are not to be removed subsequently and that the image of the backing strip does not interfere with the interpretation of the resultant radiographs.

(b) If the weld reinforcement and/or backing strip is not removed, a shim of material radiographically similar to the weld metal should be placed under the penetrameter.

3-48 What two imperfections in a weld are considered unacceptable by the Power Boiler Code and which appear on a radiographic film?

Ans. Any type of crack or zone of incomplete fusion or penetration is considered unacceptable.

3-49 How long must a manufacturer retain in file radiographs taken on a power boiler?

Ans. A set of radiographs taken for each power boiler must be kept in file by the manufacturer for at least 5 yr.

3-50 *(a)* To what type of material may magnetic-particle examinations be applied?

(b) In prod magnetic-particle testing, what is the maximum prod spacing permitted?

(c) In magnetic-particle testing, what direction is the magnetic field in relation to current flow?

Ans. *(a)* Only to ferromagnetic materials.

(b) A maximum of 8 in.

(c) At right angles, or perpendicular, to current flow.

3-51 *(a)* When are the best results obtained in magnetic-particle inspection—when the magnetic field is parallel to a discontinuity or when the magnetic field is at right angles to the discontinuity?

(b) Differentiate or describe the difference between the continuous method and residual method as used in magnetic-particle inspection.

Ans. *(a)* The best results are obtained when the magnetic field is at right angles to the discontinuity or when the magnetizing current is flowing parallel to the discontinuity.

(b) In the continuous method, the dry powder is applied to the surface of the work under test while the magnetizing current is flowing. In the residual method, the dry powder is applied to the surface of the work after the magnetizing current has been switched off. This method requires the work to be one that retains the magnetism, whereas the continuous method can be used on material which has low magnetic retentivity.

3-52 How many ultrasonic examinations are required of an area to be examined in a part under test?

Ans. At least two separate examinations shall be carried out on each area. The second test should produce lines of magnetic flux approximately perpendicular to the first test. This is necessary in order to pick up discontinuities which might be parallel to the magnetic field and thus not become visible if only one test were performed.

3-53 Name the three types of penetrant recognized by the Code in making liquid-penetrant examinations.

Ans. Water-washable, postemulsifying, and solvent-removable.

3-54 How does liquid penetrant work so that it can be used for examination for discontinuities?

Ans. Liquid-penetrant examination is used to detect discontinuities which are open to the surface of nonporous material. A liquid called a penetrant is first applied to the surface to be examined. The penetrant is allowed to stand for awhile on the part so that it can enter any discontinuity. The part is then wiped free of surface penetrant. A drier is then applied to the surface. Finally, a developer is sprinkled over the part. Any penetrant trapped in a discontinuity below the surface will now wet the developer so it sticks out and thus exposes the discontinuity that needs to be evaluated for acceptance of the part or rejection unless it is repaired.

3-55 What is type A and type B liquid-penetrant inspection?

Ans. Type A uses fluorescent penetrants while type B uses a visible-to-the-eye dye penetrant.

3-56 Name the two types of ultrasonic principles used in nondestructively testing material to be used for pressure-vessel parts.

Ans. The first type of ultrasonic principle used in nondestructive testing is pulse echo. In this system, sound pulses are sent through a material, and any reflecting surfaces echoes this sound. The signals are displayed on a CRT screen, and with proper adjustment, the echo principle is used to detect flaws at the interface of materials which produce echoes.

The second ultrasonic principle used in ultrasonic testings is the resonance principle. In this system, the frequency of sound from an instrument is varied until it matches the material under test or is in resonance with it. By electronic and electric circuit principles, the resonant frequency can be converted to thickness measurements of the material under test. This is the chief application for resonant ultrasonic-sound instruments.

3-57 *(a)* What method of ultrasonic examination of welds is permitted by the Code?

(b) When is it permissible to use ultrasonic examination on welded joints for power boilers?

Ans. *(a)* Ultrasonic examinations shall be conducted with a pulse-echo type of system with rated frequency ranges of 1 to 5 MHz.

(b) When it is impractical to use a combination of radiography parameters so that a geometric unsharpness of 0.07 will not be exceeded, then ultrasonic examination shall be used to check the weld and heat-affected zones of those joints requiring radiographic examination by Section I rules.

3-58 Describe how the following surfaces shall be prepared for ultrasonic examination: *(a)* contact surfaces, *(b)* welded surfaces, *(c)* base material.

Ans. *(a)* The finished contact surfaces shall be free from weld spatter and any roughness that would interfere with free movement of the search unit or impair the transmission of ultrasonic vibrations.

(b) The finished weld surfaces, where accessible, shall be of adequate smoothness to prevent interference with the interpretation of the examination. The weld surface shall merge smoothly into the surfaces of the adjacent base material.

(c) After the weld is completed but before the angle-beam examination, the area of the base material through which the sound will travel in angle-beam examination shall be completely scanned with a straight beam search unit to detect reflectors which might affect the interpretation of angle-beam results. Consideration must be given to these reflectors during interpretation of weld examination results, but their detection is not a basis for rejection of the base material.

3-59 *(a)* Name three types of waves used in ultrasonic inspections.

(b) What does the term CRT mean as applied to ultrasonic testing?

(c) What does a piezoelectric element do as applied to ultrasonic testing?

Ans. *(a)* Straight-through waves, angle-beam waves, and surface waves.

(b) Cathode-ray tube. This is the tube on which the ultrasonic signals appear. If it is adjusted properly, they will show "beeps" where sound enters and where it bounces off either a fault or an opposite surface.

(c) The piezoelectric element is a carefully manufactured crystal which has the property of converting electric signals into mechanical vibrations or sound signals. The crystals also have reverse capability, namely, of converting vibrations or sound signals into electric signals.

3-60 *(a)* What nominal frequency should be used to calibrate by the ultrasonic angle-beam method?

(b) What is considered the primary reference response of a pulse echo using the angle-beam method when one is calibrating ultrasonic equipment?

(c) Describe the ultrasonic transfer method used to correlate the response from the basic calibration block and from the production material when the angle-beam method is used.

Ans. *(a)* The nominal frequency shall be 2.25 MHz unless a variable, such as production-material grain structure, requires the use of other frequencies to ensure adequate penetration.

(b) Seventy-five percent of full screen on the cathode-ray tube.

(c) Transfer methods are used to correlate or check the response from the basic calibration block and from the component under examination. Readings are taken on the calibration block, and differences in readings from the component under test and the reference block are then corrected by recalibrating the instrument.

3-61 *(a)* When one is examining welds by the ultrasonic method, what shall serve as basic calibration reflectors and what are the material requirements for these calibration reflectors?

(b) In ultrasonically examining welds by the angle-beam method, how often must the transfer method be used to check technique on vessel welds and pipe welds?

Ans. *(a)* Drilled holes shall be used as basic calibration reflectors, and these holes may be located in the base metal or extension of it or may be located in a calibration block of similar metallurgical structure as the metal being welded.

(b) For vessels, the transfer method must be used at least once for each 10 ft of weld or less per plate, and shall be performed at least twice for each type of welded joint. For pipe, the transfer method must be used, as a minimum, once for each welded joint on pipe sizes of 10-in. diameter and over and once for each 5 ft of weld for pipe less than 10 in. in diameter.

3-62 *(a)* Who has the responsibility of preparing an ultrasonic examination report for power boilers?

(b) What information shall the ultrasonic examination report contain?

(c) What minimum length of discontinuity is not permitted by the Power Boiler Code on welds examined by the ultrasonic method and where the base plate being welded is 3 in. thick?

Ans. *(a)* The manufacturer.

(b) (1) All procedures and equipment shall be identified sufficiently to permit duplication of the examination(s) at a later date. This shall include initial calibration data for the equipment and any significant changes in subsequent recheck. (2) A marked-up drawing or sketch indicating the weld(s) examined, the item or piece number, and identification of the operator who carried out each inspection or part thereof are required. (3) A record of repaired areas shall be kept as well as the results of the re-examination of the repaired areas.

(c) Discontinuities with lengths of ¾ in. or over for plate thickness over 2¼ in.

3-63 *(a)* In ultrasonic examination of steel castings by the angle-beam method, how often must the transfer method be used to recheck the calibration of the ultrasonic instrument?

(b) When must both angle-beam and straight-beam methods be used in ultrasonically examining welded joints?

(c) When may flat basic calibration blocks be used for calibrating ultrasonic instruments that are to be used in examining circumferential welds on curved contact surfaces?

Ans. *(a)* The transfer method shall be used for calibration checks against the standard at least each ½ hr.

(b) If the geometry of the welded joint does not make it possible to make examination by the angle-beam method for both sides of a weld from a single surface or a combination of surfaces, then both the angle-beam and the straight-beam methods shall be used to examine the welded joint.

(c) For contact surfaces with curvatures greater than 20-in. diameter, flat basic calibration blocks may be used or blocks of essentially the same curvature as the part to be examined.

3-64 *(a)* Name the two types of induced electric or magnetic fields that a part under examination may be subjected to when being tested by eddy current.

(b) Describe the term *inductor* as applied to *eddy-current* testing.

(c) When the *eddy-current* tests are applied, how are the magnitude and direction of the induced eddy currents altered?

Ans. *(a)* The effect of the eddy current on the part being tested may be to induce electromagnetic *currents* or, if the material is magnetic, to induce or set up *magnetic* fields.

(b) The inductor is an electromagnetic coil to be operated with alternating current which is positioned close to the article under test and which, when energized, induces eddy currents or sets up magnetic fields or does both in the part under test.

(c) The magnitude and direction of the induced eddy currents are altered by discontinuities in the metal being tested.

3-65 *(a)* How is the change in the induced eddy current noted?

(b) How is the change in induced magnetic flux noted when eddy-current testing is used?

(c) What frequency of alternating current is used in eddy-current testing?

Ans. *(a)* The change in the induced eddy current is detected by a detector coil which is connected to electronic circuitry. This circuit registers the discontinuity in several different electrical variables, depending on the design, but these may be voltage, current, or impedance. These then are related in magnitude to the discontinuity through reference blocks.

(b) The distribution of the induced magnetic flux is affected by discontinuities, too, and this is revealed by the change in magnetic-flux distribution as it is affected by the defect.

(c) The frequency chosen to excite the electromagnetic field depends on the depth of penetration beneath the surface of the part being tested. Frequencies vary from 500 to 20,000 Hz (cycles per second). The higher frequencies are used for small-diameter high-wall nonmagnetic tubing.

3-66 How would you define a leak test as applied to checking welds on a pressure vessel?

Ans. Leak tests of welds of pressure vessels are used to determine discontinuities that extend through a weld and thus result in a loss of contents—either liquid or gas. The tests used in leak testing may range from a hydrostatic test using water to sophisticated methods using low-pressure gas and electronic equipment for the detection of the leaking gas through a discontinuity.

3-67 *(a)* Who is responsible for drafting a nondestructive examination (NDE) procedure?

(*b*) What shall it contain?

(*c*) How is the procedure proved?

(*d*) To whom shall the procedure be available?

Ans. (*a*) The manufacturer or assembler per Section V, ASME Codes; for nuclear in-service inspection, the owner of the nuclear facility is responsible for qualifying the NDE personnel.

(*b*) For nuclear work, all nondestructive examinations shall be performed to a written procedure, which will clearly indicate the method and technique to be used in order to properly inspect the vessel under test. The exception to this rule on power boilers is detailed in article 3 of Section V for radiographic examination in which procedure qualification is not required as long as density and penetrameter image requirements are met. As an example for good radiographic procedures for nuclear vessels, the following must be detailed: (1) material to be radiographed; (2) range of thickness to be radiographed; (3) type of radiation, source, voltage rating, and manufacturer of the radiation source equipment; (4) film brand or type to be used; (5) type and thickness of intensifying screens and filters to be used; and (6) minimum source-to-film distance.

(*c*) The procedure must be such as to prove the method to be used will show defects not permitted by the Code. In all cases, the procedure must demonstrate this capability to the satisfaction of the inspector. This is a good rule to follow in inspection work.

(*d*) To the nondestructive examination personnel of the manufacturer and the authorized inspector.

4

Strength of Materials and Stress Analysis

BOILERS AND STRESSES

The potential loss of life and damage to property caused by the explosion of boilers have been brought to public attention in news accounts. Pictures of such disasters are shown and causes and preventives discussed in detail later in this book. See Fig. 4-1.

Boiler failures are preventable; and, since every failure involves stress, a working knowledge of stresses should be had by all concerned with boiler operation, design, inspection, or maintenance.

INTERNAL PRESSURE VERSUS STRESSES

The cause of stresses in boilers should be understood before an attempt is made to calculate or analyze them.

Internal pressure is the first cause to consider. The U.S. measure of steam pressure is in units of pounds per square inch as read on the pressure gauge; it is above the pressure of the atmosphere (14.7 psi at sea level). This unit measurement is known as *gauge pressure.* In some operation formulas and steam tables, absolute pressure is used. This means the pressure in pounds per square inch above a perfect vacuum, that is, the gauge pressure plus approximately 14.7 at sea level.

The foregoing facts relate to unit pressure, that is, the force acting on 1 in.^{2}. If the area of a surface in a steam boiler is 20 in.^{2} and if it is exposed to

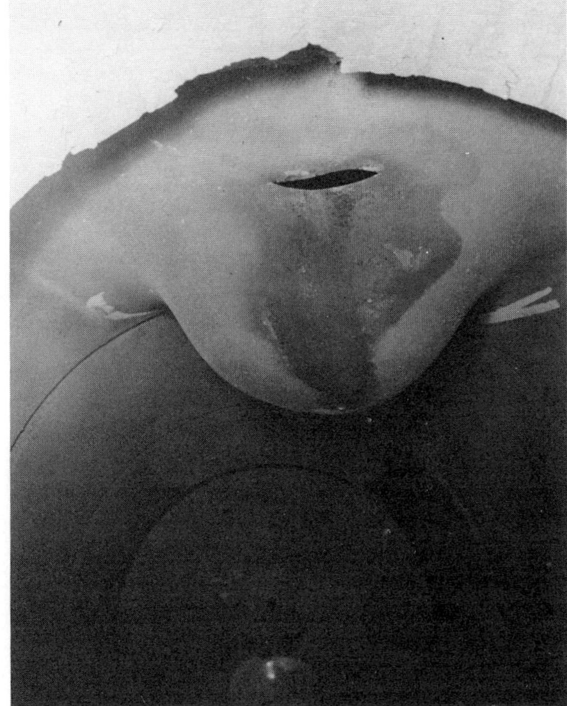

Fig. 4-1 Collapse of scotch marine boiler furnace from overheating. *(Courtesy Royal Insurance Co.)*

a steam pressure of 50 psi, the total pressure, or load, on that surface will be 20 × 20 × 50 = 20,000 lb. Thus, it is easily seen that the pressure acting on a large exposed area may cause a tremendous load, even with a comparatively low unit pressure. Also, this example makes clear the fallacy in feeling that there is no danger possible with a low-pressure boiler.

In addition to pressure acting on the structural components of a boiler, other factors to consider are temperature, cycling, corrosion, creep of the material used in high-temperature service, and a multitude of factors involving the properties of the materials used in the fabrication of the boiler. With the advent of nuclear power, vast amounts of technical progress have been made in systematically analyzing the forces that can act on a pressure vessel's material and the methods or stress analysis that can be used to ensure a safe operating life. This includes the use of fracture mechanics to predict crack growth rates and determining cycles of operation before repairs are needed.

The properties of a material that are important in analyzing its strength to resist loadings must be first reviewed.

Any body or material subject to external forces on it will resist these external forces. This resistance of the material comes from *within* the material. The

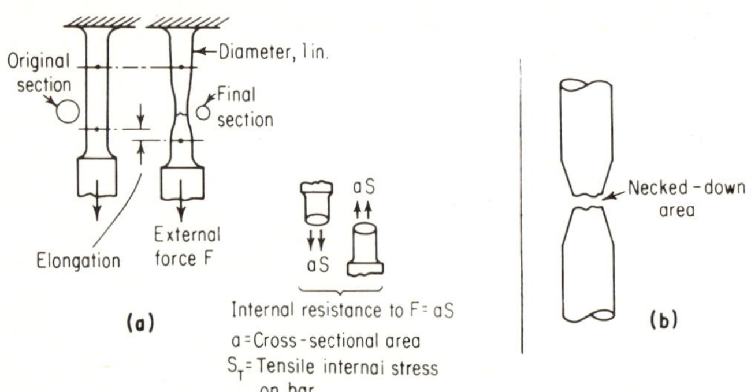

Fig. 4-2 Determining tensile stress of a bar: *(a)* specimen is stretched until failure occurs; *(b)* necked-down area after rupture.

internal structure of the material is subject to intercrystalline loading when an external force is applied. Thus, *stress* is defined as the internal force per unit area on the material which resists external forces on the material. It is expressed in pounds per square inch, but the notation *psi* is *not* used for stress as it is in pressure notation. Stress is always expressed by engineers with the notation $lb/in.^2$ to differentiate it from the psi designation of pressure, which is an *external force* per unit area on the material.

There are several general classifications of stresses that affect materials. A *normal stress* is a *stress* on an area of a material produced by a force at a right angle to the area acted on. Normal stresses are further classified as *tension stresses* or *compression stresses.* In Fig. 4-2*a*, a 1-in.-diameter bar is pulled by a force *F*. This force produces a normal stress at a right angle to the cross-sectional area of the bar. Since the force tends to pull (stretch) the bar apart, it is called a *tensile* stress. Within the material, the intercrystalline structure (assuming it is steel) is also being stressed. The tensile stress of the bar is found by the following equation, based on the definition of stress as pounds per square inch, or internal force per unit area within the material:

$$F = aS_T$$

where

F = external force
a = cross-sectional area of material resisting F
S_T = tensile (internal) stress on the material

From this, the equation for stress S_T is

$$S_T = \frac{F}{a}$$

This equation shows that the tensile stress S_T is found by dividing the external force F, acting normal to the cross-sectional area, by the cross-sectional area of the material resisting the force.

If the force were acting in the opposite direction, a compressive stress (known also as a *bearing stress*) would be imposed on the material. But the compressive stress would be found by the same equation,

$$F = a \times S_C$$

where S_C = compressive stress.

Assume that in Fig. 4-2*a* the tensile force is 15,000 lb and the rod is of 1-in. diameter. What would be the tensile stress on the bar?

$$S_T = \frac{F}{a}$$

$$F = 15{,}000 \text{ lb}$$

$$a = \frac{\pi\,(1)^2}{4} = 0.7854 \text{ in.}^2$$

Substituting, we get

$$S_T = \frac{15{,}000}{0.7854}$$

$$= 19{,}099 \text{ lb/in.}^2$$

If a force acts tangent (sideways) to the area of a material, a shear stress is produced. This is illustrated in Fig. 4-3, in which a force F acts on the rivet area tangent to its cross-sectional area, thus producing a shear stress on the rivet. The shear stress is found by the equation $F = a \times s$, but a is the cross-sectional area resisting shear. In Fig. 4-3*a*, since only one area of the rivet is resisting the external force F, it is in *single shear.*

In Fig. 4-3*b*, a butt-riveted joint is shown with butt straps on each side of the butting plates. The rivets in this illustration are in double shear, as *two*

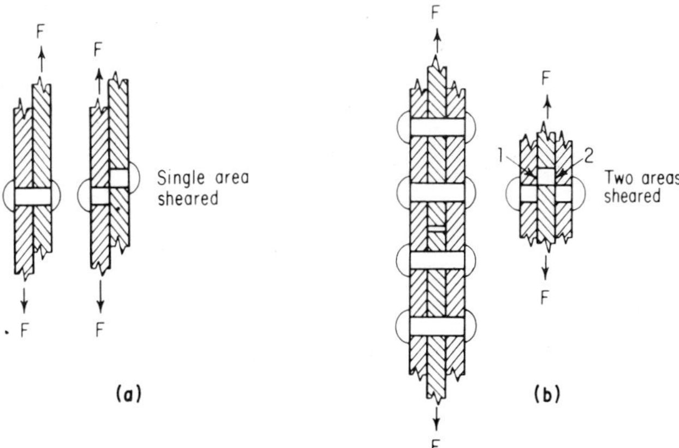

Fig. 4-3 *(a)* Rivet in single shear; *(b)* rivet in double shear.

areas of the rivet are resisting the load F. An example will illustrate the significance of single shear and double shear.

Assume that in Figs. 4-3*a* and 4-3*b* only one rivet is being considered for analysis, but one is in single shear, the other in double shear. Assume the rivet to be 1 in. in each case and the load F to be 15,000 lb on each rivet. What is the shear stress for each rivet?

Single-shear rivet

$$S_S = \frac{F}{a}$$

$$F = 15{,}000 \text{ lb}$$

$$a = \frac{\pi D^2}{4}$$

$$= \frac{\pi (1)^2}{4}$$

$$= 0.7854 \text{ in.}^2$$

$$S_S = \frac{15{,}000}{0.7854}$$

$$= 19{,}099 \text{ lb/in.}^2$$

Double-shear rivet

$$S_S = \frac{F}{a}$$

$$F = 15{,}000 \text{ lb}$$

$$a = 2\frac{\pi D^2}{4}$$

$$= 2\frac{\pi (1)^2}{4}$$

$$= 1.5708 \text{ in.}^2$$

$$S_S = \frac{15{,}000}{1.5708}$$

$$= 9549.5 \text{ lb/in.}^2$$

This shows that a rivet in double shear is *twice* as strong as a rivet in single shear.

Another stress is that due to bending. A beam supported on each end and loaded in the middle will develop a bending stress. The beam, when bending, will actually be under a tension stress on one side and a compression stress on the opposite side. This is illustrated in Fig. 4-4. Flat plates and stayed surfaces in boilers are some elements that are subjected to bending stresses.

Note: Stress due to torsion, such as in an axle being rotated and transmitting power, is another stress considered in an analysis of resistance of materials. Stress analysis is beyond the scope of this book. However, a knowledge of *tension, compression, shear,* and *bending* stresses is essential in understanding how pressure is contained in a pressure vessel, pipe, or any other apparatus made of material designed to confine that pressure within safe limits.

When a body or material is subjected to external forces, internal stresses resist these forces, but there is always some deformation with load. For exam-

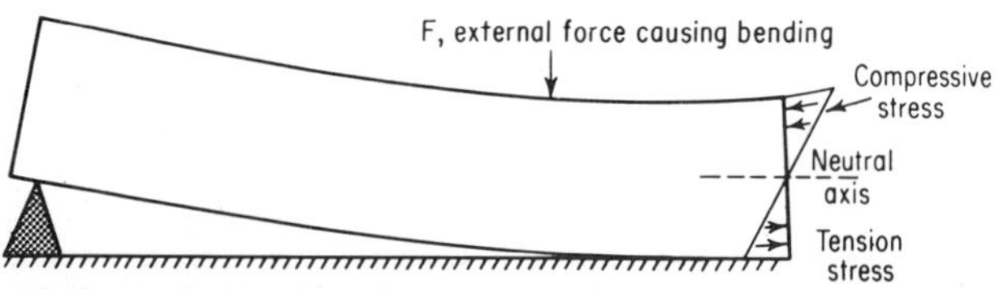

Fig. 4-4 Bending causes compressive stress on top of the beam and tension stress on the bottom of the beam.

ple, a steel rod will stretch when pulled on by an external force. The *total stretch* is expressed in a length measurement such as inches or centimeters. *Strain* is defined as the stretch *per unit length,* or deformation of a body per unit length, and is always expressed as inches per inch, centimeters per centimeter, etc. For example, assume that a steel rod 10 in. long stretches 0.010 in. with load; then the unit strain will be 0.010/10 = 0.001 in./in.

Some of the fundamental properties of all structural materials that must be determined are found by means of *stress-strain diagrams.* From the stress-strain diagram, structural properties determined are proportional limit, yield point, ultimate strength, and modulus of elasticity.

Modern engineering practice requires testing of materials so as to specify and identify their physical properties. This is particularly true for materials intended to be used in boilers, pressure vessels, and nuclear reactors. Steel manufacturers' laboratories run tests, and the Boiler Code shows sketches and requirements for preparing samples to run tensile tests and bending tests on boiler materials.

A test specimen of a specified grade of steel is cut out from a rolled stock, then machined to fit a test machine. A test specimen for a tensile test is shown in Fig. 4-5*a*. The square ends are clamped in jaws of a tension-test machine. The ½-in. round center piece is marked off in a 2-in. gauge length. An extensometer is attached to the 2-in. gauge length. This tension-test machine has a dial gauge indicating the force F applied to pull the rod apart. Incremental loading is applied. At each increment, the force F is recorded and the total amount of length l is taken at that incremental loading. This procedure is followed until rupture occurs. The data are tabulated, and then by the relationship F/a, where a is the original ½-in.-diameter area, stresses are found for each increment of load.

The strain, or stretch, is found by calculating the amount of stretch from the original 2-in. length to obtain the unit strain ϵ. These values are then plotted as in Fig. 4-5*b*, showing a stress-strain diagram for ductile steel.

The stress-strain diagram shows a sloping straight line extending from zero upward to the point marked *PL*. The reason is that as the load is increased, the imposed stress S increases, as does the strain ϵ. The increase for both is in the same ratio, meaning that if the load is doubled, the stress is doubled, and so is the strain. This is why a straight line is drawn, not a curve. The *proportional limit* of a material is thus the maximum unit stress that can be developed in the material without causing a deviation from the law of proportionality of unit stress to unit strain. Of even more significance, it implies that if the load is decreased, the material will return to its original length without having a permanent *set* as a result of the loading. The material will not be permanently deformed, but will return to its original shape as long as the proportional limit is not exceeded.

As the load on our test specimen is increased further, causing a stress greater than the proportional limit, a unit stress is reached at which point the material continues to stretch *without* an increase in load, assuming it is ductile steel. The unit stress at which this stretch-without-load occurs is called the *yield*

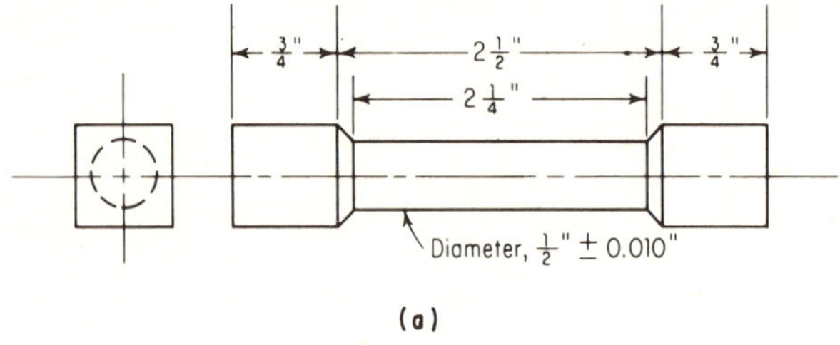

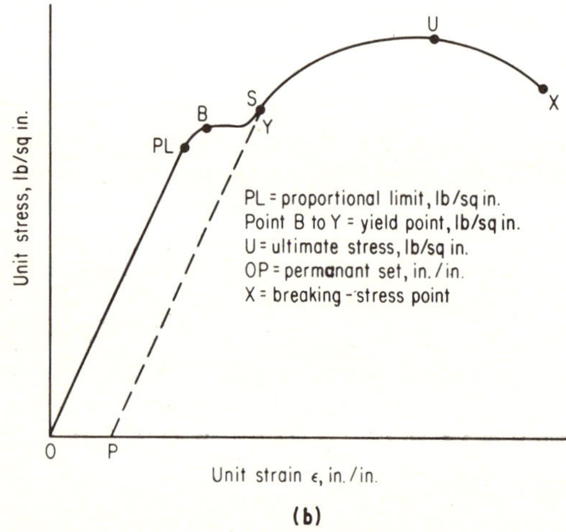

Fig. 4-5 *(a)* Specimen is prepared for tensile testing. *(b)* Stress-strain diagram is obtained from tensile test for a ductile material.

point and is represented by the short horizontal line *B* to *Y* on the stress-strain diagram. The *yield point* of a material is defined as the minimum unit stress in the material at which the material deforms or stretches appreciably without an increase of load.

If a material is stretched or loaded slightly beyond the yield point, a permanent set or deformation occurs in the material. For example, in Fig. 4-5*b*, if the load is reduced to zero after just passing the yield point, the extensometer will show a permanent stretch or deformation. This is found by drawing a line parallel to the proportional-limit line, and the set will be length 0 to *P* per inch of test specimen.

If the loading on our test specimen is increased, as indicated by the curve *S* to *U*, a point of maximum unit stress is reached. Then the unit stress declines with slight additional loading and stretches until it breaks. This is particularly true of ductile material, which *necks down* very rapidly after reaching its maximum unit stress because of reduced area in the neck section, which requires less load to cause rapid stretching to complete breakage. See Fig. 4-2*b*.

The *ultimate strength* of a material is defined as the maximum unit stress that can be developed in the material as determined from the *original* cross section of the material. It is point U in the stress-strain diagram. The curve from U to X is a rapid, unstable testing condition, with point X called the *breaking-load point.* The maximum unit stress is at point U, and this is the ultimate stress designated for the material.

Let us take an example. Data supplied on a tested specimen are: total length, 19 in.; sections at each end of a specimen are 1 in. × 1 in. × 6 in. long; center section 7⁄16-in. diameter by 7 in. long with center section concentric with the square sections at each end; center 7-in. section has a 2-in. gauge length marked.

The specimen being tested here broke when the load reached a maximum of 11,274.75 lb, and the break was through the original 7⁄16-in. diameter. But the diameter was now ¼ in. The cross-sectional area of a 7⁄16-in.-diameter bar is 0.15033 in.^{2}. The length of the 2-in. gauge section had stretched to 2.55 in.

1. Find the ultimate strength of the material.
2. What is the ultimate strength in pounds per square inch of the 1 in. × 1 in. section at each end?
3. What is the percentage elongation of the 2-in. gauge section?

1. In this problem the break was in the 7⁄16-in.-diameter section, so the stress S is

$$S = \frac{F}{a}$$

where a = original cross-sectional area.

$$S = \frac{11{,}274.75}{0.15033}$$
$$= 75{,}000\text{-lb/in.}^2 \text{ ultimate stress}$$

2. The ultimate stress at the 1 in. × 1 in. section is the same (75,000 lb/in.^{2}) because it is still the same material. It did not break at this section because the cross-sectional area is larger than at the 7⁄16-in.-diameter section.

3. The percentage elongation is found as follows:

$$\frac{(\text{Final length} - \text{original length}) \times 100\%}{\text{Original length}} = \%\ \text{elongation}$$

$$\frac{(2.55 - 2) \times 100\%}{2} = \frac{55\%}{2} = 27.5\%$$

The modulus of elasticity, also known as Hooke's law, states that the unit stress in a material is proportional to the accompanying unit strain, provided that the unit stress does not exceed the proportional limit. In different words, it states that the ratio of stress to strain for a certain material is always a constant, called E, the modulus of elasticity, or in equation form,

$$E = \frac{\text{stress}}{\text{strain}} = \frac{S}{\epsilon} = \text{constant}$$

For steel, the modulus of elasticity is usually taken as 30,000,000 and written 30×10^6 lb/in.^{2}. This is the modulus of elasticity for normal or axial loads. There is also a shear modulus of elasticity. For steel it is 12,000,000 lb/in.^{2}.

Strain gauges are used to determine stresses at critical areas of boilers, nuclear reactors, and pressure vessels for which exact calculations cannot be made. With the following relationship among stress, strain, and the modulus of elasticity of the material, the stress can be calculated. It is much easier to measure strain, or the deformation of a material under load, than to measure stress.

We explained that stress for normal loads is F/a, where F = imposed load and a = original area of material resisting the load. We also explained that unit strain ϵ is e/l, where ϵ = strain in inches per inch, e = amount of strain from original length l, and l = original length.

Now,

$$E = \frac{\text{stress}}{\text{strain}}$$

Substituting the above values gives

$$E = \frac{F/a}{e/l}$$

$$E = \frac{S}{e/l}$$

as

$$S = \frac{F}{a}$$

Rewriting this in terms of stress S,

$$S = \frac{Ee}{l} = E\epsilon$$

as

$$\frac{e}{l} = \epsilon$$

It can be seen that if strain is measured, stress can be calculated by knowing the modulus of elasticity of the material, which is usually a constant for the class of material being considered.

The modulus of elasticity is a measure of the *stiffness* of a material. For example, if one material has a modulus of elasticity twice as large as that of another material, the elastic unit strain in the one material for a given unit stress is one-half as large as that in the other material. Thus, one material is considered twice as stiff as the other. Some common E values are steel, 30 million; cast iron, 15 million; aluminum, 12 million, concrete, 3 million.

The *elastic limit* is the maximum unit stress that can be developed in the material without causing a permanent set. Test results show that for most structural metals the elastic limit of the material has about the same value as the proportional limit, and in most technical literature the elastic and proportional limits are considered identical. A small difference is apparent in testing work, but for practical purposes they can be treated as identical quantities.

LONGITUDINAL VERSUS CIRCUMFERENTIAL STRESSES

Internal pressure in a cylindrical shell closed at each end tends to burst the vessel along two distinct axes. First, the total pressure acting on the shell tends to cause rupture along a longitudinal axis. The total pressure acting on the heads tends to cause fracture of the shell around its circumference.

Thin-walled cylinders, meaning those where the thickness of the shell does not exceed one-half the inside radius, have two stresses, *longitudinal stress* and *circumferential stress*. The latter sometimes is called the *transverse stress*. Thick-walled cylinders have these stresses also, but they are determined differently. Both stresses are known by these names because of the loading they resist in a cylinder. Both are fundamentally tensile stresses.

Figure 4-6*a* shows a seamless cylinder with an inside diameter D, shell thickness t, length L, and with a uniform pressure P acting inside the cylinder. Pressure acts on the cylinder walls, so the resultant force created tends to split the cylinder along its long axis. Thus the first stress to be considered is the longitudinal stress resisting this force tending to split the cylinder along this axis. The pressure acts in all directions. But if we cut the cylinder as in Fig. 4-6*b*, which shows the external force on one side and also the internal material stress resisting this external force, the following is developed for a condition of equilibrium to exist.

The force tending to split the cylinder is area times pressure. This is

$$D \times L \times P = \text{force acting on one side}$$

where $D \times L$ = projected effective area. The internal force of the material resisting this force is

$$\text{Stress} \times \text{material area}$$

or

$$S_L \times t \times L \times 2 = \text{resisting force}$$

where $t \times L$ = one area of the material. But since there are two material areas resisting the force, it is multiplied by 2. Equating the two forces gives

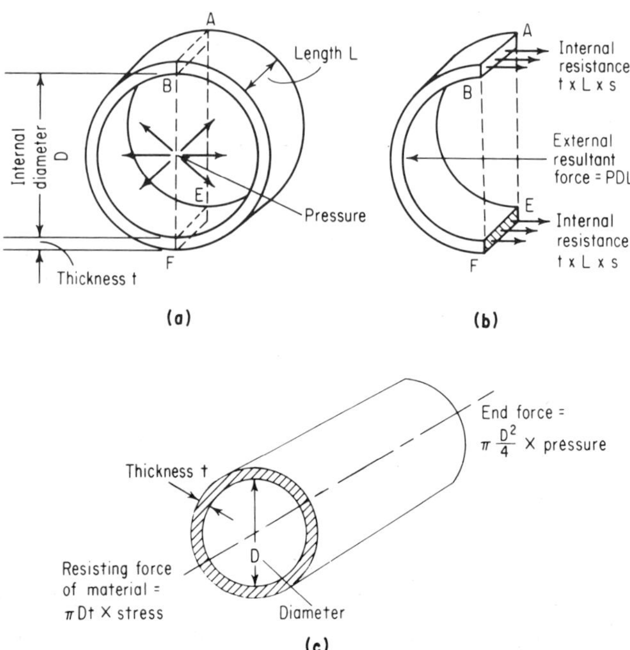

Fig. 4-6 *(a)* and *(b)* Longitudinal force with pressure acting on the side of the cylinder tends to split the cylinder lengthwise. *(c)* Pressure on the end of the cylinder tends to split the cylinder circumferentially.

$$D \times L \times P = S_L \times t \times L \times 2$$

From this,

$$\text{Longitudinal stress } S_L = \frac{DP}{2t}$$

The force tending to split the cylinder endwise, or around its circumference, is shown in Fig. 4-6*c*. Pressure acting on each end creates a force which is equal to the end area (circle) times the pressure, or

$$\frac{\pi D^2}{4} \times \text{pressure} = \text{end force}$$

The material resists this by a force equal to the end area of the material times the stress, or

$$\pi D t S_c = \text{resisting force}$$

where S_c is circumferential stress. Equating the two forces for equilibrium gives

$$\frac{\pi D^2}{4} P = \pi DtS_c$$

By elimination, solving for S_c, circumferential stress, gives

$$S_c = \frac{DP}{4t}$$

If we compare this with the longitudinal stress, we find the circumferential stress is one-half the longitudinal stress.

The two equations for longitudinal and circumferential stresses are fundamental strength-of-material equations. They are modified somewhat by the Boiler and Pressure Vessel Code to take into account manufacturing and experience factors.

The equations developed are for seamless construction, meaning that no welded or riveted joint is present. Later chapters show how the joint efficiency has to be considered to modify these equations. Note that equations for both longitudinal and circumferential stresses (due to pressure) are independent of the length of the vessel. But if a vessel is very long, the bending stress will have to be added to the stress due to pressure. This is especially true of a vessel filled with a substance of considerable weight.

The significance of the circumferential stress being one-half the longitudinal stress in a cylinder enters many problems in boiler design and calculation. For example, riveted circumferential joints do not have to be as strong in this direction as they do longitudinally. But in many calculations it is extremely important to check a cylinder both longitudinally and circumferentially, so as to make sure that the strength circumferentially is at least one-half the strength longitudinally. This is brought out in other chapters.

Temperature above designed limits has the immediate effect of lowering the permissible stress on a material. For example, SA-30 grade-A quality carbon-plate firebox steel has an allowable stress of 12,000 lb/in.^{2} for temperatures from −20 to 400°F. At 900°F the allowable stress is only 5000 lb/in.^{2}. By assuming the same pressure at both temperatures, it can be seen that a boiler designed for 12,000-lb/in.^{2} normal stress will be weakened to 5000/12,000, or 41.7 percent of its original strength with a temperature increase to 900°F.

Certain parts of boilers, particularly tubes, tube sheets, furnaces in scotch marine boilers, and cast parts in cast-iron boilers are very susceptible to temperature or overheating damages. A large temperature increase in a material, with accompanying *lower permissible stress levels,* is one of the most common causes of boiler damage. Low water, poor circulation, and scale are some causes of overheating of the material beyond safe stress levels. Let us not forget that the firing side of boilers is hot enough to melt steel. And with existing pressure on the water or steam side, it does not require much overheating to cause ruptures, bulges, and other deformation. Thus if the material is stressed well beyond the yield stress at high temperatures, permanent deformation will take place. In severe cases, the ultimate stress of the material is reached at the elevated temperature level, leading to complete rupture of the affected parts of the boiler.

Stress on boiler parts can also be caused by expansion due to temperature rise, and it is pronounced if the parts are restrained because of restrictions or uneven metal thicknesses being joined abruptly with no transition sections. Temperature causes expansion of steel, which can be calculated as follows:

$$e = nl\,(T_2 - T_1)$$

where

e = change in length
l = original length
T_1 = original temperature, °F
T_2 = final temperature, °F
n = coefficient of expansion (change in length per unit of length per degree change in temperature)

Example: Steel has a coefficient of thermal expansion of 0.0000065 in. per in. per °F. To show the possible rate of expansion to be considered, assume that a stay in a horizontal-return-tubular (HRT) boiler running from tube sheet to tube sheet is 30 ft long. How much will this rod expand with a temperature change from 70 to 300°F, assuming free expansion?

Substituting into the equation, we get

$$\begin{aligned} e &= 0.0000065\,(30)(12)(300\text{–}70) \\ &= 0.538\text{-in. stretch, which is over } \tfrac{1}{2} \text{ in.} \end{aligned}$$

If we assume that the stay rod was fixed at each end and that the tube sheets would *not give*, what compressive stress would be imposed on the rod, neglecting the column effect of a long rod?

This is calculated from the modulus-of-elasticity equation

$$S = E_\epsilon$$

where

$$\epsilon = \frac{\text{stretch}}{\text{inch}}$$

So,

$$\begin{aligned} S &= 30{,}000{,}000\,\frac{0.538}{30 \times 12} \\ &= 30{,}000{,}000\,(0.001494) \\ &= 44{,}820 \text{ lb/in.}^2 \end{aligned}$$

This example illustrates the importance of considering temperature effects in boiler design and the rapid stress buildup when a part becomes accidentally overheated above design conditions. Remember that the stress developed is not calculated as simply as shown by the illustration. For example, we assumed that the shell and tube sheets would *not* expand because of temperature.

This is obviously *not* true. If the shell is fixed or anchored, some relief will still be obtained from the expansion of the tube sheet. It does illustrate, however, the high stresses possible on stays and tubes. This is one of the chief reasons tubes start to leak around rolled joints or become bowed when a low-water condition develops in a boiler.

Also, on the long stay rod we ignored the column effect. But long, thin structures have to be treated as columns, which involves the ratio of the length to the radius of gyration. Stay rods that bow from temperature effects are also influenced by the strength-of-material equations involving columns.

If a tube leaks at the rolled joint but does not become bowed, it is an indication that the expansion force is greater than the rolled joint's holding power. Rolled joints are equivalent to *press fits,* depending on the friction of contact areas to hold the tubes tight in a tube sheet. The exception, of course, is welded-in tubes, where a shear stress is imposed by expansion.

If a structural material has an abrupt change in a section, for example, a flat plate containing an opening or a sharp corner as shown in the rod in Fig. 4-7, the stress distribution is not uniform over the cross-sectional area

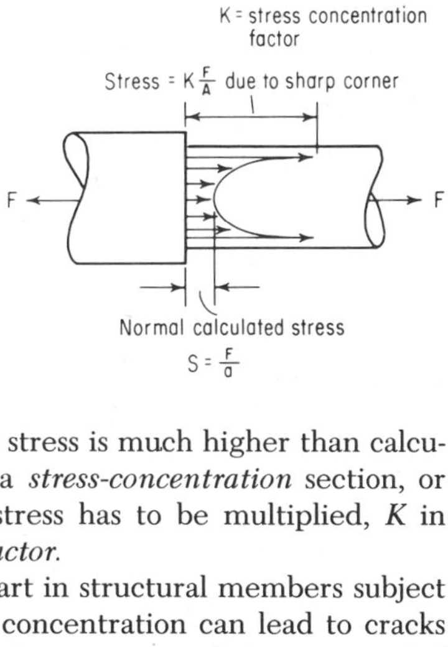

Fig. 4-7 Sharp corners can cause stress concentration which can increase normal stresses.

of the material. Near the abrupt change the stress is much higher than calculated. The affected section is said to have a *stress-concentration* section, or area, and the ratio by which the normal stress has to be multiplied, K in Fig. 4-7, is called the *stress concentration factor.*

Stress concentration plays an important part in structural members subject to repeated type of loadings, for the stress concentration can lead to cracks and fatigue failures. If the stress concentration is severe enough (even in normal loading), stresses may be induced far above the normal expected stress. Sharp corners in welded joints and other sharply formed shapes must be avoided. Thus openings cut into plates must be reinforced to strengthen the edges around the opening against stress concentration.

The Boiler Code specifies permissible joint connections to avoid stress concentrations. Fillet radii are specified on formed shapes. Openings must be calculated by Code rules. In analytical and design work, stress concentrations are determined by the *photoelastic method, stress-coat method,* and *strain-gauge method* using the electrical resistant wire gauge. In nuclear vessels one must carefully design the elements by the endurance limit and other stress-analysis methods. The *endurance limit* (also known as *fatigue limit*) is the maximum unit stress that can be imposed and repeated on a material through a definite cycle, or range of stress, for an indefinitely large number of times without causing the material to rupture.

How is the endurance limit determined? By testing a material through a complete reversal of stresses. When stressed nearly to its ultimate strength, the specimen will rupture after a few cycles. If a second sample of the same material is again tested but stressed slightly less than before, a larger number of reversals, or cycles, can be imposed. This is continued until a stress value, known as the *endurance limit,* is reached where an almost indefinite cycle of stress can be imposed without causing rupture.

Figure 4-8 shows an *S-N* diagram, where stress to rupture is plotted on one side and number of cycles to failure on the other. The horizontal line obtained is the endurance stress for the material. In Fig. 4-8 this is 22,500 lb/in.^{2} Endurance limits are widely used in machine design work, and with the adoption of the Nuclear Vessel Code, Section III, it will receive increasing attention in the Code. Endurance limits of materials can be modified considerably by environment, temperature swings, corrosion effects, hydrogen embrittlement, and similar other conditions which generally involve the study of the fatigue of metals. For example, discontinuities or stress concentration factors such as sharp corners or cracks can lower or modify endurance limits.

Fatigue crack growth rate has been quantified as a result of intense developmental work in the space and nuclear stress analysis field. If a crack is assumed to exist in a material, under stress this material, if it is steel, will undergo

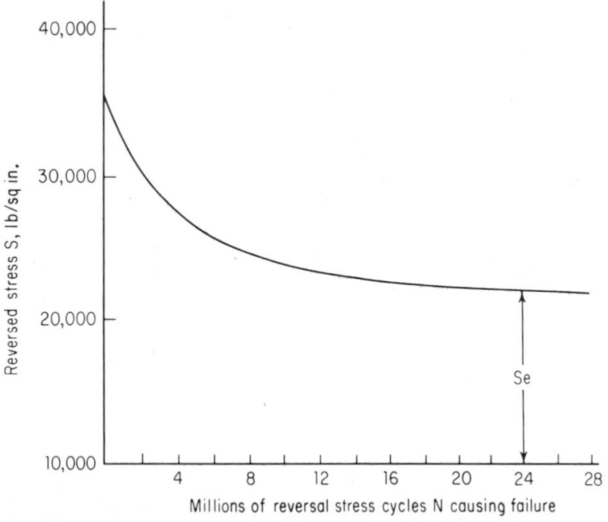

Fig. 4-8 *S-N* diagram to determine the endurance limit.

plastic deformation about the crack tip. The crack can grow under plastic deformation as the applied stress or load is increased. The distance that the crack front advances with each cycle of loading is a function of the stress intensity applied to the tip of the crack, which is expressed as the stress intensity factor ΔK. The crack growth rate, in inches per cycle of stress application, is expressed as $d(a)/d(N)$, where a = crack size in inches and N = fatigue life or number of cycles of stress before failure. An equation for crack growth rate used in failure analysis is

$$\frac{d(a)}{d(N)} = C(\Delta K)^m$$

where

C = material constant determined by test for class of material (for steels at room temperature, 4×10^{-24} is used)

m = material constant determined from tests (4 is usually used for ferrous material)

See Fig. 4-9. Crack propagation data can be used to determine the stress intensity ΔK that can be tolerated for a particular design. Also, the number

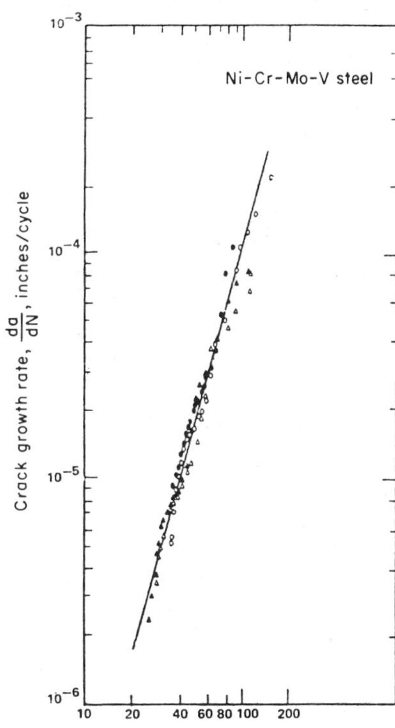

Fig. 4-9 Fatigue crack growth rate as a function of stress intensity at the tip of the crack for a nickel-chrome-moly steel.

of cycles required to extend an initial crack size a_1, to what is considered a critical crack size, a_2, can be established by using the fatigue crack propagation equation, $d(a)/d(N)$.

Fracture mechanics and NDT methods are being used to determine if a flaw needs immediate repair or whether many more cycles of stress can be still applied until the defect has grown to a size that requires repair or replacement. In the last few decades, there has been a great deal of research activity directed to this failure-prediction method as applied to the basic mechanism of fracture phenomena in solids. The determination of rate of crack growth helps in establishing the number of cycles that can be tolerated before corrective actions will be needed. Nondestructive testing methods are used to periodically check a developing defect in conjunction with fracture-mechanics methods.

Prevention of fatigue failures relies on containment of crack growth within what are considered safe limits. Where in-service inspections are possible, a safe service interval is established between inspections. Where in-service inspections are not feasible, a more conservative design is needed so that any anticipated defect will not grow to a dangerous size during the expected life of the component.

The simultaneous action of a part being subjected to repetitive stresses and some form of attack from the medium in which the part must operate can cause a part to fail as a result of *corrosion fatigue.* The combined effect of these two factors is much greater than the effect of either one alone. The cracking usually begins at surface defects, pits, or irregularities. These points act as stress-concentration points with the corroding medium intensifying the pitting action. The more repetitive the stress, the faster will be the pitting action, which again will increase the stress concentration. A cumulative effect can materialize that will cause a failure well below a predicted endurance stress.

Pits formed under simultaneous stress and corrosion are always sharper and deeper than pits formed in the same time under stressless conditions. The more repetitive the stress, the faster will be the pitting action. At low-cycle repetitive stress, pitting will proceed entirely by normal corrosion, and the section will fail by the normal tension failure or when the resisting area is thinned down so the stress rises proportionately, assuming a constant load.

Stress corrosion, then, is primarily caused by repetitive stresses of high frequency in a corroding medium, leading to pitting caused by stress and corrosion. The pitting then develops high-stress concentration points which lead to fatigue failure.

No experimental data are available for determining the extent to which the endurance limit is reduced for most materials in combination with corroding solutions or media. But for boilers and pressure vessels, the obvious precaution to take against stress corrosion is to make sure the water or medium being confined is free of any corrosive tendencies. This is determined by analysis of the water at regular intervals by personnel experienced in water analysis. This also points out the value of periodic internal inspections of pressure-containing parts and examination of surfaces for evidence of pitting and corrosion.

Brittle materials are those which are comparatively weak in tension. Brittleness is a characteristic of a material that is opposite to ductility and toughness. A material is generally considered brittle if its elongation at rupture from a tension test is less than 5 percent in a test specimen of 2 in. Brittle materials will fail with very little give or stretching before failure because of their lack of ductility. Brittle materials have low toughness or low resistance to impact loading, sometimes called lack of resiliency. As a result, impact testing is used as a measure of brittleness or toughness (described as the ability to absorb energy). When a material has a notch and is brittle, failure can be unexpected and well below calculated allowable stresses.

Charpy V-notch tests are used to measure resistance to impact loading or brittleness. The test uses a pendulum-type apparatus to strike a specimen, usually notched, and the foot-pounds of energy needed to cause a fracture are correlated to whether the material is considered brittle or not. A complete description of notched-bar impact testing can be secured from studying the E-23 standard as developed by the American Society for Testing and Materials (ASTM).

The effect of temperature altering or changing the brittle nature of a material was dramatically displayed in World War II when merchant ships cracked unexpectedly. It was noted that this occurred in cold waters, but not in warm waters. It has been established since these failures that materials may exhibit ductile behavior in a normal environment but act in a brittle manner if the environment, such as temperature, is changed. Stress corrosion cracking can also be of a brittle nature as a result of the combined action of stress, susceptible material, and a corrosive surrounding.

Low-carbon and low-alloy steels frequently exhibit a transition zone from ductile behavior to brittle failure over a small temperature range called the *nil-ductility temperature.* (See Fig. 4-10.) Any flaws in the material will aggra-

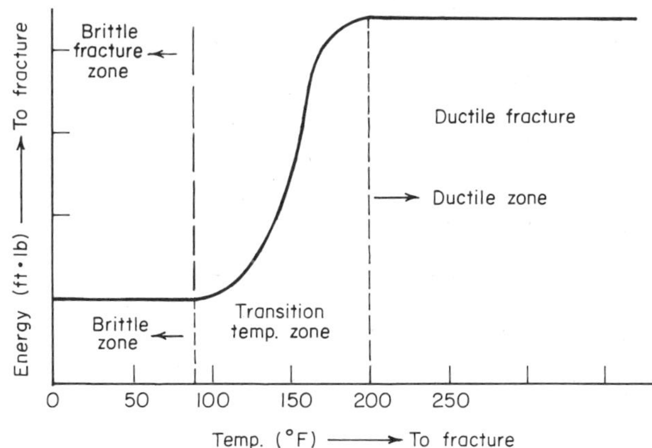

Fig. 4-10 Transition-temperature zone is important on certain steels to determine brittle and ductile fracture possibilities.

vate the tendency of a material with a nil-ductility characteristic to act brittle below the transition temperature.

Figure 4-11 shows how a Charpy V-notch impact test is made. A V notch is cut out per standards on one face of a small specimen. When the heavy pendulumlike weight strikes the specimen's face opposite the notch, the foot-pounds needed to cause fracture are calculated by weight–lever arm relationships. Carbon steels with cold-weather resistance to impact failure are usually

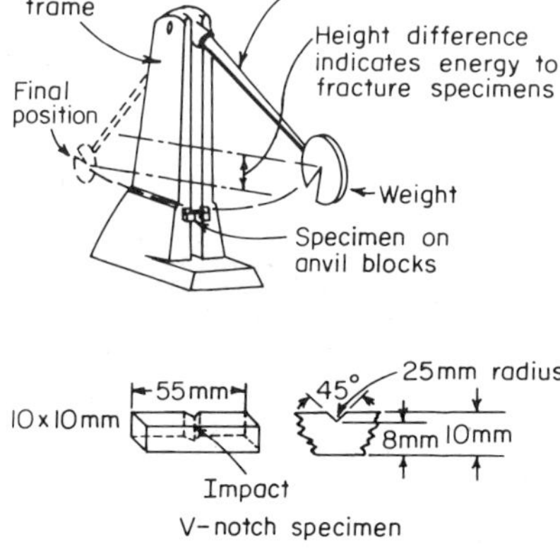

Fig. 4-11 Charpy V-notch impact test is used to determine material toughness against brittle fracture. *(Courtesy Power magazine.)*

killed steels, or a steel that was deoxidized in manufacture so that no evolution of gas occurred during pouring and solidification. Nickel steels of 3½ or 9 percent are considered superior impact-resistance steels, as are austenitic stainless steels.

Welding can also cause changes in a material that may make it act brittle in the weld metal or heat-affected zone (HAZ). It is thus important to determine and consider nil-ductility transition temperatures of material to be used in boilers and pressure vessels in order to make sure this temperature is not experienced during any service temperatures that the boiler or pressure vessel may be exposed to, even perhaps under possible test or operating abnormalities. One such possibility is hydrostatic testing with water that will be below the nil-ductility temperature. The other is testing when the surrounding temperature may be too low, as may be the case in winter. Failures due to brittleness from not considering the effect of temperature transitions on the property of the material have occurred during hydrostatic tests of pressure equipment and others such as water towers.

Creep is defined as the slow deformation of material with time, with the deformation taking place at elevated temperatures with no increase in stress. Modern material testing includes the determination of how much creep or permanent deformation can be expected in what length of time so that design can reflect these factors in life prediction for the pressure vessel. Creep properties are obtained by imposing a constant tensile load on a specimen at a constant temperature and then measuring the strain at intervals of time. Data are plotted to obtain a variety of constant-stress creep-strain curves or constant-time creep-stress curves. In using creep data, the designer tries to establish expected service life and the corresponding amount of creep deformation that can be tolerated. With these established, a stress can be selected that will satisfy these conditions.

Hydrogen embrittlement can occur under high-pressure and -temperature operations of boilers and pressure vessels. Hydrogen attack on steels can result in severe loss of ductility with cracks developing that can lead to unexpected brittle fracture. The effect is most severe on high-strength steels. The hydrogen under high pressure and temperature diffuses into the metal as atomic hydrogen. It recombines in the metal as molecular hydrogen in grain boundaries and causes high pressure in these "voids," resulting in bulging and blistering. Hydrogen also combines with carbon in steel under high pressure and temperature and decarbonizes the steel, with embrittlement taking place. Frequently chromium-molybdenum steels are used where resistance to hydrogen embrittlement is required.

Nuclear pressure vessels may be subjected to *neutron irradiation* of steel. It is necessary to avoid impurities in the steel as well as large grain sizes in order to limit any vacancy or voids in the material. It is theorized that the radioactivity present in nuclear vessels promotes the formation of helium which, with time, collects at metal grain boundaries to weaken it by a decrease in ductility. Quality control is thus more stringent on nuclear vessels, so cavities may be eliminated.

Allowable stresses are used to design structures or machine components. The allowable stress is also sometimes called the *allowable working stress.* It is the maximum stress considered to be safe when the material is subjected to resisting loads assumed to be applied in service. In boiler applications, the term *allowable pressure* is often used. Actually, the allowable pressure is determined by applying the forces acting on a material and then calculating the allowable pressure from the allowable stress on the material.

The allowable stress is determined from the ultimate strength of the material, which is divided by a safety factor. The safety factor used in modern boilers is 4. However, certain elements of a boiler, such as rivets, have to be designed with a safety factor of 5. Other parts have to be designed with a safety factor as high as 12.5, such as the rivets holding lugs on brackets on an HRT boiler to be suspended from a beam.

In boiler usage, the factor of safety is the ultimate strength divided by the allowable loadings, or the ultimate stress S_u divided by the allowable stress Sa. In equation form,

$$\text{Safety factor} = \frac{S_u}{Sa}$$

Another method of expressing the safety factor is by dividing the bursting pressure by the allowable pressure. This method is used on state inspection reports and on ASME data reports. In equation form, it is

$$\text{Safety factor} = \frac{\text{bursting pressure}}{\text{allowable pressure}}$$

In boiler design and usage and other critical structures where life and property may be at stake, there is a definite need for selecting working stresses and loadings considerably less than the ultimate, or the yield stress, for these reasons:

1. There is always some uncertainty in materials being used, how they were made, how they were assembled, and how they were joined or fabricated with other materials.
2. There is always some uncertainty as to the exact loading that a structure, or part of it, may have to resist and how it is abused in operation.
3. Calculations of all stresses possible in a fabricated structure are never that exact when one considers the variables to be encountered in service through the years.

Let us remember that the ASME Code was drawn up when yield stresses for many materials were not available. Then again, certain brittle materials such as cast iron do not have a *definite* yield point. Thus it was much easier to work from ultimate stress and apply a safety factor to this to obtain an allowable stress. In some countries of Europe, the *yield point* is the basis of design. The Nuclear Vessel Code uses the yield stress or endurance stress or both as a basis of design. With increasing technological changes, the present Boiler Code may also be changed on this matter in time.

On existing installations, the question of safety factor and allowable pressure arises quite often. For example, can a boiler that was originally designed with a safety factor of 5 be operated with a safety factor of 4, since the latest Power Boiler Code allows a safety factor of 4? Usually, the original safety factor of 5 remains. The safety factor of 4 was drawn up principally for seamless-steel or welded boilers that met stiff quality-control and inspection requirements on welding. Older boilders may not meet this requirement; thus the original safety factor of 5 should govern the allowable pressure.

If a boiler is stamped for an allowable pressure of 275 psi and the safety valve is set at 150 psi, what is the safety factor? Assume a Code-welded boiler meeting latest Code requirements and an original design safety factor of 4. The bursting pressure of this boiler would be $4 \times 275 = 1100$ psi. So,

$$\text{Safety factor} = \frac{\text{bursting pressure}}{\text{allowable pressure}} = \frac{1100}{150} = 7.33$$

This question brings up an important consideration as to which pressure to use in calculating safety factors on existing boiler installations. For example, in this question, should the safety-valve setting be used or the stamped allowable pressure? If the safety-valve setting is *below* the stamped allowable working pressure, use the safety-valve *setting.* If by chance the safety-valve or valves are set *higher* than the allowable pressure (above Code limits), then a dangerous condition exists because safety valves must always be set *at* or *below* the allowable, or working, pressure stamped on the boiler.

The safety-valve setting is used where the stamped allowable pressure is above the safety-valve setting in calculating the safety factor because the allowable pressure on the boiler is the *safety-valve setting.* The boiler is *not* supposed to operate above this safety-valve setting. If an increase in pressure is needed, other items will have to be checked before the pressure can be raised to the maximum allowable pressure stamped for the boiler. Then, a new safety valve will be required. Also, Code specifications on valve ratings and water-column connections will have to be checked to see if they meet the Code requirements for the new pressure.

The question of a safety factor continuing for the life of the boiler must be considered quite often, especially where the expense of repair or alteration must be evaluated in comparison to new equipment. The safety factor continuing on a boiler depends on the type of boiler, the condition of the boiler, and even the state in which it is located. Assuming that internal and external inspections are satisfactory, the National Board regulations state:

1. Lap-riveted longitudinal-joint boilers operating over 50 psi can be operated at this pressure for 20 yr. After that, 50 psi or less is permissible; but if the boiler is relocated, only low-pressure service is permitted.
2. For boilers of butt construction, at the end of 25 yr and every 5 yr thereafter, the safety factor must be increased by 0.5 unless a hydrostatic test of 1½ times the allowable pressure is imposed; if this test is satisfactory, no increase in the safety factor is necessary.

The best rule to follow on any safety-factor changes is to check with an insurance company boiler inspector or legal jurisdiction inspector. *Never* assume that the boiler can be operated at a higher pressure by just changing safety valves, even if the boiler is stamped for the higher pressure. There are other requirements to be met on feedwater, blowdown, water-column connections, low-water fuel cutoff, and service-valve ratings that must be considered. In determining the allowable pressure on a boiler, stress calculations must be made as shown in later chapters.

Any pressure vessel and parts confining pressure must be analyzed per component by carefully considering the strength of the material being used, its physical characteristics as to type and grade, allowable stress, thickness, etc. The forces acting on this material must then be analyzed. This force is usually created by pressure, but may also include temperature, the weight it is supporting, and stress concentration, such as around an opening. The problem then evolves to comparing the forces acting on the material and determining whether the material is being stressed beyond the allowable stresses governed by the Boiler Code rules. Elements to be considered depend on the type of boiler but will generally include shells or drums, tubes, tube sheets, heads, flat surfaces, stays, stay bolts, openings, furnaces, rivets, welded joints, structural supports, and connected piping and valves. Each of these is governed by Boiler Code rules as to allowable material, allowable stresses, and method of calculating forces to obtain the allowable pressure. Finally in boiler and pressure-vessel application, the weakest element producing the lowest pressure then determines the *allowable pressure* for the boiler.

In the equation for longitudinal stress developed earlier in the chapter,

$$S=\frac{DP}{2t}$$

Existing installations in many states require the allowable pressure on a cylinder to be calculated by the following equation:

$$P=\frac{TS\times t\times\%}{FS\times R}$$

where

TS = ultimate tensile strength of material of shell, lb/in.^{2}
t = plate thickness, generally minimum, in.
% = efficiency of joint in shell
FS = factor of safety
R = inside radius of shell or drum

The factor of safety can range from 4 to 6, depending on whether the boiler is welded or riveted, lap- or butt-constructed if riveted, the age of the unit, requirements for periodic hydrostatic test to prove that the boiler is tight, and similar considerations. A licensed or commissioned boiler inspector of the jurisdiction in which the boiler is located should be consulted.

Chapter 9 provides typical calculations that are made on Code-constructed boilers, incorporating some of the stress considerations of this chapter.

Questions and Answers

4-1 How would you define the ultimate tensile strength of a material mathematically if S = ultimate tensile strength, in pounds per square inch, F = maximum load when failure occurs, a_f = final cross-sectional area of a round specimen at the time of failure, and a_i = initial cross-sectional area of a round specimen?

Ans. $S=\frac{F}{a_i}$

The initial area is used in material testing for the ultimate tensile strength.

4-2 Define ductility mathematically, given the following:
Percentage reduction in area = A
Initial area of specimen = a_i
Fractured area of specimen = a_f

Ans. $A = \dfrac{a_i - a_f}{a_i} \times 100\%$

4-3 *(a)* A boiler drum has a 40-in. diameter; its tensile strength is 55,000 lb/in.², its plate thickness is ½ in., and its factor of safety is 5. What is the maximum safe working pressure?

(b) If the pressure is increased 10 percent, what is the factor of safety *FS*?

(c) If corrosion causes a general reduction in thickness of ⅛ in., what is the maximum safe pressure if joint efficiency is 94 percent?

Ans. *(a)* $P = \dfrac{55{,}000 \times 0.5 \times 0.94}{20 \times 5} = 258 \text{ lb/in.}^2$

Note the existing installation equation is used with a safety factor of 5 because of the riveted construction.

(b) $$258 \times 1.10 = 284 \text{ lb/in.}^2$$

Transposing the formula and solving for the factor of safety, we have

$$FS = \frac{55{,}000 \times 0.5 \times 0.94}{20 \times 284} = 4.56$$

(c) A reduction in thickness of ⅛ in. will be 25 percent, which will leave 75 percent of the solid plate. Since this is below the efficiency of the longitudinal seam (94.0 percent), the maximum safe pressure should be reduced as follows:

$$P = \frac{55{,}000 \times 0.375 \times 1.0}{20 \times 5} = 206$$

4-4 What is stress?

Ans. Stress is the resistance to an external force that a material provides within its crystalline structure; it is expressed as pounds per square inch of the material's cross-sectional area.

4-5 What is the test most commonly used for measuring a metal's strength, elasticity, and ductility?

Ans. The tension test is the mechanical test usually applied to determine these properies of a material.

4-6 What is strain as applied to material testing?

Ans. Strain is a measure of deformation of a material under load, and it is expressed as the increase in length per inch of original length.

4-7 What is the proportionality constant called that exists between stress and elastic strain?

Ans. This constant for a class of material is called the modulus of elasticity *E*.

4-8 How would you define ultimate tensile strength?

Ans. The ultimate tensile strength is the stress which a material experiences in a testing machine under maximum load during the test. The ultimate stress is obtained by dividing this maximum load by the original cross-sectional area of the material.

4-9 What is notch sensitivity?

Ans. Notch sensitivity or notch toughness of a metal is its resistance to the start and propagation of a crack at the base of a standardized notch. Notch sensitivity is measured by the amount of energy absorbed in foot-pounds by a specimen as it fractures under impact of a hammer blow that is delivered by a standard weighted pendulum. Brittle materials will absorb little energy in fracture.

4-10 What is meant by fatigue?

Ans. Fatigue is the tendency of a metal to fracture under conditions of repeated cyclic-type stresses that are considerably below the ultimate strength of the material.

4-11 How is ductility defined?

Ans. The ductility of a metal is the amount of permanent deformation or strain that it can undergo before fracture occurs. Two methods are used to define or measure ductility: the percentage of elongation and the reduction in area of a specimen tested under tension. Welds are tested for ductility by a bend test. This consists of bending the sample by a specific amount about a plunger of given radius. The increase in distance between gauge marks on the tension side of the specimen is noted, and the change is expressed as the percentage of elongation.

4-12 What are brittle materials?

Ans. These are materials that can be deformed very little without rupture taking place. This is usually characterized by a sudden, shattering type of failure. Cast iron, concrete, brick, and glass are examples of brittle materials.

4-13 What are resilient materials?

Ans. These are materials that can absorb large amounts of energy without experiencing permanent deformation. To express it another way, they return to their original shape after the load is reduced. Materials with capability of low modulus of elasticity and high elastic limit would produce high resilience.

4-14 Define tough materials.

Ans. Tough materials can absorb large amounts of energy before rupturing. This quality is related to a material's high strength, high ductility, or flexibility. Toughness is a useful measure of the ability of a material to absorb shock loading or sudden blows without rupturing.

4-15 Name some environmental impurities that can enter a boiler's steam or water system which could produce stress corrosion cracks on a high-strength turbine steel.

Ans. The following have been identified as possibly causing stress corrosion cracks on turbine blades: (1) use of coordinated phosphate treatment with improper control, which can produce free caustic that can attack metals; (2) sodium getting into the boiler water from a leaking condenser tube; and (3) untreated water used in attempering sprays for the control of superheat or reheat.

4-16 What causes thermal stress?

Ans. Thermal stress is caused when a material is not free to expand or contract because of temperature changes to which a material may be subjected in service. The resultant stress that is developed can be found by the equation

$$s = Ea(t_2 - t_1)$$

where

$$E = \text{modulus of elasticity of material}$$
$$a = \text{coefficient of expansion for material}$$
$$t_1,\ t_2 = \text{initial and final temperatures}$$

4-17 What can the effect of creep be on a structural metal?

Ans. Creep causes planes of slow movement in a material's crystalline structure that are also a function of time. This slow movement or slip can cause sufficient deformation to cause a sudden fracture even when the applied stress is much lower than that which normally could produce a failure under normal loadings.

4-18 The term *concentration factor* defines what ratio?

Ans. The stress concentration factor is the ratio of the actual stress existing on a given plane of a member under loading to the calculated stress that is needed to resist that loading without taking into account the discontinuity which causes normal stresses to be magnified by the stress concentration factor.

4-19 How is Poisson's ratio described?

Ans. Below the proportional limit, a material under load will stretch in tension lengthwise and be thinned at right angles to the length. The ratio of the unit strain at right angles to the stress to the unit strain in the direction of the stress is called *Poisson's ratio,* expressed as follows:

$$u = \frac{e_t}{e_s}$$

where

$$e_t = \text{unit strain in tranverse direction}$$
$$e_s = \text{unit strain in direction of stress}$$

4-20 A ½-in. round bar is stretched 0.00195 in. in a length of 2 in. and decreased 0.000162 in. in diameter when a load in tension of 2000 lb is imposed. What is Poisson's ratio?

Ans. Unit strain in direction of stress $= \dfrac{0.00195}{2} = 0.000975$

$$\text{Unit strain in transverse direction} = \frac{0.00016}{0.5} = 0.00032$$

$$\text{Poisson's ratio} = \frac{0.00032}{0.000975} = 0.328$$

4-21 Calculate the tensile strength of the specimen in the above question.

Ans. $s = \dfrac{P}{a} = \dfrac{2000}{\pi(0.25)^2} = 10{,}191 \text{ lb/in.}^2$

4-22 A specimen has a modulus of elasticity of 6.5 million. What is the maximum load axially that may be applied on a ½-in. rod without stretching the rod $\frac{1}{16}$-in. in a length of 6 ft?

Ans. Use $E = \dfrac{F/a}{e/l}$ and solve for F

where

F = load
a = rod area
$e = \frac{1}{16}$
$l = 6 \times 12 = 72$ in.

$$F = 6{,}500{,}000 \times \frac{0.0625}{72} \times 0.0625 = 325.6 \text{ lb}$$

4-23 How much will the diameter of a steel rod 2-in. diameter change if an axial load of 60,000 is imposed and Poisson's ratio is 0.30?

Ans.

$$\text{Axial stress} = \frac{60{,}000}{\pi(1)^2} = 19{,}091$$

Use $E = 30$ million for steel.

$$E = \frac{\text{stress}}{\text{strain}}$$

$$\text{Strain axially} = \frac{19{,}091}{30{,}000{,}000} = 0.000636 \text{ in./in.}$$

$$\text{Transverse strain} = 0.3 \times 0.000636 \text{ in./in.} = 0.0001808$$

$$\text{Change in 2-in. diameter} = 2 \times 0.0001808 = 0.0003616 \text{ in.}$$

4-24 How is the endurance limit of a material determined?

Ans. The endurance limit of a material is determined by applying to several test specimens repeated loads, which cause complete reversed stresses of known values, and recording the number of stress reversals each specimen endures before it fails or breaks. Each specimen is subjected to a lower stress than the preceding one and also breaks after a large number of cycles of stress. Finally, a stress is reached that does not cause failure regardless of how many reversed cycles are applied. This stress is called the endurance limit for the material under test.

4-25 What loading besides internal pressure is required by the ASME high-pressure Boiler Code?

Ans. The loadings that must be considered include the effect of the weight of water and stresses imposed on a boiler during an hydrostatic test. As well as any additional loading that will increase the average stress above the working pressure and induced stresses by more than 10 percent of the allowable working stress. These loads can include wind, snow, and earthquake loads where specified as in the nuclear Codes. It also includes the reaction of supporting lugs, rings, saddles, and similar types of support.

4-26 Who must prepare stress reports for nuclear pressure vessels?

Ans. The manufacturer of the pressure vessel is responsible for the preparation of the stress report, and it must be reviewed by the owner of the plant and in most cases by an inspection agency.

4-27 What broad outline is stipulated for making a stress report?

Ans. The stress report must be prepared and signed by professional engineers experienced in pressure-vessel design. Generally it has three sections: thermal analysis, structural analysis, and fatigue evaluation. Of interest is the fact that areas of severest stress condition at any transient must be listed in the report, including the values of the stresses.

4-28 What do the following tests mean as applied to welding? *(a)* free-bend test; *(b)* root-bend test; *(c)* face-bend test.

Ans. *(a)* When a specimen is bent cold, the outside fiber of weld shall elongate at least 20 percent before failure occurs.

(b) The specimen is bent against the bottom of the weld.

(c) The specimen is bent against the surface of the weld.

4-29 What is the purpose or use of these tests? What is a reduced-section tensile test?

Ans. These tests are required to qualify welding and operator procedures per ASME Code requirements:

(a) A reduced-section tensile test is a qualifying procedure. When the section is broken in tension, it shall have a tensile strength at least that of the minimum of the range of the plate which is welded, and elongation or stretch shall be 20 percent minimum in 2 in.

(b) A free-bend test is a qualifying procedure. It consists of bending a specimen cold; the outside fibers of weld shall elongate at least 20 percent before failure occurs.

(c) A root-bend test is for qualifying operators. It consists of bending a specimen against the bottom of the weld.

(d) A face-bend test is for qualifying operators. It consists of bending a specimen against the surface of the weld.

(e) A side-bend test is for qualifying operators. It consists of bending a specimen against the side of the weld.

4-30 Name three hardness tests applied to metals that are welded.

Ans. Brinell, Rockwell, and Vickers.

4-31 What type of fracture normally is analyzed by the notched-bar impact test?

Ans. Notched-bar impact tests are used to analyze ferritic steels for possibilities of failure by *brittle fracture.*

4-32 Name the destructive test normally carried out on ferritic steels in order to determine the nil-ductility transition temperature in order to determine the low-temperature ductility properties of the ferritic steel.

Ans. The drop-weight test is used to determine the *fracture resistance* of ¾-in. and over thick steel at *different temperatures* with the steel having a notch that can start a fracture.

4-33 Name some typical items to be observed in welding a joint to be used for power boilers.

Ans. The following details should be checked:

1. Welding process being used versus specifications or written procedure
2. Extent of cleaning of joint prior to welding in order to ensure a good weld
3. Preheat and interpass temperatures maintained and how these compare with the Code requirements for the material being welded
4. Joint preparation and if this meets Code specifications and print dimensions
5. Filler metal being used and whether this is suitable per welding procedure and/or material being welded per Code requirements
6. Chipping, grinding, or gouging that is being carried out after each welding pass in order to remove slag and impurities prior to making the next welding pass
7. Bracing of plate during weld in order to control distortion
8. Postweld heat treatment being conducted and whether this meets Code requirement for temperature and time held per Code requirements
9. Check if welders doing the job are qualified and have the necessary papers on file with the manufacturer or contractor

5
Fire-Tube Boilers

HEAT TRANSFER

Boiler configurations are influenced by heat-transfer requirements so that as much of the heat released by a fuel may be extracted as material and economic considerations permit. The effect of shape is of immense importance to the strength and bracing requirements of a boiler. It is a well-known law of science that pressure of a fluid exerts itself in an equal amount in every direction. Following this law, an irregularly shaped vessel subjected to internal pressure always *tends* to be forced into a perfect spherical shape. The first tendency of an oval-shaped drum or shell would be to change its cross-sectional shape to a true circle.

With regard to these facts, one may look at a flat head of a vesssel operated at above very low internal pressure and see that bracing is used. However, if a semispherical or dished head is used, the bracing is dispensed with. Heads for high-pressure boilers are practically always dished or elliptical in form. No braces are required for these heads.

A boiler is a heat-transfer apparatus that converts energy to a working medium, such as steam, hot water, or organic fluids. The basic laws of heat transfer stipulate that when energy is transferred from one body to another, a temperature difference must exist. Another fundamental law is that heat may be transferred from a high-temperature region to one of lower temperature, but *never* from a lower-temperature region to one of higher temperature. And this flow of heat may occur in one of, or a combination of, these three ways: *conduction, convection*, and *radiation*. These three methods of heat transfer are utilized in boiler design to convert fuel energy to a useful heat medium.

Conduction is the transfer of heat from one part of a material to another or to a material with which it is in contact. Heat is visualized as molecular activity—crudely speaking, as the vibration of the molecules of a material. When one part of a material is heated, the molecular vibration increases. This excites increased activity in adjacent molecules, and heat flow is set up from the hot part of the material to the cooler parts. In boilers, considerable surface conductance between a fluid and a solid takes place, for example, between water and a tube and gas and a tube, in addition to conductance through the metal of a tube, shell, or a furnace.

While surface conductance plays a vital part in boiler efficiency, it can also lead to metal failures when heating surfaces become overheated, as may occur when surfaces become insulated with scale. The surface conductance, when expressed in Btu per hour per square foot of heating surface for a difference

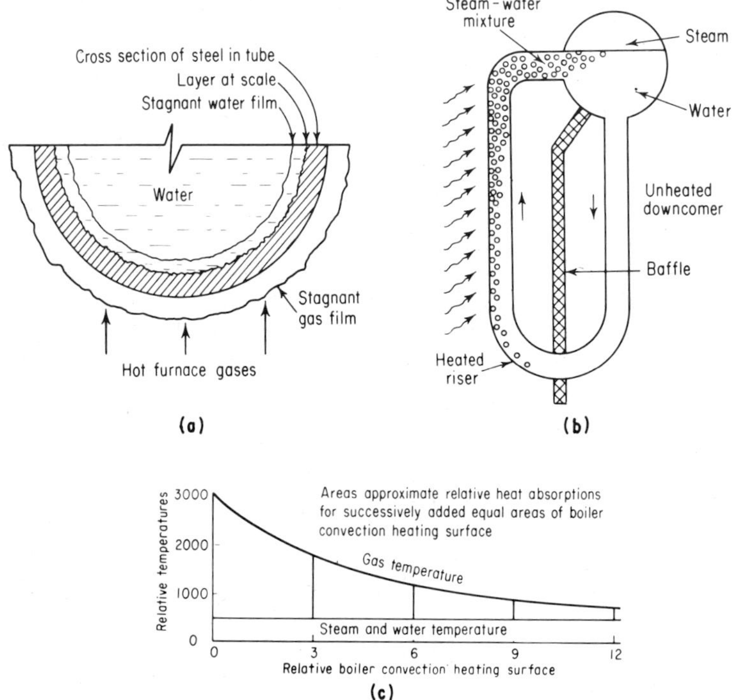

Fig. 5-1 Factors affecting heat transfer in boiler components: *(a)* Scale and stagnant gas and water near the tube affect heat transfer across the tube. *(b)* Circulation depends on heated water rising while cooler water descends to replace it. *(c)* Adding boiler heating surface increases heat absorption but at a reduced rate.

of one degree Fahrenheit in temperature of the fluid and the adjacent surface, is known as the *surface coefficient* or *film coefficient.* Figure 5-1*a* shows stagnant areas near the tube where the film coefficient will reduce heat transfer.

The coefficient of thermal (heat) conductivity is defined further as the quantity of heat that will flow across a unit area in unit time if the temperature gradient across this area is unity. In physical units it is expressed as *Btu per hour per square foot per degree Fahrenheit per foot.* Expressed mathematically, the rate of heat transfer Q by conduction across an area A, through a temperature gradient of degrees Fahrenheit per foot T/L, is

$$Q = kA\frac{T}{L}$$

where k = coefficient of thermal conductivity.

Note that k varies with temperature. For example, mild steel at 32°F has a thermal conductivity of 36 Btu/(hr/ft²/°F/ft), whereas at 212°F it is 33.

Convection is the transfer of heat to or from a fluid (liquid or gas) flowing over the surface of a body. It is further refined into *free* and *forced convection.* Free convection is *natural* convection causing circulation of the transfer fluid due to a difference in density resulting from temperature changes.

For example, in Fig. 5-1*b* the heated water and steam rise on the left and are displaced by cooler (heavier) water on the right. This causes free convection of heat transfer between heat on one side of the U tube and cooler water on the other side. Actually, conduction has to take place first between the gas film and metal of the tube, then the water. But if the water did not circulate, eventually equal temperatures would result. Heat transfer would then cease.

Forced convection results when circulation of the fluid is made positive by some mechanical means, such as a pump for water or a fan for hot gases. The heat transfer by convection is thus aided mechanically.

Adding boiler surface may increase the heat absorption, but as shown in Fig. 5-1*c*, the temperature gradient will drop more and more. Then at some point the gain in efficiency will be far less than the cost of adding heating surface. Further, the mechanical power required for forced circulation will also increase with the addition of heating surface by convection.

Note that in Fig. 5-2*a* more tube area is required at lower pressure than higher pressure for the same circulation to exist. But the force producing circulation is less at high pressure than at low pressure. This involves the change in the specific weight of water and steam as pressures increase. The mixture actually weighs less in pounds per cubic foot at higher pressures. For example, in the sketch in Fig. 5-2*b* at the critical pressure (3206.2 psia), water and steam have the same specific weight. Friction losses due to flow are generally less at higher pressure. This is primarily due to more laminar, or streamlined, flow and less turbulent flow in the tubes.

When boiling occurs in a tube, bubbles of vapor are formed and liberated from the surface in contact with the liquid. This bubbling action creates voids (Fig. 5-2*c*) of the on-again-off-again type, because of the rapidness of the action. This creates a turbulence near the heat-transfer surfaces, which generally in-

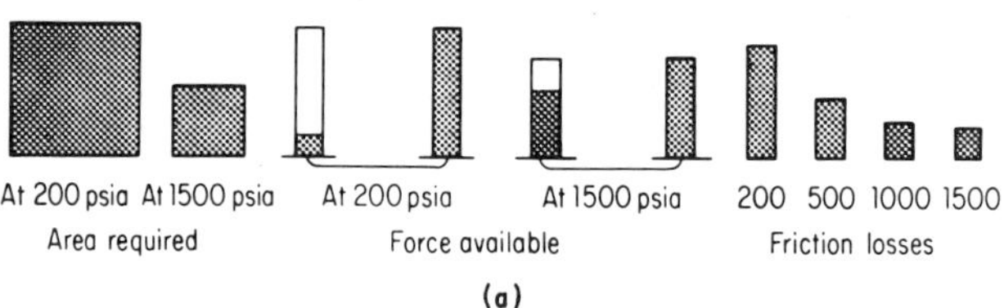

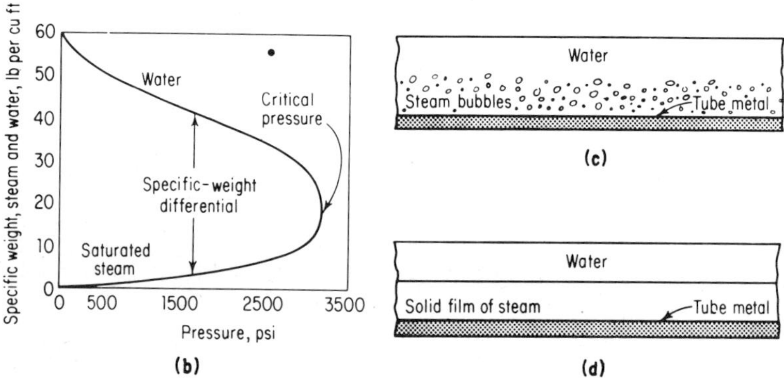

Fig. 5-2 The effect of pressure on circulation rate: *(a)* The tube area needed is higher at low pressure; the force to produce the circulation is less at high pressure; friction loss is greater at low pressure. *(b)* At critical pressure, water and steam have the same specific weight (3206.2 psia). *(c)* At low pressure, steam bubbles form near the tube metal. *(d)* At high pressure, a solid film or layer of steam is formed at the tube metal.

creases the heat-transfer rate. But the loss of wetness as the bubbles are formed may diminish heat transfer.

Pressure has a marked effect on the boiling and heat-transfer rate. With higher pressures (Fig. 5-2*d*) bubbles tend to give way to what is called *film boiling,* in which a film of steam covers the heated surface. This phenomenon is very critical in boiler operation, often causing watertube failures due to starvation, even though a gauge glass may show water. It is further compounded by the formation of scale and other impurities along the boiling area of a tube.

Radiation is a continuous form of interchange of energy by means of electromagnetic waves without a change in the temperature of the medium between the two bodies involved. Radiation is present in all boilers. In fact, all boilers utilize all three means of heat transfer: *conductance, convection,* and *radiation.*

FIRE-TUBE BOILERS

Types and Arrangement Fire-tube boilers are classified into horizontal-return-tubular (HRT), economic or firebox-type, locomotive firebox-type, scotch-marine-type, vertical tubular, and vertical tubeless boilers. The HRT boiler now represents only about 5 percent of the boilers in service of the total fire-tube boilers still being operated. The scotch marine (SM) design is the dominant fire-tube type for both heating and industrial process use up to about 35,000-lb/hr capacity. Above this capacity watertube boilers are generally used.

General Construction Details The fire-tube boiler has tube ends exposed to the products of combustion and has other flat surfaces that require staying with structural steel in order to avoid excessively thick plates. Tubes in all fire-tube boilers must be rolled and beaded (Fig. 5-3*a*) or rolled and welded. If they are rolled and welded in high-pressure boilers, see Power Boilers, Section I, ASME Boiler Code. Tubes are beaded to prevent the ends from being burned off by the hot gases in this area. Beading also increases heat transfer near the tube sheet and tube juncture. The edges of tube holes are chamfered about 1/16 in. after the holes are drilled so that there will be no sharp edges to cut into the tube when it is expanded. Fire-tube holes are finished 1/16 in. larger in diameter than the outside diameter of the boiler tube so as to permit a tube with scale to be drawn from the tube sheet without damage to the tube-sheet hole.

Fire tubes are normally under external pressure; thus they may collapse but not burst. The biggest problems are loosening of tubes in the tube sheet; cracking, burning, and corrosion of tube ends; waterside pitting and corrosion leading to leakage; and fireside corrosion and pitting leading to leakage or pulling out of the tube sheet (as a result of poor rolling). Also scale buildup on the waterside leads to overheating and possible sagging and loosening in the tube sheet.

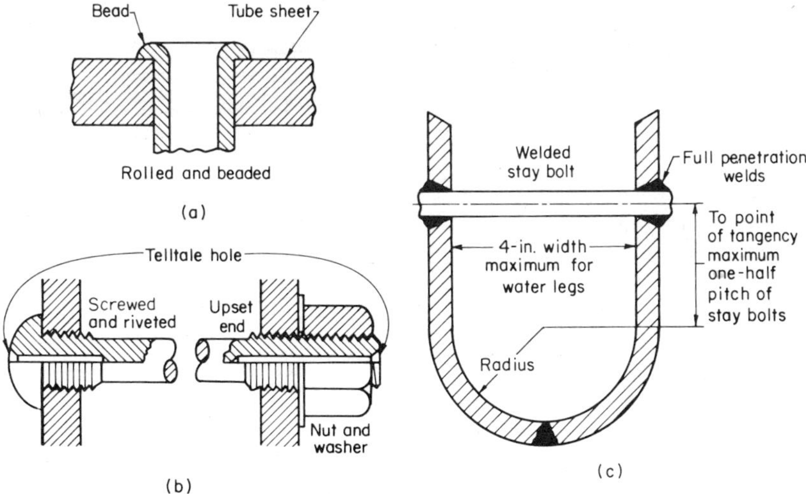

Fig. 5-3 *(a)* Tubes in a fire-tube boiler are rolled and beaded. *(b)* Threaded stay bolts have telltale holes. *(c)* Welded stay bolts do not require telltale holes.

Flat surfaces are stayed with either stay bolts or stays. Stay bolts can be threaded, welded, and hollow. They are generally used to strengthen short spans, such as the inner and outer plates of water legs.

See Fig. 5-3*b* and *c*. The ends of the stay bolts, or stays screwed through the plate, must extend beyond the plate not less than two threads when installed, after which the ends must be riveted over without excessive scoring of the plate. They may also be fitted with threaded nuts, provided the stay bolts, or stays, extend through the nut.

If stay bolts are solid, 8 in. in length or less, and threaded, they must be drilled with telltale holes (Fig. 5-3*b*) at least $\frac{3}{16}$ in. in diameter to a depth extending at least ½ in. beyond the inside of the plate. If the stay bolt is reduced in diameter between the ends, the telltale hole must extend ½ in. beyond the point where the reduction in diameter begins. For bolts over 8 in. in length, telltale holes are not required, nor if the stay bolt is attached in by welding and Code rules have been followed. See Fig. 5-3*c*. Since stay bolts usually break near the plate supported, warning is given by water flushing from the telltale hole.

The stay bolts and stays must be inserted in countersunk holes through the plate, except for diagonal stays, which can be fillet-welded to the shell but not the head, provided the weld is not less than ⅜ in. and the fillet weld continues the full length of the stay. Stay bolts inserted by welding cannot project more than ⅜ in. beyond the plate exposed to products of combustion. The welding must be stress-relieved after it is completed. Radiographing (x-ray photograph) is not required.

Staying of Tube-Sheet Segments In fire-tube boilers the segment of the tube sheets above the top row of tubes requires staying. For this, there are three common methods. In a boiler that does not exceed 36-in. diameter or 100-psi working pressure, structural shapes such as angle irons or channel irons may be riveted to the segment. The outstanding legs are proportioned to have sufficient strength in bending to resist the pressure load.

For boilers exceeding 36-in. diameter or 100-psi working pressure, the flat segment of the tube sheets requires staying either by diagonal stays between the tube sheet and shell or by through-stays running the entire length of the boiler. The former are usually preferable, for they leave more room inside the boiler for cleaning and inspection. The through-stays may make it quite difficult for a worker or an inspector to move around over the tubes in the boiler.

There are three general types of diagonal stay: the Huston, the MacGregor, and the Scully (Fig. 5-4). It is important to have them all under tension, and this effect is accomplished as follows: The tube sheets are marked and drilled for the "crowfoot" of the stays before riveting up. Each stay then has its crowfoot tack bolted to the tube sheet. The holes in the "palm" of the stay are located on the shell. The stays are then removed, and the shell is drilled about $\frac{1}{32}$ in. farther from the tube sheet than the marks. The stays then have their crowfeet riveted to the tube sheets or heads. In order that the holes in the stay palms line up with the $\frac{1}{32}$-in. offset holes in the shell, the stays are

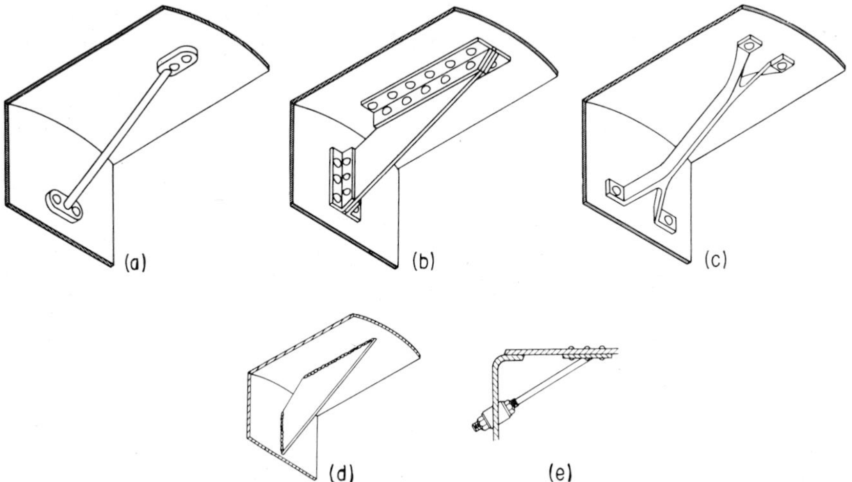

Fig. 5-4 Stays used in fire-tube boilers: *(a)* Scully diagonal stay, *(b)* riveted gusset stay, *(c)* crowfoot diagonal stay, *(d)* welded gusset stay, *(e)* palm stay.

heated to expand them this amount. While hot, the palms are riveted, and, on cooling, the stays contract enough to provide proper tension. Some shops do not take the trouble to heat the stay to line up the holes but "stretch" the stay into position with a driftpin.

The diagonal stays are stretched into position before the tubes are installed in the boiler. In order to prevent distortion of the head through the tension of the stays, one or more heavy steel bars are clamped across the tube sheet as a beam, the tube sheet thus being held against the tension of the stays until some of the tubes are installed. This beam is known as a *strongback.*

The staying of the section of the tube sheet below the tubes of HRT and similar boilers is usually effected by through-to-head stays. These stays are "spooled off" at the rear tube sheet. The front ends of the through-to-head stays—usually two in number—pass through the front tube sheet with inside and outside nuts and washers. They are usually made tight against leakage by grommets or soft metallic packing of various types beneath the outside nuts and washers.

The reason why the rear ends of the stays do not pass through the rear tube sheet but are spooled off inside is that the heat of the fire would damage the nuts and threaded ends.

Manholes and Handholes ASME Code specifies that "all boilers or parts thereof must be provided with suitable manhole, handhole, or other inspection openings for examination or cleaning, except special types of boiler where such openings are manifestly not needed or used. . . ."

A manhole is required in the upper part of the shell or head of fire-tube boilers over 40-in. diameter. A handhold above the tubes will suffice for very small boilers.

If the boiler is 48-in. diameter or larger, a manhole is required in the front head below the tubes. A manhole or handhole is required at this location in smaller boilers. When a handhole is used in the front head below the tubes, it is preferable to provide also a handhole in the rear head below the tubes—whether required by a local code or not. Elimination of this rear handhole may cause considerable difficulty in cleaning the lower internal surfaces of the shell.

HORIZONTAL-RETURN-TUBULAR BOILERS

The HRT boiler (Fig. 5-5*a* and *b*) consists of a cylindrical shell, today usually fusion-welded, with tubes of identical diameter running the length of the shell throughout the water space. The space above the water level serves for steam separation and storage. A baffle plate (or dry pipe) is ordinarily provided near the steam outlet to obtain greater steam dryness.

The HRT boiler is simple in construction, has a fairly low first cost, and is

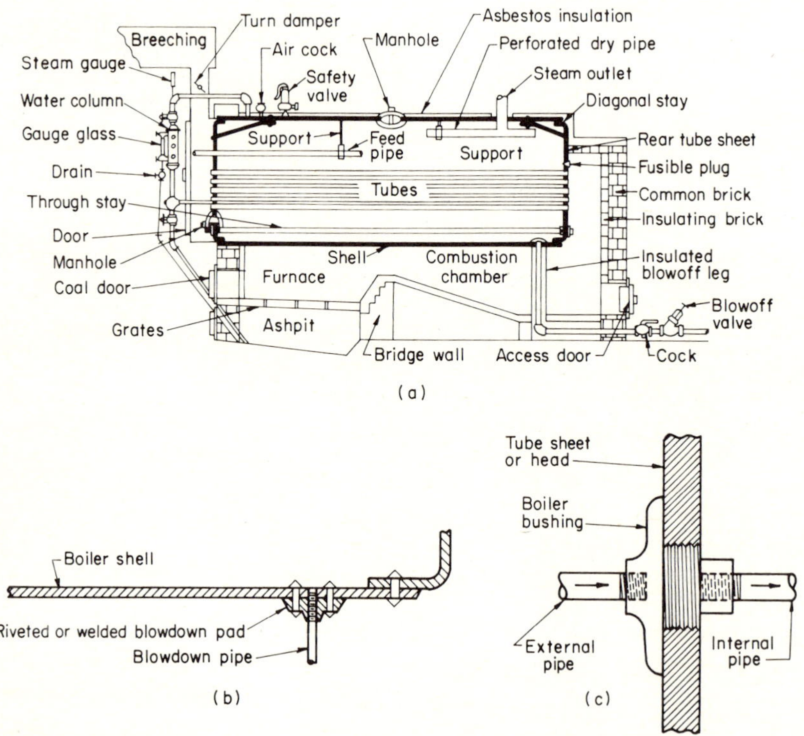

Fig. 5-5 Horizontal-return-tubular boiler details: *(a)* major components of the boiler, *(b)* blowdown connection requiring reinforcement, *(c)* bushing detail as inserted through tube sheet.

a good steamer. It is more economical than the vertical tubular or locomotive types, but the scotch marine boiler is replacing it. One disadvantage is that hard deposits of scale are difficult to remove from water surfaces of inner rows of tubes. Another disadvantage is the danger of burning the shell plates above the fire if thick scale or deposits of mud form on the waterside on these plates. But difficulty of cleaning scale from the tubes holds true for all other types of fire-tube boilers.

Horizontal-return-tubular boilers are not very practical in shell sizes over 96 in. in diameter or for pressures exceeding 200 psi. Thick plates for higher pressures exposed directly to the flames would deteriorate very rapidly from overheating, because of poor heat transfer to the water. This could lead to bags (bulges) in the shell.

Bags also result when mud or other sediment from water settles on the bottom of the shell. For this reason an HRT boiler should be pitched 1 to 2 in. toward the blowdown connection. Also, proper blowdown and water chemical treatment must be used to avoid scale or sediment formation and to keep solids to a minimum.

On older HRT boilers the blowdown line is in the path of the hot gasses and thus must be protected from overheating. A V-shaped pier of firebrick, usually in front of the blowdown pipe, deflects the hot gases. If not properly protected, the blowdown pipe may melt under certain conditions and drain the boiler, possibly leading to a serious explosion.

Blowdown Connection The blowdown pipe cannot be merely screwed into the shell, for there would not be a sufficient number of threads in contact to give proper support. Instead, a pad is riveted to the bottom of the shell, and the blowdown pipe is screwed into this pad (Fig. 5-5*b*).

The minimum size of blowdown pipe for boilers over 10 horsepower (HP) (100 ft^2 of heating surface) should be 1 in. A minimum size of ¾ in. is permitted by most states for boilers of less than this size. The maximum-size blowdown pipe in any case should be 2½ in.

The blowdown pipe of the HRT boiler is carried through a cast-iron pipe sleeve in the rear or side wall with a clearance of ½ to 1 in. all around to allow for expansion. If the boiler settles an appreciable amount, the clearance will become unequal. In severe cases, the clearance on one side or the tip or bottom may disappear, and the blowdown pipe will be pressed against the brickwork. Here, too, severe stresses in the pipe may result in failure. The clearance space is usually sealed gastight with a plastic refractory.

Allowance should be made for expansion of the shell, that is, its unrestricted elongation caused by bringing it up to operating temperatures. Restriction of expansion may cause serious stresses, and it may be produced in several ways. If the boiler is suspended from overhead cross girders, the longitudinal expansion will be practically unopposed (Fig. 5-6). However, if the boiler is supported by brackets on the brick walls, it is important that the points of contact do not prevent free movement. The proper installation places the front brackets stationary, with the rear brackets resting on rollers of short lengths of ¾- or 1-in. steel pipe, free expansion thus being allowed. Flat steel

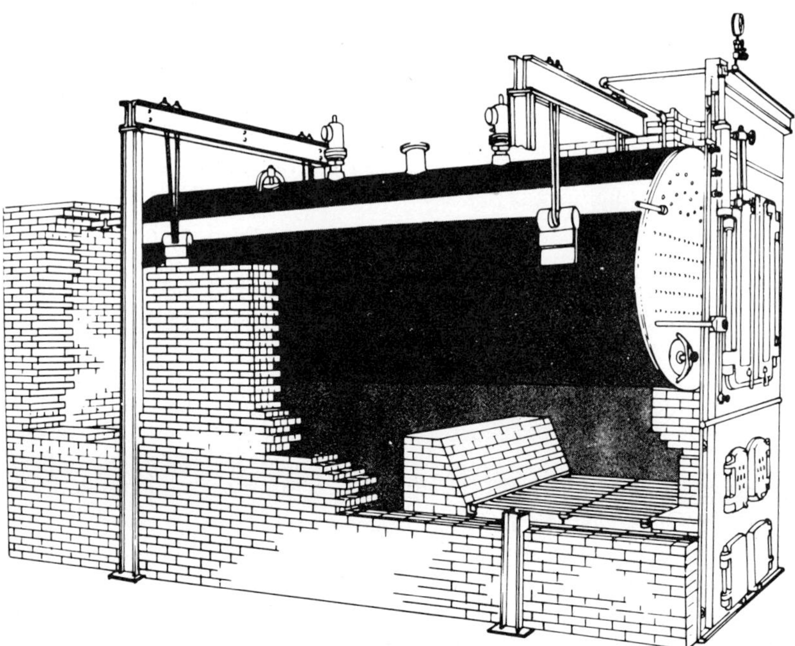

Fig. 5-6 An HRT boiler is supported by a steel structure independent of the brick walls. *(Courtesy Power magazine.)*

bearing plates are used under all points of bracket contact so as not to concentrate the total bracket load on a small area of brickwork but to distribute it over sufficient area on the brick to result in a conservative unit load pressure.

See Fig. 5-6. This type of steel structural support is required on all HRT boilers 72 in. in diameter and over, since brickwork cannot support this heavier weight. On older boilers, the pad on the shell was riveted to the shell, with rivets designed with a safety factor of 12.5. But if the pads are welded, full peripheral fillet welds are required, and these must be stress-related on high-pressure boilers. Radiographing is not required.

Feedwater Inlet The feedwater should enter through the upper part of the shell or front head, and it should discharge clear of any riveted seam or part of the boiler exposed to radiant heat or high temperatures.

If the boiler is over 40-in. diameter, a manhole is provided above the tubes. Then the feed piping should enter through a boiler bushing (Fig. 5-5*c*). An internal feed pipe should discharge the water approximately three-fifths the length of the boiler from the hottest end—usually from the front—so that solids in the feedwater will not be precipitated onto the hot shell plate at the front, where overheating and damage might result.

Circulation The circulation of water in the HRT boiler at operating temperatures is as follows: The radiant heat of the fire on the shell causes the water to rise around each side of the shell; since the tube temperature is somewhat less, the water then returns downward between the tubes. Also, there is some difference in circulation at different longitudinal locations. The hot gases enter the rear ends of the tubes, and as the gases progress toward the front, some of the heat is absorbed by the boiler water. Although the temperature of the gases at the rear ends may be over 1000°F, the exit temperature at the front end may be 500°F. Thus, the circulation in a longitudinal axis will tend to rise between the tubes at the rear of the boiler and to be downward at the front end.

Restricted circulation in the HRT boiler is uncommon, though it is possible if extremely heavy scale deposits build up and block the space between tubes. In such cases, it is probable that damage would occur to the boiler by overheating resulting from restricted circulation in the affected parts.

Rating The conventional means of rating an HRT boiler is according to its heating surface. Most authorities accept the standard of 10 ft² of water heating surface per boiler horsepower. The water heating surface of this type of boiler is taken as one-half the area of the shell plus the total area of all tubes, based on their inside diameter, plus two-thirds the area of the rear tube sheet minus the aggregate area of the tube holes.

$$HS = \frac{(L \times \frac{1}{2}\pi D) + (N \times d \times \pi \times l) + (0.5236D^2 - N \times 0.7854d^2)}{144}$$

where

L = length of boiler shell exposed to products of combustion, in.
D = shell diameter, in.
N = number of tubes
l = length of tubes, in.
d = inside diameter of the tube, in.
HS = water heating surface, ft²

The area of the front tube sheet may be neglected, for the gas temperature is usually too low to cause much evaporation at this point. One-tenth of this result will be the boiler horsepower at 100 percent rating. The evaporation equivalent of a boiler horsepower, 34.5 lb/hr of steam with feedwater at 212°F, may be exceeded by forcing the boiler at a higher rating than that found by the heating-surface standard, but about 125 to 150 percent is the usual practical limitation for this type of boiler. Forcing the output at above this point may result in difficulties.

During an inspection, some of the areas to check carefully on an HRT boiler are the following: Internally, on the section above the tubes, check for corrosion and pitting. Look for grooving on the knuckles of heads, shells, welds, rivets, and tubes. Check the seams for cracks, broken rivet heads, porosity, and any thinning near the waterline of the shell plate. Check all stays for soundness and proper tension. Examine the internal feed pipe for soundness and support, and see that it is not partially plugged. Check the openings to the water-column connections, safety valve, and pressure gauge for scale obstruction. Also check shell and tube surfaces for scale buildup. Follow the same procedure internally below the tubes. Then check the opening to the blowdown connections and make sure that the bottom of the shell is pitched toward blowdown and that it has no blisters or bulges.

Externally, remove the plugs from the crosses of water-column connections and make sure they are free of scale. Examine the blowoff piping pad and the blowoff pipe to make sure it is protected from the fire, that the pipe is sound, and that the blowoff valves are in good order. Examine tube ends and rivets or welds for cracks and weakening of the tube to the tube-sheet connection. Check for fire cracks around the circumferential seam and for leakage at the caulked edge. Then examine the setting and supports for soundness.

ECONOMIC BOILER

The economic-type boiler (Fig. 5-7) is an adaptation of the HRT boiler, giving somewhat greater heating surface per square foot of floor space. An added

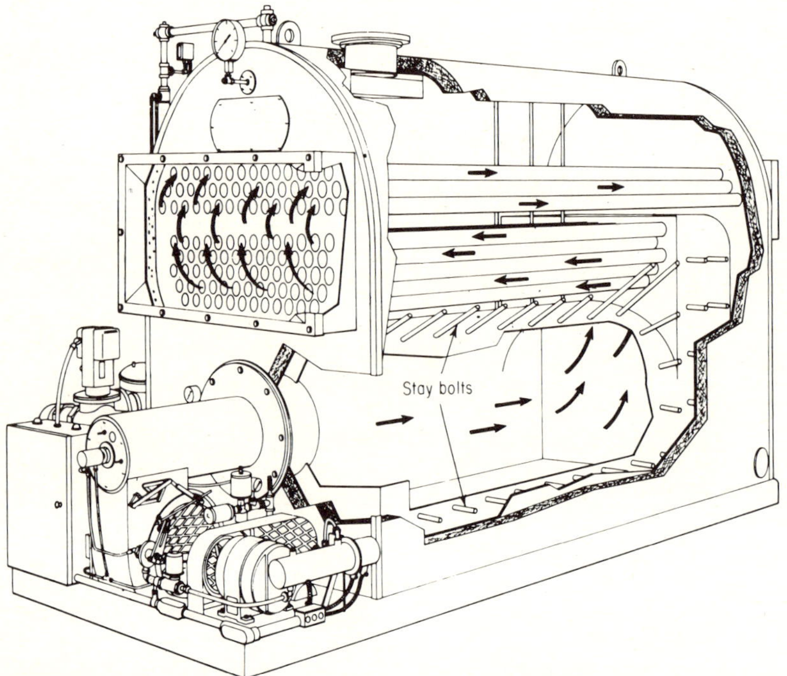

Fig. 5-7 Economic-type fire-tube boiler consists of three passes with arched crown sheet and stayed surfaces.

advantage is that the required amount of brickwork is much less since the boiler is self-supporting in its special casing. The rear end is supported by an iron cradle under the shell. The boilers of this type may be shipped as a unit with furnace walls held in position by a permanent steel casing. The construction principles, piping connections, and circulation are essentially the same as those of the HRT boiler. Also, the practical limitations for pressure and capacity are similar.

Because the rear course of the shell plate is oval, the flat surface on each side requires bracing as a stayed surface. This requirement is met by through-braces passing between horizontal rows of tubes from side to side. The ends of these braces are riveted over. The pitch and size of these braces are calculated by the same method as for diagonal stays in an HRT boiler.

In the economic boiler, the fusible plug, if used, is located in the rear head. If the distance between the top row of tubes and the top of the shell is over 13 in., the fusible plug should be at least 2 in. above the top row of tubes. Otherwise, if may be located not lower than the upper part of the top row of tubes.

The boiler is self-supported in its special casing and thus requires little brickwork. But this type has the same size and pressure limitations as the HRT boiler. The flat surfaces on each side require staying. The front water legs are stayed to each other by means of stay bolts. In the back, the side sheets are stayed by through-braces passing between the horizontal tubes.

The economic-type boiler is considered to be an externally fired, fire-tube design because its steel-encased combustion chamber is not a pressure part of the boiler. Boilers of this type are usually shipped as a unit and thereby qualify as being among the first so-called compact designs.

LOCOMOTIVE FIREBOX BOILERS

See Fig. 5-8. Like the vertical tubular and scotch marine types, the locomotive firebox boiler is an internally fired fire-tube unit. But its shell is horizontal, and the firebox is not contained within the cylindrical portion of the boiler. The firebox is rectangular with a curved top known as a crown sheet. This crown sheet is supported by radial stays screwed into the crown sheet and the outer wrapper sheet. Ends of radial stays are riveted over. The inner sheets of the firebox are connected to the outer side sheets by stay bolts. The space between these sheets is called the *water legs*. The fire tubes are within the barrel and run from the firebox tube sheet to the smokebox head of the barrel. This head in the smokebox is formed by extending the barrel beyond the tube sheet.

The firebox front sheet above the crown sheet and the smokebox tube sheet above the tubes are supported by longitudinal stays. In some cases diagonal stays also are used for this purpose. The steam dome provides additional steam storage space and allows the main steam outlet to be taken off at a considerable height above the waterline, thus reducing the possibility of water carrying

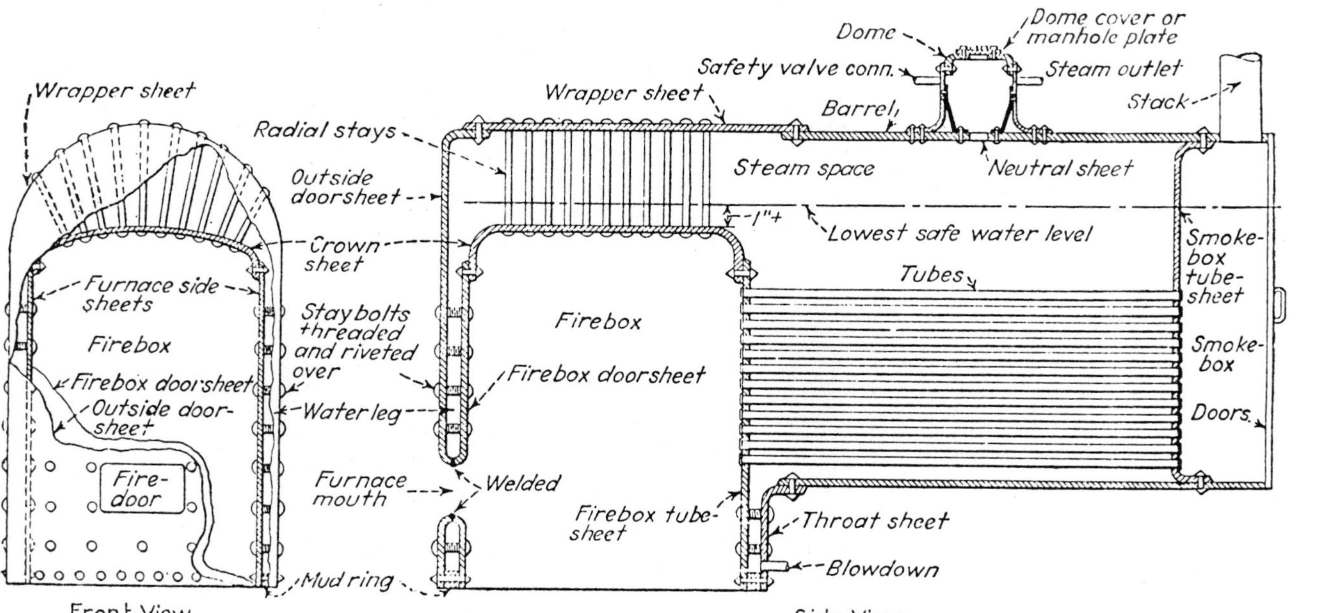

Fig. 5-8 Locomotive boiler sheets and parts. The outside door sheet and smokebox tube sheet are stayed to the shell.

over with steam. The steam space extends over both the furnace and the barrel, which usually has 3-in. tubes. All tubes are of one diameter and length.

As is true of most internally fired fire-tube boilers, some water spaces are very difficult to clean, either mechanically or manually. Also, the locomotive firebox boiler is limited as to pressure and capacity, just as is the HRT boiler.

Water Leg The water space between the outside sheets and the firebox sheets is known as the water leg, as used on all horizontal or vertical firebox boilers which have this type of furnace construction. The inner and the outer sheets are fastened together at the bottom by either a mud ring or an "ogee" flange (Fig. 5-9). The former is more expensive but is usually more satisfactory, for it is easier to keep clean. An inherent disadvantage of the ogee flange is that the breathing action due to expansion and contraction of the furnace may cause fatigue stress and cracking. Modern construction would flange the inner and outer sheet together and butt-weld them.

Most locomotive boilers are equipped with a steam dome because the steam space and the water surface are limited and more space for the steam flow to disengage drops of water is desirable. An additional service for the steam dome in railroad service is to house the throttle valve. The steam dome is usually installed with its vertical seam on one of the two long sides of the dome—this, of course, on one side of the boiler shell. Thus, the longitudinal seam of the dome will not have to be flanged so sharply.

It is important to protect the furnace sheet from overheating at the mud ring. Also, the bottom few inches of the water leg may become filled with sediment, and that area, too, should be protected.

In hand-fired installations, the grate is well above this zone. But with mechan-

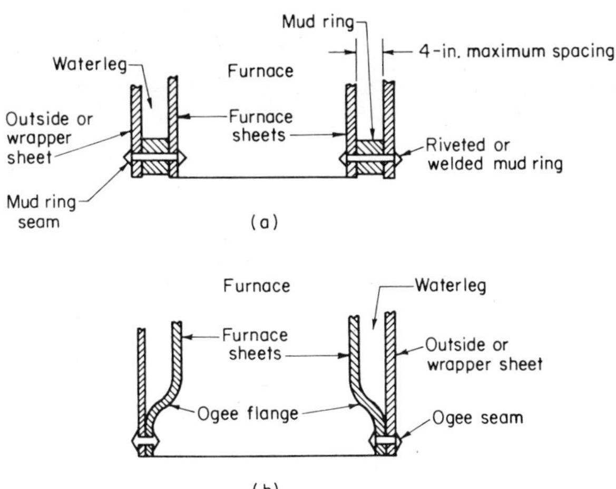

Fig. 5-9 Water-leg construction details: *(a)* water-leg bottom, firebox-type boilers, mud-ring style; *(b)* water-leg bottom, firebox-type boiler, ogee-ring style.

ical firing, a firebrick wall should be built up about 6 in. above the mud ring on all sides of the firebox.

Inspections For internal corrosion, check: (1) waterline and top row of the tubes because of oxygen and other impurities released when boiling; (2) top of the crown sheet, on and around the ends of the radial stays and stay bolts; (3) water legs because of oxygen release and the presence of corrosive sediment; and (4) bottom of the barrel where pitting occurs if the boiler is improperly laid up when out of service.

External corrosion on a locomotive firebox boiler should be especially looked for around handhole plates and at the bottom of the first head. Rainwater wetting the sooty smokebox will corrode its bottom and also the handholes in this area. The bottom of the barrel should be watched for corrosion because of possible dampness or leakage. Corrosion due to sediment deposits forming at the bottom of water legs and their handholes may eat through the plate and gaskets of handholes. Leaking tubes or stay bolts cause corrosion at the furnace end of the tube sheets.

SCOTCH MARINE BOILERS

The largest number of boilers in use today for commercial and small industrial plants is the scotch marine (SM) boilers. This boiler was originally used for marine service because the furnace forms an integral part of the boiler assembly, permitting very compact construction that requires a small space for the capacity produced. The SM boiler is sold as a package consisting of the pressure vessel, burner, controls, draft fan, draft controls, and other components assembled into a fully factory-fire-tested unit. Most manufacturers test their models as a unit before it is shipped to the site, basically delivering a product that is pre-engineered and ready for quick installation and connection to services such as electricity, water, and fuel. Some manufacturers provide starting service as part of the purchase price. A factory-trained specialist starts up the unit, readjusts controls, checks out the unit in operation, makes necessary adjustments, and trains an operator to troubleshoot on controls.

The SM boiler is built either as a wet-back furnace, as shown in Fig. 5-10*b*, or as a dry-back, shown in Fig. 5-10*a*. This boiler is an adaptation to stationary practice of the well-known SM wet-back boiler. It consists of an outer cylinder shell, a furnace, front and rear tube sheets, and a crown sheet. The hot gases from the furnaces pass into a refractory-lined combustion chamber at the back (sometimes built into a hinged or removable plate) and are returned through the fire tubes to the front of the boiler and then to the uptake. This boiler is suitable for coal, gas, and oil firing.

In the wet-back design (Fig. 5-10*b*), the shell, tube, and furnace construction are similar to the dry-back type, but the combustion chamber, being inside the shell, is surrounded by water. Thus no outside setting or combustion-chamber refractory is needed. The dry-back type is a quick steamer because of

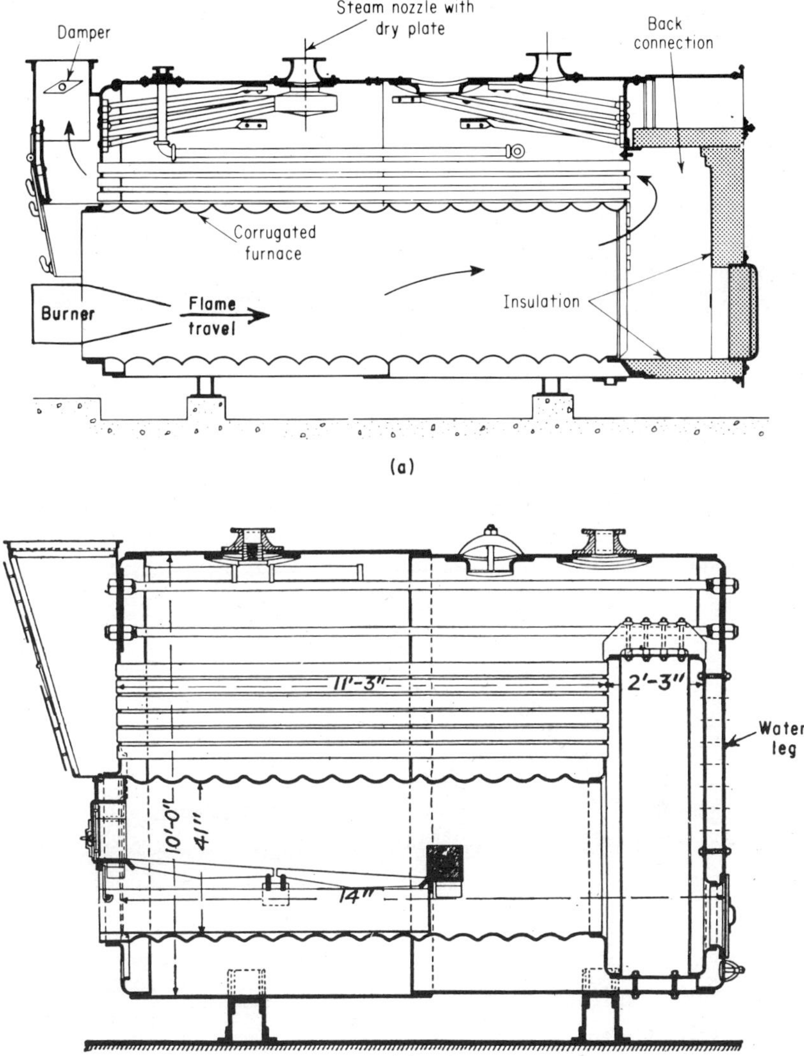

Fig. 5-10 Scotch marine boilers: *(a)* dry-back type, *(b)* wet-back type. Both shown have corrugated furnaces.

its large heating surface. It is also compact and easily set up and shows fairly good economy.

The internal furnace is subject to compressive forces and so must be designed to resist them. Furnaces of relatively small diameter and short length may be self-supporting if the wall thickness is adequate. For larger furnaces, one of these four methods of support may be used: (1) Corrugating the furnace walls; (2) dividing the furnace length into sections with a stiffening flange (Adamson ring) between sections; (3) using welded stiffening rings; and (4) installing stay bolts between the furnace and the outer shell. If solid fuel (coal, wood, etc.) is to be fired, a bridge wall may be built into the furnace at the end of the grate section.

The scotch boiler may be by far the largest in diameter of any fire-tube boiler, being built up to about 15-ft diameter. Since the area of the segment of heads above the tubes is large, diagonal stays are usually precluded because of the great number that would be required. Instead, it is customary to use a smaller number of head-to-head through-stays of 2- to 3-in. diameter.

In boilers of large diameter, it is the practice to use more than one furnace. Two, three, or even four furnaces are used in the large boilers of this type. Figure 5-11 is a cut-away view of a four-pass model. This unit maintains a continuously high gas velocity. As the hot gases travel through the four passes, as shown in Fig. 5-12, they transfer heat to the boiler water and thus cool and occupy less volume as the gases pass through the different tube passes. The number of tubes is reduced proportionately to maintain the high gas velocity and thus keep producing as much as possible a constant heat transfer.

While more passes can be used, today four is the practical limit. The two-

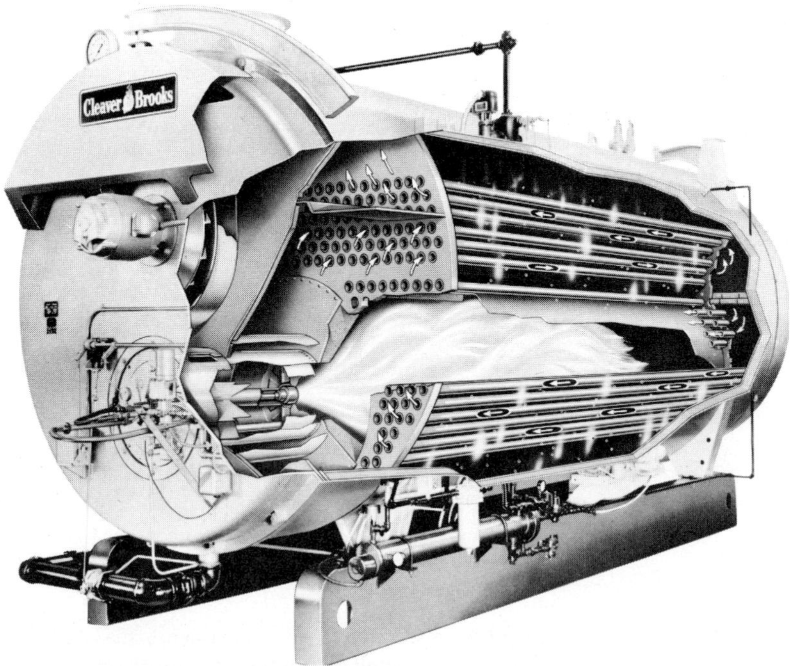

Fig. 5-11 Four-pass SM boiler. *(Courtesy Cleaver-Brooks Co.)*

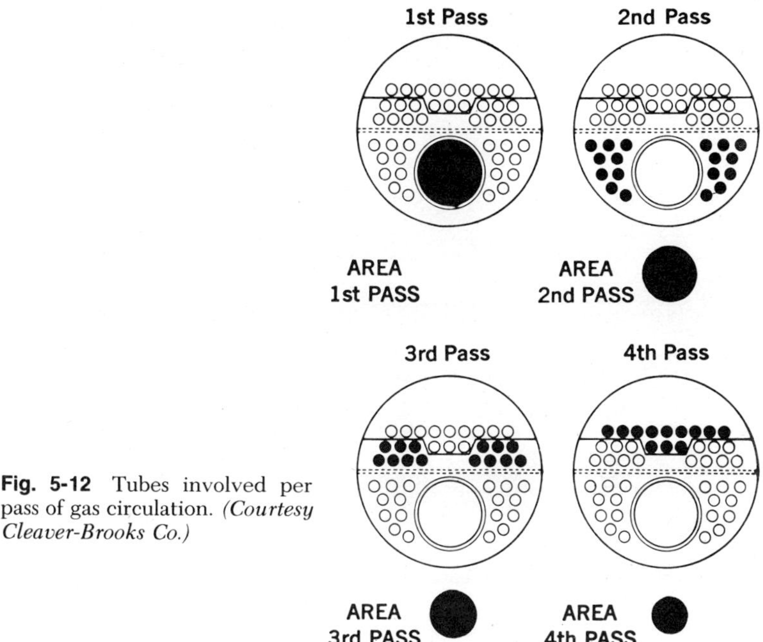

Fig. 5-12 Tubes involved per pass of gas circulation. *(Courtesy Cleaver-Brooks Co.)*

pass design is the most popular for average plants, whether for high- or low-pressure service. Designers also give attention to the number and arrangement of the second-pass tubes. For example, waterside inspection and cleaning are easier when the tubes are aligned uniformly both vertically and horizontally. But a staggered layout of tubes tends to give a more circuitous flow of water around these tubes, thus promoting increased heat transfer. Improved transfer from gas to water is also obtained by using slip-in spiral fittings (retarders) at the tube inlet to impart a swirl to the hot gases. Some boilers have tubes finned on the fireside known as extended-surface tubes.

The furnace of an SM boiler may provide up to 65 percent of the boiler output even though it may have only about 7 to 8 percent of the total heating surface. In the furnace, most of the heat is transferred by radiation. The furnace should have sufficient volume to permit complete combustion of the fuel-air mixture before the flue gases reach the tube passes. Most designers try to limit the heat-release rate in the furnace to below 150,000 Btu/(hr/ft^3) of furnace volume; otherwise, the ratio of air to fuel becomes critical. Rates above 150,000 Btu/(hr/ft^3) of furnace volume can cause the fuel to be still burning at entry to the first pass of tubes, and this can cause cracking of tube ends or cracking of welds between the furnace and the rear tube sheets. Any sacle or other deposits can aggravate this cracking with high furnace heat-release rates. Good feedwater treatment is thus essential for SM boilers with high heat-release rates in the furnace.

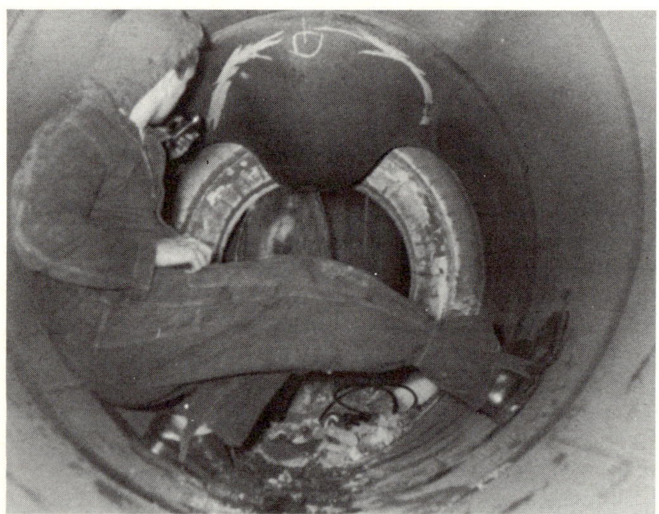

Fig. 5-13 Collapsed furnace of SM boiler from low water. *(Courtesy Royal Insurance Co.)*

There is also the problem of corrosion on the fire side when sulfur-burning fuels are used. Corrosion can occur when the plate or tube temperatures drop below the acid dew point, such as may occur from on-off firing, and reaches a maximum rate when the temperatures fall to less than the water dew point. On-off firing usually requires purging of the furnace, and this can also produce thermal gradients in the boiler which might produce cracking from expansion and contraction effects.

Compact designs also tend to make surfaces less accessible for inspection and cleaning. Thus today's higher heat-transfer rates can easily cause overheating, especially if forced. See Fig. 5-13. This results in loose tubes in tube sheets, cracks between ligaments on tube sheets, weld cracks in high-heat zones, bulged furnaces, and low water. Even more dangerous is the complete reliance on automatic controls to safely cycle a boiler, without periodically checking the controls for (1) conditions of electric contacts; (2) electric connections; (3) water-column connections; (4) waterside plugging of pressure switches; (5) low-water fuel cutoffs; (6) soot accumulation in tubes; (7) operation of solenoid valves in fuel-cutoff lines; (8) firing-equipment timing and operation of flame-failure devices; and (9) operation of safety valves.

VERTICAL TUBULAR BOILER

The vertical fire-tube boiler is used where floor space is at a premium and the pressure and capacity requirements come within the scope of this type of boiler. The vertical tubular (VT) boiler is an internally fired fire-tube unit.

It is a self-contained unit requiring little or no brickwork. Requiring little floor space, it is popular for portable service, such as on cranes, pile drivers, hoisting engines, and similar construction equipment. Vertical tubular boilers are used for stationary service where moderate pressures and capacity are required for process work, such as pressing, drying-roll applications in various small laundries, and in the plastics industry.

The coil-type watertube boiler is a competitor of the VT boiler for small capacities and lower pressures to 150 psi. But the VT boiler is limited in capacity and pressure even more than the horizontal fire-tube boiler. For this reason, most VT boilers of the fire-tube type seldom exceed 300 HP, or about 10,000-lb/hr capacity with a maximum pressure of 200 psi. There are five general classifications:

1. Standard straight shell type with dry top (Fig. 5-14*a*)
2. Straight shell type with wet top (Fig. 5-14*b*)
3. Manning boiler with enlarged firebox
4. Tapered-course bottom with enlarged firebox
5. For even smaller capacity, the vertical tubeless unit

The Manning and the tapered-course types provide a greater grate area and furnace volume, which permit somewhat better combustion efficiency.

The submerged-head type was developed to prevent overheating of the upper ends of the tubes, which are in the steam space in the standard vertical

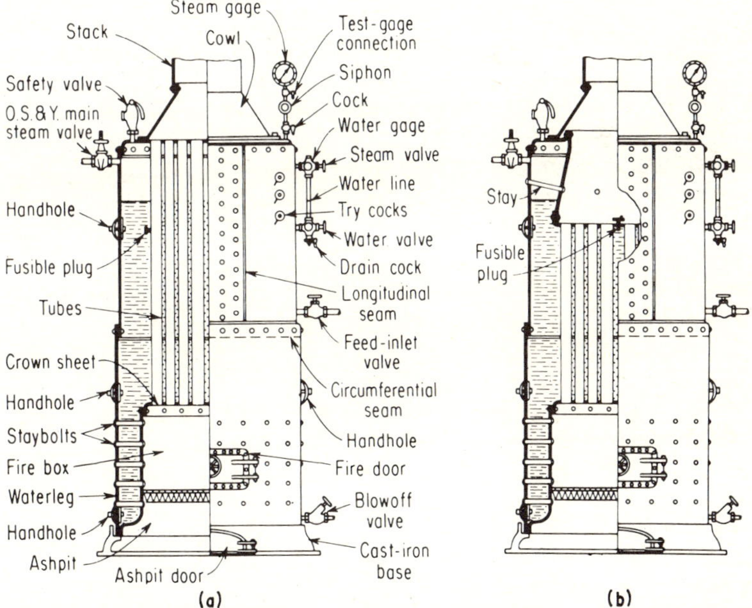

Fig. 5-14 Vertical tubular boilers: *(a)* dry-top design, and *(b)* wet-top design.

fire-tube boiler. Overheating and damage result sometimes when the latter boiler is forced or too hot a fire is maintained when starting up. A slow fire is essential until steam generation starts. Then the upper ends of the tubes may be "cooled" by steam.

The advantages and disadvantages of VT fire-tube boilers are (1) compactness and portability; (2) low first cost; (3) very little floor space required per boiler horsepower; (4) no special setting required; and (5) quick and simpler installation.

Disadvantages are that (1) the interior is not easily accessible for cleaning, inspection, or repair; (2) the water capacity is small, making it difficult to keep a steady steam pressure under varying load; (3) the boiler is liable to prime (carry over with steam) when under heavy load because of the small steam space; and (4) the efficiency is low in smaller sizes because hot gases have a short, direct path to the stack, so much of the heat goes to waste.

Deposits settle in the 4-in.-wide water legs (maximum per Code) because they have restricted circulation. Cleanout openings should be periodically opened around the circumference in the water legs and bottom tube sheet so all areas are accessible for cleaning. Some units have a continuous drain in the bottom of the water legs. When this boiler is opened for cleaning, a length of chain can be pulled around inside to get the sludge and scale to a cleanout opening for removal.

The Code indicates no specific water level for the dry-top boiler, except to state: "It shall be at a level at which there shall be no danger of overheating any part of the boiler when in operation at that level." But the level is generally taken as a minimum at a point two-thirds the height of the shell, above the bottom head, or crown sheet or tube sheet. This is the same minimum requirement as for miniature vertical fire-tube boilers.

For the submerged wet-top type, the minimum water level must be at least 2 in. above the top tube ends, except for miniature boilers, where it can be 1 in.

Internal and external corrosion usually take place in a VT boiler.

1. At the waterline, usually on the tubes. This is due to oxygen and organic materials being released there during the process of boiling.
2. In the vicinity of the feedwater discharge, as a result of oxygen release.
3. On top of the lower tube sheet, because of scale formation.
4. On and around the ends of the stay bolts, as a result of the stresses imposed and the subsequent expansion and contraction (this leads to stress corrosion).
5. In the water legs, especially at the bottom, where grooving may occur in addition to the usual pitting under the scale.

External corrosion occurs:

1. On top of the top tube sheet, bottom tube sheet, and tube ends, because of acid formed by the damp soot in contact with the products of combustion.
2. Around all manhole, handhole, and washout openings. Here corrosion is from leakage due to poor gaskets, improper handhole-cover installation, and thermal expansion and contraction which loosen opening closures.

3. At the bottom of shell, water legs, and furnace sheet, as a result of soot attack.

4. Around all openings including gauge connections, safety-valve connections, steam-outlet connection, feedwater connection, blowdown connection, and water-column connection, because of leakage.

Questions and Answers

HORIZONTAL-RETURN-TUBULAR BOILERS

5-1 Why, instead of through-to-head stays, are diagonal stays not used below the tubes, as above the tubes, to brace the tube sheet?

Ans. Because there is usually insufficient room for the proper number without placing them too close together or using sizes larger than is practical. Also, they would tend to hold loose scale and sludge and prevent its free movement to the blowdown.

5-2 How much space should there be in front of the boiler in planning installation?

Ans. Sufficient room for replacement of tubes.

5-3 If, on looking in the furnace, the bottom of the shell is found to be bulged, what should be done?

Ans. The boiler should be shut down immediately and then inspected by an authorized boiler inspector. The inspector's recommendations should be followed before the boiler is returned to service.

5-4 Why are boilers over 72-in. diameter required to be supported by the outside-suspension type of setting?

Ans. Because the weight of the larger boilers may be in excess of the safe load on brickwork. Crushing or buckling of the walls might result from the load of a large boiler full of water.

5-5 What are two dangerous conditions offered by a weakened rear arch?

Ans. (1) If the arch collapses wholly or in part, the upper part of the rear tube sheet may become overheated and damaged. (2) Anyone walking on top of the arch may cause it to collapse, and that person will fall through to a horrible death.

5-6 What precaution is required with a flush-front-set HRT boiler?

Ans. The front arch protects the dry-sheet and front head seam from damage by overheating. This arch should be kept in good condition.

5-7 Why should the hole in the brick wall through which the blowdown pipe passes be inspected?

Ans. To see that the pipe is not resting on the bottom of the opening. If it does, it is an indication that the boiler is probably settling and that it may not have sufficient clearance to allow free expansion and contraction.

5-8 Would loose bricks lying in the firebox be of any interest? If so, why?

Ans. They might be from the closing-in line or from some point where they are supposed to protect part of the boiler from overheating. Also, they might be from a point where their loss would weaken the walls.

5-9 Name three points where brickwork should protect the boiler from overheating.

Ans. The rear arch, the closing-in-line along the sides of the shell, and the front arch in the flush-front-set boiler.

5-10 How and under what conditions may the segments of the heads above the tube dispense with diagonal or through-stays?

Ans. In boilers not exceeding 36-in. diameter or 100-psi maximum allowable pressure, the segments may be braced by stiffening with channel irons or angle irons (riveted back to back) riveted to the tube sheet. The specifications for structural form should comply with Code rules.

5-11 What is a strongback?

Ans. It is a bar bolted to the tube sheet to prevent the sheet from buckling when the stays are put in tension before the tubes are installed.

5-12 If the boiler is pitched forward slightly, what should be done?

Ans. Unless the installation was designed in this way with the blowdown in front, the boiler should be reset with the proper pitch toward the rear as soon as possible. In the meantime, the boiler should be shut down frequently and the manhole or handhole in the front head below the tubes removed. All sediment should be washed out. The frequency of these washouts depends on the rapidity with which sediment accumulates.

5-13 What is the difference between a through-brace and a through-to-head brace? Where is each used? Why?

Ans. The through-brace has washers and nuts on each end. The through-to-head brace has nuts and washers on one end; the other end is forged into an eye that is held clear of the rear head by a pin or other construction. The through-stays may be used above the tubes, for the rear outside nuts are protected from burning off by the rear arch.

The through-to-head braces are used below the tubes where the rear ends have to be protected from the high-temperature gases.

5-14 If a fusible plug is used, where should it be located?

Ans. Near the centerline of the rear tube sheet not less than 1 in. above the top row of tubes.

5-15 How may the proper height of a water column be checked?

Ans. Fill the boiler with water to a level 3 in. above the top row of tubes at their highest end (usually the front). The water column should be at a level where water just shows in the bottom of the gauge glass. Or the height of a water column may be checked by measurement from any common point of elevation.

5-16 Why do the flat segments of the tube sheets or heads require bracing?
Ans. Because internal pressure tends to bulge these areas outward into a semispherical shape.

LOCOMOTIVE BOILERS

5-17 What are the sheets of a locomotive boiler?
Ans. Barrel or shell, smokebox tube sheet, dome shell, dome head, wrapper sheet, throat sheet, firebox side sheets, firebox tube sheet, crown sheet, inside-door sheet, and outside-door sheet.

5-18 Where is the most dangerous part of the boiler in case of low water?
Ans. The crown sheet.

5-19 From which end are tubes removed and replaced?
Ans. The smokebox end.

5-20 How are the firebox sheets supported?
Ans. The side, door, and throat sheets are supported by stay bolts. The crown sheet is supported by radial or girder stays.

5-21 How may expansion of long firebox side sheets be accommodated?
Ans. Vertical corrugations in the sheets are used between several of the vertical rows of stay bolts.

5-22 What are required with sling stays that are not needed with radial stays?
Ans. Crown bars and girder stays.

5-23 What are the advantages and disadvantages of radial stays and girder stays?
Ans. The radial stays are more flexible and tend to hold less scale from circulation than girders. About the only advantage for the girder stays is that they pass straight through the sheet rather than at an angle.

5-24 If the segment of the smokebox tube sheet below the tubes requires bracing, what type of stay may be used?
Ans. Any approved type of diagonal stay.

5-25 Where is the usual location for the manhole?
Ans. In the dome head, or in the shell if no dome is used.

5-26 How many manholes are used in a locomotive boiler? Explain your answer.
Ans. One. The part of the barrel over the tubes is the only part of the interior that is accessible.

5-27 What should be the lowest level for the bottom nut of the gauge glass?

Ans. The lowest visible part of the glass should be at least 3 in. above the highest part of the crown sheet.

5-28 How far through the sheets should the ends of stay bolts extend before they are riveted over?

Ans. Not less than two threads.

5-29 Why should the feedwater not discharge into the water leg?

Ans. The cooling action of the water against the hot furnace sheets would cause serious stresses and probable damage.

5-30 What purposes do smokebox doors serve?

Ans. Inspection, cleaning, and tube replacement.

OTHER FIRE-TUBE BOILERS

5-31 Name four types of firing-door construction in a VT boiler.

Ans. Flanged together and riveted, flanged together and butt-welded, both sheets flanged out and riveted (sucker-mouth), riveted doorframe ring.

5-32 What are the four types of VT boiler?

Ans. Standard straight-shell, Manning, tapered-shell, and submerged-head.

5-33 What are the reasons for or advantages in the deviations from the standard type of the other types named in the answer to question 5-32?

Ans. The submerged head protects the upper ends of the tubes from overheating. The Manning and tapered-shell types allow a larger grate area and furnace volume.

5-34 On what point of construction is the maximum allowable pressure based in the standard VT boiler?

Ans. On the ability of the shell or furnace, whichever is the weaker, to resist collapse.

5-35 What additional method is used in calculating the maximum allowable pressure of a Manning boiler?

Ans. It is necessary to figure the strength of the shell at its point of greatest diameter, between the reverse flange and the top row of stay bolts.

5-36 Is the economic boiler internally fired?

Ans. No. Its combustion chamber is steel-encased, but the casing is not a pressure part of the boiler.

5-37 Name four internally fired boilers.

Ans. Vertical tubular, locomotive, scotch marine, and scotch dry-back.

5-38 What boiler has a crown sheet that is braced similarly to the locomotive crown sheet?

Ans. The scotch marine boiler.

5-39 What internally fired boiler frequently has more than one furnace? How many has it?

Ans. The scotch marine or the scotch dry-back often has two to four circular furnaces when the boiler is of large diameter.

5-40 How are internal furnaces braced?

Ans. If flat, they are stay-bolted. If circular, they may be thick enough to be self-supporting for moderate pressures. If they are not thick enough, they may be stay-bolted, corrugated, or braced by Adamson rings.

5-41 If used, where should the fusible plug be located in any boiler?

Ans. Not lower than the lowest safe water level.

GENERAL QUESTIONS APPLYING TO ALL TYPES OF FIRE-TUBE BOILERS

5-42 What would cause a fire tube to burst or explode?

Ans. Fire tubes are normally under external pressure. They may collapse, but do not burst.

5-43 How should a boiler be prepared for inspection?

Ans. Extinguish the fire, and allow the boiler to cool slowly. When it is cool, open the blowdown and drain, venting the boiler to atmosphere. The soot and ashes should then be blown and swept clear of all tubes, shell plates, heads, seams, and every accessible external surface.

The blowdown valve should be shut if any other boilers feed this line. The manhole, handhole plates, and inspection plugs should then be removed. All loose deposits of sludge or other sediment should be washed out. The blowdown valve should be opened only when it is certain that there is no pressure in the line and no one inside the boiler. Attached scale or oil deposits should be left for the inspector to see.

A boiler properly prepared for inspection should be cool, clean, and dry.

It is advisable to attach a red tag marked "Person in boiler" to the steam, blowdown, feed, and fuel valves and also to the manhole plate whenever anyone is in the boiler.

5-44 How are diagonal stays installed?

Ans. When the crowfoot is against the head, the holes in the shell should be placed so that the holes in the palm of the stay are about $\frac{1}{32}$ in. shy of lining up. Often, the crowfoot is bolted to the head, and the shell or palm holes are marked off and drilled to meet this requirement. The crowfoot is then riveted to the head. The stay is elongated so that the shell and palm holes line up. This effect may be accomplished by heating, but a driftpin is often sledged in one hole to serve the purpose. The palm is riveted in position, and the stay becomes in tension to support the head. Diagonal stays are installed before the tubes are. The tube sheet must be held from buckling by a strongback until the tubes are put in.

5-45 May a safety-valve nozzle or steam nozzle be inserted flush and butt-welded to the boiler shell?

Ans. Yes, if all requirements for fusion-welded boilers are met.

5-46 Briefly, what are these requirements?

Ans. The joints should be fusion-welded by certified welders. They should be subjected to stress relieving, radiographic examination, and a hydrostatic and a hammer test by an authorized inspector.

5-47 Of what value is a steam dome?

Ans. It is designed to give drier steam, and it adds very slightly to the steam-storage capacity of the boiler.

5-48 How large a steam dome may be used?

Ans. A steam dome may be as large as 0.6 times the shell diameter unless the dome head or shell is stayed fully to the neutral sheet of the boiler shell. In the latter case, the dome may not exceed 0.8 times the boiler-shell diameter.

5-49 How should a dome be attached to the boiler shell?

Ans. If the dome is 24-in. diameter or over, it should be double-riveted or double-full-fillet-welded. If it is under 24-in. diameter, it should be double-full-fillet-welded or single-riveted unless the dome diameter times the maximum pressure in pounds per square inch exceeds 4000. Double riveting is then required. If the double full-fillet welding is used, the weld must meet all specifications for fusion welding except that the x-ray examination may be omitted.

5-50 Where and why are drain holes required in a steam-dome installation?

Ans. A drain is required on each side of the dome in the lowest part of the neutral sheet so that condensate will not collect and possibly cause corrosion.

5-51 May a dome have a lap-riveted longitudinal seam?

Ans. Yes, if the dome is under 24-in. diameter and the maximum working pressure is calculated with 8 as a factor of safety.

5-52 What other types of seam in question 5-51 may be used?

Ans. Butt-and-double-strap or butt-welded if all requirements for fusion welding are complied with.

5-53 What precautions should be taken before entering a boiler shell?

Ans. If the blowdown enters a common line with other boilers in operation, all valves on the line to the open boiler must be closed. If other boilers are operating on the same steam header, *both* stop valves must be closed and the drip valve between them must be open. Any other valves on lines under pressure leading to the boiler must be checked. The engineer in charge and the operator on duty must be told that someone is going inside. A responsible person should be stationed at the manhole or entrance doors while anyone is inside the boiler.

5-54 *(a)* Why do the upper ends of tubes in VT boilers which are not covered with water not burn out?

(b) Why do the rear ends of tubes on a high-temperature boiler leak in case of low water?

Ans. *(a)* Because the products of combustion at this point are at such a temperature that cooling of the tube ends is provided by the steam and prevents the upper ends of the tubes from burning.

(b) The products of combustion at the front tube sheet are at a much lower temperature than at the rear tube sheet, so that when low water occurs, the tubes overheat and expand; at the same time, the rear tube sheet overheats and the tube hole expands and gets larger, so that the seating of the tube produced by the expander is destroyed and, therefore, leakage occurs at the rear ends.

5-55 Which type of boiler, fire-tube or watertube, has the most heating surface in the tubes?

Ans. The Code states heating surface shall be computed for that side of boiler surface exposed to products of combustion. Therefore, a watertube contains more heating surface than a fire tube of the same diameter.

5-56 What is the minimum-size feed pipe required for a VT boiler with *(a)* 50 ft^2 of heating surface, *(b)* 125 ft^2 of heating surface?

Ans. *(a)* ½-in. pipe size. *(b)* ¾-in. pipe size.

5-57 *(a)* What is the minimum thickness required for a steel boiler tube in order that a fusible plug may be installed?

(b) What is the maximum allowable pressure and temperature allowed on copper tubes or nipples in fire-tube boilers?

Ans. *(a)* 0.22 in. *(b)* 250 psi and 406°F.

5-58 A VT fire-tube boiler has 95 ft^2 of heating surface. What is the minimum size of blowoff piping that can be used on this boiler?

Ans. ¾ in.

5-59 *(a)* State the reason for beading the ends of tubes of fire-tube boilers.

(b) State the reason for flaring the ends of tubes of watertube boilers.

Ans. Both beading and flaring increase the holding power of tubes. Tests have indicated that a flared tube has more holding power than a beaded tube.

(a) Beading increases the holding power of the tubes and eliminates to a large degree the burning and erosion of tube ends. Therefore, beaded tube ends are used in fire-tube boilers.

(b) Flaring of tube ends increases the holding power of the tubes. Since fire is not in contact with a watertube end, tubes in watertube boilers are usually flared.

5-60 What fusible metal is used in fire-actuated fusible plugs? What is its required melting point?

Ans. Tin; 445 to 450°F.

5-61 Explain the meaning of externally and internally fired boilers.

Ans. Externally fired boilers have a separate furnace built outside the boiler shell. The HRT boiler is probably the most widely known example of the externally fired boiler. In internally fired boilers, the furnace forms an integral part of the boiler structure. The VT, locomotive firebox, and the scotch marine are well-known examples of internally fired boilers.

5-62 Where are stay bolts most likely to break in the locomotive firebox boiler?

Ans. Usually the top row of stay bolts and the first row of radial stays, with fracture occurring close to the inner surface of the outer sheet (wrapper sheet). This area is a high-heat zone, causing large expansion and contraction movements.

5-63 What are the most common staying methods used on an HRT boiler?

Ans. Four methods are used:

1. Through-stays are used below the tubes because there is insufficient room for other stays, which would accumulate deposits more readily.
2. Diagonal stays are used above the tubes to support the flat, unstayed tube sheet above the tubes.
3. Gusset stays are used above the tubes.
4. If not over 36 in. in diameter and not over 100 psi, structural shapes can be used if they are arranged according to ASME Code requirements.

5-64 Name three causes of bagging (bulging) of the shell of an HRT boiler.

Ans. 1. Oil (from lubricated pump rods, etc.) getting into the boiler feedwater and being carried to the lower part of the shell, which is exposed to fire, and thus causing overheating

2. Scale or mud (from sediment in the water) deposits on the lower portion of the shell, restricting heat transfer

3. Excessively localized flame on a portion of the shell, causing overheating

5-65 Name three kinds of angle brace (for segment of tube sheet over tubes in HRT boilers not exceeding 36-in. diameter or 100 psi maximum pressure).

Ans. Angle irons, channel irons, T irons.

5-66 In a boiler shop, why should you examine punchings from rivet holes in boiler plate?

Ans. To see if the allowable tolerances were exceeded. If the plate does not exceed 5⁄16-in. thickness, the holes may be punched to within 1⁄8 in. of full diameter. If the plate exceeds 5⁄16-in. thickness, the holes may be punched not over 1⁄4 in. less than full size.

5-67 How are tube holes made?

Ans. They may be punched to not over 1⁄2 in. less than finished diameter and finished by machining. They are often cut with a pilot drill and rotating cutter.

5-68 How far should the caulking edge of a longitudinal seam be from the center of the nearest row of rivets?

Ans. Not less than 1½ or more than 1¾ times the rivet-hole diameter.

5-69 *(a)* How much larger than a fire tube may the tube hole be? *(b)* How much larger for a watertube?

Ans. *(a)* 1/32 in.-larger diameter at the fire end, 1/16 in. at the opposite end. *(b)* 1/32 in. for either end.

5-70 How are the ends of shell plates, and butt straps, formed for the longitudinal seam?

Ans. They should be formed by rolling or pressing, and not by blows.

5-71 What are the maximum- and minimum-size openings for a blowdown connection in a power boiler?

Ans. Maximum, 2½ in.; minimum, 1 in.; miniature boiler, ¾ in.

5-72 *(a)* When may cast-iron headers be used? *(b)* When may malleable-iron headers be used?

Ans. *(a)* If the pressure does not exceed 160 psi and the cross-sectional area falls within a 7 in. × 7 in. rectangle. *(b)* If the pressure does not exceed 350 psi and the cross-sectional area falls within a 7 in. × 7 in. rectangle.

5-73 What objection is there to shell plates of an HRT boiler exceeding ⅝ in. at the girth seams, and what should be done if they do?

Ans. The combined thickness of metal at the seam might cause the fire side to overheat and crack. It is required that plates over ⅝ in. thick be machined down to not over 9/16 in. with a radius at least 1 in. at the edge of machining at girth seams.

5-74 How should tubes be expanded when there is a reinforcing plate?

Ans. If the plate is caulked on the outside, the tubes should be expanded into both the shell and reinforcing plates so that the tubes will aid in holding both together.

5-75 When may a lap-riveted longitudinal seam be used in a shell over 36-in. diameter?

Ans. When the shell is in a section stay-bolted as a furnace sheet.

5-76 When is a manhole required in the front tube sheet below the tubes in HRT boilers?

Ans. When the boiler is 48-in. diameter or larger.

5-77 When is a manhole required above the tubes of a horizontal fire-tube boiler?

Ans. If externally fired, for boilers of 40-in. diameter or over; if internally fired, for boilers of 48-in. diameter or over.

5-78 *(a)* What is the minimum face width of a manhole flange for a gasket bearing surface? *(b)* What could be done if the flange were only 7/16 in. thick?

Ans. *(a)* 11/16 in. *(b)* A steel ring is shrunk onto the flange, and the double face is machined for a bearing surface.

5-79 Name four ways of supporting a circular furnace subjected to external pressure.

Ans. It may be: (1) self-supporting; (2) stay-bolted; (3) corrugated; or (4) equipped with Adamson rings.

5-80 What is the maximum permitted length of a course in an HRT boiler of riveted construction?

Ans. 12 ft.

5-81 Why are submerged top tube sheets used in a VT boiler?

Ans. A submerged top tube sheet of a VT boiler protects the upper tube sheet and tube ends from overheating and possible fire cracks.

5-82 What size limitations must not be exceeded for a fire-tube boiler to be classified as a miniature boiler?

Ans. The following cannot be exceeded: inside diameter, 16 in.; gross volume, 5 ft^3; water heating surface, 20 ft^2; allowable working pressure, 100 psi.

5-83 Name three advantages in using a corrugated furnace in a scotch marine boiler.

Ans. The corrugated SM type of furnace offers the following advantages over a plain furnace: (1) The corrugations stiffen the furnace so it does not collapse as easily from external pressure as does a similar thick, plain furnace; (2) the corrugated furnace permits more expansion and contraction to take place as a result of the bellowslike action of the corrugations; (3) there is a slight increase in the heat-absorbing area over a plain furnace of the same diameter and length.

6
Watertube Boilers

INTRODUCTION

The chief differences between the watertube boiler and the fire-tube boiler is that in the former the water circulates *through* the tubes instead of around them. The hot gases pass around the tubes.

Fire-tube boilers are designed with the tubes contained in the shell. The tubes of most watertube boilers are located outside the shell or drum. There are two advantages to this feature of the watertube boiler: (1) Higher capacity may be obtained by increasing the number of tubes independent of shell or drum diameter. (2) The shell or drum is not exposed to the radiant heat of the fire.

The biggest advantage over fire-tube boilers is the freedom to increase the capacities and pressures. That is impossible with fire-tube boilers because the thick shells and other structural requirements become prohibitive over 30,000-lb/hr capacity and over 300 psi. The large capacities and pressures of the watertube boiler have made possible the modern, large utility-type steam generators.

Water-Column Connections The water column is connected to the upper and the lower part of the front head of the main steam drum. If more than one drum is installed at the same level, it is not necessary usually to provide a separate water column for each drum, for the water level should equalize in all the drums.

Manholes and Handholes A standard-size manhole is provided in at least one head of each drum. Tube caps are used to close the tube holes which serve the important purpose of allowing for tube insertion, replacement, and inspection. Handholes are provided in external mud drums to permit proper cleaning and inspection.

Rating The rating of all watertube boilers, as of fire-tube boilers, is based on the square feet of heating surface. This is the sum of the areas of drum surface exposed to the products of combustion, the area of all watertubes so exposed based on their outside diameter, and the aggregate projected area of all headers so exposed.

For high-pressure service, the ASME Code stipulates the following attachment of water tubes to tube sheets, headers, and drains: (1) Expanded or rolled and flared (Fig. 6-1); (2) flared not less than ⅛ in., rolled, and beaded;

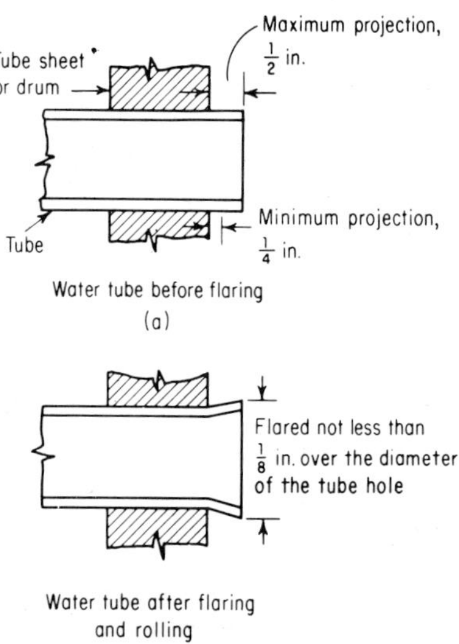

Fig. 6-1 Watertube boiler-tube rolling and flaring requirements.

(3) flared, rolled, and welded; (4) rolled and seal-welded, provided the throat of the seal weld is not more than ⅜ in. and the tubes are rerolled after welding; (5) superheater, reheater, waterwall, or economizer tubes may be welded without rolling or flaring, provided that the welds are heat-treated after welding and the welding is done according to Code requirements.

Large watertube boilers do have complex circulation loops consisting of waterwalls, screen walls, radiant platen superheaters which are designed for mass-flow passes through tubes at different loads with the aim of keeping the tube metal within a safe temperature limit. *Mass flow* is defined as the weight of fluid mixtures in pounds per hour passing through the tube. As boilers become larger, it is obvious that extensive calculations must be made in design to consider the variables involved with mass flows, such as velocity of flow, pressures, temperature zones, specific heat of fluid, conductivity, viscosity, tube diameter, and internal surfaces that could affect the friction of flow through a tube. In comparison, the early watertube boilers were relatively simple in design.

SINUOUS- AND BOX-HEADER BOILERS

Figure 6-2*a* shows an early straight-tube watertube boiler. The tubes are placed in the furnace, and the shell above is used primarily as a storage tank for water and steam. Circulation from the drum is down the back headers, through the watertubes, and through the front header. With this arrangement, the tubes on boilers began to be separated from the internal shell, in contrast to fire-tube boilers. The design shown has one drum; larger boilers had two or three drums. The drum runs from the front to the rear of the boiler. The inclined straight steel tubes, usually of 4-in. OD, are connected with the drum by pressed-steel headers of the sectional type, the tubes being staggered in pairs. A mud drum below the rear headers was used to collect sediment and was blown out from time to time. Tube headers are in one piece for each vertical row of tubes. Header handholes (for tube cleaning) are closed by bolted covers with machined joints.

The sectional header, which is also referred to as a sinuous, or serpentine, header (Fig. 6-3*a*), is either a casting or a forging. In older low-pressure boilers, the headers were also constructed of cast iron. But for high pressure, they were limited by the ASME Code to a maximum design pressure of 160 psi and 350°F. In contrast, headers constructed of forged steel have been used for pressures up to 1200 psi.

A box header (Fig. 6-3*b*) is constructed of flat plates, referred to as a tube sheet and tube cap sheet. But these surfaces must be stayed to prevent deformation. The sides, top, and bottom are flanged and riveted (welded on new boilers) to the tube sheets and cap sheets. The staying of sheets limits pressure for a box header to about 600 psi.

Feedwater Connections The feedwater pipe enters through the front head in the standard longitudinal-drum boiler. An internal feed pipe, several feet long, discharges the feedwater below the waterline to the rear of the front cross box. As in other boilers, the water should not discharge against a riveted seam or directly against the drum surfaces. The internal feed pipe permits the water to be heated somewhat and then discharge in the same direction as the normal circulation. Solids are carried back and are deposited in the mud drum.

The longitudinal-drum type (Fig. 6-2*a*) has the drum, or drums, parallel to the inclined tubes and above the headers. In the longitudinal-drum box-header type, the water leg at the high end of the tubes has a flanged semicircular throat, which is welded directly to the drum (or drums). At the low end of the tubes, the rear box header is connected with the drum (or drums) by tubes expanded into the top of the headers and into a throat connection welded to the drum (or drums).

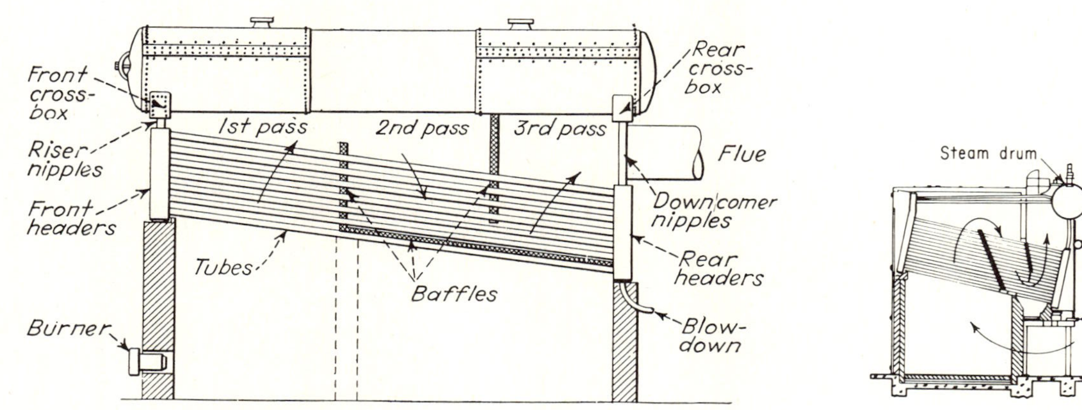

Fig. 6-2 Early straight-tube watertube boilers: *(a)* longitudinal-drum type, *(b)* cross-drum type.

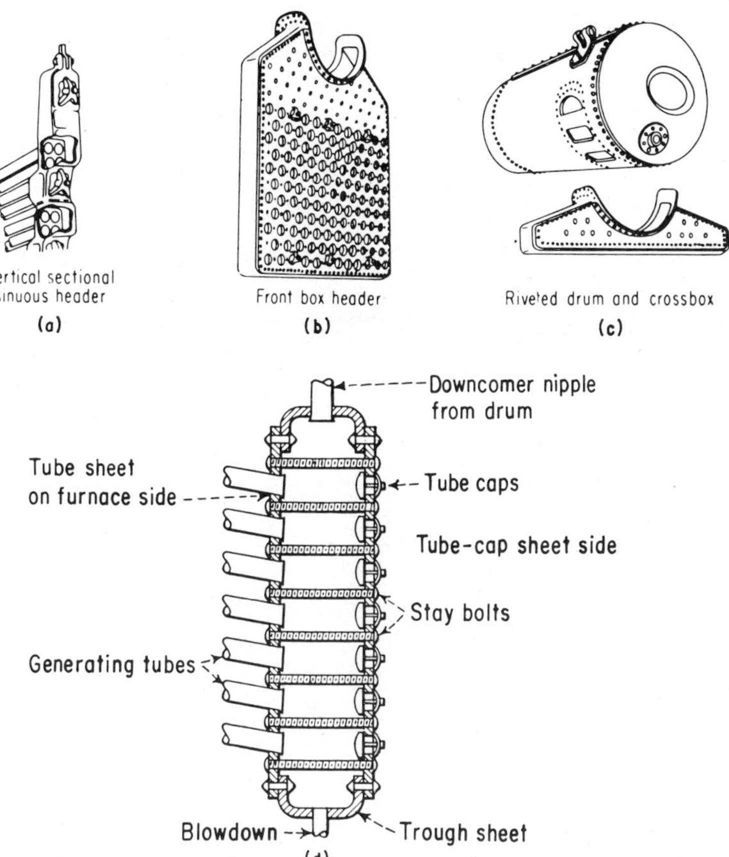

Fig. 6-3 Sinuous and box headers were used on straight watertube boilers: *(a)* vertical sectional sinuous header, *(b)* front box header, *(c)* cross box connecting shell to box header, *(d)* box-header detail.

Cross-Drum Boilers Cross-drum watertube boilers are constructed with either sinuous or box headers, as with the longitudinal-drum types. (See Fig. 6-2*b*.) The rear headers are supplied with water by a row of downcomer nipples connected to the lower part of the cross drum. These are often the same diameter as the generating tubes. The upper ends of the front headers are connected to the front side of the cross drum by one or more rows of horizontal circulating tubes which carry the mixture of steam and water (delivered to the front headers by the generating tubes) into the drum.

A baffle plate is used in the drum to deflect the steam-and-water mixture from the horizontal circulating tubes downward; otherwise, a considerable amount of water might be carried over with the steam flow, especially in operating at high ratings.

Both the sinuous-header and box-header types of boilers have tube caps, which is a favorite source of leakage if gaskets and the caps are not kept tight. On sinuous-header boilers, corrosion can take place on the tube-entrance outer plate as a result of leakage from handholes or tube caps. Leakage and corrosion also take place between the sinuous headers, which are usually packed with asbestos. On box-header boilers, leakage at handhole plates causes corrosion on the wrapper or external sheet where the closing caps are located. Leakage of these handholes finds its way down to the bottom of the headers (Fig. 6-3*d*), which is usually concealed in brickwork. Thus leakage from above causes undetected corrosion. When inspecting box-header boilers having bricked-in bottom headers, always remove the brickwork and check this surface for corrosion.

Some of the manufacturers who built header-type watertube boilers are Heine, Union Boiler, Murray Boiler, Keeler Co., Springfield Boiler Co., Babcock and Wilcox Co., and the Wickes Boiler Co.

VERTICAL WATERTUBE BOILERS

The vertical watertube boiler is a type requiring moderate headroom and small floor space per unit of capacity. The watertube boiler in Fig. 6-4*a* represents this type. These boilers make use of a Dutch oven or furnace extension to secure proper combustion space. Some installations use the front row of generating tubes to form a roof waterwall in the furnace extension.

The upper drum is known as the steam drum; the lower, as the mud drum. The tube sheets are braced by sling stays of proper size and pitch. The heads

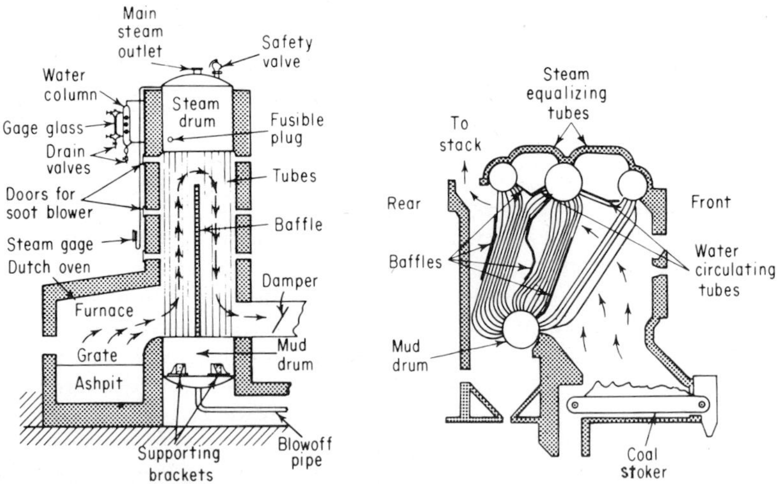

Fig. 6-4 *(a)* Vertical watertube boiler details; *(b)* early bent-tube boiler of the Stirling type.

opposite the tube sheet in the steam drum and in the mud drum do not need staying, for they are dished to the proper radius to be self-supporting at the design pressure.

Tube replacement may be accomplished by insertion through a circumferential row of handholes provided in the upper head of the steam drum for this purpose. Access to these drums is by a manhole in the dished head of each.

BENT-TUBE WATERTUBE BOILERS

Bent tubes are more flexible than straight tubes. Boilers can be made wide and low, where headroom is limited, or narrow and high, where floor space is at a premium. Also, bent-tube boilers allow more heating surface to be exposed to the radiant heat of the flame. Drums serve as convenient collecting points in the steam-water circuit and for separation of steam and water. Thus boilers with two, three, and sometimes four drums have been used. As boilers grew in size (made possible by bent-tube design), the demand for more active furnace cooling increased. It was then that waterwalls and other improvements in design were made. A better knowledge of fluid dynamics resulted in simpler and much safer methods for the circulation of waterwall fluids, on both the gas and the steam sides.

Two-drum boilers, even boilers with but one drum on top and one or two large headers at the bottom, became commonplace. Thousands of boilers of the vertical-header type and of multidrum bent-tube design are still operating today. The tubes are bent for several reasons:

1. Heat-transfer reasons make it impossible to use straight tubes.
2. The bent tube allows for free expansion and contraction of the assembly, usually on the lower mud-drum end, since the upper drum (or drums) is separated or suspended by steel structures.
3. The bent tubes enter the drum radially to allow many banks of tubes to enter the drum.
4. Bent tubes allow greater flexibility in boiler-tube arrangement than is possible in straight tube boilers.

STIRLING BOILERS

The Stirling boiler was one of the first types of bent-tube boiler to come into common use. Figure 6-4*b* shows a unit with front and middle boiler banks connected to both the front and the middle upper drums. This arrangement equalizes the discharge of the steam-and-water mixture to improve circulation and reduce carryover with the steam. Boilers of this general type were usually designed for pressures from 160 to 1000 psi and capacity range from 7500 to 350,000 lb/hr of steam. Both the top and the bottom drums have brickwork built partly around them. Because the Boiler Code required the longitudinal joint to be away from furnace heat, the joint was usually under this brickwork. Even if it means removing some brickwork, always check the condition of longitudinal and circumferential rivets, including the caulked edges, and look for caustic embrittlement affecting the drums.

Feedwater Entrance and Circulation The feedwater pipe enters through the top of the rear steam drum or through the upper rear part of its manhole head. Feedwater discharges into a long trough riveted or welded along the rear part of this drum. It then spills over the front of the trough to enter the boiler-water circulation.

The water circulates downward through the rear bank of tubes to the mud drum and supplies the middle and front banks of tubes through which it rises to the respective steam drums. It circulates from the front to the middle steam drum through the short circulating tubes, thus tending to equalize the water level in these two drums. As the rear rows of tubes from the middle drum pass back to join the rear tube bank where lower gas temperatures permit a downward circulation, the water level in the middle and rear steam drums tends to equalize through the mud drum.

The mud drum is large in diameter, and all the riser and downcomer generating tubes are remote from its bottom. Sediment is deposited here, for this section is not disturbed by rapid circulation. The blowdown connects into a forged-steel fitting riveted or welded to the bottom of the mud drum. In the case of large boilers with long mud drums, more than one blowdown may be provided.

Modern packaged boilers have grown in popularity and size since their inception in the 1940s. Today, most packaged watertube boilers follow one of the three designs shown in Fig. 6–5. These are known as A, D, and O types.

A typical factory-assembled D-type watertube boiler consists of two drums and comes equipped with a low-pressure air-atomizing burner. The furnace walls are water-cooled at the front and rear walls, and the outside walls, floor, and roof are cooled with tangent tubes. The boiler can be equipped for either top or side flue-gas outlets. Both the inner and outer casings are constructed of 10 gauge steel. The inner casing is completely seal-welded and pressure-tested at the factory. The heavy 10 gauge outer seal-welded casing is suitable for outdoor installation. The inner furnace walls are made of tangent tubes.

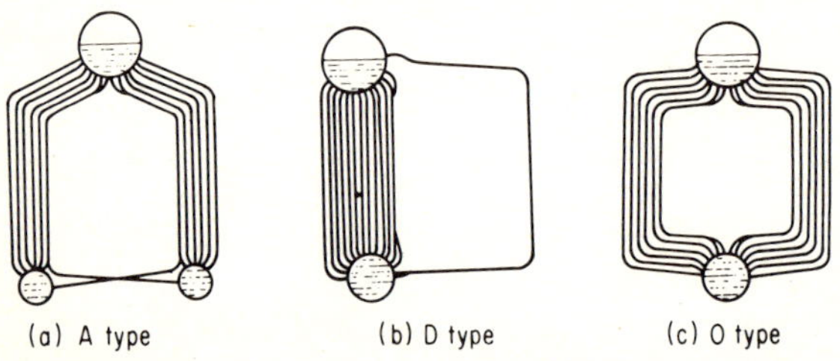

Fig. 6-5 Bent-tube boilers are arranged for packaging at the factory in A, D, and O configurations.

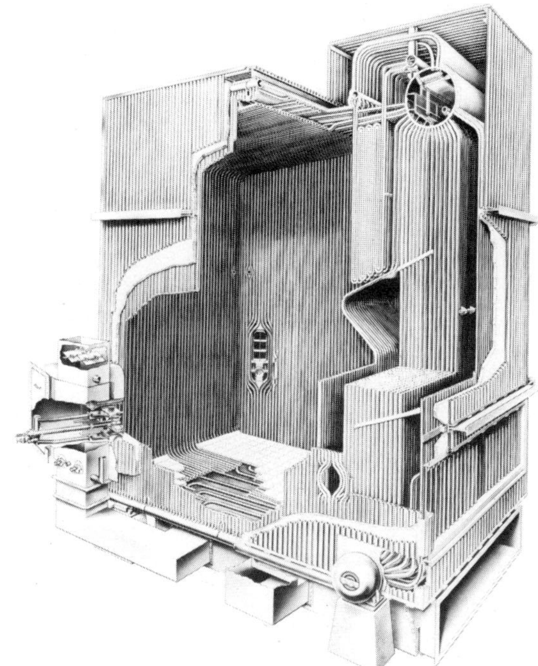

Fig. 6-6 Combustion Engineering VU steam generator is a two-drum bent-tube boiler equipped with water walls. *(Courtesy Combustion Engineering, Inc.)*

Packaged watertube boilers usually are equipped with safety combustion-ignition and flame-failure equipment. Both the pilot and the main burner are monitored by a flame-sensitive scanner (see later chapters). Safety shutdown, safety interlocked starting, and main-flame operation features are provided by most manufacturers. Limit switches are included with interlocks for low water, high steam pressure, fan failure, low oil pressure and temperature for units firing oil, high and low fuel gas pressure for those firing gas, and flame-failure devices suitable for either fuel.

The combustion engineering VU steam generator is a cross-drum bent-tube boiler (Fig. 6-6) with self-contained furnace. Great flexibility of operation, as well as compactness and high efficiency, is afforded by the ample combustion space and adequate heating surfaces offered to the radiant heat. This boiler is designed especially to burn fuels in suspension with intertube-type burners as shown in the figure.

Coal-Fired Packaged Boilers Stoker-fired, watertube boilers generally are built up to about the 250,000-lb/hr rating. Above this rating, pulverized-coal and cyclone-fired units are generally used. Modern stoker-fired boilers are usually of the two-drum type. Long-drum and cross-drum designs are used

Fig. 6-7 A 500-HP, CP steam generator, long-drum on top and short drum on bottom, fired by underfeed coal stoker. *(Courtesy E. Keeler Co.)*

with bent tubes. On a long-drum boiler, the flue gases flow lengthwise to the drums, while on a cross drum they flow across or perpendicular to the drum. Figure 6-7 is an artist's drawing of a 500-HP steam generator fired by a single-retort underfeed stoker with the design patented by the E. Keeler Co.

Boilers burning combination fuels, such as shown in Fig. 6-8, require special considerations. Among these are the following. When oil and/or gas capability is added to a solid-fuel boiler, consider (1) superheater and attemperator to be used; (2) excess air requirements; (3) NO_x emission standards to be observed; (4) tube spacing needs to avoid fouling and draft restrictions; (5) air-heater metal temperatures to be expected; (6) final flue-gas temperatures to be obtained.

When solid fuel is burned on a grate, air cooling must be provided for oil or gas burner parts to withstand the heat released by radiation from the furnace interior. This may affect efficiency by burning with excess air. When oil or gas from wall burners is burned, a flow of cooling air is required through the grates in order to prevent overheating. For this reason, it is wise to consult

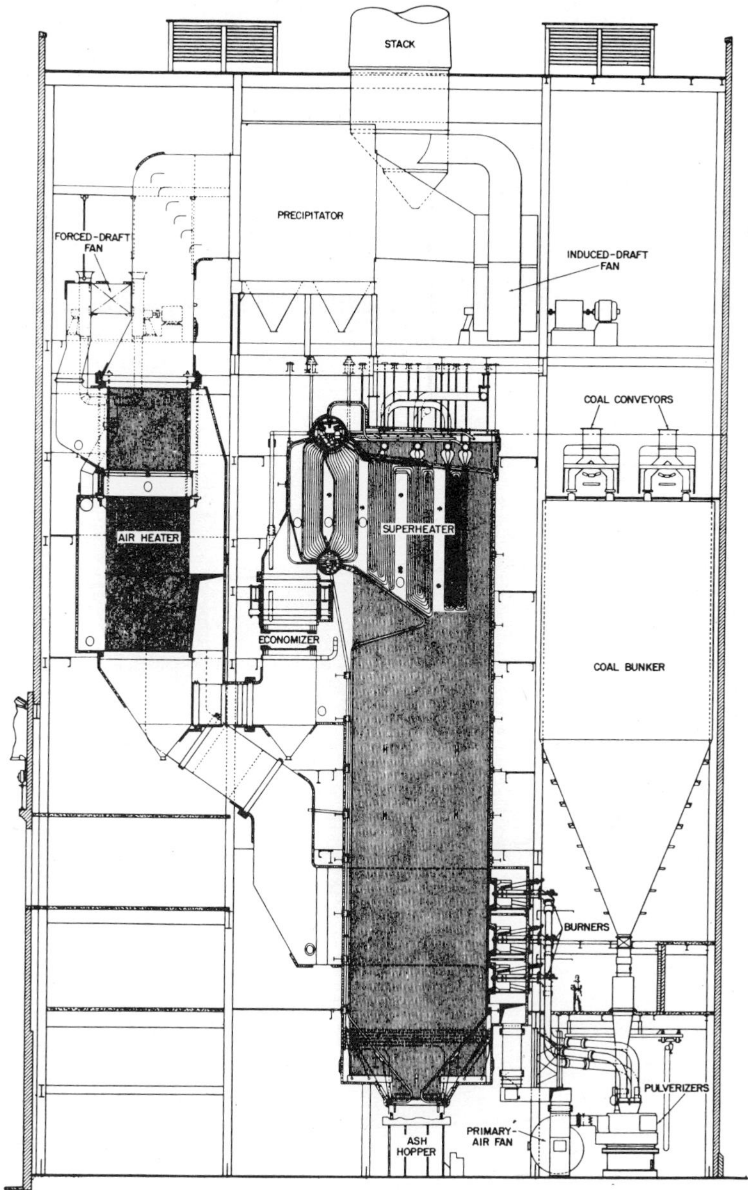

Fig. 6-8 Boiler equipped to burn pulverized coal, gas, and oil must be designed accordingly. *(Courtesy Babcock & Wilcox Co.)*

with the original boiler manufacture for design recommendations when considering converting a boiler to a multifuel burning boiler.

UTILITY BOILERS

Utility boilers are used to generate electric power at the lowest heat rate possible, which is generally defined as the net Btu of fuel needed to generate one kilowatt of electricity. (See Fig. 6-9.) The trend has been to install single-boiler–steam-turbine arrangements that include a need to coordinate boiler-turbine startup and loading. The starting and loading coordination is needed because the process involves the heating and cooling of large metal parts, and this can cause high thermal stresses.

High reliance is placed on engineered integrated control systems on new

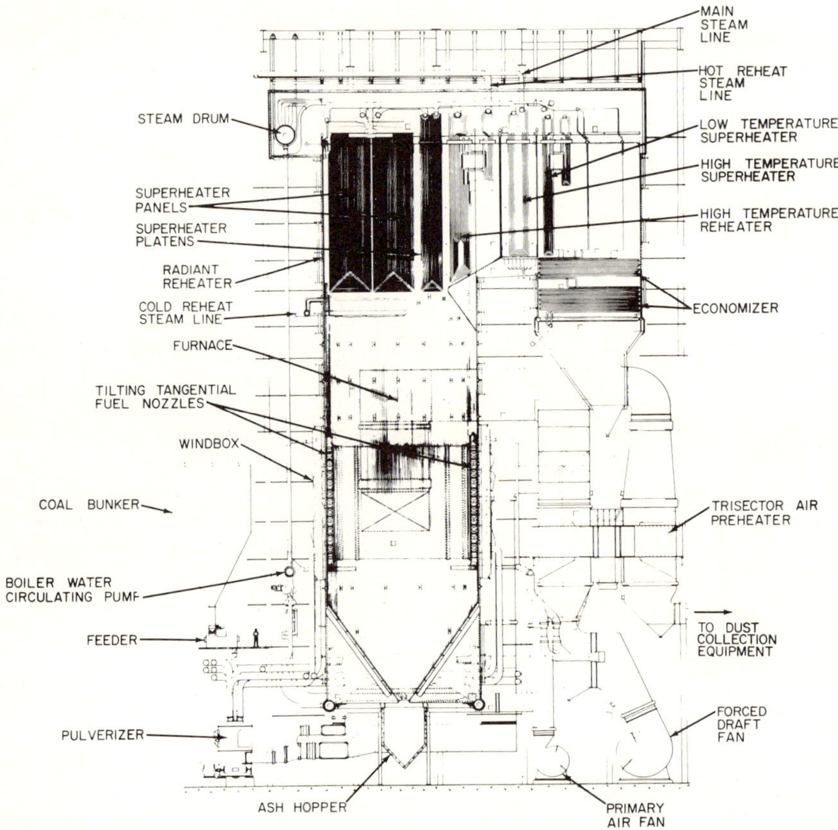

Fig. 6-9 A 650-MW controlled subcritical steam generator. *(Courtesy Combustion Engineering, Inc.)*

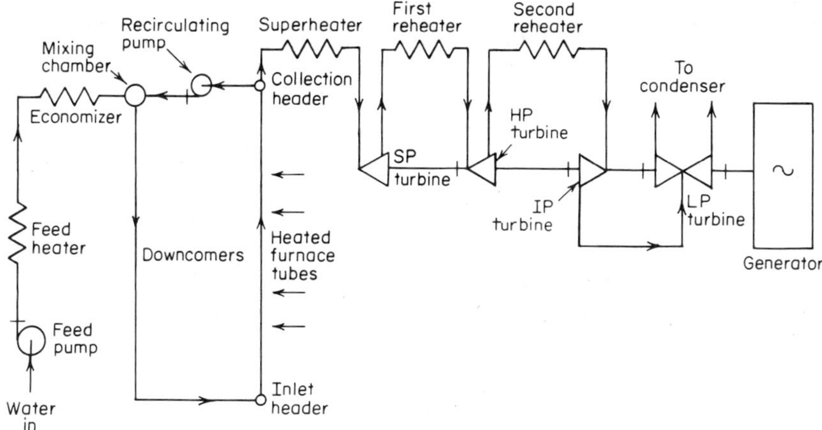

Fig. 6-10 Boiler-turbine arrangement (LaMont type), using supercritical pressure and forced recirculation.

units with the operators acting as overseers. Turbine controls are generally digital-hydraulic and are integrated with boiler controls. Computers are used quite often for bulk storage of data and to program starting and shutdown cycles.

Depending on the design for pressure, capacity, and final temperature, a modern, large steam generator will have waterwalls, superheater, economizer, reheater, and air-heater tubes. The basic purpose of these tubes is to extract all possible heat from the fuel input, thus lowering the exit temperature going up the stack. Utility boilers are classified as subcritical and supercritical, and this is determined by whether an unit is operated above or below the critical pressure of 3206.2 psia. Further classification is also made to indicate whether the boiler has natural circulation or forced (controlled) circulation.

Figure 6-10 is a schematic diagram of a subcritical controlled-circulation utility boiler equipped for tangential firing. The unit is designed to produce 3.980 million lb/hr of steam with superheater outlet pressure of 2680 psig and temperature conditions of 1005°F at both superheater and reheater outlets. Fuel is pulverized coal, but gas- and oil-fired units can be made available by the manufacturer. The steam generated by this unit is used to drive a 650,000-kW turbogenerator.

Controlled circulation permits water to be carefully apportioned to furnace walls, boiler-tube sections, parallel tubes or tube, in accurate, predetermined amounts. Water can even be changed in total flow or distribution within a certain range at any subsequent time in operation. This is usually done by installing circulating pumps between the boiler drum and water inlet to the heat-absorbing surfaces. The result is positive flow in one direction at all times, regardless of heat application. See Fig. 6-10.

By using forced circulation, small-diameter, thin-walled tubes can be used where natural-circulation design would not be possible because of the high temperature which the natural circulation may have to operate against. This is especially true of large-flow and high-pressure units (supercritical) and once-through designs.

The LaMont-type boiler shown in Fig. 6-11 is called a controlled-circulation boiler because the quantity of water passing through the boiler is from 3 to 20 times the amount evaporated. Thus two pumps are required, one for circulating the high rate of flow through the tubes (no natural circulation), the other as a conventional boiler feed pump. The feed pump operates on the same principle as most boiler feed pumps by maintaining a constant level of water in the drum.

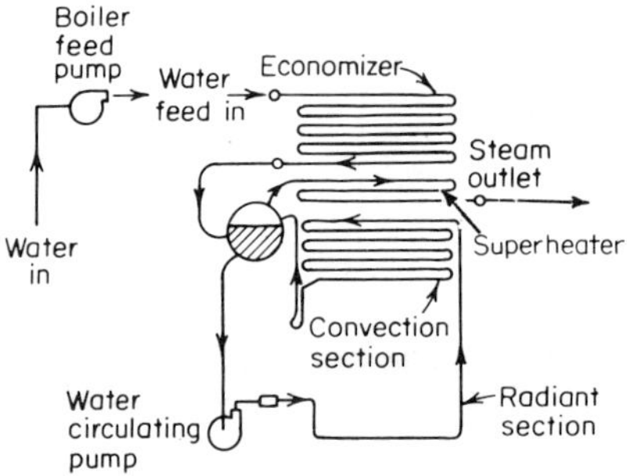

Fig. 6-11 LaMont forced-circulation principle uses a pump in the cycle to overcome frictional resistance to flow.

The function of controlled circulation is to establish a flow through the first section of inlet to the boiler to prevent the water in the tubes from evaporating to complete dryness. Instead, evaporation is only to the extent that dissolved solids and salts will remain in solution. This solution (a mixture of water and steam) passes to the steam-and-water drum, where the steam is separated while the excess water is removed. The separated water, along with feedwater, is returned to the pumps through downcomers.

By means of continuous or intermittent blowdown, some water and solids in solution are removed. The separated steam is passed through the superheater for final usage.

Special ASME Code exceptions or rules have been made in reference to once-through forced-flow steam generators which have no fixed waterlines and steam lines. For example,

1. It is permissible to design the pressure parts for different pressure levels along the path of water-steam flow.
2. No bottom blowoff pipe is required.
3. If stop valves are installed in the water-steam flow path between any two sections of line, certain safety valves or power-actuated pressure-relief valves, with control impulse interlocks, are required. (This is different from the typical boiler with safety valves protecting the entire boiler.)

4. Pressure gauges required are more numerous at various sections. No water-gauge glass or gauge cocks are required.

WATERTUBE BOILER COMPONENTS

Large watertube boilers have far more extensive heat-absorbing components than do fire-tube-type boilers, as well as other construction details that need to be reviewed.

Older boilers, especially the bent-tube type, used to be set up in pairs with a common parting wall between them which covered the two heads next to each wall. This could be a steam drum or mud drum. Because the external surfaces of the heads were covered, the term *blind head* was applied to these drum heads.

There is a tendency for soot, ashes, and moisture to collect in the space between the brickwork and external surfaces of the blind heads. And this condition is not directly observable. In older inspection days, it was common for an inspector to insist that the brickwork be removed from around these blind heads. Then the external surface was available for inspection and drill-tested for thickness. Today, ultrasonic instruments can be used to get a profile of the thickness of the head. These checks on deterioration are made without disturbing the brickwork unless the ultrasonic check indicates it is necessary.

Firing doors on watertube boilers are required to be of the inward-opening type or a type provided with self-locking door latches of a style omitting springs or friction contact, so that the door will not be blown open from pressure in the furnace in case of a tube rupture or furnace explosion and thus possibly burn or scald an operator standing by the boiler. Explosion doors, if used and if located in the setting walls within 7 ft of the firing floor or operating platform, must be provided with substantial deflectors to divert any blast.

Baffles deflect the hot gases back and forth between the tubes a number of times to enable greater heat absorption by the boiler tubes. They also permit designing for better temperature differences between tubes and gases throughout the boiler. Baffles help maintain gas velocity, eliminate dead pockets, deposit flyash and soot for proper removal, and prevent high draft losses. When a furnace baffle breaks, the gases short-circuit one or more passes, causing excessive flue-gas temperatures and a loss in efficiency and capacity. Overheating and damage might result in those parts of the boiler designed for low gas temperatures. Thus on any outage inspection, the baffles should always be carefully checked for erosion, breaks, leakage (around tubes), or dislocation, for tube failure may result.

Waterwalls consist of relatively close-spaced vertical tubes forming the four walls of the furnace. They were originally developed to cool and protect the furnace lining. One design of large power-generating boilers has 144-ft-high, 0.340-in.-thick tubes at the hottest furnace zone (below 85-ft elevation) but only 0.320-in.-thick tubes above.

Depending on the type of boiler, the waterwall heating surface may account for only 10 percent of the boiler's total heating surface, yet represent as much as 50 percent of the total heat absorption. Waterwalls perform three basic functions: (1) protect the insulated walls of the furnace; (2) absorb heat from the furnace to increase the unit's generating capacity; and (3) make the furnace airtight (on pressurized furnaces with tangent-welded tubes).

Heat is transferred to the waterwall tubes as radiant heat from the zone of highest temperature in the furnace. Because of the great amount of heat absorbed by that part of the boiler, feedwater must be of the best quality. Also, the circulation of water must be rapid and plentiful to ensure positive flow through each tube at all times.

Blowdown valves are required for each header at the bottom of a series of waterwall tubes, for the same reasons that the boiler itself needs them. Sediment accumulation in a header supplying wall tubes might cause interruption of circulation, with consequent overheating and failure of the tubes.

Warning: Under no circumstance should waterwall headers be blown down while the boiler is operating. If they are blown, the boiler's normal circulation will be upset, and the overheated tubes will bulge or rupture.

Figure 6-12 shows some typical arrangements. The waterwall in Fig. 6-12*a* is designed for moderate cooling. This design has the tubes spaced apart and the wall surface composed of part firebrick. The brick is backed with several layers of insulation and strong steel casing. Reinforcing metal lath is often used in wall construction. Figure 6-12*b* shows how tangent tubes are placed in the furnaces of many large and small boilers. The staggered-tube arrangement offers a high heat-absorbing surface that is backed by solid block, or plastic, insulation and a strong steel external casing. Figure 6-12*c* shows steel lugs or longitudinal fins welded to nontangent wall tubes. In some designs the lugs protrude from the tube into the furnace and are covered with a chrome-base refractory or slag. To ensure furnace tightness, adjacent fins are often welded.

Figure 6-12*d* shows newer gastight casings, known as membrane walls. Here tightness is obtained by welding a flat strip of metal between the tubes. This eliminates the casing and many of its problems. Insulation is applied directly to the outside of the tubes, while metal lagging is attached to give the outer surface durability and good appearance. Figure 6-12*e* shows how the outer casing, insulation, and steel-skin casing are often constructed.

Superheaters Each pressure of saturated steam has a corresponding temperature. Heat added to the dry steam at this pressure is known as superheat and results in a higher temperature than that indicated on the curve for the corresponding pressure.

The advantage of superheated steam in prime movers is twofold: (1) Work may be done down through the superheat range before condensation starts to take place. This represents an increase in steam-utilization efficiency. (2) This period of work performed with dry steam eliminates corrosive and erosive effects of condensate. The deterioration of high-speed turbine blades caused by impingement of drops of condensate may be considerable.

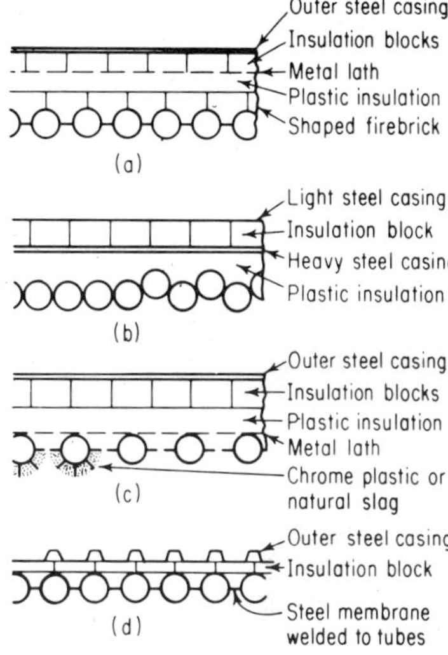

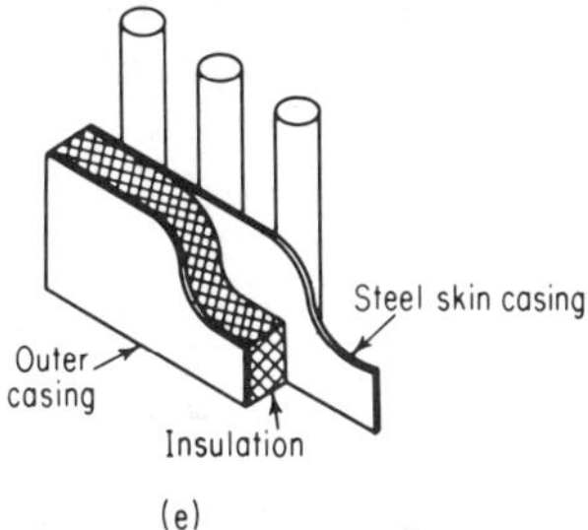

Fig. 6-12 Arrangements of waterwall tubes in a watertube boiler.

Superheat is produced by passing the flow of saturated steam from the boiler through a superheater of one or both of two types, radiant and convection.

Figure 6-13 shows some typical superheaters and their locations in the boiler. The general classification includes radiant and convection types, depending on whether they absorb radiant or convection heat. The interdeck type has tubes arranged between banks of primary boiler tubes. The pendant type is a suspended series of coils, usually shielded against radiant heat by a screen of boiler tubes. It is often arranged as the first steam heater before the steam goes to the superheater outlet header. The platen type is similar to the pendant type, but the tubes are in one plane. Usually the steam goes through the platen superheater before it enters the pendant superheater.

Constant temperature of superheated steam is the desire of most designers, for a steam turbine is designed for the particular steam temperature at which it will operate most efficiently. Characteristics of the convection-type superheater may produce a drooping temperature curve with increasing combustion rates, whereas the reverse may be true with radiant types. Thus a combination

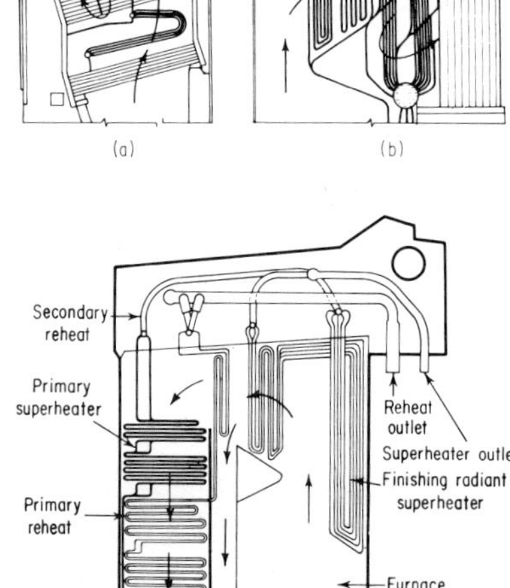

Fig. 6-13 Superheater arrangements in different types of watertube boilers.

of the two types in certain installations, in order to obtain a practically constant superheated steam temperature with varying loads, is employed. Damper control of gas through the passes is another development for control of superheat temperature.

Economizers serve as traps for removing heat from the flue gases at moderately low temperature, after they have left the steam-generating and superheating sections of the boiler. The general classification is:

1. Horizontal- or vertical-tube, according to the direction of gas flow with respect to the tubes in the bank
2. Parallel-flow or counterflow, with regard to the relative direction of gas and water flow
3. Steaming or nonsteaming, according to thermal performance
4. Return-bend or continuous-tube types
5. Plain-tube or extended-surface types, according to the details of design and the form of the heating surface

The tube bank may be further designated as of the staggered or in-line arrangement, with regard to the pattern and spacing of tubes which affect the path of gas flow through the bank, its draft loss, heat-transfer characteristics, and ease of cleaning.

Theoretically, economizer heating surface could be added to a boiler until the exit temperature neared the outside air temperature. But an abnormally high heat-surface area would be needed. Further, each fuel burned has a dew-point temperature which can cause moisture accumulation on the economizer and corrode the surface in a short time. The amount of heating surface which should be used in the economizer is limited by the final gas temperature at the exit. If the gas temperature is cooled below the dew point, condensation (sweating) may result. Sulfur in the soot then unites with moisture to produce a sulfurous acid, which is extremely corrosive to all steel construction contacted between the economizer and the stack.

Feedwater of low oxygen content is recommended in using the steel-tube type of economizers. Relief valves (water safety valves) are required on economizers to protect them against excessive pressure that might be built up by the feed pump if the regulator or feed valve to the boiler were closed.

A *reheater* is essentially another superheater used in modern utility boilers for boosting plant efficiency. While the superheater takes steam from the boiler drum, the reheater obtains used steam from the high-pressure turbine at a pressure below boiler pressure. This lower-pressure steam passing through the reheater is heated to 1000°F and then is introduced into the intermediate, or low-pressure, turbine. Reheaters, like superheaters, are also classified according to their location in the boiler as convection or radiant. Convection superheaters and reheaters may be of the horizontal or pendant type.

Air heaters make the final heat recovery from boiler flue gases with which they preheat the incoming furnace air for its combustion with fuel. Thus some fuel is saved which would otherwise be used in heating the air-fuel mixture to its ignition point. But the temperature of the flue gas must not be reduced below its dew point since moisture would condense out of the flue gas. That would cause water to combine with sulfur and possibly carbon dioxide, also carbon monoxide, to form highly corrosive sulfurous and carbonic acids.

A regenerative air heater is shown in Fig. 6-14. This type of air heater

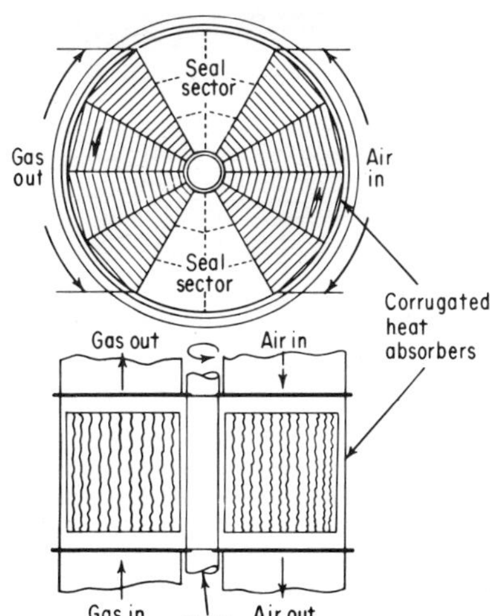

Fig. 6-14 Rotating regenerative air heater.

offers a large surface of contact for heat transfer. It usually consists of a rotor which turns at about 2 to 3 revolutions per minute (r/min) and is filled with thin, corrugated metal elements. Hot gases pass through one half of the heater; air passes through the other half. As the rotor turns, the heat-storage elements transfer the heat picked up from the hot zone to the incoming-air zone.

Figure 6-15 illustrates the *internals* of a typical drum, which performs two essential functions: separates steam from water to provide the downcomer system with steam-free water necessary for proper and safe circulation, and it separates moisture from steam to provide high steam quality. The drum internals shown provide both functions by means of two stages of separation. The normal water level is 1½ in. below the horizontal center line of the drum. Vortex eliminators separate the steam and water passages in the drum.

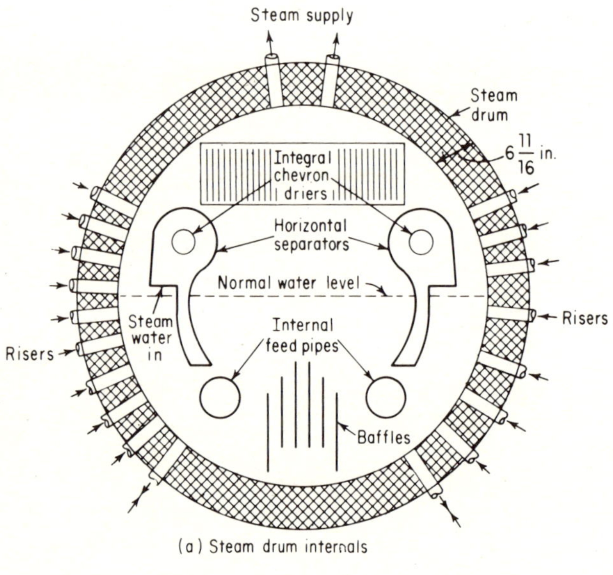

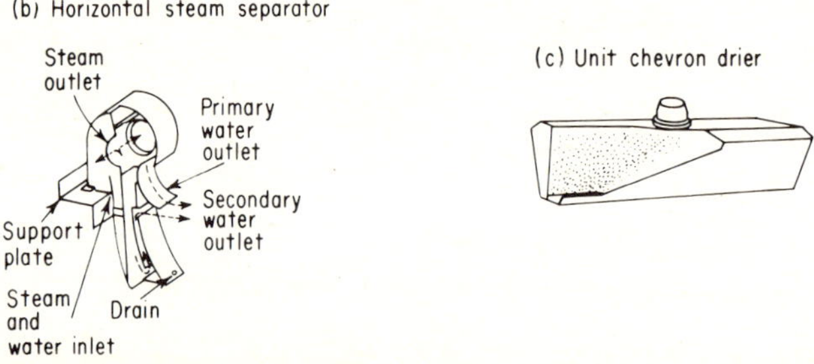

Fig. 6-15 Internals of a high-pressure steam drum showing steam separator and chevron drier.

The total circulating steam-and-water mixture from the steam-generating tubes is directed to the horizontal turboseparators (Fig. 6-15*b*). The steam-and-water mixture enters the separators and is centrifuged by following the curved contour of the separator dome. Most of the separated water is discharged horizontally at the water level of the drum. The separated steam flows to the chevron driers at the top of the drum. The steam flows from the chevron through the dry box, then out the top of the drum via steam tubes across the roof to the partition-wall superheater.

The opening to the main steam line will usually be adequate to handle the design flow capacity. The ASME Boiler Code has definite rules on the opening to a safety valve. For example, internal collecting steam pipes, splash plates, or pans are permitted to be used near safety valve openings, provided "the total area for inlet of steam thereto is not less than twice the aggregate areas of the inlet connections of the attached safety valves. The holes in such collecting pipes must be at least ¼ in. in diameter."

In the case of steam scrubbers or driers, the ¼-in.-diameter opening does not apply, provided "the net free steam inlet area of the scrubber or drier is at least 10 times the total area of the boiler outlets for the safety valves." Therefore, when inspecting the internals of drums, check the condition of the ¼-in.-diameter holes in the collecting pipes. They must be free and clear. The same applies to the openings on driers, because plugged driers on collecting pipes could lead to restrictions of the safety-valve openings. That would reduce relieving capacity flow, which could be dangerous.

Questions and Answers

6-1 Why are tubes flared in a watertube boiler?

Ans. To add to the holding power of the tubes after rolling and to prevent the tubes from pulling out of the tube holes if the holes should become enlarged from overheating caused by low water or other reasons.

6-2 What is the difference between a watertube and a fire-tube boiler?

Ans. In the fire-tube boiler, the products of combustion pass through the tubes and the water surrounds them. The reverse is the case in watertube boilers.

6-3 How is a new tube placed in position for installation in a vertical watertube boiler?

Ans. Through one of the handholes provided for this purpose in the top head of the steam drum.

6-4 What type firing door is required on watertube boilers? Explain your answer.

Ans. The inward-opening type or a type provided with self-locking door latches of a style omitting springs or friction contact, so that the door will not be blown open in case of tube rupture or furnace explosion.

6-5 Where, other than as tube caps, are handholes required on sinuous-header type of watertube boilers?

Ans. In the external mud drum.

6-6 Name three types of mud drum used in straight-tube watertube boilers.

Ans. The internal mud drum (Heine), the external mud drum (Babcock and Wilcox), and the lower water drum (Wickes vertical boiler).

6-7 Are watertube boilers supported by the brick sidewalls?

Ans. Not as a rule. They are supported by independent steel structure.

6-8 Where is the manhole in horizontal watertube boilers?

Ans. In at least one dished head.

6-9 What is the method of forming the manhole frame?

Ans. The head is flanged in. If it is not at least $^{11}/_{16}$ in. thick for a gasket bearing surface, a steel band is shrunk onto the flange to bring the combined thickness to at least $^{11}/_{16}$ in.

6-10 What would happen if a baffle broke down?

Ans. Gases would short-circuit one or more passes, excessive flue-gas temperatures and a loss in efficiency and capacity thus resulting. Overheating and damage might result in parts of the boiler designed for low gas temperatures.

6-11 What is the purpose of an external mud drum (Babcock and Wilcox type)?

Ans. It provides a chamber out of the rapid circulatory system for deposition of sediment, which is then removed by blowing down. It also equalizes circulation between headers.

6-12 Of what material were mud drums of the external Babcock and Wilcox boiler type?

Ans. Wrought or cast steel on modern boilers. Cast iron was used on some old installations.

6-13 Can a ligament be strengthened on a boiler in the field?

Ans. Yes, by removing the drum tubes and riveting or welding on a doubling plate.

6-14 Are tubes beaded or flared in the straight-tube watertube boiler? Explain your answer.

Ans. They are flared, for the ends are not contacted by hot gases and nothing would be gained by beading.

6-15 To what points are water-column connections made in most watertube boilers?

Ans. To the upper and lower parts of one head of the main steam drum.

6-16 How many gas passes are there in most standard-type watertube boilers?
Ans. Three.

6-17 Is the volumetric capacity of each gas pass the same? Explain your answer.
Ans. No, it decreases in each succeeding pass. The gases contract as they cool, and in order to maintain the high gas velocity necessary to sweep off stagnant gas films to effect good heat transfer, the cross-sectional area of the passes must decrease as the gases require less space.

6-18 In what type of watertube boiler are stay bolts used?
Ans. The box-header types.

6-19 How are stay bolts used in connection with soot blowing?
Ans. The holes through hollow stay bolts in Heine box headers are of sufficient size to permit a steam of air lance to enter and blow soot from the tubes along the "horizontal" gas passes.

6-20 What is a downcomer nipple?
Ans. It is a short length of boiler tube between the steam drum and header carrying downward circulation of boiler water.

6-21 What type of cross-drum boiler often has inclined downcomer nipples? Explain your answer.
Ans. The standard Springfield boiler, because the steam drum may be located forward of the line of rear headers.

6-22 Where are interdeck nipples used?
Ans. In double-deck boilers where two sets of headers are installed one above the other. Interdeck nipples connect them vertically.

6-23 Name the two main sheets of a box header.
Ans. The tube sheet and the handhole (or tube-cap) sheet.

6-24 What is the narrow plate sometimes used to form the bottom of a box header called?
Ans. The header trough.

6-25 How many courses are there in a cross-drum boiler? Explain your answer.
Ans. Usually one, for a girth seam might unnecessarily interfere with the tube ligament.

6-26 Is it more difficult to withdraw a tube from a watertube boiler or from a fire-tube boiler? Explain your answer.
Ans. It is usually more difficult in a fire-tube boiler, for any scale will be on the outside of the tube.

BENT-TUBE BOILERS

6-27 How many drums are there in bent-tube boilers?
Ans. Usually two, three, or four. A few types have one; a few, more than four.

6-28 What is an advantage of the multidrum type? What may be a disadvantage?

Ans. The greater steam- and water-storage capacities enable the multidrum type to meet peaks in load fluctuations with less pressure drop. The added cost of the additional drums is the disadvantage.

6-29 How is the lower drum of large bent-tube boilers, such as the Stirling boiler, supported? Give the reason for this method of support.

Ans. It is suspended by the tubes to permit free expansion and contraction.

6-30 Which is the shortest drum in a Stirling boiler? Explain your answer.

Ans. The lower (mud) drum is the shortest. It is suspended by the tubes, and the heads of the drum are inside of the boiler sidewalls.

6-31 Is the water level equal in the steam drums of a bent-tube boiler of a type having the drums at the same level?

Ans. It is when the boiler is idle or operating at low ratings. At high ratings, the water level in the drum supplying the downcomer tubes is often lower than that in the other drums.

6-32 Name three methods of supporting the upper drums of bent-tube boilers (cross type).

Ans. (1) U bolts around each end of the drums and tied to overhead cross beams; (2) cast or forged L pads attached to each head and resting on cross beams; and (3) cast cradles under each end of the drums and resting on cross beams.

6-33 For what reasons are bent tubes used in Stirling-type boilers?

Ans. To allow for expansion and contraction; to permit tube replacement; to allow tubes to enter drums perpendicular to surface tangent; to allow flexibility in design.

6-34 How are Stirling-type boilers designed in regard to tube arrangement to allow removal and replacement of tubes?

Ans. The tube may be pitched equally at a greater spacing than the tube diameter, or the tubes may be grouped in twos with closer pitch in the pairs and the wide pitch between the pairs.

6-35 What is the purpose of a bridge wall in a Stirling boiler?

Ans. It protects the mud drum from exposure to the direct heat of the fire.

6-36 What might happen if the Stirling bridge wall collapsed?

Ans. The longitudinal seam might be overheated and damaged.

6-37 Where is the manhole in bent-tube boiler drums?

Ans. In at least one head of each drum.

6-38 What is the purpose of circulating and equalizing tubes in bent-tube boilers?

Ans. Circulating tubes connect the water space of adjacent steam drums (usually at the same level) and aid in equalization of water level. Equalizing tubes connect the steam space of such drums to equalize the steam pressure.

6-39 In riveted drums of bent-tube boilers, why is it customary to select a longitudinal seam of comparatively low efficiency (45 to 75 percent), whereas in straight-tube types longitudinal seams are designed usually with an efficiency of 85 to 95 percent?

Ans. Because the bent-tube type has a tube-ligament efficiency below 55 percent usually, whereas the straight-tube type has no ligament. It would be a waste of money to design a riveted longitudinal seam of high efficiency when the pressure on the boiler were limited by the tube ligament.

6-40 What is the difference between a mud drum in a Stirling and in a straight-tube type of boiler?

Ans. A Stirling mud drum is one of the main drums in the circulatory system. In straight-tube boilers, the mud drum may be a trough in the steam drum (Heine boiler) or an external box at the bottom of the headers (as in the Babock and Wilcox boiler).

6-41 Where are the blowdown connections on bent-tube boilers, and how many are there?

Ans. In the bottom of the lowest, or mud, drum; one or more, depending on the length of the drums.

6-42 Why are circulating tubes bent in Stirling-type boilers?

Ans. To permit replacement and allow for expansion.

6-43 What is the difference between the tube sheet in a bent-tube boiler and in an HRT boiler?

Ans. In the bent-tube boiler, the tube sheet is that portion of the drum into which the tubes are rolled. In the HRT boiler, the tube sheet is a flat head.

6-44 *(a)* Define a superheater as used with a steam boiler.

(b) Name two separate ways of producing superheated steam.

Ans. *(a)* A superheater is a form of heater used to raise the temperature of saturated steam above the temperature due to its pressure. It usually consists of an inlet and an outlet header with interconnecting tubes.

(b) One separately fired and the other ones located in the boiler setting or enclosures and sometimes in the walls of the furnace.

6-45 *(a)* Define an air preheater and mention some advantages gained in using it as part of the steam-generator unit.

(b) Define air-cooled walls used in boiler installations and what is gained by their use.

Ans. *(a)* An air preheater is a heat exchanger usually installed in the smoke stack or in the boiler breeching. The cool incoming air is heated by heat in the gases that would otherwise be wasted, and the heated air is used in the furnace again, saving fuel or heat needed to heat the air.

(b) Air-cooled walls are in the furnace setting. Cool air keeps the furnace brickwork from burning, and the air acts as an insulating blanket to prevent heat loss through the brickwork of the setting.

6-46 How do water and steam circulate in a straight-tube watertube boiler?

Ans. This type of boiler has natural circulation with the water and steam rising along the inclined tubes to the higher front header, then through the headers to the drum. The water then circulates through the downcomers to the rear header to the inclined tubes to complete the cycle.

6-47 Name two classifications used to describe furnace enclosures for watertube boilers.

Ans. Two classifications are furnace enclosures consisting primarily of refractory to keep the heat within the furnace and furnaces with water-cooled walls.

6-48 What is usually the weakest part (from the viewpoint of construction) of a bent-tube boiler?

Ans. The tube ligament.

6-49 How many courses are there in bent-tube boiler drums? Explain your answer.

Ans. One, because a girth seam would unnecessarily interfere with the tube ligament. Also, different-length tubes would be required for each row.

7
Coil, Electric, Cast-Iron, and Special-Application Boilers

COIL BOILERS

Coil boilers were developed to satisfy industry's need for a compact, fast-steaming, factory-assembled packaged boiler. They find special application where a process requires high-pressure steam in one part of the process flow and the capacities required are moderate. A packaged unit is placed where the load need exists, and this makes it unnecessary to operate large, centralized boilers at reduced capacity during periods of operation when other parts of the plant may have low demand. Several packaged boilers can be placed close to the steam loads of a plant at widely separate locations, thus avoiding long steam-line losses that may exist with a centralized steam plant. Coil-type boilers are used over packaged fire-tube types when high pressures and capacities may be required. Pressures up to 900 psi are possible with coil-type watertube boilers. Capacities generally are below 10,000 lb/hr, but units of greater size are available.

The generating tubes of coil-type boilers consist of small-diameter helical, spiral, or horizontal coils of tubing. Some large units consist of a series of bundles connected to make a continuous coil. In case of a coil failure, the affected coil section need only be removed and connected to the remaining coil sections. Figure 7-1 shows the steam-and-water flow of a coil-type unit. Note the boiler uses forced circulation and forced recirculation, which permits smaller tubes to be used with high steam velocities and high heat-transfer rates.

The flow of water and steam in the system shown in Fig. 7-1 starts with

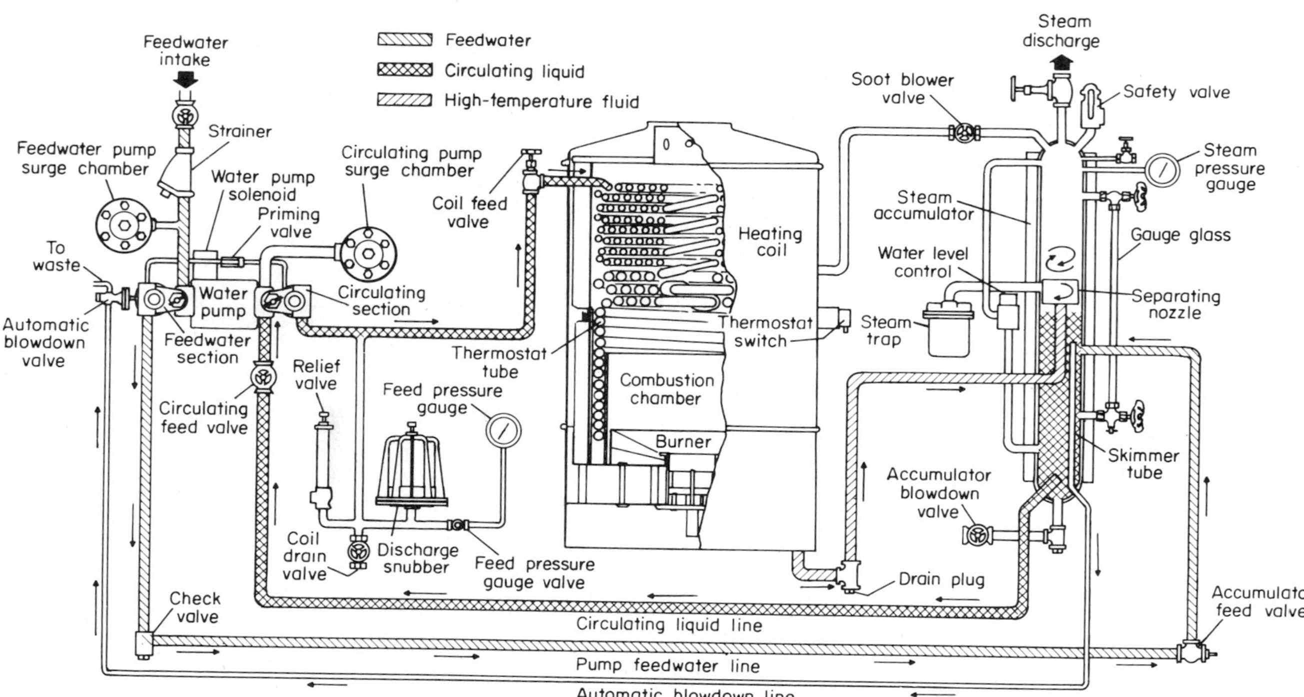

Fig. 7-1 Water-and-steam flow for a coil-type watertube boiler. *(Courtesy Clayton Mfg. Co.)*

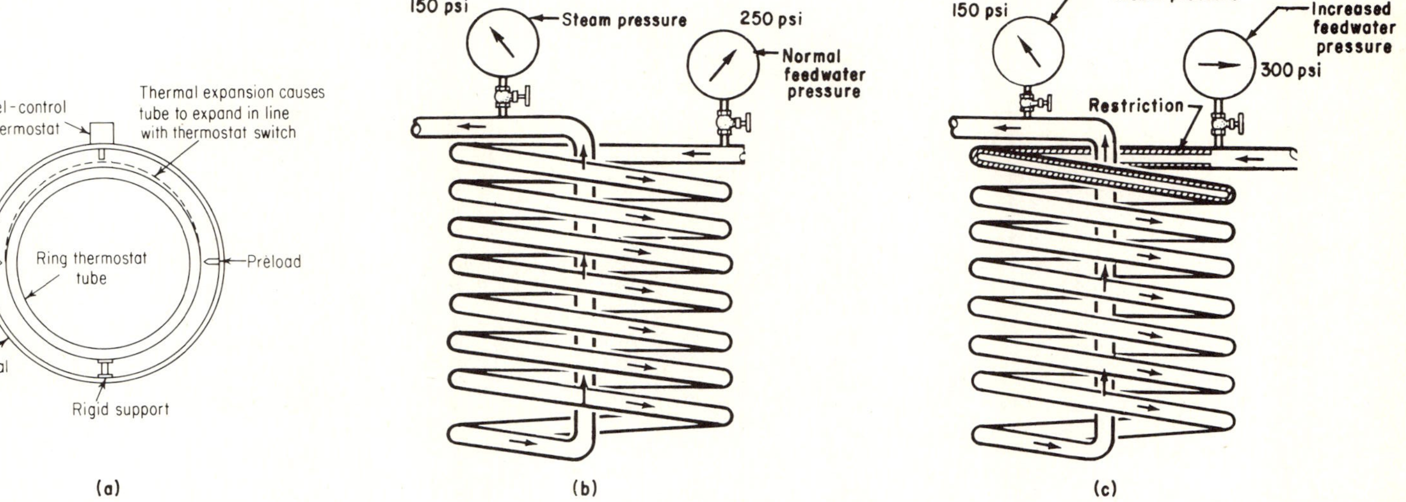

Fig. 7-2 Coil-type boilers may show scale or obstruction inside coils by the following: *(a)* The ring thermostat shuts the burner off from overheating; *(b)* the normal feedwater pressure when coils are new establishes bench marks; *(c)* abnormal or increased pump pressure may be a sign of flow restriction.

water entering the feedwater section of the water pump with the water pumped directly to the steam accumulator. The feedwater rate is controlled by the water-level control, which in turn responds to the liquid-level control in the accumulator. The circulating liquid from the accumulator is then pumped to the heating coils of the boiler. Flow of liquid in the coils is downward through the coils spiral-wound in a counterflow direction to the combustion gases. When leaving the spiral generating section, the fluids pass through the ring thermostat and the helically wound waterwall section and then to the separating nozzle in the accumulator. The centrifugal action of the nozzle separates dry steam from the liquid and allows excess liquid to return to the lower section. The dry steam from the accumulator is discharged on top through the discharge valve.

An unusual feature is the ring thermostat tube shown in Fig. 7-2*a*. This tube is an integral part of the heating coil, actuated directly by combustion heat. In case of water shortage or any excessive-heat condition, the thermostat control will expand beyond the set point and thus shut off the fuel supply.

Another design has three nests of coils and inlet and outlet headers. It is widely used in small sizes up to 300 HP and pressures to 250 psig.

Malfunctions in the loop may cause a coil failure from overheating. This could come about from pump failure, partial blockage of inlet and outlet lines, blocked tubes (scale), fireside soot accumulation in concentrated heat zones, malfunction of controls, etc. Thus it is essential to keep both the waterside and fireside of the coils clean; proper feedwater treatment is vital. A pressure-differential chart is often used to show whether the difference between suction and discharge pressure on the recirculation pump exceeds a certain pressure differential. See Fig. 7-2*b* and *c*. If so, it is a sign of tube or flow obstruction. In a conventional boiler of multitube design, a leaky tube can be plugged, then the boiler operated until it is convenient to replace the affected tube. With a coil-type boiler, the entire coil must usually be replaced. All coil-type packaged boiler manufacturers supply excellent instruction manuals with their units which include maintenance and feedwater treatment sections as well as a description of the controls used. Prominent coil-type boiler manufacturers are Clayton, Besler, Francis, Mund-Parker, Vapor-Clarkson, and Sid Parker.

ELECTRIC BOILERS

Two basic types of electric boilers are available. First, units for low capacity and voltage generally consist of the resistance type. In the resistance type, current generates heat by flowing through resistance elements. This is wire encased in an insulated metal sheath, and these are submerged in water to generate usually moderate pressure steam at low capacities. These types of units do not depend on the conductivity or resistance of the water for generating heat. Second, in electrode boilers, the current flows through the water, as shown in Fig. 7-3*a*, *b*, and *c*, and not through wires. The liquid in the boiler converts electric energy to heat energy.

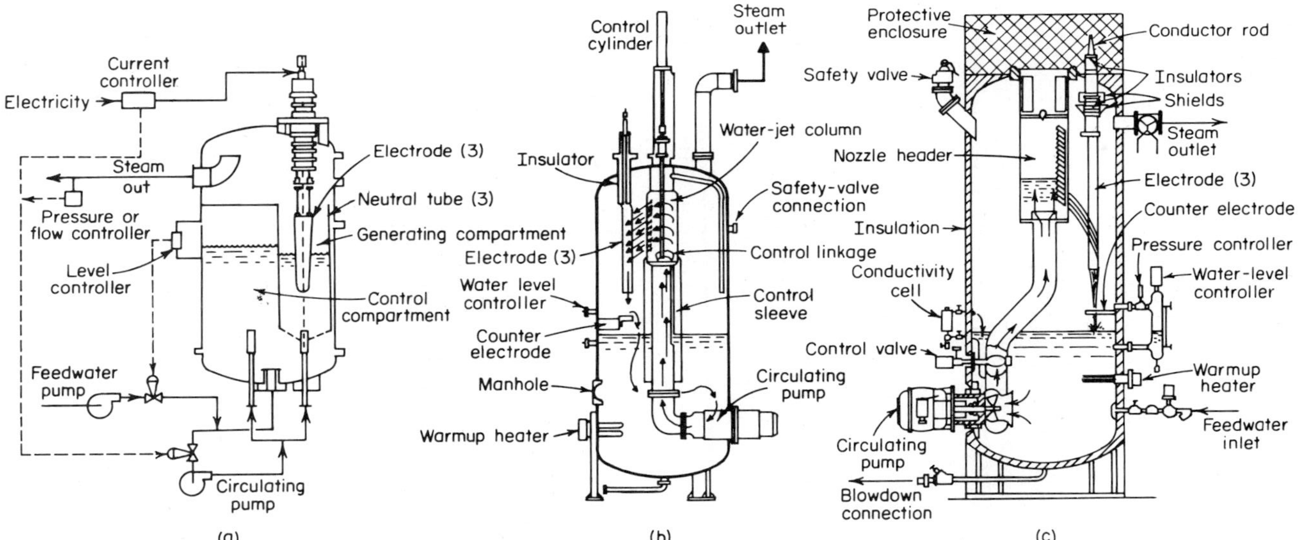

Fig. 7-3 Types of electrode boilers: *(a)* current flows through fluid to the wall of the generating compartment; *(b)* high-voltage sprayed electrode boiler with control sleeve to regulate water flow; *(c)* high-voltage sprayed electrode boiler with water flow regulated by a rotating distributor.

The energy crisis and air pollution regulations have created a demand for higher-output electric boilers above the 10,000-lb/hr ratings. Where electric power is economical to buy, high-voltage units are now available for increased capacities. A high-voltage unit is classified as boilers with energy input between 2300 and 15,000 volts (V). The high-voltage steam electrode boiler shown in Fig. 7-3*a* has three generating compartments, one for each phase or electrode. Steam is generated as current flows from the electrodes to the neutral walls of the cylindrical compartments that contain them.

High-voltage sprayed electrode boilers, as shown in Fig. 7-3*b* and *c*, control the output by regulating the amount of water sprayed on electrodes that are suspended in the steam space. The current path is created when the water flows through ports in a storage compartment to the electrodes. Water that is not converted to steam runs down the electrodes and falls to a counterelectrode. This creates a second current path for steam production—from the electrodes to the counterelectrode. The remaining water returns to the reservoir.

Electrode boilers are packaged; they come fully equipped with controls and safety devices. Most boilers have controllers to maintain conductivity within the manufacturer's limits by monitoring the water and adding prescribed chemicals as needed, as well as blowing down the boiler when necessary. With electrode boilers it is necessary to control the conductivity within specified ranges of the manufacturer's specifications. If the conductivity is too low, the boiler output may not be attained; if it is too high, short circuits may occur. Water treatment is also important with these boilers because faulty treatment can affect conductivity as well as steam problems such as foaming.

Most states include electric-type boilers in their rulings. The ASME Boiler Code states that electric boilers of a design employing a removable cover, which will permit access for inspection and cleaning of the shell, and having a normal water content not exceeding 100 gallons (gal) need not be fitted with washout or inspection openings.

The usual access for inspection and cleaning is the electrode cover connection to the inside of the boiler. This must be pulled out to check the internal conditions of the boiler. At the same time, check the condition of the electric elements.

The capacity of the safety valve is determined by the kilowatt input. The minimum safety-valve capacity must be at least 3.5 lb/(hr/kW) input. This is true whether it is a high- or low-pressure boiler. On a Btu-per-hour basis, the requirement is 3500 Btu/hr for each kilowatt input.

While low-water-level controls are required for the electric-resistance boilers, they are not needed for the electrode type. The reason is that they cannot generate steam when the water is low or when no steam is required. To ensure that the boiler carries the proper salinity, a conductivity control is generally supplied. This means that salts are added or the unit is blown down, depending on the condition of the water. All electric boilers should be built in accordance with the ASME Boiler Code and also approved by Underwriters Laboratories.

CAST-IRON BOILERS

The cast-iron boiler is basically a watertube type because the water is inside the cast sections (no tubes) and the products of combustion are on the outside. But because of the limitations of cast iron, the Boiler Code treats cast-iron boilers as a special type, without considering the heat-transfer method. Many cast-iron boilers are stamped by the manufacturer as cast-iron watertube boilers, which should not be misinterpreted as a Code classification.

Cast iron is a term applied to many iron-carbon alloys which can be cast in a mold to make a particular shape. But for cast-iron boilers, gray cast iron is generally used. When the casting is cooled slowly in the molds, part of the carbon separates out as graphite. This makes the gray cast iron less brittle and easier to machine. Also, when it is alloyed with nickel, chromium, molybdenum, vanadium, or copper, considerable tensile-strength properties can be achieved. The general practice is to classify cast iron by class:

Class no.	*Ultimate tensile strength, lb/in.*^{2}
20	20,000
25	25,000
30	30,000
35	35,000
40	40,000

Cast-iron boilers are built to various shapes and sizes, but can be grouped into the three following broad classifications:

1. Round cast-iron boilers (Fig. 7-4*a*) consist of a firepot (furnace) section with base, a crown-sheet section, one or two intermediate sections, and a top or dome section. The sections are held together by tie rods or bolts with push nipples interconnecting the waterside of the section. Thus water circulates freely through the nipples from section to section. Fuel is burned in the center furnace, with the flue gases rising and flowing through the various gas passages of the water-filled sections, then out to the stack. The round cast-iron boiler used to be popular for hot-water-supply service and was often stamped "hy-test" to indicate an allowable pressure of 100 psi for domestic hot-water service. But it is rapidly being replaced by fired steel-welded water heaters or by electrically heated water heaters.
2. Sectional boilers (vertical) consist of sections assembled front to back, with sections standing vertically and assembled by means of push nipples or screwed nipples.
3. Sectional boilers (horizontal) consist of assembled sections stacked like pancakes. Here each section is laid flat in relation to the base. This type of vertical stacking may be supplemented by having three vertically stacked boilers side by side and interconnected to gain additional capacity. In this arrangement a common supply and return header is used with no intervening

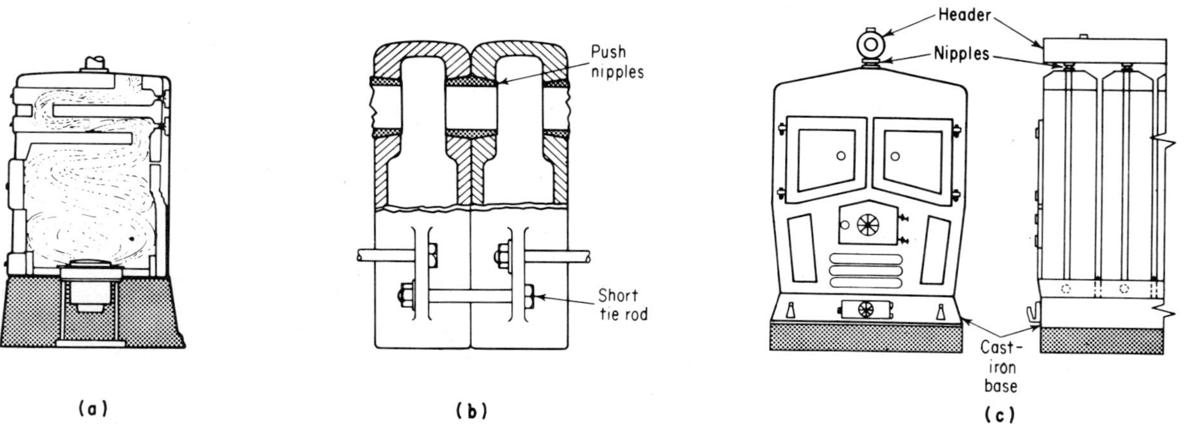

Fig. 7-4 Types of cast-iron boilers: *(a)* round unit for hot-water supply, *(b)* sectional boiler with sections held together by tie rods and push nipples, *(c)* screwed-nipple type of sectional header boiler.

valves between the vertically stacked boilers. These units are usually gas-fired, with a burner for each vertical stacking.

Two methods used to aid in assembly of vertical cast-iron boilers are the following:

1. Internal, tapered push nipples are inserted into holes of the vertical section. Then by means of through tie rods, or short tie rods (Fig. 7-4*b*), the sections are pulled together by tightening nuts against washers on the tie rod. That interconnects the sections on the waterside, enabling them to withstand pressure as assembled sections.
2. In the external header type (Fig. 7-4*c*), the sections are individually assembled to headers (supply drum and return drums) by means of threaded nipples, locknuts, and gaskets. This type of assembly allows replacing an intermediate section, because only the locknuts, gaskets, and threaded nipple have to be removed on each header to slide a section out. In contrast, with through tie-rod construction, all the sections in front of the intermediate section to be replaced have to be removed first to get at the affected section.

When checking new boilers, it is necessary to check the type of nuts securing the tie rods. If solid-steel or brass nuts are used, they should be only hand-tight or backed off a few threads. The first choice on new boilers is collapsible washers, with shallow split brass nuts. The second choice is split shallow nuts backed off hand-tight (or cloverleaf-type nuts that fail when a slight expansion takes place). On an older unit idle during summer months and located in a damp basement, make sure the holding nuts are not tight. Also determine whether the tie rods are rusted into their holes, which may have the same binding effect as tight securing nuts. Obviously, if the rods are free and nuts slacked off, there should be no problem of expansion cracking. If rust growth and tight tie rods are the cause of cracking, the boiler must be dismantled and the rust buildup removed by chipping.

Header-type boilers have no tie rods or tapered nipples, so nothing can be adjusted to allow for abnormal expansion. In addition to rust depositing between the sections, rapid start-up can cause serious damage to these units. Make sure older boilers without good blowdown facilities do not develop scale buildup, because it causes cracking. Restriction of water supply and circulation can also be caused by scale buildup in these units, resulting in overheating.

When controls are not operating properly, or if the safety relief valve is stuck or of the wrong size, a cast-iron boiler will explode. Many have. The big difference between a shell-type steel boiler explosion and a cast-iron boiler explosion is that a cast-iron boiler will usually fragment into smaller pieces of the affected sections. A steel boiler, on the other hand, rips and tears along the sheet of a drum or shell. Then it flies apart in the form of curved panels of steel. But in each explosion, the danger to life and property is very great.

In cast-iron boilers of different poured shapes, unpredictable stress-concentration areas, geometry, service factors, waterside fouling, soot and carbon accumulation on the fireside, and rapid temperature changes are all causes for cracking. Thus they may never fail twice in an identical manner. Careful investigation will usually point to one or more of the following causes for unexpected cracking failures:

1. Rapid introduction of cold water into a hot boiler, as may occur with poorly operated manual makeup or an automatic-feed device
2. Makeup line connected to a section instead of to the return line to temper the cold water with the returning condensate or hot water (the Code requires makeup to *enter return lines*)
3. Controls not functioning and thus overstressing the boiler by either pressure or temperature
4. Insufficient water in the boiler
5. Internal concentrated deposits, blanking off proper heat transfer or obstructing circulation from section to section
6. Defective material or casting which does not become noticeable until after several years of service
7. Poor assembly or work or improper installation of the boiler as to pitch, alignment, tie-rod tightening, screwed-nipple fitting, etc.

HIGH-TEMPERATURE HOT-WATER BOILERS

A high-temperature hot-water (HTHW) system shown in Fig. 7-5 is a heat-utilization boiler that uses water at an elevated temperature, usually over 300°F, with no specific pressure limitations. However, the ASME Boiler and Pressure Vessel code, Section I, defines a HTHW boiler system as one using water at a temperature over 250°F and at a pressure over 160 psi. The HTHW systems usually are not designed for temperatures exceeding 500°F. Operating temperatures within the range of 350 to 450°F are the most frequently encountered. Above 500°F, other fluids are used to obtain high temperature, such as Dowtherm and similar inorganic fluids.

At these temperatures, water in its liquid form can exist only at pressures above the corresponding saturation pressures. The HTHW systems have widespread use for supplying the heating needs of large airports, military bases, office buildings, hospitals, colleges, and other large multibuilding complexes. The application of HTHW systems has been extended to many industrial processes in the fields of chemistry, plastics, rubber, metal plating, paper, textiles, etc.

Advantages of HTHW Systems For many applications, HTHW systems are claimed to have distinct advantages over steam systems. The major advantages are:

1. Because of its large heat-storage capacity, an HTHW system permits very close control of temperature, which is important with many process applications.
2. The large quantity of heat in the system forms a heat reserve so that

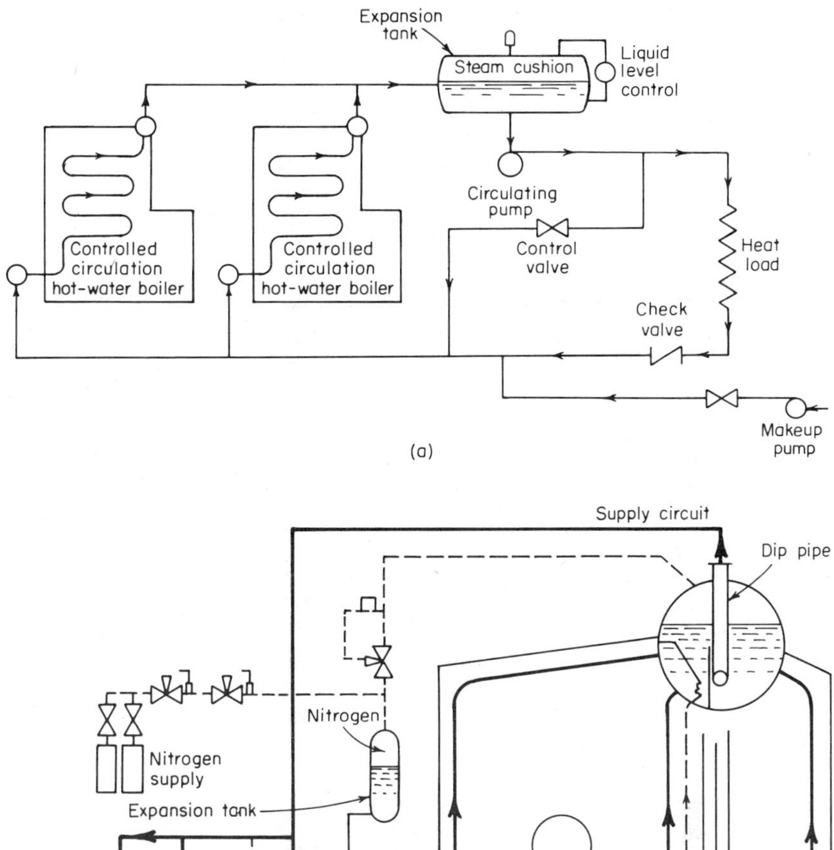

Fig. 7-5 High-temperature hot-water boilers use: *(a)* expansion tank *(b)* drum of boiler, which acts as an expansion tank.

fluctuating loads have a minimal effect on the boiler, permitting use of a lower-capacity boiler unit.

3. The absence of return-line corrosion in a HTHW system, which is frequently a problem in a steam system, is advantageous.

4. No traps, pumps, receivers, vents, or other condensate-return equipment are needed on HTHW systems, thus producing lower first cost, less maintenance, and no steam losses.

5. A small amount of makeup is required; thus there is no need for an expensive feedwater-treatment system.

Disadvantages of HTHW Systems There are several disadvantages, one of the most important being the large water content of the system. This produces the following disadvantages:

1. The system takes longer to heat up initially.
2. More time is needed for cooldown when a repair or alteration has to be made and much water has to be discharged.
3. If a pipe or pressure-vessel break occurs, the water content, being above the atmospheric boiling point, will partly flash into steam with a powerful disruptive effect. Even a relatively small leak can result in considerable water damage before the system can be depressurized.

Both fire-tube and watertube boilers are used for HTHW systems. The main advantage of fire-tube boilers for this application is that they are generally less expensive in the smaller sizes below 600 HP. Above this size, watertube boilers are used. Surprisingly, the box-header boiler is prominently used for lower capacities; however, the forced-circulation unit (LaMont type) is used extensively.

Watertube hot-water generators are available in packaged ratings up to 150 million Btu/hr for water temperatures up to 650°F. The combustion equipment used is the same as that used for steam boilers. All types of fuels can be used.

Pressurization and Flow Control Three considerations in the design of an HTHW system, which are not factors in steam-system design, are:

1. How to maintain the circulating-water pressure above the saturation pressure for the water temperature required
2. How to provide for expansion of heating-system and generator water
3. How to ensure uniform flow rates under all operating conditions

There are three pressurization methods used—steam cushion, gas, and mechanical—and two ways to accommodate water expansion—in a separate accumulator or in the upper drum of the boiler for drum-type units serving small heating systems.

The steam-cushion arrangement generally is used for large-capacity, continuously operated hot-water systems, such as those used at many airports, because it maintains tighter control of water temperature than the other two. See Fig. 7-5*a* and *b*.

High-temperature hot water from the boiler discharges into a separate expansion drum, where a small amount of the fluid flashes to steam to maintain system pressure. The boiler's firing rate is controlled by drum pressure. The drum is located high enough above the boiler to permit free vapor release and to ensure a reasonable suction head for the circulating pump. It is necessary to avoid the introduction of large slugs of cold return water into the drum to prevent the cushion from collapsing. Such a condition causes flashing and water hammer in the piping system.

In the inert-gas pressurization system, a nitrogen blanket is maintained in an expansion vessel, which is partially filled with water from the circulating system. The pressure of the nitrogen is kept higher than that of the saturated liquid.

High-temperature hot-water boilers should have a low-water fuel cutout in case the water level drops below a safe level. On forced-circulation units, a pressure-differential switch or flow switch is advisable, so that the fuel to the burner is cut off in the event no water circulation is occurring. High temperature hot-water boilers may not require a water gauge glass or gauge cocks if the boiler is completely filled with water and has an external expansion tank. However, if the boiler is a natural-circulation boiler with a drum, then the Code requires a gauge glass and cocks, because the drum may have fluctuating water levels that must be displayed to the operator who notes if it is proper. Figure 7-5*b* shows a D-type, natural-circulation boiler in which the drum acts as an expansion tank. This boiler would require a gauge glass per Code rules.

High-Temperature Heat-Transfer Liquid and Vapor Organic Fluids Thermal fluids such as oils, silicates, glycols, and similar liquids with high boiling points are used where higher temperatures are demanded by process requirements but at low operating pressures. Uses include drying of fabrics, clay, wood, paint, etc., which previously may have been done by direct firing of natural gas. Temperatures up to 750°F are now practical with thermal fluids; a 1000°F limit reportedly is not far off.

Where HTHW systems can be used, they are preferred to thermal liquids, because they have about twice the heat capacity, do not deteriorate in use, and are cheap.

Thermal-liquid systems also can be used in conjunction with steam systems. Plants that require high-pressure steam for specific applications and that can apply thermal liquids economically for other uses as well can install a thermal-liquid-to-steam heat exchanger and use part of its output to generate steam. This eliminates installation of a separate gas- or oil-fired boiler.

When an organic substance is in the liquid state, the heating unit is referred to as a heater, and if it is vaporized, it is called a vaporizer. The heating may be performed in electric-fired units, up to about 1 million Btu/hr, with units above this size generally being fired by gas, oil, and even coal. Units are generally packaged for automatic operation to a maximum capacity of 300 million Btu/hr. Multiple units are used for greater capacities as a rule.

The systems used can be classified as follows:

1. Vapor systems are generally of the fire-tube type using either gravity returns or pumped condensate returns. See Fig. 7-6*b*.
2. Liquid systems use watertube boilers with either natural circulation or the more pronounced forced circulation, as shown in Fig. 7-6*a*. Figure 7-7 shows a horizonal, helical watertube type. The once-through forced-circulation La Mont design is used for larger capacities.

Among the organic fluids used are Dowtherm, Aroclor, Therminol FR, and Tetralin. Each of these substances has similar properties, but complete details should be obtained from the manufacturer of the fluids. Since many of these fluids are hydrocarbons, a tube failure can create a fire hazard in the unit. Some fire insurance inspection departments require a steam smothering device

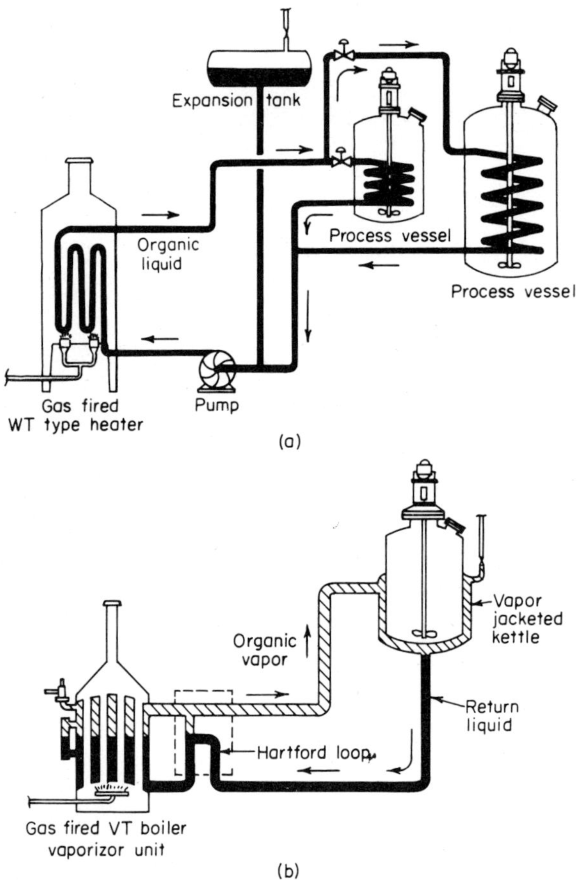

Fig. 7-6 Organic heaters: *(a)* liquid heater with forced circulation, *(b)* vapor heater with gravity condensate return. *(Courtesy Dow Chemical Co.)*

or an inert-gas smothering system to be installed to smother a fire in the unit from a leaking pressure part. Safety controls should include a high-temperature cutout, pressure cutout, and safety relief valves set to the maximum allowable pressure of the unit.

Dowtherm is one of the better known organic fluids, and a review of some

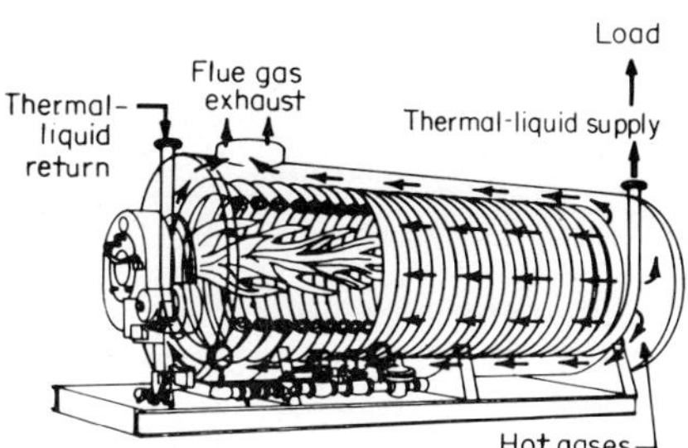

Fig. 7-7 Coil-type watertube boiler heats organic liquid to 20 million Btu/hr. *(Courtesy Power magazine.)*

of its characteristics is somewhat a review of other organic fluids. In Fig. 7-8 it will be noted that, at a temperature of 650°F, a Dowtherm pressure of but 53 psig is required. To attain this temperature with saturated steam, a pressure of 2196 psi would be required. The comparative constructural costs of boilers for these pressures are obvious.

Dowtherm is an organic material. Dowtherm A, which boils at 500°F and is recommended for temperatures up to 750°F, is composed of 26½ percent diphenyl and 73½ percent diphenyloxide. Dowtherm C is used in only the liquid phase and for temperatures up to about 800°F. Dowtherm is nontoxic.

It is more difficult to maintain pressuretight joints with Dowtherm than with steam or water, and so joints of the boiler proper are welded instead

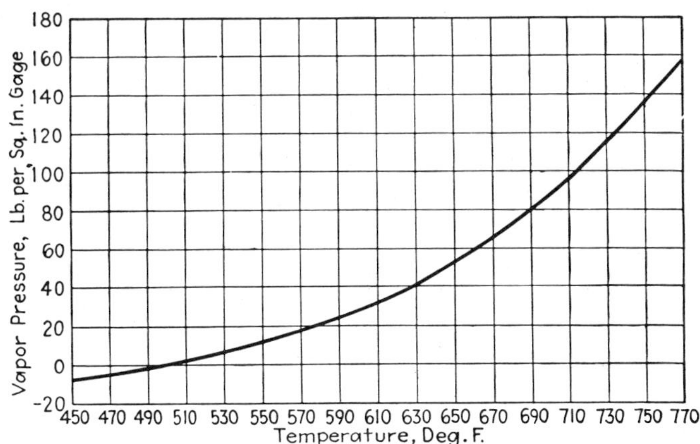

Fig. 7-8 Dowtherm A, vapor-pressure curve. *(Courtesy Dow Chemical Co.)*

of riveted and tubes are welded into the drums. Whereas a hydrostatic test under cold-water pressure may show leakage at defects in a steam boiler, no indication would be shown for leaks that hot Dowtherm might find. The most satisfactory test for Dowtherm boilers is to introduce ammonia gas into the system at a pressure of up to about 1 psi or to fill the boiler half full of aqua ammoniae. Air pressure at about 50 percent of the rated boiler pressure is then applied. A lighted sulfur candle or a dilute hydrochloric acid swab is passed in front of seams, joints, or points where leakage is considered likely. If leakage is occurring, a white smoke will result from the indicator.

Some decomposition may occur if the recommended temperatures are exceeded greatly, owing to faulty operation or localized overheating. Such decomposition may produce a fixed gas, which may be vented during operation, and soluble triphenyls. If the triphenyl content becomes excessive, purification of the Dowtherm may be effected by a distillation or fractional crystallization method.

Blowdown connections are provided on Dowtherm boilers, but are used only when the boiler is off the line and cool and for the purpose of emptying the boiler or to withdraw liquid for purification. Since Dowtherm is a valuable chemical, no waste is permissible. All vents, the discharge of safety valves, and so on are conducted back through a condenser to the storage system. Welding or burning on equipment with organic fluids generally requires complete draining of the fluid and then flushing with steam until traces of the fluid are no longer noted before welding can be performed. Tube failures occur from coking of the organic fluid due to overheating; contamination of the fluid, causing it to break down chemically to carbon; flame impingement, leading to hot spots; and poor circulation for the heat input.

WASTE-HEAT, COMBINED-CYCLE, AND COGENERATION SYSTEMS

Waste-Heat Boilers There are a number of manufacturing processes that give off considerable quantities of high-temperature gases. Common among these are the exhaust from gas turbines or diesel engines. The value of heat recovery depends primarily on three considerations:

1. The cost of producing an equivalent amount of heat by other means
2. The cost of heat-recovery equipment
3. The operating and maintenance cost of the waste-heat-recovery equipment

Steam boilers may be designed to use waste heat as all or part of the steam-generating medium. Since the gas temperature is usually 500 to 800°F, whereas combustion products in the conventionally fired installation may enter generating passes at about 2000°F, some means of compensating for the lower gas temperature must be employed. Otherwise, to have an appreciable steam-generating capacity, the boiler would have to be beyond all reason in size.

Source of Gas	Temp, F
Ammonia oxidation process	1350-1475
Annealing furnace	1100-2000
Cement kiln (dry process)	1150-1500
Cement kiln (wet process)	800-1100
Copper reverberatory furnace	2000-2500
Diesel engine exhaust	1000-1200
Forge and billet-heating furnaces	1700-2200
Gas turbine exhaust	850-900
Garbage incinerator	1550-2000
Open-hearth steel furnace, air blown	1000-1300
Open-hearth steel furnace, oxygen blown	1300-2100
Basic-oxygen furnace	3000-3500
Petroleum refinery	1000-1100
Sulfur ore processing	1600-1900
Zinc-fuming furnace	1800-2000

Fig. 7-9 Waste-heat gases with temperatures available from industrial processes.

Other factors to consider besides pressures and temperatures of available waste gases are the physical and chemical properties of the gas, their effect on boiler parts, the effect on the heat-recovery system by plant-process disturbances, and similar considerations involving continuity of service. Recovery of exhausted heat from industrial processes and combustion equipment can often reduce overall plant fuel consumption with minimal capital investment.

Supplementary firing is used when the waste-heat gases do not have sufficient heat to produce the desired final pressure or temperature of the steam. Figure 7-9 shows some typical waste-heat gases and temperatures usually available from processes.

Depending on the properties of the waste gases and the pressure and capacity needed, the following waste-heat boilers are used:

1. Fire-tube boilers, both the vertical and horizontal types, if waste gas is relatively clean
2. Straight-tube watertube boilers, for clean or moderately dust-laden waste gas
3. Watertube of the bent-tube (Stirling) type, for very heavy dust loadings
4. Positive circulation boilers, for clean, low-temperature gases
5. Pressurized or supercharged boilers, for gas turbine exhaust (Velox type)

Because waste gases quite often have inert gases and solid entrapped particles in the mixture, special material and other design factors must be considered in the application of a waste-heat boiler. These factors and operation and maintenance practices will prevent the unexpected type of failures that can occur on these units also. Among the possible failures are the following:

1. Tube failures from scale and mud buildup, erosion of tubes from particles in the gases, corrosion of tube material from chemical attack from the gas side
2. High-heat-content mass gas input, causing the equivalent of overfiring damage
3. Fireside plugging of gas passages, resulting in localized overheating
4. Malfunction of controls or equipment, causing a low-water condition, with tube and other damage

Fire-tube boilers can be designed for waste-heat application. For example, a two-pass wet-back SM design with no refractory-lined surfaces and equipped with dampers for flue-gas control is shown in Fig. 7-10. The pressure controller PC_s modulates the fan-inlet damper according to the need for steam. The pressure controller PC_a is at a low-limit setting in the illustration to lower the flow through the boiler if it is reducing furnace pressure. The pressure

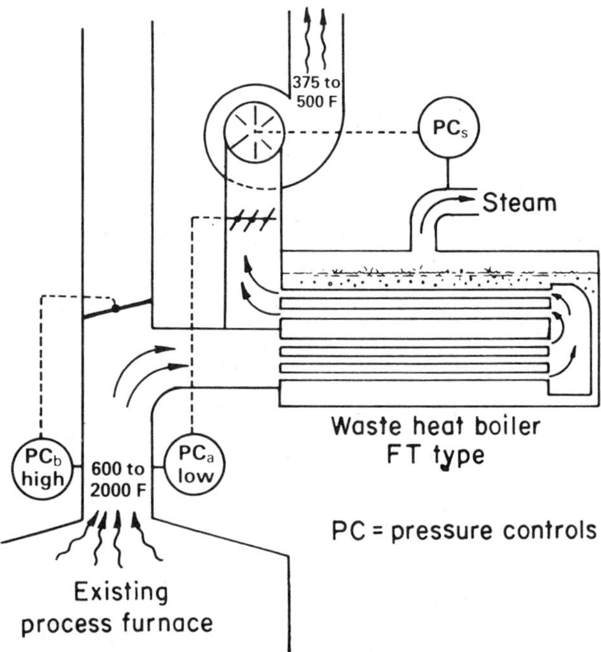

Fig. 7-10 Control schematic diagram for waste-heat boiler uses damper regulation. *(Courtesy North American Mfg. Co.)*

controller PC_b as shown is at a high-limit setting to bypass the boiler if the furnace pressure becomes too high. The boiler also comes equipped with a low-water fuel cutoff of the float type as well as of the electrode type, operating and high-limit pressure cutouts in addition to other auxiliary devices.

A *combined-cycle* generating system which utilizes a gas turbogenerator or diesel with the exhaust heat going to either an unfired or a partially fired waste-heat boiler can be used to recover sensible heat and thus lower the Btu's required to generate one kilowatt of electricity. The boilers that can be used are generally watertube types of either natural- or forced-circulation design. *Cogeneration* produces electric power from a combined cycle usually, but also provides process steam needs with the use of an extraction steam turbine.

A factor or condition often neglected on waste-heat steam boilers is the possibility of developing a dangerous low-water condition which, if undetected, has caused tubes in the convection passes to be melted. This is an extremely important design consideration. It is also a very important operating check that should be made periodically to ensure that the unit is in working condition. This test should be the same as when low-water fuel cutoffs are tested on a suspended fuel-fired standard boiler. Basically the design should include a mechanism, either a heavy damper or some other quick-closing mechanism. In this way in case of low water, the waste gases can be cut off from the boiler and bypassed to the stack. Then the basic process will not be interrupted, and the boiler will also be saved from serious damage due to overheating.

A preferred method is a device that automatically diverts the gases as soon as the water level drops to a predetermined dangerous level. *Time* is of extreme importance in a low-water condition, so the heat input must be quickly removed.

Waste-Fuel Boilers Fuel characteristics must be considered when the many available waste fuels are burned such as wood and trimmings from lumber, paper, furniture, and similar industries using lumber as a basic raw material. Liquid and gas wastes offer another burning problem. The furnace design must be ample so that these waste fuels are burned efficiently in the furnace without causing obnoxious fumes to be emitted from the stack. Regulations on permissible emissions now govern many designs, including whether the boiler will be of the fire-tube or watertube type. Escalating fuel costs and shortages that may develop in the future have caused industry and governments to re-examine the potentials of waste fuels of all types as a combustive alternate to the once-plentiful supply of fuels such as oil and gas.

Solid by-product fuels are many. Among them are wood chips, sawdust, hulls from coffee and nuts, corn cobs, bagasse (waste product from sugar cane), coal char (residue from low-temperature carbonization of coal), and petroleum coke (final solid residue from a refinery). Each product must be handled in a special manner because of differences in moisture content, consistency, specific weight, and heat content.

The furnace rather than the steam generator is affected when these special fuels are used. Products like bagasse, which has about 50 percent moisture,

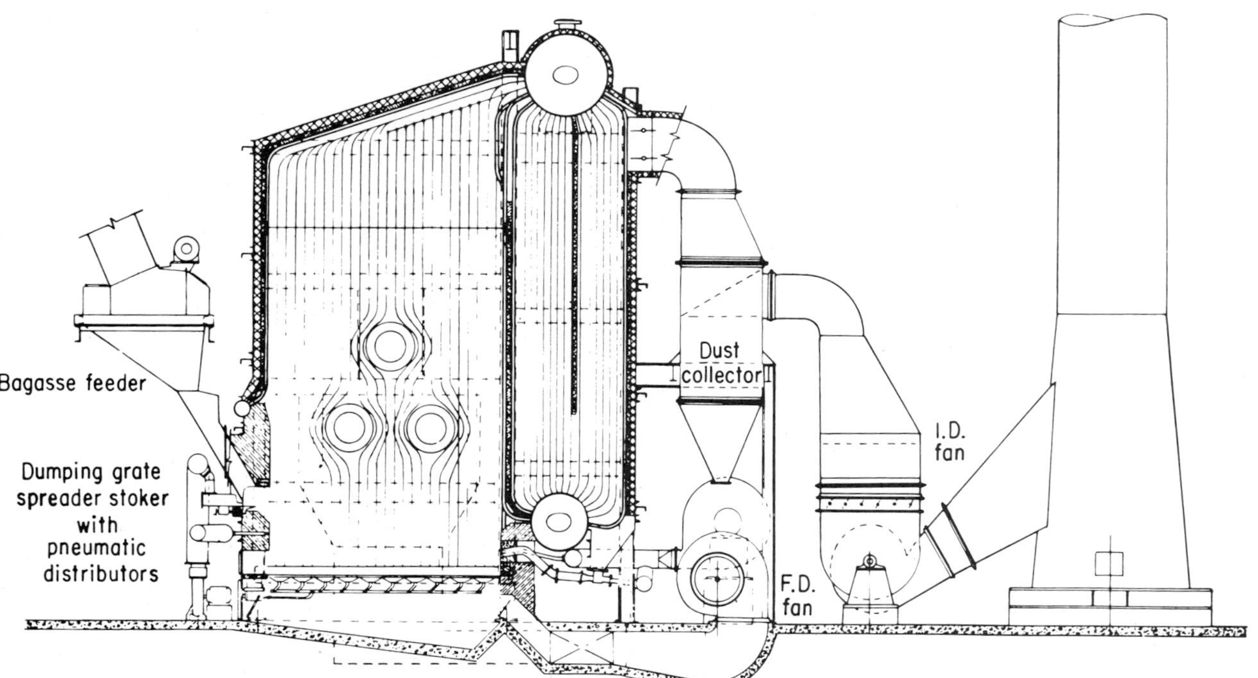

Fig. 7-11 Bagasse-burning watertube boiler with dump-type spreader stoker and auxiliary oil burners; capacity: 100,000 lb/hr, 250 psi. *(Courtesy Riley Corp.)*

require a Dutch oven. The Ward furnace is a popular design for bagasse, both in the United States and abroad. Here bagasse fuel is partly dried and burned in refractory cells below a radiant arch. The combustion of gases is completed above the arch. Spreader stokers can also be used as shown in Fig. 7-11.

In refining sugar from cane, the juice squeezed out of the cane eventually is processed into sugar. The remaining fibrous, tenacious, and bulky crushed cane is called bagasse. It is also moist. Depending on where it is grown and the efficiency of the juice extractor, bagasse contains 30 to 50 percent wood fiber and 40 to 60 percent water. Heating value is 8000 to 8700 Btu/lb as a dry solid, with a yield of about 4500 Btu/lb at around 45 percent moisture content.

The steel industry has large quantities of gaseous by-product energy available. Heat content varies from less than 100 Btu/ft^3 for blast-furnace gas to 525 to 600 Btu/ft^3 for coke-oven gas. The main problem is getting this gas clean enough to avoid fouling the burners.

Oil refineries (catalytic cracking of crude petroleum) produce large volumes of gas as a by-product of catalyst regeneration. This gas contains 5 to 8% CO (carbon monoxide), about twice that much CO_2 (carbon dioxide), and air. The gas temperature is around 500°F, with a heat content of about 145 Btu/lb. Increasingly, refineries reclaim this energy by burning the gas, together with oil or gas and additional air, in a carbon monoxide steam generator.

Liquid by-products include residue from chemical processes such as tar and pitch. These can be handled in conventional oil burners, but for satisfactory results, they must be heated to maintain viscosity at the proper level. Filtration is also required to remove any solid contaminants. High moisture content can be poured off (decantation) or emulsified. In designing heat-recovery apparatus, consideration must be given to the dew point of the flue gas that is created in order to reduce the possibility of fireside corrosion and tube-passage plugging, especially when fuels that contain sulfur are burned.

Black-Liquor Boilers Black liquor is a by-product of wood-pulp processing in the papermaking industry. See Fig. 7-12. Chips of wood are cooked by steam in a solution of sodium sulfide and sodium hydroxide in a large tank known as a "digester." Strong liquor from the digester flows to a storage tank where it is joined by weak liquor from pulp washers. In order that this liquor may sustain combustion, it is then concentrated by evaporation and crushed salt cake is added until it contains over 58 percent solids.

The concentrated liquor is pumped at about 220°F through oscillating burners which spray it onto the furnace walls; deposits of combustible char build up until they are heavy enough to drop to the furnace floor, where combustion is assisted by primary air nozzles. The gases and a small percentage of fuel particles rise to the upper part of the furnace where secondary air is admitted to complete the combustion process.

A considerable percentage of chemical, a form of soda ash, is recoverable from the ash of this process. Thus, the combustion of this black liquor is twofold, namely, steam generation and for soda-ash recovery. The capacity of a black-

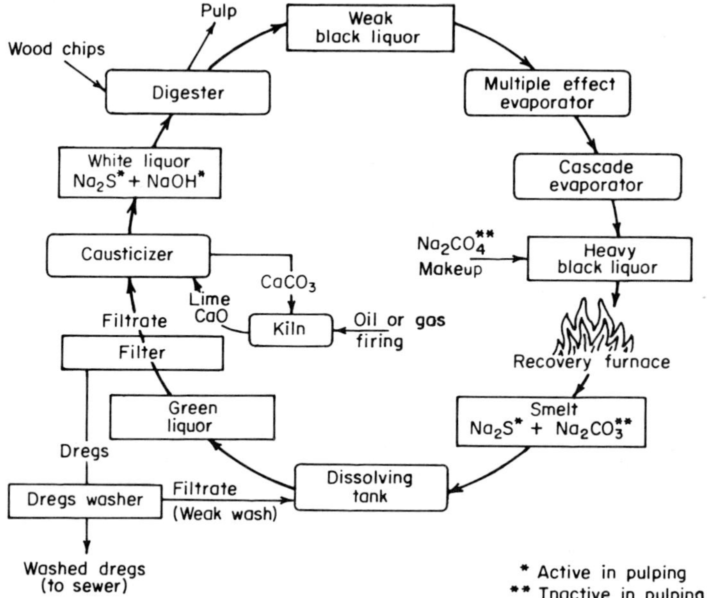

Fig. 7-12 Kraft recovery cycle in paper mills recovers chemicals used to break up wood fibers in digesters by burning black liquor in a recovery boiler.

liquor recovery boiler is also expressed in the amount of tons of dry solids it can burn in 24 hr. As an example, a unit designed for 1600 psi with a capacity of 392,000 lb/hr of steam is rated at the same time as being able to burn 800 tons/day of black liquor. The two major manufacturers in the United States are B&W and Combustion Engineering (CE). The B&W units use a slanting-bottom furnace (Fig. 7-13). The oscillating burners spray the black liquor on the furnace walls. The deposited liquor dries, forms char, and falls to the hearth where sodium chemicals are smelted and the organics in the char are turned to gas which burns in the furnace. The smelted chemicals drain down the sloping floor through the water-cooled spout and then into the dissolving tank.

The CE unit shown in Fig. 7-14 sprays the black liquor into the center of the furnace. The bottom of the furnace in the CE unit is flat and is called a decanting furnace. Most other boiler details are similar for the two manufacturers' designs.

Serious furnace explosions on recovery boilers have instituted research as to the cause. It is generally believed that the smelt in the furnace can chemically combine with water to produce a chain-reaction type of detonation in the furnace. This has produced rapid pressure buildup in the furnace with shock-wave results to the structural components of the furnace. The paper industry, insurance companies, and the boiler manufacturers have drawn up

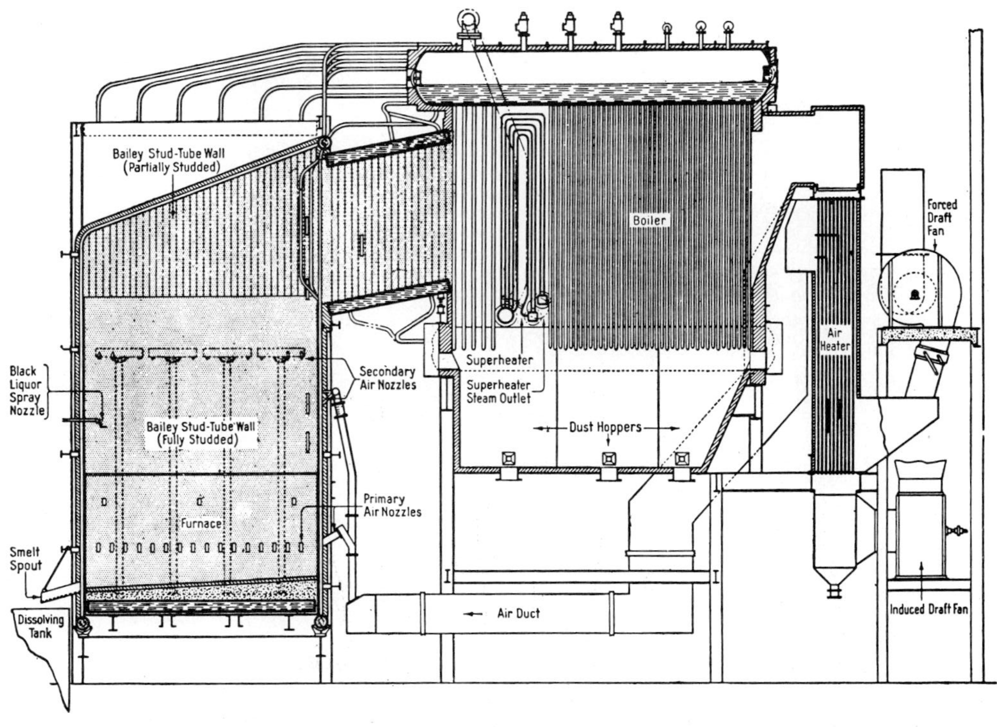

Fig. 7-13 Black-liquor recovery boiler has slanted-bottom furnace for smelt removal to dissolving tank. *(Courtesy Babcock & Wilcox Boiler Co.)*

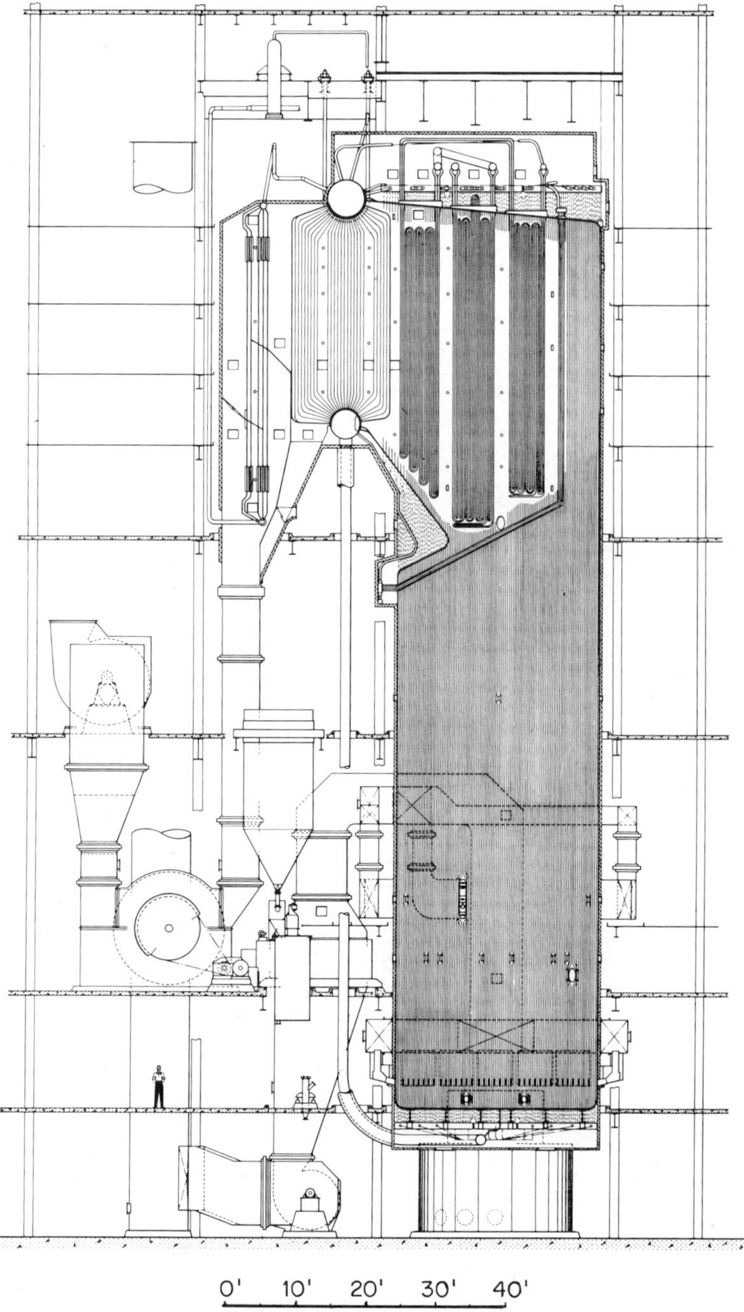

Fig. 7-14 Combustion Engineering black-liquor recovery boiler with flat-bottom furnace. Fuel is sprayed into the center of the furnace for burning. *(Courtesy Combustion Engineering, Inc.)*

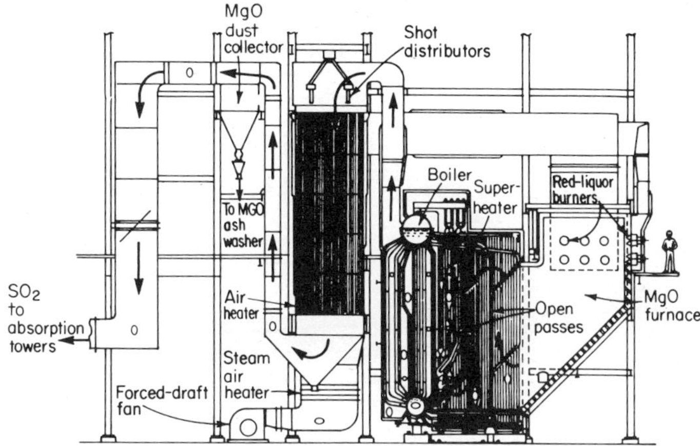

Fig. 7-15 Red-liquor-burning watertube boiler used in the acid process of chemical recovery and steam generation in the paper industry. *(Courtesy Power magazine.)*

guidelines, including emergency procedures, in order to prevent smelt-water explosions. Tube integrity is stressed, and NDT inspections of welds in the furnace area are recommended, as is black-liquor-concentration monitoring. Among the novel procedures recommended is rapid draining of the boiler to a level 8 ft above the low point of the furnace floor at any time when water is suspected of entering the furnace area.

The black-liquor recovery boiler is used in the papermaking industry where hydrogen sulfite is used as the chemical to break up the lignin in wood that is cooked in digesters. The process is considered alkaline. Another process is called the *red-liquor acid process.* Red liquor with a concentration of about 50 percent solids is fired with steam-atomizing burners, as shown in Fig. 7-15. The red liquor burns in suspension and forms little or no slag in the furnace. Magnesium oxide ash is removed from the furnace floor as well as from the flue gas. Sulfur dioxide is also recovered from the flue gas by passage of the flue gas through absorption towers where a magnesia oxide slurry absorbs the sulfur dioxide. The result is a cooking acid that is reused in the digesters.

Black and red liquors do not burn alike; thus the steam generators used are not alike. Black liquor in the kraft process is very difficult to burn. Large furnaces are needed to keep the temperature relatively low because the liquor has a high content of low-fusion-temperature ash. Smelt collects on the refractory sloping hearth, and a reducing atmosphere must be maintained in the lower part of the furnace for chemical conversion. Also, since superheater and boiler surfaces have a tendency to coat with slag, they operate at low absorption rates. Thus frequent soot blowing and shot cleaning of heating surfaces is necessary.

Red liquor in the MgO (magnesia oxide) process, on the other hand, burns completely in suspension, making little or no slag. Thus a smaller steam generator can be used for an equivalent amount of steam production.

Questions and Answers

7-1 What limitation does the Code place on hot-water coil boilers that do not have a steam space?

Ans. The Code does not cover this type of boiler if the following restrictions are complied with: (1) tubing of ¾-in. diameter with no drums or headers attached; (2) maximum 6-gal water-containing capacity; and (3) no steam generated in the coil.

7-2 Name two reasons why a coil-type boiler may develop excessive coil temperatures.

Ans. Low water and insufficient water being circulated through the coil because of partial obstructions or pump failure. For this reason, coil-type boilers should have a high-temperature cutout and possibly a flow switch that will shut off the firing mechanism when high coil temperature is developed. This is usually done by a thermostatic switch.

7-3 How is the feedwater rate controlled in a Clayton-type coil boiler?

Ans. The feedwater is controlled by the water-level control which in turn responds to the liquid level existing in the accumulator. When the accumulator water level drops, the water-pump solenoid de-energizes to allow the feedwater pump to operate.

7-4 What procedure is recommended in checking a coil-type for leaks?

Ans. Most manufacturers recommend the boiler be hydrostatically tested; if failure is small, repair the leak by an approved welding process. The coil should be replaced if difficulties are encountered in the repair.

7-5 What is the minimum thickness of plate permitted for shells or heads subjected to pressure on an electric boiler?

Ans. The plate must be at least 3⁄16 in. thick per Section I (ASME Code) rules.

7-6 When are postweld heat treatment and radiographs not required on a welded joint of an electric boiler?

Ans. When the following limitations are not exceeded: 16-in. inside shell diameter, 100-psi pressure.

7-7 When must an electric boiler have a built-in inspection opening?

Ans. Electric boilers over 5 ft^3 in volume and not having a manhole must have an inspection opening in the lower portion of the shell or head so that the unit can be periodically checked for scale or other deposits and cleaned accordingly. The opening cannot be smaller than 3-in. pipe size.

7-8 If an electric boiler has a power input of 750 kW, what Code requirements on capacity and number of safety valves must be met?

Ans. Since the input is over 500 kW, two or more safety valves are required. Their total capacity must be at least equal to $750 \times 3.5 = 2625$ lb/hr.

7-9 How small can the feedwater connection be on an electric boiler?

Ans. It cannot be smaller than ½ in.

7-10 What type of electric boiler does not require a low-water fuel cutoff?

Ans. Electrode-type boilers do not require a low-water cutoff because the water in the boiler completes the electric circuit from electrode to electrode. Low water would interrupt the electrical flow, thus serving the same purpose as a low-water cutoff.

7-11 When are two gauge glasses required on electric boilers?

Ans. Non-electrode-type boilers operating at pressures over 400 psi require two water gauge glasses.

7-12 What is the minimum-size safety valve permitted for a steam-heating boiler?

Ans. ½ in. and no larger than 4½ in. (Section IV of ASME Code).

7-13 What are the sizes for a hot-water-heating boiler?

Ans. ¾ in. and no larger than 4½ in.

7-14 What pressure rise is permitted on a boiler before the safety-valve capacity may be considered inadequate for heating boilers?

Ans. For steam-heating boilers, the safety-valve capacity shall be sufficient at maximum firing rate that the pressure does not rise more than 5 psi. For hot-water-heating boilers, the relief-valve capacity must be sufficient to prevent the pressure from rising more than 10 psi with the burners operating at the maximum firing rate.

7-15 Name the two pressure controls required on a steam-heating boiler and the temperature controls required on a hot-water-heating boiler.

Ans. Automatically fired steam-heating boilers require an operating-pressure cutout switch set lower than the allowable pressure and an upper-limit cutoff switch set at a pressure no higher than 15 psi. For hot-water-heating boilers, an operating-temperature cutout switch set less than the maximum allowable temperature is required as well as an upper-limit temperature cutout switch set no higher than 250°F.

7-16 When does the Heating Boiler Code require a low-water fuel cutout for hot-water boilers?

Ans. When the boiler is automatically fired with heat input greater than 400,000 Btu/hr.

7-17 What hydrostatic tests are required on sections of a steam-heating boiler made of cast iron?

Ans. The individual sections must be tested with a hydrostatic pressure not less than 60 psi. The assembled boiler is tested at a hydrostatic pressure not less than 45 psi.

7-18 What hydrostatic tests are required for a cast-iron boiler to be used for hot-water heating?

Ans. For boilers with working pressure not over 30 psi, the hydrostatic test pressure must be at least 60 psi for each individual section. Those with working pressures over 30 psi require a test pressure 2½ times the maximum allowable pressure on each section. In both of the above, the assembled boiler requires another hydrostatic test not less than 1½ times the maximum allowable working pressure. These requirements apply to the manufacturer of the cast-iron boilers. The test pressures required must be controlled within a 10-psi range.

7-19 Name the three methods or types of cast-iron boilers.

Ans. Horizontal sectional, vertical sectional, and one-piece types.

7-20 What is a fluid-vaporizer generator?

Ans. This is a closed vessel in which a heat-tranfer medium, other than water, is vaporized under pressure by the application of heat. Here the heat-transfer medium is used externally to the closed vessel.

7-21 Do fire insurance companies and local fire regulations have requirements on liquid-phase heaters or on vapor-phase heaters?

Ans. Yes, because a distinct fire hazard may exist with some of the organic fluids, including mineral oil.

7-22 What kind of gauge glass is required on an organic-fluid vaporizer generator?

Ans. It must be of the flat gauge-glass type with forged-steel frames. Gauge cocks cannot be used in order to avoid spilling or draining the organic fluid.

7-23 Can a safety valve have a lifting lever when installed for overpressure protection on an organic-fluid vaporizer?

Ans. No lifting lever is permitted in order to avoid discharging accidentally the organic fluid. The safety valve should be removed at least once per year and tested off the unit for pressure setting and capacity. The discharge from an installed safety valve should be directed to a condenser or, as a minimum, to a safe point outside a building.

7-24 To what pressure must a vaporizer be designed?

Ans. At least 40 psi above the operating pressure at which it will be used with the usual safety factor applied to this design per Code rules.

7-25 What two requirements govern the selection of safety relief valves for liquid- and vapor-phase organic generators?

Ans. The pressure setting can be no higher than the maximum allowable pressure of the unit, and the capacity must equal at least the maximum Btu-per-hour output of the generator.

7-26 What kind of valves should be used for HTHW application?

Ans. Valve seats, plugs, and bodies should be made only of cast steel, forged steel, or steel alloys. Valve seats should be stainless chrome-nickel steel, to avoid corrosion and erosion by flow of water. Pressure ratings should follow the Power Boiler Code rating on stop valves, feed valves, and blowdown valves, which is 125 percent of boiler allowable working pressure (AWP).

7-27 Do HTHW generators require a water and gauge glass?

Ans. Only if a natural-circulation boiler has a drum which is used as an expansion tank. If the boiler is completely filled with water under pressure and has an external expansion tank, the Power Boiler Code does not require a water gauge glass or gauge cocks.

7-28 On forced-circulation boilers for HTHW systems, what provision should be made to prevent hot spots, or lack of circulation, in case the pump fails?

Ans. A pressure-differential or flow switch should be installed to shut off the fuel-burning equipment in case no water is flowing.

7-29 Should a low-water fuel cutout be used on an HTHW generator of the suspended fuel-fired type?

Ans. Some states now require a low-water fuel cutout on low-pressure hot-water-heating systems. Thus it follows that this safety device is even more necessary on an HTHW boiler. It should be installed on the boiler to shut off the fuel in case the water level drops below a dangerous level.

7-30 Name three methods for pressurization of an HTHW system.

Ans. Three methods are: (1) steam cushion; (2) gas pressurization such as use of a nitrogen gas cushion; and (3) mechanical pressurization which makes use of an expansion tank and a pump.

7-31 When chemical water treatment is applied to an electrode boiler to minimize scale formation and similar water considerations, what other factor in the treatment must be considered?

Ans. The effect to conductivity of the treatment. If the conductivity becomes too high (low resistance to current flow), short circuits may occur that may damage the electrodes and the insulation. If the conductivity is too low (high resistance to current flow), the unit may not be able to reach its rated output.

7-32 Name some typical advantages in burning waste fuels in a steam generator.

Ans. Advantages to a plant that has waste fuel as a by-product of a manufacturing process are: (1) an inexpensive or economical means is available to produce steam in comparison to fossil-fuel-fired boilers which require the purchase of fuel; (2) a fuel is available that may not be affected by shortages of oil or gas; and (3) the by-product can be disposed of in the steam generator thus waste-disposal regulations may be avoided when the waste has to be dumped in a landfill or similar disposal means that could affect the environment.

7-33 Name three burning requirements in the burning of wood refuse in a steam generator.

Ans. In the burning of wood refuse, one must consider the design of the furnace, the type of burning equipment to be used, and similar factors to accomplish the following: (1) evaporation of the moisture in the wood; (2) distillation and combustion of the volatile components in the refuse; and (3) burning of the remaining carbon material in the wood.

7-34 What is cogeneration?

Ans. Cogeneration is the coincident generation of electricity and process-steam needs for a manufacturing complex. Usually high-pressure steam is produced in on-site steam generators. This high-pressure steam is passed through a turbogenerator to obtain electric power and process steam, by extracting the steam from the turbine at a pressure needed for the process. Several extraction pressures can be used. The overall heat balance for a cogeneration plant can thus be attractive in comparison to just buying power and generating only process steam.

7-35 Name three reasons why waste-heat boilers are usually of the watertube type.

Ans. The watertube boiler type is favored for the following reasons: (1) Large mass flows of waste heat are possible; (2) solid entrapped particles are more readily recovered in hoppers and similar equipment; and (3) a better tube arrangement is possible to avoid slagging and erosion problems.

7-36 Name three possible sources of a furnace explosion in a recovery boiler of the black-liquor type.

Ans. (1) Reaction between water and smelt from a tube failure and a similar source of water entering the furnace; (2) reaction between a weak or low solid black liquor that is sprayed into the furnace and then because of its high water content, reacts with the smelt in the furnace; and (3) conventional ignition of unburned fuel from auxiliary burners using gas or oil as a fuel, with the ignition leading to a fireside explosion when the accumulation of unburned fuel is ignited.

8
Nuclear Power Plant Reactors and Pressure Vessels

Nuclear power plant equipment consists of many pressure vessels, including those that generate heat under pressure, called reactors, because in its core a nuclear reaction is taking place. Steam or high-pressure hot water is also made in nuclear reactors; by means of heat exchangers, the steam produced then follows a conventional flow to a turbogenerator. This chapter briefly reviews some of these systems. However, the reader is referred to more advanced texts to more fully study nuclear energy.

A review of the structure of an atom will reveal that the nucleus is made of protons and neutrons defined as follows:

A *proton* is a *positive-charged particle* of matter. There is usually one proton in the nucleus for each negative-charged particle, or electron, circling the nucleus. The positive-charged proton attracts the negative-charged electron whirling around the nucleus, and this prevents the electron from flying off by centrifugal force just as the earth is held in orbit around the sun by gravitational forces.

A *neutron* is matter in the nucleus with no electric charge, and it weighs about the same as a proton. There is a strong binding force holding together protons and neutrons. The number of neutrons is not fixed, but varies with atoms, although this number does affect the atomic weight. Neutrons do not affect the chemical properties of the elements, as protons and electrons do. Since neutrons have no charge, they are very useful in nuclear-bombardment applications. They travel in straight lines until they come in contact with other matter. This makes them very useful in splitting the nucleus of an atom.

Electrons are negatively charged particles, quite a bit lighter than protons.

In fact, they weigh about $1/1800$ the weight of a proton. Electrons are arranged in rings around the nucleus, with definite numbers of electrons per ring. That is, the first ring holds 2 electrons; the second, 8; the third, 18; until a maximum of 32 is reached. The total number of electrons in an atom is equal to the number of protons and neutrons in the nucleus, which is also the atomic weight of the atom. Uranium has the highest atomic weight at 92 and also has the greatest number of electrons surrounding the nucleus.

Material that is capable of capturing neutrons, and therefore splitting an atom into two or more particles, is called a *fissionable material.* Uranium 235, uranium 233, and plutonium 239 are fissionable materials. When an atom is split by a neutron traveling at high speed so that two or more fragments are split from the atom, it is called *fission.* During fission, two to three neutrons are released from the split atom. If a chain reaction is to result, one of these free neutrons must be captured by another atom and cause fission of that atom; then in turn it must produce another free neutron to establish the chain reaction of fission. The ratio of the number of neutrons which cause fission to the number of neutrons initially produced is called the *effective multiplication.*

The fundamental law of physics on which a nuclear reaction is based is Einstein's law of the interrelationship of mass and energy, namely, energy equals mass times velocity squared ($E = MV^2$). One must realize, though, that the velocity term is near the speed of light. It has been established, for example, that 1 lb of uranium 235, when hit by a neutron bullet, will *explode* to form lighter atoms and spare neutrons. Some of the mass, about one-thousandth of a pound, is lost in the form of energy which is equivalent to about 11.4 million kilowatthours (kWhr) of energy. Many other principles of nuclear physics are involved in a nuclear reaction, but they are beyond the scope of this book.

The present method of nuclear power generation is by fission, or the splitting of the nuclei of certain heavy atoms by bombarding them with neutrons. Hitting the uranium-isotope ^{235}U nucleus, these neutrons split the atom to form new elements, such as krypton and barium, release energy in the form of heat, and liberate additional neutrons from the nucleus that can bombard another nucleus to keep the reaction going. Other atoms that fission are ^{233}U, thorium, plutonium, and ^{238}U.

Reactor vessels are used to house the fuel elements in a nuclear plant where the chain reaction of fission takes place. Reactor vessels, fuel elements, coolants, moderators, and control rods are hardware common to most types of reactors.

Reactors are generally classified by the type of coolant used to extract the heat from the fission reaction. Most common are the pressurized-water reactor, boiling-water reactor, heavy-water reactor, gas-cooled reactor, and metal-cooled reactor (sodium).

Figure 8-1*a* shows a boiling-water reactor (BWR) where the nuclear-reaction-heated water in the reactor evaporates in the same pressure vessel and then is directed to a steam turbogenerator. The steam is condensed, passed through feedwater heaters, and returned to the reactor by a feed pump.

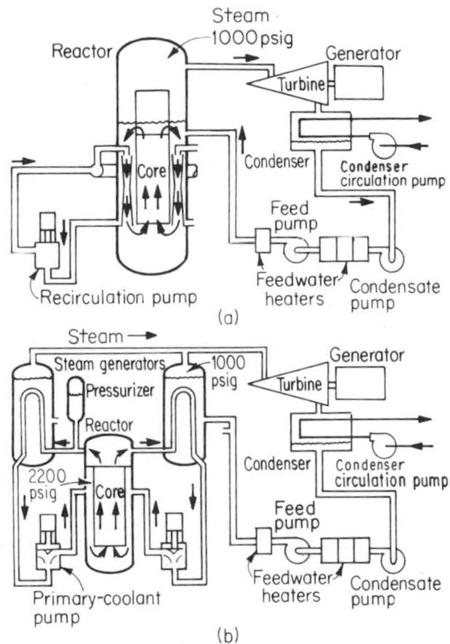

Fig. 8-1 *(a)* Boiling-water reactor: The steam to turbine comes directly from the reactor. The recirculation pump promotes circulation. *(b)* Pressurized-water reactor: The reactor is cooled by pressurized water. The steam to turbine comes from the heat exchanger. Coolant pumps maintain hot-water circulation to the reactor core. The pressurizer maintains water under pressure and permits expansion of water with a rise in temperature.

Figure 8-1*b* illustrates a pressurized-water reactor (PWR). The coolant is high-pressure water which is pumped through the core of the reactor to remove heat by primary coolant pumps. The pressurized hot water is passed through water-to-steam heat exchangers, with the tubes carrying the radioactive water within the shells containing the steam going to the turbogenerator. The water must be at a higher temperature in the heat exchangers than the steam, with the water pressure in the reactor being about 2200 psi to prevent boiling at high water velocities. A pressurizer is used to keep the water under constant pressure; it is similar to an expansion tank on an HTHW system.

Figure 8-2*a* illustrates a heavy-water reactor, which resembles the PWR type except that in this system, heat transfer is from fuel rods to heavy-water coolant and then to light water in steam generators or water-to-steam heat exchangers. This design is popular in Canada.

Figure 8-2*b* illustrates a gas-cooled reactor. This system uses a large compressor for circulating the helium through the reactor core and the steam generators. Helium is at 700-psia pressure and a temperature of 1430°F, with the steam produced at higher temperatures (1000°F) and pressure than is possible with the BWR or PWR types.

Under development are breeder reactors for power generation. The breeder reactor is so named because it produces more fissionable fuel material than it consumes. The now wasted uranium 238 and low-grade thorium ores would be converted in a breeder reactor, through neutron bombardment, to a fissionable fuel from the dwindling supply of the present fuel used, uranium 235.

Fusion power plants are still in the conceptual and development stages. Fusion requires temperatures to 100 million degrees Fahrenheit which no

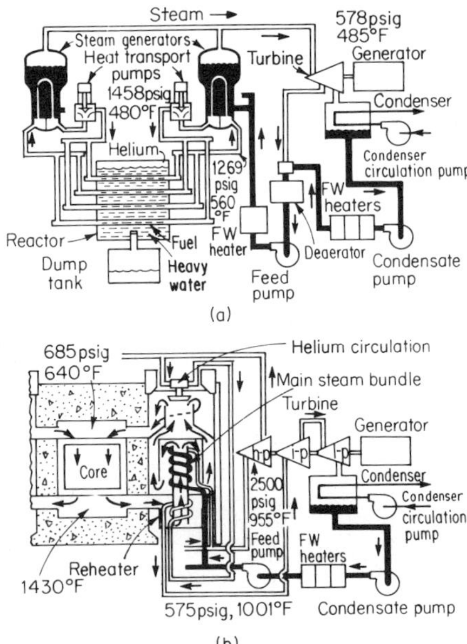

Fig. 8-2 *(a)* The heavy-water reactor has water flowing through fuel-rod tubes with the moderator being heavy water. *(b)* The gas-cooled reactor system uses helium as a coolant through a graphite core and helium-to-steam heat exchanger to make steam.

known material can withstand. It is necessary to develop intense magnetic fields to confine and also to obtain the hydrogen-gas plasma that is the fusion fuel. In the fusion process, hydrogen or its isotopes, deuterium and tritium, are fused to form helium, the next higher element in the atomic scale. The joining or fusion takes place in nuclear ovens which generate intense heat that must be extracted for the production of useful energy. Intense development is going on all over the world in an effort to find an economical method of extracting usable energy from a fusion nuclear reaction.

The heat produced in a nuclear reactor is usually expressed in so many megawatts of power. The heat produced depends on the thermal or neutron flux developed in the nuclear chain reaction per unit of volume, the energy produced per fission, and similar nuclear-physics criteria. One concern in nuclear reactors that is similar to the case of fired steam generators is the possibility for overheating damage. The limiting factors are the temperature limits imposed on fuel assemblies in order to avoid burnout of the fuel cladding and/or melting of the oxide fuel. These factors also influence coolant flow rates in order to avoid overheating damage of the fuel rods. This also requires maintaining proper coolant levels in the reactor, just as one has to maintain proper water levels in a steam generator.

RADIATION HAZARD

The radiation hazard in nuclear plants has received wide publicity, but the extent of exposure becomes significant only when the pressure-containing boundary is pierced. Waste disposal problems are under control, but the long-term effect of storage is a matter of concern to many environmentalists. The main fear of radiation is its ability to cause cancer and genetic defects; however, the threshhold limit required for this to occur is a medical matter.

Three distinct kinds of radiation are emitted from radioactive material. Alpha and beta rays have little penetrating power, but gamma rays can penetrate great thicknesses. Neutrons, too, have great penetrating power and are the primary radiation hazard in an operating reactor.

To safeguard personnel against neutrons, gamma rays, and heat, the reactor, and much of its auxiliary equipment, must be enclosed within thermal and biological shielding. Gamma rays can be absorbed by a number of materials, particularly those with the greatest density, such as lead or steel. The same effect is obtained by using a much greater thickness of water or concrete. For land-based reactors space and weight limitations are not major considerations. The cost of material and construction is usually more important. The cheapest and most widely used shielding material is concrete. When several feet thick, a concrete biological shield is an excellent neutron absorber. The addition of some percentage of denser material to the mix, such as iron or barytes, will improve local gamma shielding. Alternatively, the use of more expensive magnetic concrete affords protection against both gamma rays and neutrons.

The intensity of radiation follows an inverse-square law as the distance from the source increases. Therefore, the amount of shielding required can be reduced by building the shield at a greater distance from the core. Since concrete is not able to withstand high heat, a thermal shield is constructed between the main shield and the reactor. This may be steel or a separate, thin concrete vessel. Coolant channels may be provided in the concrete vessel to carry away heat. The major problem is preventing radiation-leaking paths where coolant inlet and outlet pipes, control rods, and other hardware leave the reactor. Any such annular paths must be stepped to avoid a direct line-of-sight path.

Nuclear safety begins with designing a plant to have numerous defenses and redundancy systems in order to avoid an unforeseen accident. See Figure 8-3. The design incorporates strict quality-control requirements that are also generally, as a minimum, legal requirements. Protective systems include instruments and measuring devices that are constantly scanning vital areas of the plant and which are capable of securing the plant from abnormalities of operation. Multiple safeguard systems are also used to protect the plant from postulated accidents. See Fig. 8-4*a* for a sketch of the system in a boiling-water reactor that provides a defense against the release of radioactive material from a reactor to the environment.

The reactor containment shell is a thin cylindrical or spherical, steel pressure vessel surrounding the shielding. Its purpose is to accommodate energy released by sudden, uncontrolled fission in the event of a reactor accident. Since personnel work inside, the shell must be provided with access locks. The pressure within is maintained below atmospheric pressure so that any leakage of radioactive gas from the reactor is retained in the shell and passes through specially filtered and monitored ducts.

The pressure-suppression containment system eliminates the need for a large

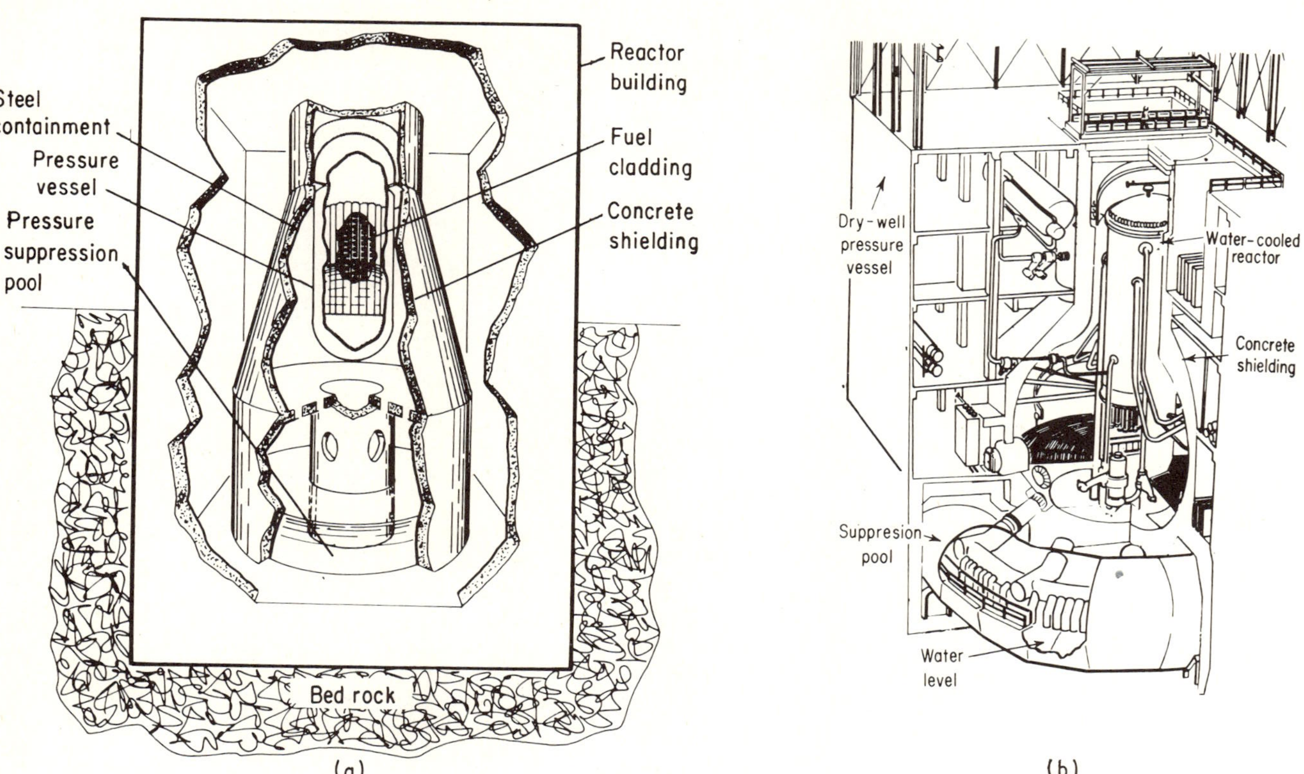

Fig. 8-3 *(a)* The reactor vessel is surrounded by a defensive in-depth containment. *(b)* Suppression pools are used to condense steam from a reactor leak into the containment vessel, thus limiting pressures.

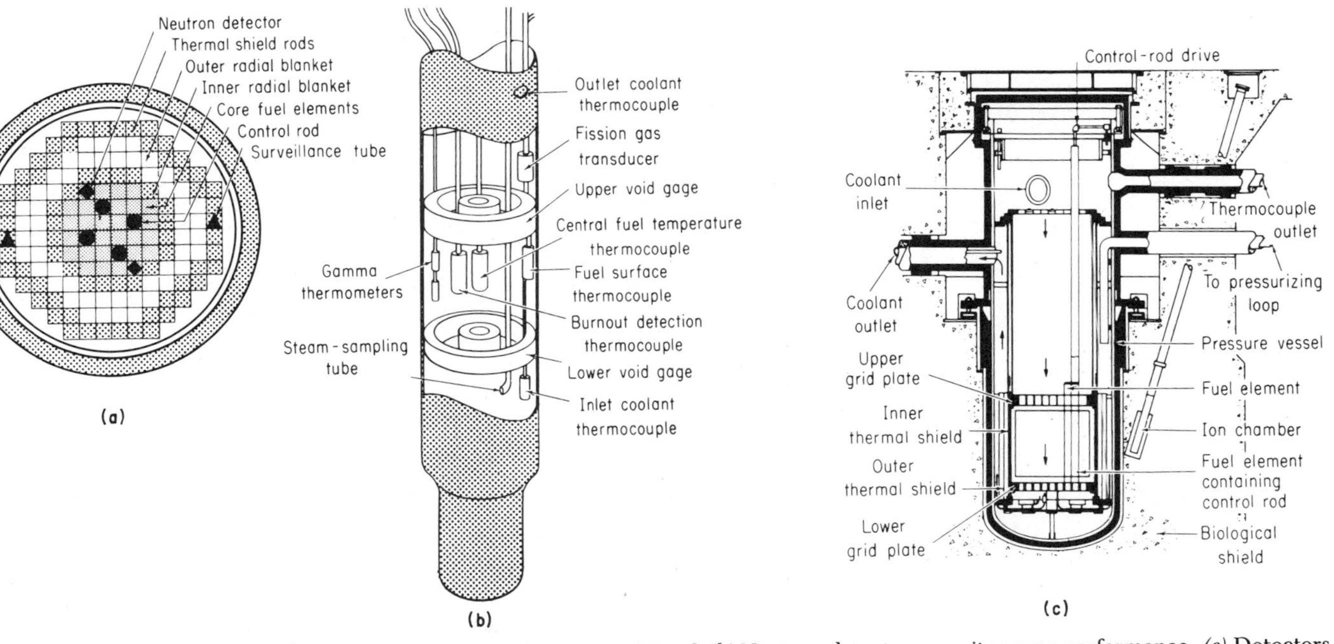

Fig. 8-4 Detectors are used to monitor reactor-core performance. *(a)* and *(b)* Neutron detectors monitor core performance. *(c)* Detectors are placed outside the reactor for core monitoring.

containment shell, at least in the case of water-cooled reactors. Instead, the reactor pressure vessel and its associated pumps and pipework are enclosed in a second pressure vessel sized to accommodate the steam formed in the event of vessel or pipe rupture. Ducts are connected to this outer vessel. They lead to an annular suppression pool in which the steam is condensed. Entrained fission products are then removed by absorption.

The energy potential of the fuel mass in a nuclear reactor and the lethal radioactivity of many fisson products demand more stringent safety measures than a conventional power plant. They may be considered under two heads: control and instrumentation for safe reactor operation and means to prevent escape of radioactivity during normal operation, possible reactor runaway, or other failure.

Reactor instrumentation can be grouped into three classes: control, safety, and monitoring. The major requirement in control instrumentation is measurement and display of the heat rate produced in the core over a complete range from subcritical to full-power operation. This is done by measuring the neutron flux level, or rate of neutron fission, with neutron detectors. Different types of instrumentation are provided for shutdown, start-up, and low-power and high-power operation.

For shutdown and start-up measurements, pulse counters are used, often taking the form of ionization chambers filled with boron trifluoride gas. Operating at a particular voltage, they give a pulse proportional to the incident radiation. Alternatively, fission chambers, coated with a fissile material, are capable of detecting neutrons by fissions inside the chamber. Each neutron produces an electric pulse. In either case, the output passes through a pulse amplifier to counting rate meters, logarithmically scaled since neutron flux increases exponentially. Start-up and similar low-power detectors retract into the biological shield during high-power operation. For power measurements in the normal operating range, where temperature effects on reactivity become important, instruments are linearly scaled. Ionization chambers act directly on high-impedance potentiometer recorders.

Reactor emergency tripping is based on set limits being executed for such factors as neutron density change, fuel-element temperature, power level, and coolant flow. Then reactor "scram" takes place: control rods are automatically inserted into the core, either by power or by releasing the drive mechanism and allowing them to fall by gravity. "Scram" time is usually 2 to 3 seconds (sec).

Containment of fission products demands several lines of defense. One sodium-cooled reactor vessel breeder plant is surrounded by a thermal and biological shield system. A stainless-steel thermal wall protects the shielding from heat; borated graphite acts as a neutron absorber. Gamma rays emitted by the absorber are captured in the outer concrete shell. A suppression pool (Fig. 8-3*b*) is a safety measure to condense steam if the water-cooled reactor pressure vessel or the piping ruptures.

Monitoring the reactor power level is the purpose of the neutron detector (fission counter or ionization chamber) placed within the core (Fig. 8-4*a*). Detectors may be alternatively placed outside the reactor, treating the core as a point source of neutrons, as shown in Fig. 8-4*c*. The extreme range of power level, as much as 10 decades, requires multiple detectors with overlapping ranges. To gain data on fuel, heat transfer, and metallurgical characteristics, extensive in-core sensing instrumentation may be applied. An instrumented fuel assembly (Fig. 8-4*b*) is used in a boiling heavy-water reactor to determine fuel power limits.

Core meltdown from a loss of coolant has received great attention by designers and regulatory agencies. To prevent core meltdowns and consequently possible release of radioactive material to the biosphere, reactor systems are equipped with numerous safety devices to forewarn of a developing incident and also to initiate backup emergency core-cooling systems if needed. This system is intended to replenish cooling water that might be lost through a rupture of the primary cooling system. All emergency core-cooling systems are designed to inject water into the pressure vessel rapidly enough to keep the core cooled. Boiling-water reactors employ a core spray and an independent low-pressure coolant injection system.

CONSTRUCTION CODES

The Nuclear Regulatory Commission of the federal government has numerous requirements and standards for nuclear power plants, including impact-on-the-environment considerations. Section III of the ASME Boiler and Pressure Vessel Code now consists of seven documents:

Division 1: Nuclear Power Plant Components, General Requirements
Division 1: Nuclear Power Plant Components, Class 1 Components
Division 1: Nuclear Power Plant Components, Class 2 Components
Division 1: Nuclear Power Plant Components, Class 3 Components
Division 1: Nuclear Power Plant Components, Class MC Components
Division 1: Nuclear Power Plant Components, Component Support
Division 1: Nuclear Power Plant Components, Core Support Structures
Division 2: Nuclear Power Plant Components, Code for Concrete Vessels and Containment

The ASME also has a Section XI Code entitled "Rules for In-Service Inspection of Nuclear Power Plant Components."

The ASME Construction Code provides for three levels of quality for design and construction, depending on the element of radioactive risk that may be involved with the component under consideration. The choice of quality level for a given component requires a knowledge of how the component functions in the process or system. Class 1 components are those that are a part of the reactor-coolant pressure boundary, i.e., those components containing coolant for the core and piping and similar components that cannot be isolated from the core and its radioactivity. Class 2 components are those involved with reactor auxiliary systems which are not part of the reactor-coolant pressure boundary, but which are in direct communication with it, such as a residual heat-removal system. Class 3 systems are those components that support Class 2 systems without being part of them.

Another document that affects the construction of nuclear power plants is American National Standard Institute (ANSI) N626, entitled "Qualification and Duties for Authorized Nuclear Inspection," which the ASME has also adopted as a requirement. This document defines authorized inspection agencies and authorized nuclear inspectors and supervisors. Great emphasis is placed on knowledge of welding and how it can affect metals in the examinations given to potential nuclear inspectors. Knowledge of NDT methods and interpretation of results are also highlighted in the requirements. Quality control procedures in monitoring a construction job are also stressed in the duties of a nuclear inspector, as are traceability of documentation on materials, welding procedures, qualification of welders, and similar areas where variables in construction may exist. The Nuclear Code has been extended to pumps and valves to ensure the integrity of the pressure containment parts and not only reactors and pressure vessels.

A separate Nuclear Code on pressure vessels had to be adopted for the following reasons:

1. Nuclear vessels have many unusual design considerations which require more stringent rules than are needed for boilers and other pressure vessels.

2. There are various types of nuclear vessels, defined as follows: *primary vessels* are those subjected to the primary coolant, which may be radioactive; *secondary vessels* are not subject to radioactivity; *pressure vessels* are basically containment vessels to guard against radioactive contamination in the event of failure of the primary vessels.

3. Since a nuclear-reactor vessel may be radioactive for years, periodic internal inspections in the usual sense as applied to boilers and pressure vessels are impossible. Sophisticated testing by nondestructive means has to be employed.

4. The hazard in a nuclear reactor vessel is not only a pressure explosion, but also a radioactive contamination hazard far more serious to life than a steam pressure explosion only. A whole area may be affected by radioactive fallout. Thus more stringent design and fabrication rules are required than for boilers or pressure vessels.

5. Nuclear reactor vessels can be subjected to very sudden heating or cooling temperature changes. This creates abnormal thermal stresses and cycling fatigue stresses. The Power Boiler Code does not have enough provisions for calculating these stresses, but the Nuclear Vessel Code requires these to be calculated.

6. The material to be used on a nuclear vessel may have to withstand radiation effects which may affect its properties. Materials for nuclear vessels thus require this consideration, whereas materials for boilers and pressure vessels may not.

7. Inspection during fabrication of a nuclear-reactor pressure vessel has to be far more thorough than of a boiler because of the hazards involved. Rigid

quality control must also be exercised far above the usual practice in boiler and pressure-vessel construction.

The design of Class 1 vessels is far more exacting in terms of Code requirements than for conventional boilers. Complete stress analysis is required on all major components of Class 1 vessels, and this must be done by a registered professional engineer. All forms of stress must be considered and combined to obtain the actual stress on each critical section of the vessel. These include normal, shear, discontinuity, bending, thermal, cycling, and fatigue stresses.

Corrosion allowance must be provided if analysis shows that it is required. All loading conditions must be considered, including: (1) internal and external pressure or combinations thereof; (2) the weight of the vessel and its contents; (3) superimposed loads, such as other vessels, insulation, piping, and cladding; (4) wind loads, snow loads, and earthquake loads; (5) reactions of supporting lugs, rings, saddles, and other supports; (6) temperature effects; and (7) environment effects due to radiation.

The allowable design stress is based on the yield strength, not on the ultimate strength, as is the case for boilers and unfired pressure vessels.

Welding requirements are *stiffer.* No backing strips are allowed to remain. Full radiographing is required on all welds subject to stress caused by pressure. Some welds require ultrasonic examination also. The material must be certified by the material manufacturer to comply with all sections of the Nuclear Vessel Code. Heat-treatment test coupons are required. Ultrasonic examination is required for steel plates on reactor vessels and all other plates 4 in. and over in thickness. Standards are established for the ultrasonic technique employed. The ultrasonic test is by means of a pulse-echo instrument.

Forgings must be inspected ultrasonically and their surface inspected by the magnetic-particle method. Forgings of nonmagnetic material must be examined on the surface by the liquid-penetrant method. Castings must be examined by radiographing, ultrasonic, magnetic-particle, and liquid-penetrant methods.

Pipes, tubes, and fittings must be examined over their full length by radiographic, ultrasonic, magnetic-particle, liquid-penetrant, or eddy-current methods, based on the material to be examined. Bolts and bolting material must be examined by the wet magnetic-particle method or by the liquid-penetrant method. Carbon steel, alloy steel, and chromium-alloy steels must be subjected to impact tests to determine ductility and brittleness.

During the fabrication phase, it is important to keep complete and detailed records of all the inspections, to be sure that the inspections were made at the proper time, in the proper sequence and not after the fact, and on all material released in fabrication.

Testing Vessel test plates must be made. The welding procedure and the nondestructive examination methods used shall be the same as those used in the fabrication of the vessel. Important test specimens must be made. A final hydrostatic test of 1¼ times the design pressure is stipulated. Pneumatic tests may be used in lieu of the hydrostatic test.

Marketing, Stamping, and Reports Design calculations and specifications must be filed with the state enforcement authority responsible, at the point of the installation, for the vessel. The design specifications must be certified as to compliance on vessel classification, detailed report on operating conditions for a complete evaluation of the design, and construction and inspection according to Code rules by a registered professional engineer experienced in pressure-vessel design. All necessary data sheets must be included. The vessels to be marked with the "N" symbol must be stamped with the following information: (1) class of vessel; (2) manufacturer; (3) design pressure at coincident temperature; (4) manufacturer's serial number; and (5) year built.

The ASME Code for nuclear power plant components now requires certificates of authorization to be obtained and retained by the following organizations that may be active in nuclear power plant work: manufacturers, engineering organizations, fabricators, installers, material manufacturers and suppliers, owners and agents of a proposed plant. The aim and purpose of certification is to have a quality assurance program from all accredited organizations that have met certain minimum Code standards. The ultimate aim is that all work performed will be conducted in rigid compliance with approved and controlled specifications and procedures, that critical activities will be verified by qualified personnel from an organization independent of work performance (third-party inspection program), that measurement and tests results will be completely documented and analyzed for conformance to established specifications and Code requirements, and that appropriate actions will be taken to preclude the recurrence of discrepancies and deficiencies. Most certificate holders are required to have a quality assurance manual. This document must be kept current and must explain in detail the controlled manufacturing system that is being followed in order to achieve Code compliance with specifications.

IN-SERVICE INSPECTION

The primary objective of Section XI of the ASME Code is to provide a means of ensuring that the mechanical integrity of the primary coolant system is maintained throughout the operating life of the facility. This objective is accomplished through the requirement for the conductance of minimum periodic inspections of critical nuclear components, such as Class 1 vessels and their weld zones and other highly stressed areas. A preservice inspection is required. Usually ultrasonic inspection techniques are used that can be duplicated later during required reinspections. This permits comparisons to be made to note changes. Automatic inspection devices are utilized as much as possible. The use of remotely operated inspection equipment under properly planned procedures has reduced the radiation hazard to examining and inspection staffs. The frequency of inspections and acceptable criteria are detailed in the ASME Code for nuclear power plant components. Fracture mechanics is used to analyze the seriousness of flaws where no Code standards exist.

Fire protection in nuclear power plants has received major attention. Fire-resistive barriers of not less than 3-hr rating are required to separate the following areas in a nuclear plant: administration building, battery rooms, boilers used for starting, cable penetrations, cable shafts, cable tunnels, computer rooms, control building or room, decontamination areas, fire pump rooms, switch gear room, and so on.

The technology exists to dispose of high-level nuclear waste, but approval by the federal government is required. Under design consideration is a system that converts high-level liquid waste to a solid and immobilizes this solid in glass for burial in deep, underground salt caves in remote areas away from the human environment.

Operator qualification and training are governed and determined by federal requirements. The criteria for the selection and training of nuclear power plant personnel are contained in American National Standard 3.1-1978, entitled "Selection and Training of Nuclear Power Plant Personnel." Most nuclear plants require a minimum presence during operation of at least one senior reactor operator, two reactor operators, and two auxiliary operators who may not be licensed as yet. Operator training is receiving increased attention, as is the need for supporting staff such as a safety engineer to supplement the operating staff in case an emergency situation develops in operation. The many valves and subsystems in a nuclear power plant as well as electrically activated controls require a broad knowledge of the interplay between systems. See Fig. 8-5. The many pumps and valves shown can lead to errors of valve opening and closing and also increase the possibility of electrical and mechanical failures. Complete reliance on automatic redundancy systems cannot anticipate all the possibilities of malfunction or sense that a malfunction is taking place; therefore, skilled operators well versed and trained in the design and operation requirements of nuclear power plants are needed.

Nondestructive testing personnel must be qualified, and these qualifications are graded as follows:

A Level 1 person must have experience or training in the performance of the inspections and tests that he or she is required to perform. This person should be familiar with the tools and equipment to be employed and should have demonstrated proficiency in their use. She or he must be familiar with inspection and measuring equipment calibration and control methods and be capable of verifying that the equipment is in proper condition for use.

A Level 2 person must have experience and training in the performance of required inspections and tests and in the organization and evaluation of the results of the inspections and tests. This person must be capable of supervising or maintaining surveillance over the inspections and tests performed by others and of calibrating or establishing the validity of calibration of inspections and measuring equipment. She or he must have demonstrated proficiency in planning and setting up tests and must be capable of determining the validity of test results.

A Level 3 person must have broad experience and formal training in the performance of inspections and tests and should be educated through formal courses of study in the principles and techniques of the inspections and tests that are to be performed. This person should be capable of planning and

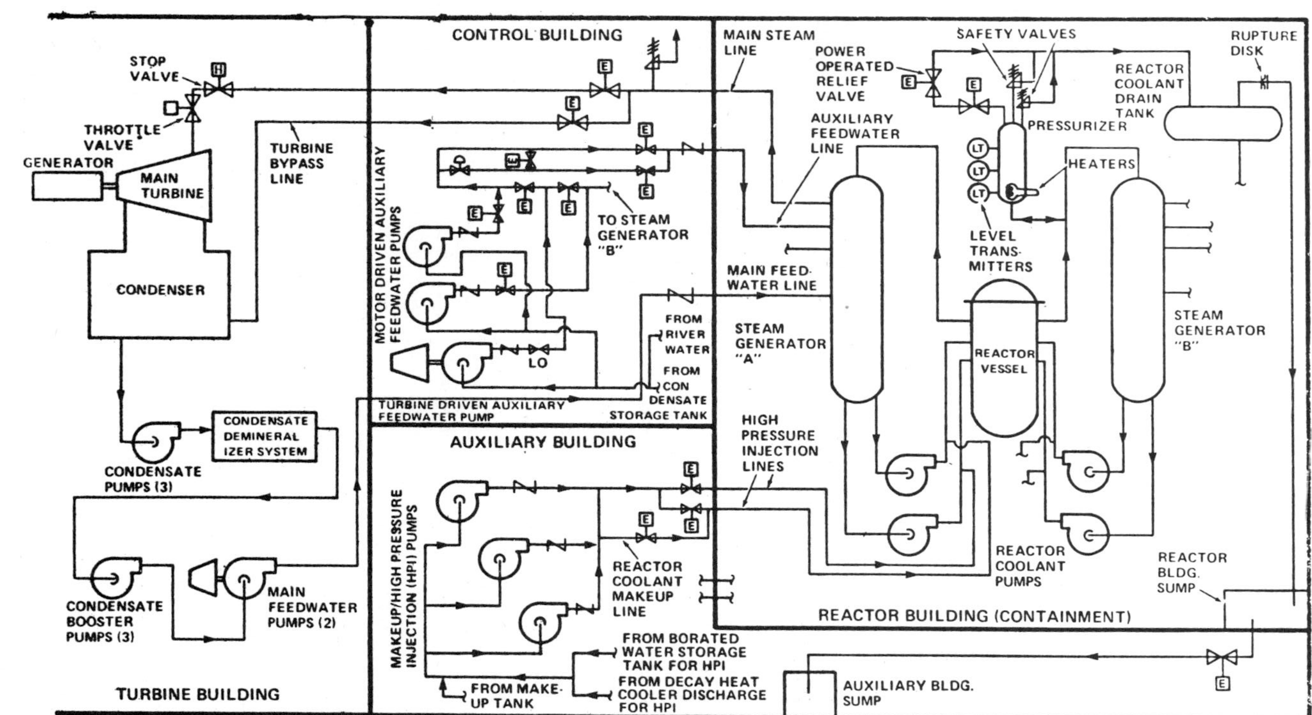

Fig. 8-5 A multitude of valves and pumps exist in a PWR nuclear power plant.

Table 8-1 Minimum Levels of Capability for Project Functions

Project function	Level 1	Level 2	Level 3
Approve inspection and test procedures			X
Implement inspection and test procedures	X		
Evaluate inspection and test results		X	
Reporting of inspection and test results		X	

supervising inspections and tests, reviewing and approving procedures, and evaluating the adequacy of activities to accomplish objectives. He or she must be capable of organizing and reporting results and of certifying the validity of results.

Personnel involved in the performance, evaluation, or supervision of nondestructive examinations, including radiography, ultrasonic, penetrant, magnetic-particle, or eddy-current methods, must meet the Level 3 qualification specified in SNT-TC-1A of the American Society for NDT and supplements. Those personnel involved in the performance, evaluation, and supervision of gas-leak test methods must meet the qualification requirements specified for a Level 2 person.

Personnel who are assigned the responsibility and authority to perform project functions must have as a minimum the level of capability shown in Table 8-1. When inspections and tests are implemented by teams or groups of individuals, the one responsible must participate and must meet the minimum qualifications indicated.

A file of records of personnel qualifications must be established and maintained by the owner. This file should contain records of past performance, training, initial and periodic evaluations, and certification of the qualifications of each person.

Questions and Answers

8-1 How is radioactivity detected?

Ans. The instruments commonly used are Geiger-Müller counters, scintillation counters, gamma survey meters, and proportional counters. The instrument to be used depends on the type and density of radiation to be measured. Geiger-Müller counters are used to detect beta and gamma radiation and are not effective for measuring alpha radiation.

8-2 What is radiation as applied to nuclear plants?

Ans. Unstable isotopes of certain chemical elements undergo spontaneous change in the atomic structure of the element. This change is called radioactive decay. While this phenomenon is going on, rays of energy are emitted; thus the term *radiation of particles* describes the radiation that is experienced in a nuclear power plant, primarily under controlled conditions in the reactor pressure vessel.

8-3 How is an isotope defined?

Ans. Isotopes are elements that have the same number of protons, but differ in the number of neutrons in their nucleus. Isotopes of an element have the same chemical properties and somewhat the same physical properties, but have different atomic weights as a result of the difference in the number of neutrons in the nucleus.

8-4 What is half-life of a radioactive substance?

Ans. *Half-life* is defined as the time required for a radioactive substance to reach one-half of its radioactive emitting strength.

8-5 What units are used to measure radioactivity?

Ans. The roentgen (R) is the unit most frequently used to measure the quantity of radiation emanating from a body. It is primarily used to measure x-rays and gamma rays. One roentgen is the production of 2.58×10^{-4} coulombs per kilogram (C/kg) of air. Exposure tolerances are usually expressed in milliroentgen (mR), or one-thousandth of a roentgen. The radioactivity of material undergoing radioactive decay is expressed in curies (Ci), which is equal to 3.7×10^{10} atomic disintegrations per second.

8-6 How are alpha, beta, and gamma radiation defined?

Ans. Alpha-radiation particles consist of two protons and two neutrons. This makes them identical to the positively charged nucleus of the helium atom. They cannot penetrate human skin, but are dangerous to human health if inhaled or ingested into the body. This could occur from inhaling alpha-bearing dust or eating alpha-contaminated food or water.

Beta particles are high-speed electrons with penetrating power sufficient to go through aluminum up to 1 in. thick. Thus they are a health hazard to the entire human body.

Gamma radiation is an electromagnetic-type ray similar to light, radio waves, and x-rays with a speed approaching that of light. Its penetrating power can reach up to 3 ft of concrete. This radiation is the most dangerous to human health in the operation of nuclear power plants.

8-7 Name some typical methods employed to protect workers from excessive radiation doses.

Ans. Some typical factors considered in the protective scheme are generally the following: (1) Control the length of an exposure; (2) control the distance between the human body and the radiation source; (3) provide shielding between the body and the source of radiation: (4) establish a strict and precise radiation-monitoring program in the work area; and (5) have medical facilities available at all times to handle any accidental exposures above normal stipulated levels.

8-8 What procedures are generally followed in decontaminating radioactive material?

Ans. Radioactive material cannot be destroyed. Therefore the process of decontamination involves one of the following methods of lessening the hazard of radioactivity: (1) Isolate the area of radioactive contamination until such time as the radioactivity is decreased to a safe level as a result of radioactive decay. The half-life of the contaminant will influence this procedure. (2) Treat the surface so that the radioactive material is absorbed, cleaned off, swept, etc., and then taken to a site where the radioactive substance will not harm people. The most difficult decontamination is where the contaminant is absorbed into porous material such as concrete. Complete removal of walls, floors, and similar contaminated areas may be necessary under this type of contamination.

8-9 What is a thermal shield in a nuclear reactor?

Ans. The thermal shield usually consists of iron or steel plates surrounding the core inside the reactor vessel, so as to conserve heat and reduce the temperature and thermal stresses in the wall of the reactor vessel. The inner layer of the shield next to the core is subject to intense neutron and gamma radiation, which is converted to heat. The thermal shield often is cooled by circulating water.

8-10 What is a biological shield?

Ans. The biological shield consists of high-density concrete or lead plates around the reactor. It prevents the escape of neutrons and radiation out of the vessel wall so as to protect personnel. Radiation is present in operation and shutdown periods because of the radioactivity of the fuel elements.

8-11 What is a containment vessel?

Ans. To prevent the release of fission products in case of a fuel meltdown or explosion of the primary reactor, most power reactors are housed in steel airtight containers. They are not normally pressurized, but are designed to withstand the maximum pressure or shock waves that could develop as a result of an accident. The primary radioactive components are usually inside the containment vessel.

8-12 Name the types of reactors possible, based on coolant used.

Ans. From the many possible arrangements of fissionable and fertile fuel, moderator, and coolant that can constitute a chain-reacting system, six types have emerged as principal contenders for full-scale electric power generation: (1) pressurized water; (2) the closely related boiling water; (3) sodium-cooled, graphite-moderated; (4) gas-cooled; (5) heavy-water-cooled; and (6) organic-cooled, heavy-water-moderated reactors.

8-13 What are a shim rod and scram rod as applied to nuclear plants?

Ans. A shim rod is a control rod used for making coarse adjustments in the reactivity of a reactor's chain reaction, whereas a regulating rod makes fine adjustments in the reactivity. Reactor control is also achieved by varying the liquid level for those reactors that use a liquid as a moderator of reactivity.

A scram rod is a safety rod that is capable of shutting down a reactor very quickly in the event the shim or control rods fail to control the reactivity within prescribed limits.

8-14 What is meant by the term *poison?*

Ans. The term *poison* applies to fission products in the fuel elements of a reactor that absorb neutrons and thus affect the reactivity of the reactor. The two most prominent fission products considered poisonous are xenon 135 and iodine 135. These are produced when a reactor uses uranium. The poison formed as fission products eventually reduces the output of the reactor; as a result, the fuel elements are spent, which eventually requires the reactor to be refueled. Poison can also be injected into a reactor to scram it or shut it down under critical emergency conditions.

8-15 What is fertile material?

Ans. This is a material that is capable of capturing a neutron and then becoming fissionable. Another term for converting fertile material to fissionable material is *breeding.* As an example, thorium 232 can be converted to uranium 233 by nuclear bombardment.

8-16 How is spent fuel processed?

Ans. The spent fuel is shipped to a reprocessing plant. The fuel cladding is chopped open to leach out uranium, plutonium, and other fission products. Uranium and plutonium are separated and put back into the fuel cycle for reuse. The remaining components are strontium 90 and cesium 137, with half-lives of approximately 30 yr. Some plutonium with a half-life of 24,000 yr is also present. The liquid waste is converted to solids and then canned in leakproof containers for eventual shipping to a federal deposit site under perpetual government control.

8-17 What is a digital audible dosimeter?

Ans. It is an instrument that provides an immediate measurement of the radiation exposure in the area as well as audibly alarming personnel that a radiation exposure exists. Set points for alarming are possible.

8-18 What are the chief functions of an authorized inspection agency in nuclear power plant work?

Ans. 1. To participate in the ASME surveys of any organization that has a nuclear code stamp and for which the authorized inspection agency will provide the Component Code inspection service.

2. To maintain a qualified staff of nuclear inspection supervisors who will monitor the shops with which an inspection agreement has been made to provide Code-required inspections.

3. Provide documented instructions to the authorized nuclear inspectors of the inspection agency on procedures to follow during routine performance and when assistance or guidance may be needed from their supervisors in order to comply with nuclear requirements.

8-19 Name the chief functions of a nuclear inspector doing work per ASME Nuclear Code requirements.

Ans. 1. Verify that work is being done by manufacturers or installers who have the ASME certificate of authorization for the work to be done.

2. Monitor the contractor or manufacturer's quality assurance program in order to note if it is being followed.

3. Verify that all material being used meets Code requirements for data, stamping, and traceability.

4. Verify that all welding is performed to qualified procedures and that the welders employed are Code-qualified.

5. Maintain a written record of inspection activity.

6. Verify that Code design calculations have been made and are available for review.

7. Verify that required heat treatments have been performed properly.

8. Verify and interpret as necessary the required nondestructive examinations required by the Code.

9. Witness required hydrostatic or pneumatic tests.

10. Witness and verify nameplate stamping as complying with specifications and Code requirements.

11. Sign the necessary documents as required by the Code to show that the component meets Code requirements.

8-20 To what part of pumps does the ASME Code on design of nuclear pumps address itself?

Ans. The ASME rules for nuclear pumps address pump casings, inlet and outlets of pumps, covers, clamping rings, seal housing, related bolting, internal pump coolers, nozzles attached to pumps, related piping, and supports.

8-21 How does the ASME define an owner of a nuclear power plant?

Ans. The Code recognizes that in the utility field there are many operating companies of facilities. The word *owner* is stipulated as being an organization that is responsible for the operation, maintenance, safety, and power generation of the nuclear plant.

8-22 To what standard must personnel performing NDT examinations be qualified?

Ans. They must be qualified for the technique and method to be used in accordance with standard SNT-TC-1A entitled "Recommended Practice for NDT Personnel Qualification and Certification" of the American Society for NDT. If this standard does not cover the proposed examination, then the NDT personnel must be qualified for the particular examination by the manufacturer or contractor.

8-23 Who audits the operators of NDT examinations?

Ans. The authorized inspector has to verify that NDT operators have been certified in accordance with SNT-TC-1A. The inspector has the prerogative to audit the NDT examination program in effect at the site. When the authorized inspector has reason to question the performance of an NDT operator, she or he can request requalification of the operator in the written procedure required for the particular examination.

8-24 How long must welded joints of a nuclear component of the Class 1 type remain exposed?

Ans. They must be left uninsulated and exposed for examination by the authorized inspector for leaks during the required hydrostatic test.

8-25 What hydrostatic test pressure is required on Section III, Class 1 components?

Ans. Completed components and appurtenances except brazed joints, pumps, and valves must be subjected to a hydrostatic test pressure not less than 1.25 times the system design pressure. Brazed joints, pumps, and valves must be hydrostatically tested at a pressure 1.5 times the system design pressure.

8-26 How is the total relieving capacity of Class 1 components' relief devices specified?

Ans. The total relieving capacity of the pressure-relief devices must be adequate to prevent a rise in pressure above 10 percent of the system design pressure and temperature within the pressure-retaining boundary of the nuclear system.

8-27 Besides spring-loaded valves, what other relief valves may be used on Class 1 systems?

Ans. Pilot-operated relief valves and power-actuated pressure-relief valves, provided they meet installation requirements of the Code.

8-28 What are considered flaw indications when one is performing NDT examination during in-service inspections of a nuclear power plant?

Ans. Evidence or signals from instruments that reveal the following: cracks, slag inclusions, segregates in the material, aligned or clustered porosity, lack of weld penetration, and plate laminations.

8-29 Differentiate between surface examination and volumetric examination.

Ans. Surface examination includes the liquid-penetrant method and magnetic-particle examination. It is used to verify surface or near surface cracks or discontinuities. Volumetric examination is for the purpose of detecting subsurface discontinuities to the extent that the entire volume of metal below the metal surface may be examined. The two principal examination methods used are radiographic examination and ultrasonic testing.

8-30 Name the five main components of a PWR nuclear power plant.

Ans. (1) Nuclear reactor; (2) steam generators or heat exchangers; (3) Piping to circulate the coolant between the reactor and the steam generators; (4) pumps to circulate the coolant; and (5) a pressurizer to prevent the coolant from boiling and to permit volume changes in the coolant without excessive pressure rise with coolant temperature changes.

9
General Construction and Stress Calculations

Chapter 4 reviewed some basic strength-of-material principles and the basic stresses that a material under load must resist. This chapter is devoted to ASME Code strength calculations as applied to high-pressure boiler components. Equations expressing the strengths of shells, stayed surfaces, and like components of low-pressure boilers and pressure vessels are similarly calculated, but specific equations are available in the appropriate section of the ASME Code. The boiler-strength calculations demonstrate methods on how to do this. It must be recognized that new materials, better manufacturing quality control, and increased knowledge of how materials behave under load have caused Code changes in the equations and allowable stresses. Therefore, the latest ASME-published Code should be consulted for specific design details.

Boiler components generally can be placed in the following categories for stress calculations: (1) tubes and pipe; (2) shells and drums; (3) heads; (4) braced and stayed surfaces; (5) furnaces and flues; (6) reinforced openings; and (7) miscellaneous.

BOILER TUBES

Three common methods of boiler-tube fabrication are used. (1) The seamless tube is pierced hot and drawn to size. (2) The lap-welded (forge-welded) tube consists of metal strip ("skelp") curved to tubular shape with the longitudinal edges overlapping. Heat is applied and the joint forge-welded. (3) The electric-resistance butt-welded tube is formed like the second type, but as its name implies, the joint is butt-welded.

It is considered good practice by some to place the weld on welded tubes away from the radiant heat of the fire. Tubes for bent-tube-type boilers are bent usually by machine.

The diameter of boiler tubes always refers to nominal outside diameter while pipe diameter refers to nominal inside diameter.

Tube Ends: Expanding, Flaring, Beading Practically all boiler tubes have the ends expanded into the tube hole of the shell or drum. This is to make the tube tight against leakage and to give it a firm grip on the tube hole so that the tube may have a definite holding or staying effect.

The edges of the tube holes are chamfered about 1⁄16 in. after the holes are drilled so that there will be no sharp edges to cut into the tube when it is expanded.

Tube holes are finished 1⁄32 in. larger in diameter than the outside diameter of the boiler tube, except in the tube sheet of fire-tube boilers. Through this, tubes must be drawn during retubing, and therefore its holes are finished 1⁄16 in. larger in diameter so as to permit a tube that is coated with scale to be removed without damage to the tube sheet.

Thick drums may be counterbored in order to have a reasonably narrow circumferential strip of tube to expand. The diameter of the counterbore should be sufficient to allow for flaring the tube end according to requirements.

The counterbore may be from either the inside or the outside. When a drum of a watertube boiler has tubes expanded in its upper side, it is best practice not to use the outside counterbore, for pockets for soot would thus be formed.

For watertube boilers the tubes and nipples should extend through the tube hole ¼ to ½ in. and be flared to at least ⅛ in. larger than the tube-hole diameter.

Fire-tube boilers have the tube ends exposed to heat and products of combustion, and therefore the tube ends might soon be burned off if they were flared. In these boilers the tube ends are driven back into a bead after expanding the tubes, in order to protect them against overheating, although the bead does not increase the holding power of the tube appreciably.

To calculate the allowable pressure for tubes in watertube boilers up to 5-in. OD, the following Code equation is used:

$$P = S\left[\frac{2t - 0.01D - 2e}{D - (t - 0.005D - e)}\right]$$

or

$$t = \frac{PD}{2S + P} + 0.005D + e$$

where

P = maximum allowable pressure, psi
D = outside diameter of tubes, in.
t = minimum required thickness, in.

S = maximum allowable stress, lb/in.^{2}
e = thickness factor for expanded tube ends

Note: For selecting the S value of *tubes,* the operating temperature of the metal shall be not less than the maximum expected mean wall temperature (the sum of the outside and inside surface temperatures divided by 2) of the tube. This in no case shall be taken as less than 700°F for tubes absorbing heat. For tubes which do not absorb heat, the wall temperature may be taken as the temperature of the fluid within the tube, but not less than the saturation temperature.

Note: Over a length at least equal to the length of the seat, $e = 0.04$ plus 1 in. for tubes expanded into tube seats. However, $e = 0$ for tubes expanded into tube seats, provided that the thickness of the tube ends over a length of the seat plus 1 in. is *not* less than the following:

0.095 in. for tubes 1¼-in. OD and smaller
0.105 in. for tubes above 1¼-in. OD and up to 2-in. OD
0.120 in. for tubes above 2-in. OD and up to 3-in. OD
0.135 in. for tubes above 3-in. OD and up to 4-in. OD
0.150 in. for tubes above 4-in. OD and up to 5-in. OD
For tubes strength-welded to headers and drums, $e = 0$.

Example: A seamless steel tube (SA-210 A-1) of 2¼-in. outside diameter, and 0.188 in. thick operates at a mean tube temperature of 650°F and is in a zone where it absorbs heat. What is the allowable working pressure of this tube when expanded into a drum of a watertube boiler? The allowable stress at 700°F is 14,400 lb/in.^{2}.

$$P = 14{,}400 \left\{ \frac{2(0.188) - 0.01(2.25)}{2.25 - [0.188 - 0.005(2.25)]} \right\} = 2455 \text{ psi}$$

For fire-tube boilers using the common SA-83 and SA-178 tube material,

$$P = 14{,}000 \left(\frac{t - 0.65}{D} \right)$$

where

P = maximum allowable pressure, psi
t = minimum required thickness, in.
D = outside diameter of tube, in.

For fire-tube boilers using copper tubes of SB-75 specification,

$$P = 12{,}000 \left(\frac{t - 0.59}{D} \right) - 250$$

Example: What pressure is allowed on a steel tube of SA-192 material, expanded into tube seats, 3 in. in diameter, with wall thickness 0.115 in., to be used in a fire-tube boiler?

$$P = \frac{14{,}000(0.115 - 0.065)}{3.0} = 233 \text{ psi}$$

SHELLS AND DRUMS

To calculate the allowable pressure on shells or drums, two factors must be taken into consideration:

1. Longitudinal-weld joint efficiency. If all weld reinforcement is removed flush with the plate, a 100 percent efficiency can be used; otherwise, a joint efficiency of 90 percent must be used. If the boiler is of riveted construction, the appropriate riveted joint efficiency is used as illustrated in App. 3. If the shell is seamless, a 100 percent efficiency is used. See Fig. 9-1*a*.
2. Since watertube boilers are arranged so that the drums have tube holes in them, the weakening effect is calculated in terms of a ligament efficiency, and this efficiency is used with the corresponding shell thickness to obtain an allowable working pressure for the shell or drum by the tube-hole area.

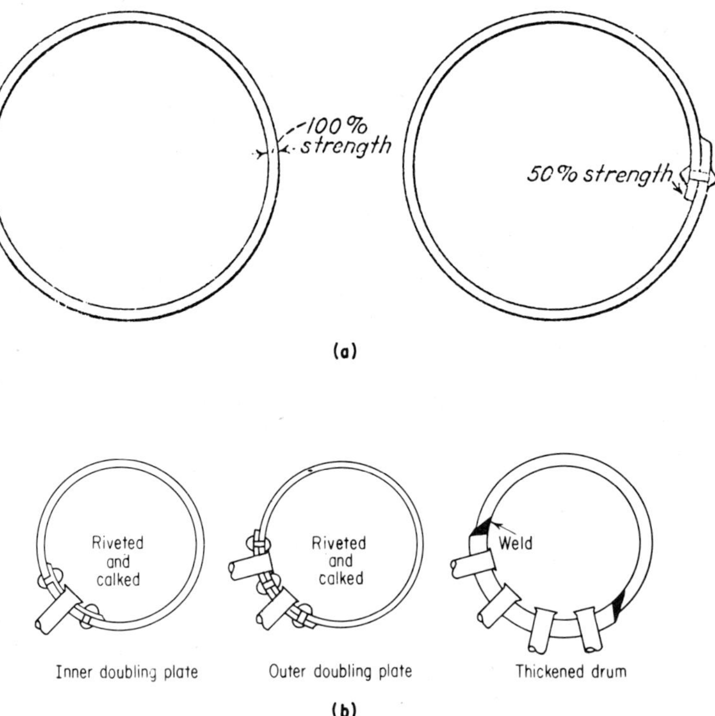

Fig. 9-1 Efficiency of joints is affected by construction: *(a)* seamless cylinder had 100 percent joint efficiency compared to 50 percent for a lap joint. *(b)* Methods of reinforcing the drum to strengthen the ligament area.

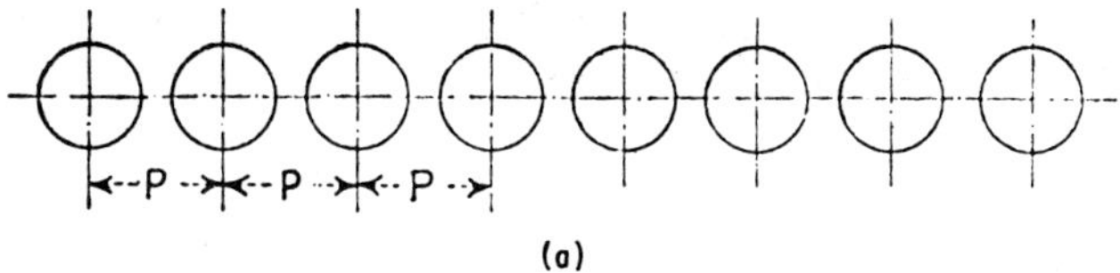

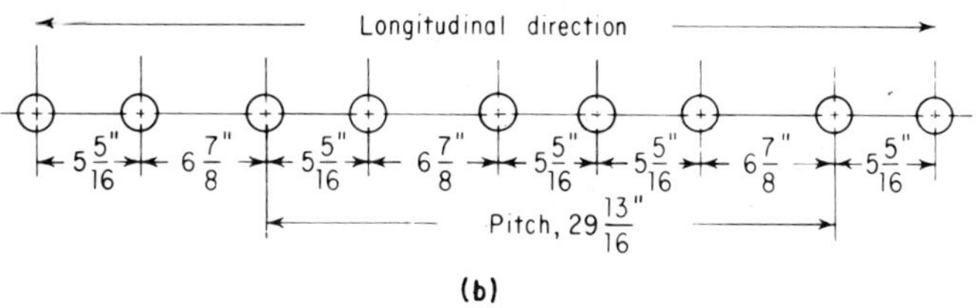

Fig. 9-2 Determining pitch to calculate ligament efficiency: *(a)* equal tube diameter, holes spaced equally; *(b)* unequal tube-hole spacing and corresponding pitch.

The efficiency of a ligament is found as follows: Where the pitch between the tube holes is equal (Fig. 9-2*a*),

$$\frac{p-d}{p}$$

where

p = pitch, or longitudinal distances between holes, in.
d = tube-hole diameter, in.

If the pitch of the tube holes is unequal (Fig. 9-2*b*), a unit longitudinal length in which all the unequal pitches are included should be selected. Then,

$$\frac{P-nd}{P} = \text{efficiency of the ligament}$$

where

P = length selected to include all variations in pitches, in.
d = tube-hole diameter, in.
n = number of tube holes in longitudinal line in selected length

The ASME Code covers methods of finding efficiency of diagonal ligaments by means of charts.

The efficiency of the usual tube ligament is quite low, usually 35 to 50 percent. On riveted boilers often the ligament was strengthened by riveting a reinforcing strap or doubling plate over the tube-hole section. The tube holes are cut through the entire section; thus, although the efficiency of the ligament is not increased because the tube spacing remains unchanged, the thickness is increased. See Fig. 9-1*b*.

Another method of increasing the strength of the ligament is to make the drum in two longitudinal halves with two longitudinal-welded seams. One half will be somewhat thinner than the other half that contains the ligament. The edges of the thicker half are machined down to the same thickness as that of the other for abutting edges of the longitudinal seams to be welded.

Two equations are provided by the ASME Code to calculate the allowable pressure of a shell or drum. First, where backing strips on the inside of drums are not removed on longitudinal-welded joints or where a drum has material that may be exposed to creep and rupture-strength considerations and all weld reinforcement is not removed, and for riveted construction, use

$$P = \frac{0.8SEt}{R + 0.6t} \quad \text{or} \quad t = \frac{PR}{0.8SE - 0.6P}$$

Second, for welded or seamless construction, with the weld reinforcement removed substantially flush and the backing strip removed, use

$$P = \frac{SE(t - C)}{R + (1 - y)(t - C)} \quad \text{or} \quad t = \frac{PR}{SE - (1 - y)P} + C$$

where

P = maximum allowable pressure, psi
S = maximum allowable stress for operating temperature of metal, lb/in.^{2}
t = minimum required thickness, in.
R = inside radius of cylinder, in.
E = efficiency of joint (E = efficiency of longitudinal welded joints or of ligaments between openings, whichever is lower. E = 1.00 for seamless cylinders, and E = 1.00 for welded joints, provided all weld reinforcement on the longitudinal joints is removed substantially flush with the surface of the plate. E = 0.90 for welded joints with the reinforcement on the longitudinal joints left in place. E = efficiency for riveted joints, E = efficiency for ligaments between openings.)
C = factor depending on whether a pipe or shell is threaded (for usual drum or shell $c = 0$; see Code for specifics)
y = factor which depends on whether steel is ferritic or austenitic as follows:

	y value	
Temperature, °F	*Ferritic steel*	*Austenitic steel*
Below 900	0.4	0.4
950	0.5	0.4
1000	0.7	0.4
1050	0.7	0.4
1100	0.7	0.5
1150	0.7	0.7

Example: A riveted boiler is 66 in. in diameter and 16 ft long. The shell plate is $\frac{7}{16}$ in. thick, the allowable stress is 13,750 lb/in.^{2}, and the allowable pressure is 125 psi. What is the least permissible circumferential efficiency of the girth joint?

Using the formula

$$P = \frac{0.8SET}{R + 0.6t}$$

$$125 = \frac{0.8(13{,}750)(E)(0.4375)}{33 + 0.6(0.4375)}$$

$$E = \frac{125(33.26)}{11{,}000(0.4375)} = 86.6\%$$

$$\text{Circumferential } E = \frac{86.6}{2} = 43.3\%$$

The circumferential strength of the shell must be always at least half that of the longitudinal joint. This was illustrated in Chap. 4.

Example: What is the maximum allowable working pressure on a welded drum that has 1.469-in.-thick shell plate and 2.406-in.-thick tube sheet? The outside diameter of the shell plate is 57.75 in. The outside diameter of the tube sheet is 58.688 in. Material is SA-515-70, and the metal temperature is not more than 650°F. Tube ligament efficiency is 0.429 and welded-joint efficiency is 100 percent. Allowable stress for this ferritic steel is 17,500. Calculations for allowable pressure must be made twice as follows:

Based on welded joint and shell-plate thickness:

$$P = \frac{SE(t - C)}{R + (1 - y)(t - C)}$$

$$R = \frac{57.75}{2} - 1.469 = 27.406$$

$$y = 0.4 \qquad C = 0$$

So

$$P = \frac{17{,}500(1)(1.469)}{27.406 + 0.6(1.469)} = 908.8 \text{ psi}$$

Based on the tube-sheet thickness and ligament efficiency:

$$R = \frac{58.688}{2} - 2.406 = 26.938$$

The same equation is used:

$$P = \frac{17{,}500(0.429)(2.406)}{26.938 + 0.6(2.406)} = 636.1\text{-psi allowable pressure}$$

The lowest pressure governs.

Example: The mud drum of a watertube boiler has a 42-in. ID. The tube sheet is ⅝ in. thick and contains 3⁹⁄₃₂-in.-diameter tube holes pitched horizontally 5⁵⁄₁₆ in., in banks of three (Fig. 9-2b) and two tubes with 6⅞ in. between banks. The shell plate is ½ in. thick. The joint efficiency between the tube sheet and shell is 67 percent. What is the allowable working pressure for this drum if the material is SA-285C and the maximum temperature is 650°F?

Allowable stress for SA-285C material is 13,750 lb/in.². This is a riveted boiler, as can be noted by joint efficiency.

The allowable pressure based on shell thickness and joint efficiency is

$$R = 21 \text{ in.} \qquad t = 0.5 \text{ in.}$$

$$P = \frac{0.8(13{,}750)(0.67)(0.5)}{21 + 0.6(0.5)} = 173 \text{ psi}$$

In the solution based on ligament efficiency and tube-sheet thickness, ligament efficiency must be calculated first:

$$E = \frac{p - nd}{p}$$

$$= \frac{29.8125 - 5(3.28125)}{29.8125}$$

$$= 0.448$$

$$P = \frac{0.8SEt}{R + 0.6t} = \frac{0.8(13{,}750)(0.448)(0.625)}{21 + 0.6(0.625)} = 145 \text{ psi}$$

DISHED HEADS

The ends of drums of most conventional-type watertube boilers are closed with a dished-out head. The shorter the radius of the curvature, that is, the nearer the dish approaches a semispherical shape of reduced radius, the greater will be the resistance to internal pressure. Conversely, the ASME Code specifies that the radius shall not be greater than the diameter of the shell or drum to which the head is attached—otherwise, the head requires bracing.

There are four types of blank unstayed heads permitted by the Power Code: segment of a sphere, semiellipsoidal, hemispherical, and flatheads. The first three are bumped heads (Fig. 9-3*a*). Some Code flatheads and methods of attachment are shown in Fig. 9-3*b*. Bumped heads are flanged, with a corner radius on the concave side of the head of not less than 3 times the head thickness and in no case less than 6 percent of the diameter of the shell for which the heads are to be attached.

Flanged-in manhole openings in dished, or bumped, heads must be flanged to a depth of not less than 3 times the required thickness of the head for plate of up to 1½-in. thickness. If thicker, the depth of the flange must be the thickness of the plate plus 3 in. The minimum width of the bearing surface for a gasket on a manhole must be ¹¹⁄₁₆ in., and the gasket thickness when compressed must be less than ¼ in.

Dished heads are calculated for allowable pressure by the following methods,

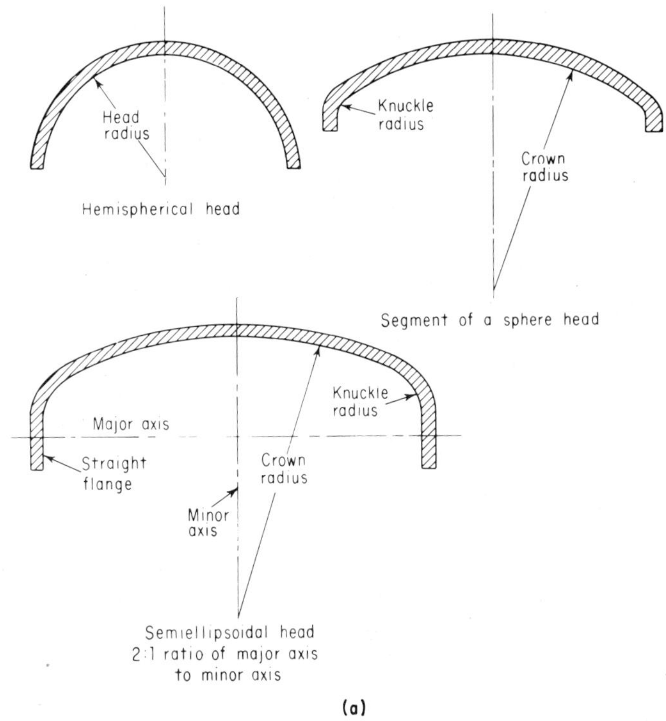

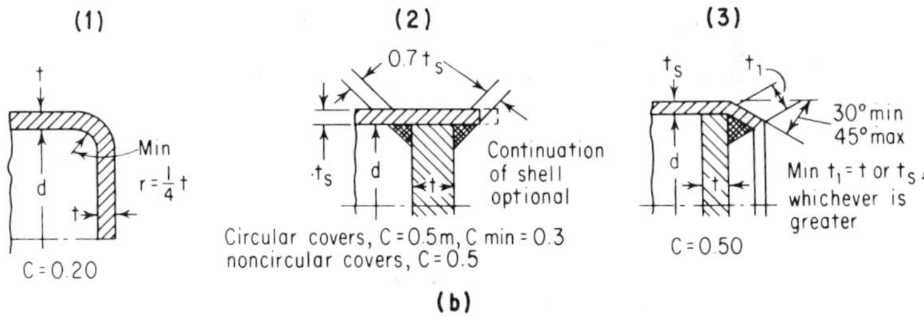

Fig. 9-3 Dished heads and flatheads. *(a)* Dished heads are hemispherical, segment of a sphere, and semiellipsoidal. *(b)* The method of flathead attachment determines the *C* factor in the flathead equation.

depending on what type of head is involved and whether it has a manhole. For segments of a sphere head without a manhole, use the formula

$$t = \frac{5PL}{4.8SE}$$

where

S = maximum allowable stress, lb/in.^{2}
t = thickness of head, in.
P = maximum allowed pressure, psi
L = radius on concave side to which head is dished, in.
E = efficiency of weakest joint, but *not* head-to-shell joint

For semiellipsoidal heads with pressure on the concave side (without a manhole), use the shell formula for cylinders, but assume that the shell is seamless (no joint efficiency). For hemispherical heads with no manhole and pressure on the concave side, use the following formula:

$$t = \frac{PL}{1.6SE} \tag{9–1}$$

or

$$t = \frac{PL}{2SE - 0.2P} \tag{9–2}$$

where

t = required thickness, in.
P = maximum allowable pressure, psi
S = maximum allowable stress, lb/in.^{2}
L = radius to which head was formed, in.
E = efficiency of weakest joint, including head-to-shell joint

Equation (9–2) is for heads integrally formed with the shell, or welded head-to-shell joints, provided that all the welding meets Code requirements, including the weld reinforcement being removed substantially flush with the plate.

When any of the heads—segment of a sphere, semiellipsoidal, or hemispherical—has a flanged-in manhole or an access opening that exceeds 6 in. in any dimension, it is computed on the following basis:

1. By the formula for a segment of a sphere head.
2. The thickness of the head must be increased by 15 percent, but in no case less than ⅛ in. after the thickness is obtained by the formula.
3. If the radius to which a head is dished is less than 80 percent of the diameter of the shell, the thickness of the head with a flanged-in manhole opening must be found (or calculated) by making the dish radius equal to 80 percent of the diameter of the shell.

At times it is necessary to calculate the dish radius based only on chord data and depth of bump. See Fig. 9-4*a*. Refer to triangle *ABO:*

R = hypotenuse of right triangle

$$R^2 = \left(\frac{C}{2}\right)^2 + (R - b)^2$$

$$2Rb = \frac{C^2}{4} + b^2$$

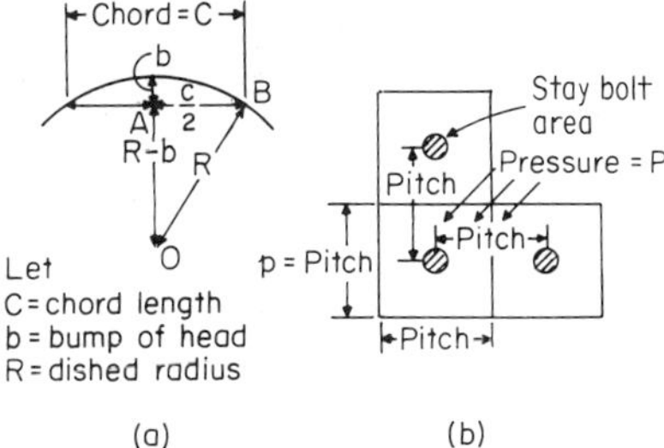

Fig. 9-4 *(a)* The dish radius can be determined by measuring chord length C and bump b. *(b)* The stay bolt must resist pressure acting on the area encompassed by the pitch dimensions.

$$R = \frac{C^2}{8b} + \frac{b}{2}$$

Example: Chord $C = 38$ in. on a head; bump $b = 4$ in. So

$$R = \frac{(38)^2}{8(4)} + \frac{4}{2}$$

$$R = 47.1 \text{ in.}$$

Examples of Dished-Head Calculations An unstayed 42-in.-diameter dished head (a segment of a sphere) with a flanged-in manhole and with pressure on the concave side is ⅞ in. thick. The radius of the dish is 36 in., and the steel is SA-285C. What pressure would be allowed on this head?

The equation for the segment of a sphere must be used, because a flanged-in manhole is in the head. Transposing the equation for P, we get

$$P = \frac{4.8SEt}{5L}$$

The allowable stress for SA-285C is 13,750 lb/in.². So

$$t = 0.875 - 0.125 = 0.75$$

(A minimum of ⅛ in. must be deducted because of the manhole.)

$$L = 36 \text{ in.} \qquad E = 1$$

Thus

$$P = \frac{4.8(13{,}750)(1)(0.75)}{5(36)} = 275 \text{ psi}$$

A fully hemispherical blank head of a 54-in.-ID drum is welded to the shell of a watertube boiler so that the pressure is applied on the concave side. Assume a working pressure of 810 psi and working temperature 650°F, the efficiency of the welded joint is 90 percent, not ground, and the steel is SA-285C. What is the required thickness of the head?

$$t = \frac{810(27)}{1.6(13{,}750)0.90} = 1.10 \text{ in.}$$

FLAT HEADS

The ASME Code has several equations for flat heads, depending on whether the head is round, rectangular, or square. For a typical round head, the equation used is

$$t = d\sqrt{\frac{CP}{S}}$$

where

t = minimum required thickness, in.
d = diameter, measured as indicated in Code
C = a factor, depending on method of attachment (Fig. 9-3*b*)
S = maximum allowable stress value, lb/in.^{2}
P = maximum allowable pressure, psi

Example: An unstayed flathead is attached as in Fig. 9-3*b*. All welding meets Code requirements. The head is circular with a 16-in. diameter and thickness of 1½ in., the material is SA-285C, the temperature is under 650°F. The shell to which the head is attached is ⅜ in., and the required shell thickness is 5⁄16 in. What is the allowable pressure on this flathead?

Using the formula below and with $S = 13{,}750$, $d = 16$, $t = 1.5$, $C = 0.5m$, $m = 0.3125/0.375 = 8.333$, $C = 0.5(0.833) = 0.4165$ or 0.417, we get

$$t = d\sqrt{\frac{CP}{S}}$$

$$1.5 = 16\sqrt{\frac{0.417P}{13{,}750}}$$

$$\left(\frac{1.5}{16}\right)^2 = \frac{0.417P}{13{,}750}$$

$$P = \left(\frac{1.5}{16}\right)^2 \frac{13{,}750}{0.417}$$

$$= 289 \text{ psi}$$

BRACING AND STAYING

The first point to remember in all problems dealing with bracing or staying is that the stress set up in a stay is due to the unit pressure in pounds per square inch acting on the area of plate supported by that stay. This *total pressure* is resisted by the internal resistance of the brace (unit stress) times the net area of the brace. These facts are the basis for all bracing formulas.

Stay bolts and stays are used in boilers to reinforce flat or other surfaces exposed to pressure loading, because the plate surfaces would have to be made too thick to resist this loading if stays or stay bolts were not used.

To calculate stay-bolt problems, it is necessary to establish the Code requirement, which states that the required area of a stay bolt at its minimum cross section shall be found by calculating the load on the stay bolt, dividing this by the allowable stress on the stay bolt, and increasing the resultant area by a factor of 1.1. See Fig. 9-4*b*. In equation form, the following can be developed to use on stay-bolt problems:

$$\frac{\text{Load on stay bolt}}{\text{Allowable stress}} = \text{resisting area of stay bolt}$$

Let

S = allowable stress, lb/in.^{2}
a = area of stay bolt (usually round), in.^{2}
p = pitch of stay-bolt spacing, in.
P = allowable working pressure, psi

Then

$$\frac{(p^2 - a)P}{S} = \frac{a - (\text{telltale hole area})}{1.1}$$

In addition to the strength of the stay bolt, the strength of the plate between the stay bolts must be adequate, or the plate might buckle between the stay bolts. The Code requires this to be checked by one of the following equations:

$$t = p\sqrt{\frac{P}{CS}} \qquad \text{or} \qquad P = \frac{t^2\,CS}{p^2}$$

where

s = maximum allowable stress, lb/in.^{2}
t = required thickness of plate, in.
p = maximum pitch, in.
P = maximum allowable pressure, psi
C = factor, depending on construction (2.1 for stays screwed through plates of not over 7/16-in. thickness; 2.2 for stays screwed through plates of over (7/16-in. thickness)

The Code has pitch and c factors depending on construction. See Fig. 9-5*a*.

Example: *(a)* What is the maximum allowable square pitch of stay bolts in the flat furnace sheet of a Code firebox boiler when the sheets are supported by 7/8-in. screwed stay bolts with 3/16-in. holes (area = 0.0276 in.^{2}) in the outer end? The stay bolts have 12-V threads per inch, and the pressure carried is 115 psi. SA-31 Grade A stay bolts and SA-285C plate under 600°F are used. Stay-bolt stress = 11,300 and plate stress allowed = 13,700.

(b). What is the least thickness of plate permissible in part *a*?

(a) a = cross-sectional area of stay bolt at bottom of threads = 0.419 in.^{2}

$$(p^2 - a)P = \left(\frac{a - 0.0276}{1.1}\right) S \text{ with } P = 115 \text{ psi}$$

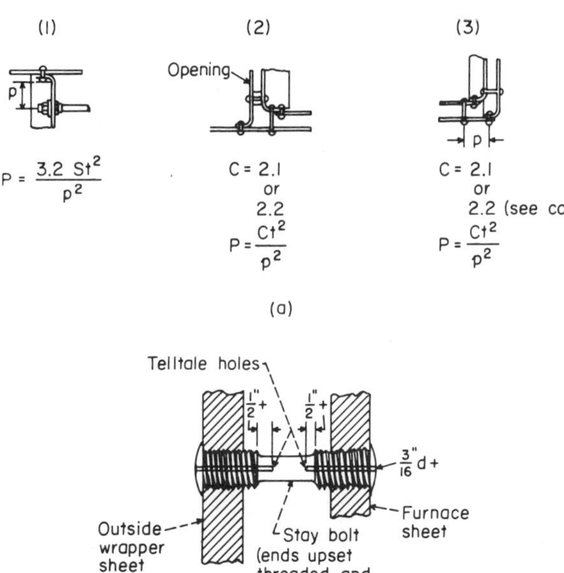

Fig. 9-5 *(a)* The pitch and *C* factor are determined by the stayed surface arrangement. *(b)* Water legs are braced with threaded stay bolts.

With SA-31 Grade A stay bolts and $S = 11{,}300$ lb/in.²,

$$(p^2 - 0.419)115 = \left(\frac{0.419 - 0.0276}{1.1}\right) 11{,}300$$

$$115p^2 = 4022.8 + 48.185$$

$$p^2 = 35.4$$

$$p = 5.95 \text{ in.}$$

(b) Use

$$t = p\sqrt{\frac{P}{CS}}$$

$$= 5.95\sqrt{\frac{115}{13{,}700(2.2)}}$$

$$= 5.95\sqrt{0.003815}$$

$$= 5.95(0.0618)$$

$$= 0.368 \text{ in.}$$

Under 7/16 in., we must recalculate with $C = 2.1$:

$$t = 5.95\sqrt{\frac{115}{13{,}700(2.1)}}$$

$$= 5.95\sqrt{0.003995}$$
$$= 5.95(0.0632)$$
$$= 0.376 \text{ in.}$$

Stay bolts may be used to stay furnaces to the outside shell or wrapper sheet (Fig. 9-5*b*). The size and the pitch of the stay bolts have much to do with the maximum allowable pressure on the boiler.

It will be noted in Fig. 9-5*b* that the ends are upset slightly so that the area of the stay bolt at the root of the threads will not be less than that of the body. However, some stay bolts are made without upsetting the ends; thus, in calculating the net area, the diameter at the root of the threads should be used.

Upset ends of stay bolts should be annealed in order to reduce any tendency to brittleness. The length of the stay bolt must be such that at least two threads extend over the plates. The ends are then riveted over.

Stay bolts sometimes break because of furnace-sheet expansion and contraction; the point of breakage is usually near the inside surface of the shell. Telltale holes are required in stay bolts not over 8 in. long, for these bolts are considered less flexible and more susceptible to breakage than the longer bolts. The telltale hole is at least 3⁄16-in. diameter and is drilled in from the outside to a depth at least ½ in. past the inner surface of the plate or, if the stay bolt is reduced in diameter, to at least ½ in. beyond this reduction. It is obvious that when the stay bolt cracks halfway through its cross section, leakage through the telltale hole should give warning.

Diagonal Stays To stay the flat portions of heads that are not supported by tubes, diagonal stays are used above the tubes. This stay is not as direct as the through-stay, and it throws stress on the shell plates as well. But the diagonal stay leaves more room above the tubes for inspection, repair, and cleaning. A common form of diagonal stay is shown in Fig. 9-6*a*. For calculating the strength of a diagonal stay, the Code requires the following to be considered:

1. What is the slant of the diagonal or its angle to the flat surface being supported?
2. Is it welded or riveted to the shell and head?
3. What is the construction on the ends of the stay where it is fastened to the shell or head: riveted, pins, split palms, or blades (crowfoot type)? See Fig. 9-6*b* and *c*.

The Code permits most diagonal stays to be calculated as straight stays similar to the stay-bolt method. This method calls for multiplying pressure times area on one side, with the holding power of the stay on the other side of the equation. For example, in Fig. 9-6*a* if the ratio of L/l is 1.15 or less (on an HRT boiler), the body of the stay is calculated as a straight stay. But the allowable stress to be used is 90 percent of that allowed for a straight stay. If L/l is over 1.15, the body of the stay is calculated by increasing the area

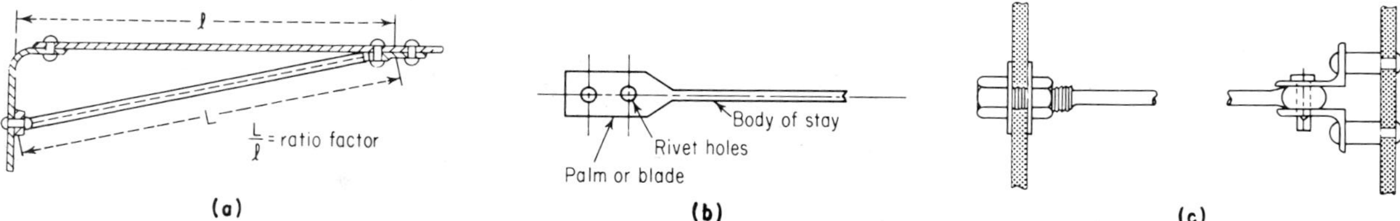

Fig. 9-6 Stays are used to brace surfaces over 8 in. apart. *(a)* Diagonal stay, *(b)* palm stay, *(c)* through-stay.

required on the body of the stay by L/l. In equation form, this is expressed as follows:

$$A = \frac{a\,L}{l}$$

where

a = cross-sectional area of direct stay body
A = cross-sectional area of diagonal stay body
l = length of right angles to area to be supported (see Fig. 9-6*a*)
L = diagonal length of stay

The Code rules on palms that are riveted on diagonal stays require the cross-sectional area of this part of the stay to be at least 25 percent greater than the body of the stay.

Example: The net area of a segment to be stayed is 504 in.^{2} and is supported by seven 1¼-in.-diameter diagonal braces (1.227-in.^{2} net area). The length of these diagonals is less than 1.15 times the length of a direct pull and does not exceed 120 diameters. SA-285C is used for weldless steel, and SA-266 Grade I is used for forge-welded steel under 400°F.

(a) What pressure is allowable if the braces are of weldless steel?
(b) What pressure is allowable if the braces are forge-welded?

(a) SA-285 Grade C material is used with an allowable stress of 13,700 for weldless construction. All temperatures are below 400°F. Thus

$$a = \text{area of brace} = 1.227 \text{ in.}^2$$
$$A = \text{area to be stayed} = 504 \text{ in.}^2$$
$$AP = 0.9(7aS)$$
$$P = \frac{7(1.227)(13{,}700)(0.9)}{504}$$
$$= 210.1 \text{ psi}$$

(b) SA-266 Grade I material is used for forge-welded steel with an allowable stress of 15,000.

$$P = \frac{7(1.227)(15{,}000)(0.9)}{504}$$
$$= 230 \text{ psi}$$

Example: The area to be stayed on the front tube sheet of a boiler is 136 in.^{2}. If it is braced with two diagonal stays of the unwelded type, what diameter of brace would be required to safely carry 165 psi, L = 29¼ in., l = 28⅝ in.? SA-285C material for stays under 600°F. The allowable stress is 13,700.

$$\frac{L}{l} = \frac{29.25}{28.625} < 1.15$$

$$AP = 0.9naS \qquad A = 136 \text{ in.} \qquad n = 2 \text{ braces}$$
$$P = 165 \text{ psi} \qquad S = 13{,}700 \text{ lb/in.}^2$$
$$136(165) = 2(13{,}700)(a)(0.9)$$

So

$$a = \frac{22{,}440}{24{,}660} = 0.91$$

Thus a 1⅛-in.-diameter brace must be used.

Staying of Tube-Sheet Segments In the HRT boiler, the segment of the tube sheets above the top row of tubes requires staying. For this, there are three common methods. In a boiler of this type that does not exceed 36-in. diameter or 100-psi working pressure, structural shapes such as angle irons or channel irons may be riveted to the segment. The outstanding legs are proportioned to have sufficient strength in bending to resist the pressure load.

For boilers exceeding 36-in. diameter or 100-psi working pressure, the flat segment of the tube sheets requires staying either by diagonal stays between the tube sheet and shell or by through-stays running the entire length of the boiler. The former are usually preferable, for they leave more room inside the boiler for cleaning and inspection. The through-stays may make it quite difficult for a worker or an inspector to move around over the tubes in the boiler.

There are three general types of diagonal stay: the Huston, the MacGregor, and the Scully. It is important to have them all under tension, and this effect is accomplished as follows: The tube sheets are marked and drilled for the crowfoot of the stays before riveting up. Each stay then has its crowfoot tack bolted to the tube sheet. The holes in the "palm" of the stay are located on the shell. The stays are then removed, and the shell is drilled about $\frac{1}{32}$ in. farther from the tube sheet than the marks. The stays then have their crowfeet riveted to the tube sheets or heads. In order that the holes in the stay palms line up with the $\frac{1}{32}$-in. offset holes in the shell, the stays are heated to expand them this amount. While hot, the palms are riveted, and on cooling, the stays contract enough to provide proper tension. Some shops do not take the trouble to heat the stay to line up the holes but "stretch" the stay into position with a driftpin.

The diagonal stays are stretched into position before the tubes are installed in the boiler. In order to prevent distortion of the head through the tension of the stays, one or more heavy steel bars are clamped across the tube sheet as a beam, the tube sheet thus being held against the tension of the stays until some of the tubes are installed. The beam is known as a *strongback*.

The first step in calculating the pressure for which a head segment is braced is to find the area that requires bracing. As specified by the ASME Code, this area is that segment enclosed by a base of a horizontal line 2 in. above the top row of tubes and a semicircular perimeter at a distance d from and parallel to the inside of the shell (Fig. 9-7a and b). Calculation of d is by the

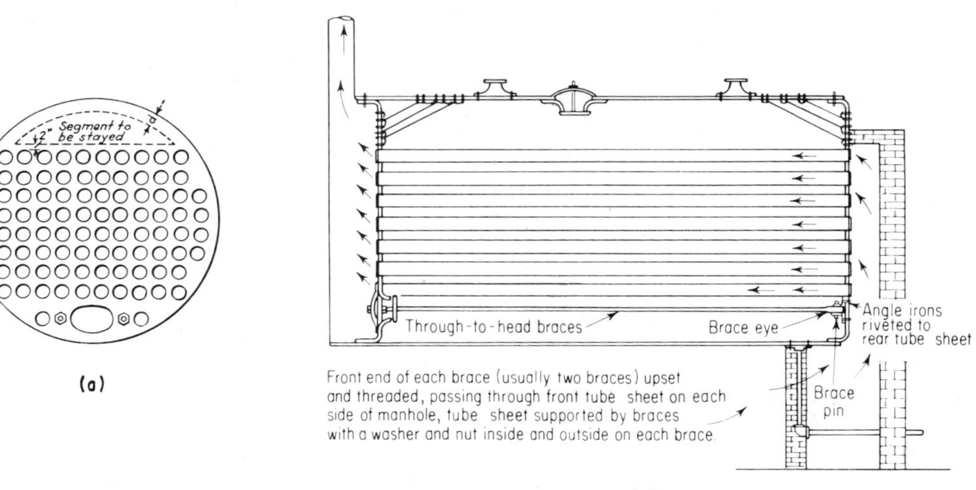

Fig. 9-7 The area above the tubes requires staying.

following methods, specified by the Code, and the larger result is used in calculating the area of the segment: (1) The outside radius of the tube-sheet flange, but not more than eight times its thickness. (2) Five times the tube-sheet thickness in *sixteenths* of an inch, divided by the square root of the maximum working pressure in pounds per square inch. This distance d is often assumed as 3 in.; a table of segment areas for boilers of various sizes is given in the ASME Power Boiler Code book, with 3 in. for d.

The distance d having been determined, it is necessary to have two more measurements for the calculation of the area to be stayed:

H = maximum distance between top row of tubes and inner surface of shell in perpendicular line at centerline of boiler, in.
R = radius of tube sheet, in.

Then,

$$\text{Area to be stayed} = \frac{4(H-d-2)^2}{3}\sqrt{\frac{2(R-d)}{H-d-2}-0.608}$$

The unit stress allowed in a weldless diagonal stay varies slightly under different codes and according to the angularity and size.

The staying of the section of the tube sheet below the tubes of HRT and similar boilers is usually effected by through-to-head stays. These stays are "spooled off" at the rear tube sheet (Fig. 9-7*b*). The front ends of the through-to-head stays—usually two in number—pass through the front tube sheet with inside and outside nuts and washers. They are usually made tight against leakage by grommets or soft metallic packing of various types beneath the outside nuts and washers.

The reason why the rear ends of the stays do not pass through the rear tube sheet but are spooled off inside is that the heat of the fire would damage the nuts and threaded ends.

If a flanged-in manhole is provided below the tubes in the front tube sheet, the stiffening effect of the flange is sufficient to allow 100-in.² deduction from the area to be stayed. However, if through-to-head stays are used, the full-sized stays required to brace the rear tube sheet must be used (unless diagonal stays between the bottom of the rear tube sheet and the shell supplement

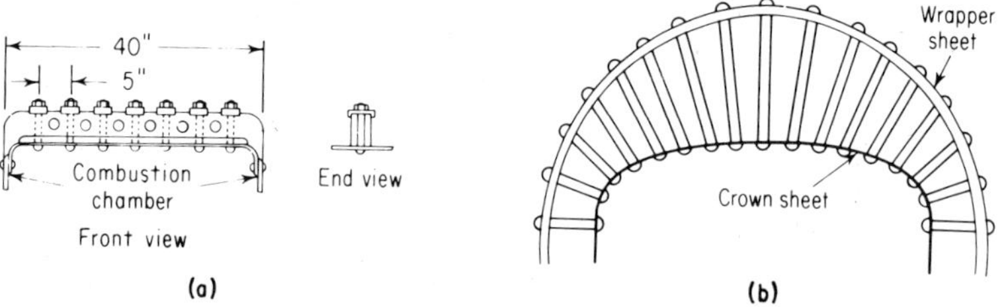

Fig. 9-8 *(a)* A girder stay is used to support the crown sheet of locomotive firebox and scotch marine boilers. *(b)* Radial stays.

the smaller through-to-head stays that would be sufficient to brace the front segment), for no deduction in area to be braced is permitted for the rear tube sheet.

Example: *(a)* A 66-in. HRT boiler is to be built for 140-psi working pressure. The flanged heads are $\frac{9}{16}$ in. thick. The distance from the upper tubes to the shell is 24 in., and $d = 3$ in. (Fig. 9-7*a*). What is the area to be stayed?

(b) If this head is to be stayed by $1\frac{1}{4}$-in.-diameter diagonal braces (weldless), how many braces will be required when L does not exceed l more than 1.15 times and 9500 lb/in.² at cross-sectional area is allowed for a straight brace?

(a)

$$A = \frac{4(24 - 3 - 2)^2}{3} \sqrt{\frac{2(33) - 3}{24 - 3 - 2} - 0.608}$$

$$= 481.3\sqrt{2.550}$$

$$= 768 \text{ in.}^2$$

(b) $768P = 0.9naS$

n = number of braces
a = area of one brace = 1.2272 in.²
S = allowable stress, lb/in.²
P = allowable pressure, psi

$$768(140) = n(1.2272)(9500)(0.9)$$

$$n = 10.2$$

Thus 11 braces must be used.

The girder stay (Fig. 9-8*a*) was formerly used very extensively to support flat crown sheets in locomotive firebox units. But it has been largely superseded by the radial stay (Fig. 9-8*b*) for this purpose. It is still used to support the tops of combustion chambers in boilers of the scotch marine type. The girder stay consists of a cast steel or built-up girder with its ends resting on the side, or end sheets, of the firebox or combustion chamber. It supports the flat crown sheet (the top of the combustion chamber) by means of bolts.

The radial stays are more flexible and tend to hold less scale from circulation than do girders. About the only advantage of the girder stays is that they pass straight through the sheet rather than at an angle. The Power Boiler Code has examples on the use of special equations for girder stays.

STAYING FURNACE AGAINST COLLAPSE

The furnace sheets of a boiler and of other internally fired boilers must resist the pressure on the external surfaces that tends to cause collapse. This collapsing tendency is resisted either by the stiffness of the furnace or by staying it to the shell with stay bolts.

A furnace not exceeding 38-in. OD may be self-supporting, and the use of stays may be eliminated provided that the thickness of the furnace is sufficient for necessary stiffness and that the span of furnace length is not too great.

The Power Boiler Code gives the following two formulas for self-supporting, unstayed circular furnaces not over 4½ diameters in length. When the length does not exceed 120 times the thickness of the furnace sheet, use

$$P = \frac{51.5(300t - 1.03L)}{D} \tag{9–3}$$

When the length exceeds 120 times the thickness of the sheet, use

$$P = \frac{1.09(10^6)(t^2)}{LD} \tag{9-4}$$

where

P = maximum allowable pressure, psi
D = outside diameter of furnace, in.
L = total length of furnace between centers of head seams, in.
t = thickness of furnace walls, in.

Example: Determine the allowable working pressure for an unstayed furnace in a vertical tubular boiler when the furnace is 26-in. OD, 7⁄16 in. thick, and 42 in. long.

Since

$$120 \times 0.4375 = 52.38$$

Eq. (9–3) applies:

$$P = \frac{51.5(300t - 1.03L)}{D} \qquad \text{where } t = 0.4375,\ L = 42,\ D = 26$$

$$= \frac{51.5[300(0.4375) - 1.03(42)]}{26}$$

$$= \frac{51.5(87.99)}{26}$$

$$= 174.3 \text{ psi}$$

Example: A plain circular furnace with 36-in. ID and 74 in. between head rivet seams is 9⁄16 in. thick. If the longitudinal joint is a single-riveted butt strap with the joint located below the grates, what is the maximum allowable working pressure on the furnace?

Since

$$120 \times \text{thickness of plate} = 120 \times 0.5625 = 67.5 \text{ in.}$$

Eq. (9–4) applies:

$$P = \frac{1.09(10^6)(t^2)}{LD} \qquad \text{where } t = 0.5625,\ D = 36 + 2(0.5625) = 37.125,\ L = 74$$

$$= \frac{1.09(10^6)(0.5625)^2}{(74)(37.125)}$$

$$= \frac{344{,}882.81}{2747.25}$$

$$= 125.5 \text{ psi}$$

If a furnace does not meet the requirements for an unstayed unit, one of the following three methods of support may be used:

1. A corrugated furnace may be used. A common type of corrugated furnace is known as a Morrison furnace. It may be used in either vertical tubular or horizontal types of firebox boiler such as the scotch marine.

2. The Adamson ring is a device used to stiffen a circular furnace against collapse under external pressure. It is used primarily in horizontal furnaces, for sediment might lodge on the flanges on the waterside if it were installed in a vertical axis which causes overheating. Figure 9-9 shows a cross-sectional

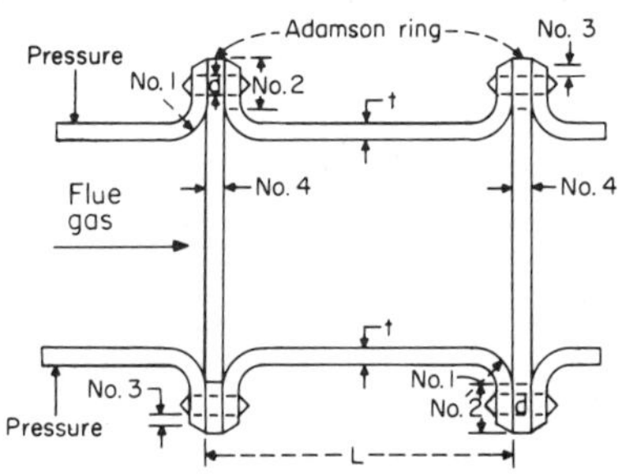

Fig. 9-9 Adamson rings are used to stiffen furnaces against collapse from steam and water pressure.

view of an Adamson-ring joint and the ASME Code specifications for proportions. The Code formula for calculating the maximum safe working pressure on a furnace subjected to external pressure and braced by Adamson rings is

$$P = \frac{57.6(300t - 1.03L)}{D}$$

where

P = maximum safe pressure, psi
D = outside diameter of furnace, in.
L = distance of furnace between Adamson rings
t = thickness of furnace sheet, in.

Example: An Adamson furnace is $\frac{9}{16}$ in. thick, 42 in. in diameter, and has furnace sections 46 in. long. What is the allowable pressure?

$$P = \frac{57.6[300(0.5625) - 1.03(46)]}{42}$$

$$= 174.2 \text{ psi}$$

3. Stay bolts may be used to stay the furnace to the outside shell or wrapper sheet. The size and the pitch of the stay bolts have much to do with the maximum allowable pressure on the boiler.

The reader should consult the Code for more complete specifications for unstayed furnaces, for it gives details not permitted by the space or intent of this book.

REINFORCEMENT OF OPENINGS IN SHELLS

In cutting a manhole in the shell of boiler, it is necessary to compensate for the metal removed. This is done by installing a manhole frame if needed.

The minimum-size elliptical manhole permitted by the ASME Code is 11 in. × 15 in. In cutting the shell for a frame having an opening of this size, the shorter dimension is placed along the longitudinal axis of the boiler so that less frame material will be required for replacement in this weaker directional axis.

In considering a cross-sectional plane of the boiler shell plate in the vicinity of a manhole "cutout," it is necessary to find the total area of metal removed, including rivet holes, and to provide a manhole frame having an equal cross-sectional area in the same plane if the shell does not have excess thickness.

The Power Boiler Code has detailed requirements on openings cut into shells or headers and how to calculate if reinforcement around the opening is necessary. The following procedure generally applies. See Fig. 9-10. The area required to be restored by the finished opening d is

$$A = d \times t_r \times F$$

where

d = diameter of finished opening in given plane, in.
t_r = required thickness of seamless shell for the pressure
F = factor that considers axis of the nozzle, usually 1.00

To determine if enough metal is available from the shell, nozzle, welds, or reinforcement, the Code provides Eqs. (9-5) and (9-6). (See Fig. 9-10 for the meaning of the symbols and limits.)

To determine the metal available from the *shell,*

$$A_1 = (E_1 t_s - F t_r) d \qquad \text{or} \qquad A_1 = 2(E_1 t_s - F t_r)(t_s + t_n) \tag{9–5}$$

The larger value is used.

To determine the metal available from the *nozzle,*

$$A_2 = (t_n - t_{rn}) 5 t_s \qquad \text{or} \qquad A_2 = (t_n - t_{rn})(5 t_n) \tag{9–6}$$

Use the smaller value.

An example will illustrate a typical reinforcement calculation. A 5-in. extra heavy pipe nozzle SA-53B is welded to a shell similar to that shown in Fig. 9-10 (½-in. welds). The shell has an inside diameter of 30 in., a thickness of

t_s = actual shell thickness
t_{rs} = required shell thickness
t_n = actual nozzle thickness
t_{rn} = required nozzle thickness

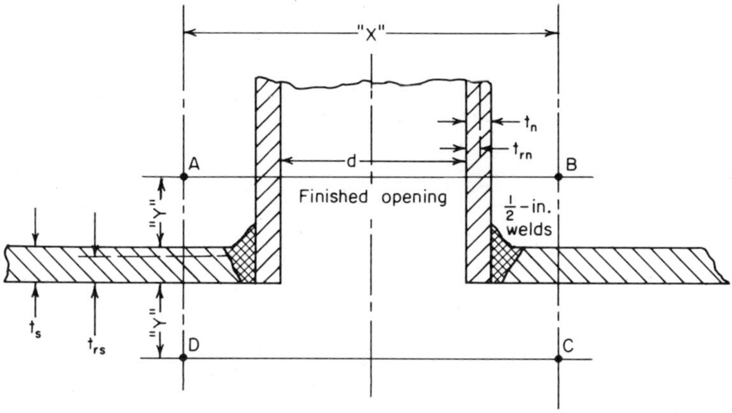

Fig. 9-10 The limits of reinforcement available in a welded nozzle attachment: (1) Excess material from shell, nozzles, and weld or reinforcement ring encompassed in rectangle ABCD. (2) The limit along shell wall (horizontally) is larger of $2d$ or $d + 2\,(t_s + t_n)$ or "X" distance. (3) The limit along nozzle wall, or "Y" distance, is the smaller of $2\frac{1}{2}\,t_s$ or $2\frac{1}{2}\,t_n$.

$\frac{7}{16}$ in., and a working pressure of 200 psi. The material is SA-285C. Assume that all welds are in accordance with the Code and that the welds are strong enough for this installation. The outside diameter of the 5-in. pipe is 5.563 in., and the thickness is 0.375 in. Does this design meet Code requirements for openings in a shell? The allowable stress for SA-285C is 13,750.

The shell thickness required is

$$t_{rs} = \frac{200(15.0)}{13{,}750(1.0) - 0.6(200)} = 0.220 \text{ in.}$$

The nozzle thickness required is

$$t_{rn} = \frac{200(2.407)}{13{,}750(1.0) - 0.6(200)} = 0.035 \text{ in.}$$

And so the area of reinforcement required is

$$\begin{aligned} A &= d \times t_{rs} \times F \\ &= 4.813(0.22)(1.0) = 1.059 \text{ in.}^2 \end{aligned}$$

The area of reinforcement provided, for $E_1 = 1.00$, is

$$\begin{aligned} A_1 \text{ (shell)} &= [1.0(0.438) - 1.0(0.220)]4.813 = 1.049 \text{ in.}^2 \\ A_2 \text{ (nozzle)} &= 2.(2.5)(0.375)(0.375\text{–}0.035) = 0.638 \text{ in.}^2 \\ A_3 \text{ (welds)} &= 1(0.50)^2 = 0.250 \text{ in.}^2 \\ \text{Total} &\quad 1.937 \text{ in.}^2 \end{aligned}$$

Construction meets code requirements because the area of reinforcement provided exceeds the area removed.

SUSPENSION OF BOILER BY LUGS OR BRACKETS

The ASME Power Boiler Code for riveted supports states the following: "Lugs or hangers, when used to support a boiler of any type, shall be properly fitted to the surfaces to which they are attached. Where it is impracticable to use rivets, studs with not less than 10 threads per inch may be used. In computing the shearing stresses, the area at the bottom of the thread shall be used. The shearing and crushing stresses on the rivets or studs used for attaching the lugs or brackets shall not exceed 8 percent of the ultimate strength." The shearing stress allowed by the ASME Power Boiler Code is 44,000 and 88,000 lb/in.^{2} for carbon-steel rivets in single shear and in double shear, respectively. These are the values used in calculating efficiencies of riveted seams, but 8 percent of these values is the maximum allowable stress for the rivets used in the hangers or brackets. This provides a safety factor of 12.5.

The Code limits the number of rivets in the same longitudinal line to two in each hanger or bracket; and, in boilers over 72-in. diameter, the circumferential distance between the top and bottom rivet of any hanger is to be not less than 12 in. After a bracket or hanger of the proper curvature and size is supplied and the rivet holes are located as specified by the Code, it becomes important to know how many rivets of a given size are required to support the boiler without exceeding the allowable stress. Modern boilers use welded lugs and brackets. See the Code for details.

Example: Assume that a horizontal boiler full of water has a total weight of 30 tons. Four hangers are to be used with ⅞-in. rivet holes drilled through the hangers and boiler shell. How many rivets are required in each hanger to meet Code specifications?

On the assumption that steel rivets are used, the rivets are in single shear, and the maximum allowable stress $= 44{,}000 \times 0.08 = 3520$ lb/in.^{2}. The load on each hanger will be $(30 \times 2000)/4 = 15{,}000$ lb.

Then, since the cross-sectional area of a ⅞-in. rivet hole is 0.601 in.^{2}, $15{,}000/(3520 \times 0.601) = 7.08$; therefore, eight rivets would be used in each hanger, arranged in two circumferential rows of four each or staggered so that not more than two per hanger would fall in the same longitudinal line.

STAMPING A COMPLETED BOILER

An identifying stamp on a power boiler is required by states and municipalities. The National Board standard form of stamping is sufficient for the boiler to pass rules and regulations for construction in practically all sections of the United States. This stamp consists of the ASME symbol above the manufacturer's serial number, the manufacturer's name or approved abbreviation, the maximum pressure for which the boiler was built, the water heating surface in square feet, and the year built.

"National Board" followed by a serial number stands for the National Board of Boiler and Pressure Vessel Inspectors. It is an enforcement body for the ASME Power Boiler Code, and a copy of the manufacturer's constructional data is filed with this board under the serial number. The National Board stamp indicates that the boiler is of ASME standard construction and that its construction was followed in the shop by a qualified inspector.

On horizontal fire-tube boilers (externally fired), the stamping should be in the middle of the front tube sheet, above the top row of tubes. On horizontal fire-tube boilers of the firebox type, the stamping should be located above the center or right-hand furnace or above a handhole at the furnace end. On vertical fire-tube boilers, the stamping should be located over the furnace door. Watertube boilers have the stamping on the drum heads above the manhole flange.

It is very important when purchase or relocation of a boiler is contemplated to ascertain the boiler laws of the state. Many states require filing a special form and following a definite procedure before relocating a boiler.

Questions and Answers

9-1 What are the major disadvantages of a lapped longitudinal joint?

Ans. The major disadvantage of a lap-riveted longitudinal seam is that the plate is "lapped"; thus, the drum or shell does not form a true circle. A certain degree of bending stress may occur along the lap of the seam when the shell is subjected to pressure. This causes a stress concentration that eventually may result in fatigue and cracking of the plate, thus producing a disastrous boiler explosion.

About the only "advantages" of the lap-riveted seam are low cost, simplicity, minimum width of space, and minimum total thickness where the seam is exposed to the fires, as in girth seams of some fire-tube boilers.

In the butt-riveted seam, the edges of the shell plate butt together so that the shell or drum is formed in a true circle. A butt strap (strip of boiler plate) is riveted on the inside and outside of the shell or drum along the abutting edges to form the seam. This is called a *butt-and-double-strap joint.*

9-2 *(a)* A boiler drum is 40 in. in diameter, the tensile strength is 55,000 lb/in.^{2}, the plate thickness is ½ in., and the factor of safety *FS* is 5. What is the maximum safe working pressure?

(b) If the pressure is increased 10 percent, what is *FS*?

(c) If corrosion causes a general reduction in thickness of ⅛ in., what is the maximum safe pressure?

Ans. *(a)* Assume a quadruple-riveted butt-and-double-strap longitudinal seam having an efficiency of 94.0 percent. Then, using existing installation equation, we have (see Chap. 4 p. 108)

$$P = \frac{55{,}000 \times 0.5 \times 0.94}{20 \times 5} = 258 \text{ lb/in.}^2$$

(b) P is 258 × 1.10 = 284 psi. Transposing the formula and solving for the factor of safety, we have

$$FS = \frac{55{,}000 \times 0.5 \times 0.94}{20 \times 284} = 4.56$$

(c) A reduction in thickness of ⅛ in. will be 25 percent, which will leave 75 percent of the solid plate. Since this is below the efficiency of the longitudinal seam (94.0 percent), the maximum safe pressure should be reduced to

$$P = \frac{55{,}000 \times 0.375 \times 1.0}{20 \times 5} = 206 \text{ psi}$$

9-3 What is the maximum diameter and pressure that can be allowed on a boiler whose longitudinal joint is fusion-welded and is not x-rayed or stress-relieved?

Ans. 16-in. ID and 100-psi pressure. Build to miniature-boiler specifications.

9-4 *(a)* When back-up strips are used, what should be done with them after welding is completed?

(b) May a welding backing ring be placed in a recess at the joint? If so, what requirements must be met?

Ans. *(a)* If 100 percent joint efficiency is desired, they must be removed on longitudinal joints.

(b) Yes. The recess must not reduce the minimum required thickness.

9-5 What is the largest opening on a shell permitted without calculation for reinforcement on a welded attachment?

Ans. 2-in. pipe size.

9-6 A dished head concave to pressure, ⅝ in. thick, and 26 in. in diameter. What is the minimum length of flange permitted where the seam is to be welded to the shell or drum?

Ans. The corner radius must be not less than 3 times the head thickness or in no case less than 6 percent of the diameter or shell, or

$$3 \times \tfrac{5}{8} = 1\tfrac{7}{8} \text{ and } 6\% \text{ of } 26 = 1.56$$

The flange must be at least 1⅞ in. (larger value).

9-7 In the building of a miniature boiler, which is of all welded construction, is not x-rayed or stress-relieved, and is used for a maximum working pressure of 100 lb, what pressure must be used in testing this boiler hydrostatically?

Ans. 300 psi.

9-8 A boiler drum is drilled for tubes in a series of rows parallel to its axis; the tubes circumferentially are in alignment. The diameter of tubes holes is 4.03125 in., and they are spaced equally in every row in groups 13½ in. center to center (tubes alternately pitched 7½ and 6 in.). Determine the horizontal ligament efficiency E and the minimum pitch circumferentially.

Ans.

$$E = \frac{13.5 - 2(4.031)}{13.5} = 0.403$$

$$\text{Pitch} = \frac{4.031}{1 - 0.202} = 5.05 \text{ in.}$$

9-9 What is the minimum required thickness for a tube sheet of a Code steam drum of the following specifications?

Working pressure, 675 psi
Inside radius, 27 in.
Longitudinal tube-hole efficiency, 46.6 percent
Circumferential tube-hole efficiency, 31.1 percent
Circumferential joint efficiency, 90 percent
Material, SA-515-70
Temperature, 650°F
Allowable stress, 17,500 lb/in.^{2}

Ans.

$$t = \frac{675 \times 27}{17{,}500(0.466) - 0.6(675)} = 2.35 \text{ in.}$$

9-10 *(a)* Why are manhole and handhole plates usually made oval?
(b) Why is water used instead of steam or air to test a boiler?
(c) What precautions are necessary to take on safety valves and steam pipes when the hydrostatic test is applied to one boiler in a battery?

Ans. *(a)* To enable the operator to remove them from the boiler and replace them. An oval manhole also provides the largest entrance opening with the removal of a minimum amount of material. The long axis of the manhole should be across, or girthwise of, the shell.

(b) Incompressibility avoids explosions.

(c) Use proper test clamp or remove safety valves and blank flanges or cap. Disconnect stop valves and blank off steam-pipe ends.

9-11 *(a)* What causes a plate formed into a cylinder to retain its shape?
(b) What is the maximum permissible distortion in a welded boiler drum?

Ans. *(a)* The stretching of the outside fiber of the plate beyond its elastic limit.

(b) 1 percent of mean diameter.

9-12 *(a)* On a boiler, why is the hydrostatic test limited to 1½ times the maximum allowable working pressure?
(b) What good does this test do?

(c) Why is there a minimum and maximum temperature set for the water during a hydrostatic test? What are the minimum and maximum temperatures?

Ans. *(a)* Because a greater pressure is likely to damage the boiler plate by causing a permanent deformation.

(b) It tests for tightness of parts and strength.

(c) So that boiler parts and water will be approximately the same temperature. The temperature must be not less than 70°F to prevent sweating. That may mask a leak through a defect, and the metal temperature cannot exceed 120°F in order to permit close examination of welds and similar joint type areas.

9-13 What is the allowable working pressure on an unstayed full-hemispherical head of welded construction with the following specifications? 48 in. ID; pressure on the concave side; thickness of plate = ⅜ in.; efficiency of weld is 90 percent; SA-285C material used; temperature 650°F; allowable stress = 13,750 lb/in.².

Ans.

$$P = \frac{1.6(13{,}750)\,(0.90)\,(0.375)}{24} = 309 \text{ psi}$$

9-14 What is the minimum required thickness permitted for a Code flat unstayed circular head forged integral with a header and complying with other requirements of the Code? The diameter D is 22 in., and the required pressure is 150 psi. The maximum allowable stress value S is 11,000 lb/in.².

Ans.

$$t = 22\sqrt{\frac{0.25 \times 150}{11{,}000}} = 1.27 \text{ in.}$$

9-15 *(a)* Name two important features to be noted in connection with a manhole opening formed by flanging inward the head plates of a boiler.

(b) Why is it preferable that heads butt-welded to shells have a skirt or flange when they are attached by the fusion-welding process?

Ans. *(a)* The plate must be flanged to a depth of not less than 3 times the thickness, and the gasket bearing surface must not be less than ¹¹⁄₁₆ in.

(b) To prevent concentration of stresses in the knuckle of the flange.

9-16 To what percentage may the knuckle of a dished head be thinned in forming?

Ans. Not over 10 percent.

9-17 The stay bolts in the firebox of a locomotive-type boiler are spaced 7 in. horizontally and 6½ in. vertically, and they are of 1¼-in. diameter with 12V threads per inch. The plate thickness is ½ in. The ends of the bolts are drilled with ³⁄₁₆-in.-diameter telltale holes, and the area of telltale hole is 0.0276 in.². What is the allowable working pressure on the furnace? These data are given: SA-285 Grade C plate under 600°F; allowable stress, 13,700 lb/in.²; area of stay bolt at bottom of threads is 0.960 in.².

Ans. The allowable pressure is 165.6 psi.

9-18 A 66-in.-diameter boiler drum has a segment of 23 in. from tubes to shell to be stayed. There are twelve 3¼ in. × ½ in. (blade dimensions) braces secured by two ⅞-in. rivets in ¹⁵⁄₁₆-in. holes supporting the segment. What pressure could be allowed on the stayed portion of the head (L is not more than 1.15 times l)? The distance d from the shell to the unstayed area is 3 in. The braces are 38 in. long. SA-285C material stays under 600°F. The allowable stress is 13,700 lb/in.²

Ans. The area to be stayed is 713 in.². We must deduct the hole from the width of the blade to get the cross-sectional area:

$$\text{Blade cross-sectional area} = (3.25 - 0.938)\ (0.5) = 1.156\ \text{in.}^2$$

The blade area to be allowed to be used is

$$1.156 \times \frac{1}{1.25} \quad \text{or} \quad 1.156(0.8) = 0.925\ \text{in.}^2$$

$$AP = 0.9naS$$
$$713P = 12(0.925)(13{,}700)(0.9)$$
$$P = 192\ \text{psi}$$

9-19 What would be the required knuckle radius of an unstayed dished head if the plate is ½ in. thick and the diameter of the drum is 48 in.?

Ans. 48 × 0.06 = 2.88 in.

9-20 Why are flat surfaces in a boiler stayed or braced?

Ans. A boiler plate under pressure tends to assume the shape of a sphere, so braces and stays are used to hold it in place.

9-21 How should telltale holes be drilled in stay bolts?

Ans. Solid stay bolts 8 in. and less in length should be drilled with telltale holes at least ³⁄₁₆-in. diameter to a depth at least ½ in. beyond the inside of the plate.

9-22 Why is the pitch of stay bolts in a vertical tubular boiler taken from the inside of the firebox rather than the outside of the shell?

Ans. The furnace sheet is the sheet that normally requires staying; therefore, the pitch is taken from inside of the firebox.

9-23 What is the maximum allowable pressure on the unstayed furnace of a vertical fire-tube boiler, with an inside diameter of 36 in., a length of 35.5 between centers of head-rivet seams, and a furnace sheet ⁵⁄₁₆ in. thick?

Ans. We have 120 × 0.5 = 60

$$P = \frac{1.09(10^6)(t^2)}{LD} \quad \text{where} \quad t = 0.5\ \text{in.},\ L = 78.25\ \text{in.},\ D = 20\ \text{in.}$$

$$= \frac{1.09(10^6)(0.5)^2}{78.25(20)}$$

$$= \frac{272{,}500}{1565}$$

$$= 174.1 \text{ psi}$$

9-24 A plain cylindrical furnace of a scotch marine boiler is 8 ft long between welded seams. It has an Adamson ring midway on its length. If the outside diameter of the furnace is 36 in., what thickness of plate is required to allow 150 psi?

Ans.

$$t = \frac{150(36)/57.6 + 1.03(48)}{300} = 0.477 \text{ in.}$$

9-25 Name three methods used to strengthen a furnace to resist collapse.

Ans. On a vertical tubular boiler by stay bolts; on horizontal furnaces by the use of Adamson rings; and by corrugations.

9-26 What is an Adamson ring and where is it used and for what purpose?

Ans. It is a flanged-type joint between two or more sections of a plain furnace. It is located somewhere in the length of the furnace; there may be one in the center or several spaced at intervals, not less than 18 in. apart. It serves as a reinforcement, stiffening the furnace against collapsing pressure. The ring between the flanges also serves as a filler for caulking.

9-27 A 16-in.-ID circular nozzle manhole (welded type) is located on a welded drum. These data pertain:

Thickness of welded neck, ¾ in.
Thickness of drum shell, 1⅛ in.
Diameter of drum shell, 48 in. inside
Working pressure, 500 psi
Working temperature, not above 600°F
Welding used complies with the Code for power boilers
Nozzle extends uniformly ¾ in. below the inside drum shell
Nozzle is welded to the drum on the inside and outside with ¾ in. × ¾ in. fillet welds
Reinforcement ring on the inside of the drum is 1 in. thick, with 21½-in. ID and 36-in. OD
Inside edge welded to the drum with 1 in. × 1 in. fillet welds
Material for the drum nozzle and the reinforcing ring have a design stress of 15,000 lb/in.^{2}

Assume the welding meets the Code requirements, and show by calculations that this reinforcement is adequate to meet Code requirements.

Ans. For the shell,

$$t_{rs} = \frac{500(24.0)}{15{,}000(1.0) - 0.06(500)} = 0.817 \text{ in.}$$

For the nozzle,

$$t_{rn} = \frac{500(8.0)}{15{,}000(1.0) - 0.06(500)} = 0.272 \text{ in.}$$

The area of reinforcement required is

$$\begin{aligned} A &= D \times t_{rs} \times F \\ &= 16.0(0.817)(1.0) = 13.072 \text{ in.}^2 \end{aligned}$$

The area of reinforcement provided is

$$\begin{aligned}
A_1 \text{ (shell)} &= [1.0(1.125) - 1.0(0.817)]16.0 &&= 4.928 \text{ in.}^2 \\
A_2 \text{ (nozzle)} &= 2(2.5)(0.75)(0.75 - 0.272) + 2(0.75)^2 &&= 2.918 \text{ in.}^2 \\
A_3 \text{ (welds)} &= 2(0.75)^2 + 1(1.00)^2 &&= 2.125 \text{ in.}^2 \\
A_4 \text{ (ring)} &= 1.0(32 - 21.5) &&= 10.500 \text{ in.}^2 \\
&\qquad \text{Total} && \quad 20.471 \text{ in.}^2
\end{aligned}$$

So construction complies with Code requirements.

9-28 A high-temperature boiler weighs 50,000 lb and is supported by columns carrying beam girders. The boiler has proper attachments at four points for the reception of U bolts, which are supplied with washers and nuts; the allowable stress is 6000 lb/in.^{2} on these bolts. What is the diameter of bolts at the weakest section?

Ans. Each bolt must carry 50,000 divided by 4, or 12,500 lb. The area of the bolt required equals 12,500 divided by 6000, or 2,083 in.^{2}.
The diameter of the bolt required equals the square root of 2.083 divided by 0.7854, or 1.6 in. The bolts should be of 1⅝-in. diameter.

9-29 Calculate the minimum number of SA-31A steel rivets required in the head joint of a drum of 72-in. inside diameter. The head is dished and flanged to meet Code requirements. It is telescoped into the shell and riveted with 1¼-in.-diameter rivets in 1⁹⁄₃₂-in.-diameter holes. The hole area equals 1.29 in.^{2}. The working pressure of the drum is 300 psi. The allowable stress on the rivets is 44,000 lb/in.^{2}.

Ans.

$$N = \frac{300(0.785)(72)^2(5)}{44{,}000(1.29)} = 108 \text{ rivets}$$

9-30 A stoker-fired 66 in. × 16 ft horizontal-return-tubular boiler, carrying 125-psi pressure, has sixty-six 3½-in. no. 11 gauge tubes. What is the total heating surface if one uses one-half the shell and ignores the heads?

Ans.

$$\text{Shell heating surface} = \frac{66(3.14)(16)}{2(12)} = 138 \text{ ft}^2$$

$$\text{Tube heating surface} = \frac{3.26(3.14)(16)(66)}{12} = 901 \text{ ft}^2$$

$$\text{Total heating surface} = 1039 \text{ ft}^2$$

Therefore, two or more safety valves are required.

9-31 Who has the responsibility to establish and maintain a quality control system for building power boilers?

Ans. The manufacturer and/or assembler holding or applying for an ASME certificate of authorization.

9-32 Briefly explain what information the manufacturer must maintain in the quality-control record file.

Ans. In general, a file should include the following:

1. Drawings or blueprints of the vessel being made or assembled
2. Calculations to show Code compliance on thickness, materials, and similar requirements
3. Identification of material to be used, including dimensions and quality to be supplied by the steel mills or other suppliers
4. Inspections made on material and fabrication during designated checkpoints
5. Qualification of any nondestructive examination (NDE) personnel who performed NDE
6. Qualification of any welders who performed welding
7. NDE records on radiographs and nondestructive testing
8. Repairs made, hydrostatic test performed, and a record of data sheets for the vessel

9-33 *(A)* Who is responsible for maintaining records for a vessel constructed in accordance with ASME Code?

(B) How long must these records be kept?

Ans. *(A)* The manufacturer or assembler holding the ASME certificate of authorization, except with nuclear vessels, where the responsibility is with the owner of the nuclear facility per NRC rules as well as ASME Code requirements.

(B) By Section I, for at least 5 yr by the manufacturer; for nuclear vessels, indefinitely by the owner of the nuclear facility.

10
Boiler Connections, Appliances, and Appurtenances

Boiler appliances, connections, and appurtenances include the following: (1) The Code requires as a minimum on high-pressure boilers a pressure gauge and test connection, a safety valve, a blowdown valve, gauge glass, gauge cocks, a stop valve in the steam line, and stop and check valves in the feed line. (2) In addition to the minimum mentioned above, boiler feed pumps and/or injectors are needed. Low-water fuel cutoffs on automatically fired boilers are now required in many jurisdictions on not only low-pressure but also high-pressure boilers. Combustion safeguards are also becoming a necessity on boilers firing fuel in suspension. These are covered in later chapters. The term *boiler fitting* is applied to valves, gauges, and other connections or devices that are attached directly to the boiler so that the unit or units can be operated safely and efficiently.

SAFETY VALVES

The function of a safety valve is to prevent excessive pressure from building up in a steam boiler. The safety valve is set at or below the maximum safe working pressure for the boiler it protects.

Safety-Valve Construction It is of the utmost importance that a safety valve be correctly constructed. Such construction may be ensured by specifying that the construction must conform to ASME or National Board approved and registered direct spring-loaded pop type, properly marked as to pressure and capacity and equipped with a testing lever. The pressure setting must match either the maximum allowable pressure for which the boiler is designed or, on older boilers, the maximum pressure allowed by state or city law. The capacity of the safety valve should be at least equal to the maximum steam that can be generated by the boiler. An ASME standard safety valve bears the following information stamped on the valve body or name plate (see Fig. 10-1*a*):

Manufacturer's name or trademark
Manufacturer's type or design number

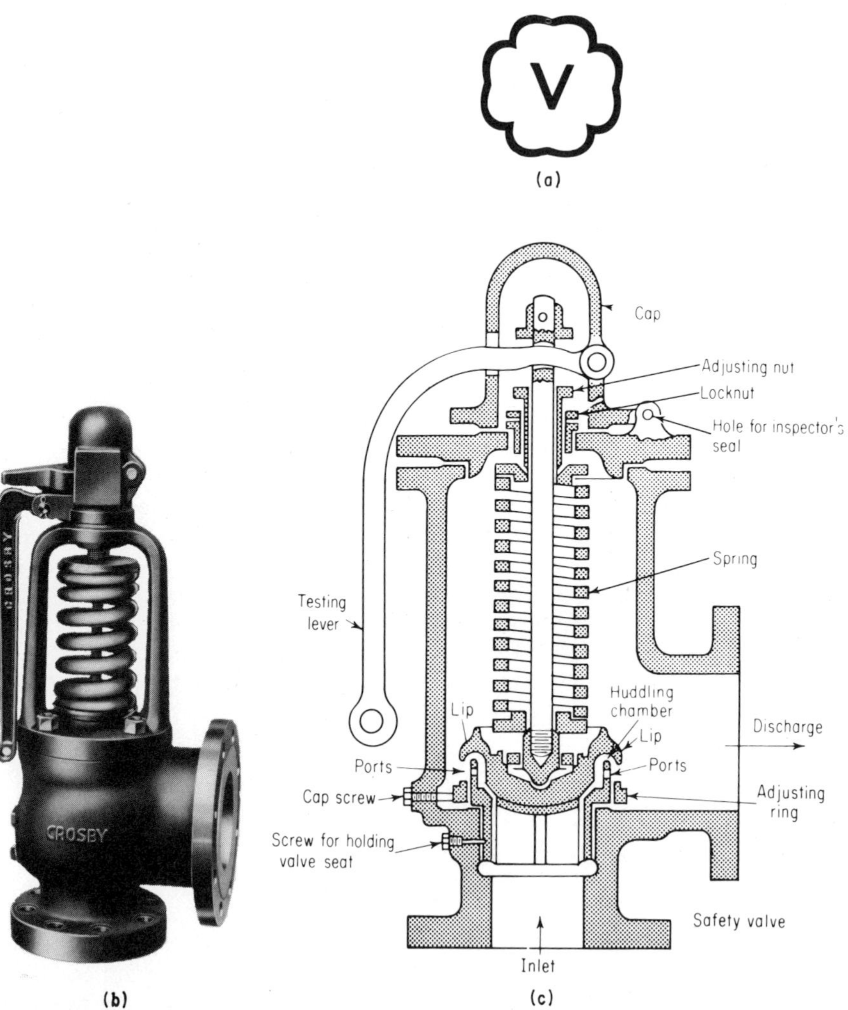

Fig. 10-1 Pop safety valves: *(a)* ASME symbol for approved safety valve. *(b)* Superheated-steam safety valve has exposed spring. *(Courtesy Crosby Valve & Gage Co.)* *(c)* Huddling chamber provides pop action for the safety valve.

Size, in.
Seat diameter, in.
Pressure at which valve is set to blow, psi
Blowdown, psi
Discharge capacity, lb/hr
Capacity lift, in.
ASME standard symbol

When a safety valve bears the ASME or National Board stamping, it is the manufacturer's guarantee that the rules of the ASME Code have been followed in the construction of the product.

In brief, the major constructional requirements are that the disk and seat be of noncorrosive material, the seat being fastened to the body so that it cannot lift with the valve disk. All parts should be constructed so that no failure of any part will interfere with full dischange capacity of the valve. The seat may be inclined at any angle between 45° and 90°.

The safety valve must be of the direct spring-loaded type. Code states do not allow the installation of weight and lever types or deadweight safety valves, for the adjustment of such valves is too easily tampered with. Their use in non-Code states is not recommended.

Safety valves should be connected directly to an independent nozzle on the boiler without any intervening valves of any description. Threaded connections may be used up to and including 3-in. diameter. For boilers operating at over 15 psi, all safety valves over 3-in. diameter should have flanged inlet connections.

Safety valves discharging steam over 450°F from superheaters should have a flanged or welded inlet connection for all sizes. Also, such valves should be constructed of steel or alloy steel throughout, suitable for heat resistance at maximum steam temperatures. The spring in superheater safety valves should be fully exposed (Fig. 10-1*b*) so that it will not be in contact with high-temperature steam.

It is important that the nozzle opening to and the escape piping from the safety valve be at least as large as the safety-valve connection. If two or more safety valves are connected on a common nozzle or fitting, the area of this nozzle or fitting should at least equal the combined areas of all safety valves served.

The safety-valve *spring* is usually of round or square stock, for maximum clearance between the coils. If the coils come in contact, the valve cannot lift. It is for this reason principally that the maximum range of adjustment permitted with a spring is 10 percent of its rated setting. This rule is for safety valves set at up to 250 psi. For higher pressures, the allowable range of adjustment is 5 percent of the spring rating. If the setting is changed to a greater deviation, a new spring and nameplate should be installed by the manufacturer's representative.

A lifting lever is required, in order to lift the valve from its seat when there is 75 percent of the popping pressure in the boiler. Lifting levers that can lock the valve in raised position are not approved.

Blowback, or *blowdown,* is the number of pounds of drop in boiler pressure from the value at which a safety valve pops to the point where the valve closes. For pressure up to 100 psi, the blowback should be not over 4 percent, but not less than 2 lb. Higher pressures call for a minimum blowback of 2 percent of the popping pressure. Safety valves used on forced-circulation boilers of the once-through type may be set and adjusted to close after blowing down not more than 10 percent of the set pressure. The valve for this special use must be so adjusted and marked, and the blowdown adjustment must be made and sealed by the manufacturer. A lesser blowback may result in a destructive chattering (rapid popping and seating) action. Too great a blowback wastes steam and fuel. Although the Code is silent regarding maximum blowback, it is usually good practice to adhere to the minimum allowed.

The ASME Code specifies that the blowback must be adjusted and sealed by the manufacturer or an authorized representative. The adjustment is effected by the *blowback ring,* or *adjusting ring,* in valves of the type shown in Fig. 10-1*c*. The adjusting-ring access screw is removed, and with a pointed tool or screwdriver the ring is rotated part of a turn on its threaded sleeve. This raises or lowers the ring, changing the area of the huddling chamber.

The action of the huddling chamber and blowback ring is to expose a greater area to escaping steam when the valve lifts slightly. The steam pressure acting on an increased area gives a greater total lifting pressure against the spring,

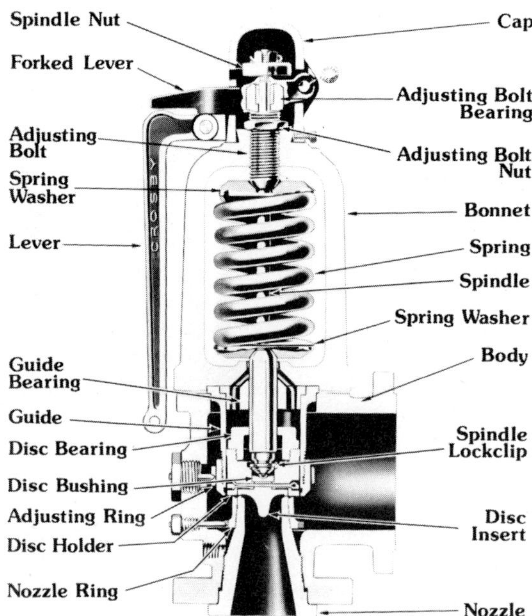

Fig. 10-2 Cross-sectional view of top-guided, full-nozzle reaction-type safety valve for temperatures to 1020°F. *(Courtesy Crosby Valve & Gage Co.)*

which results in the valve opening with a pop, with the abrasive cutting action of the steam (known as "wire drawing") which might be caused on the valve and seat by slow opening being thus eliminated.

The Crosby nozzle-type safety valve (Fig. 10-2) dispenses with the blowback-ring principle. When the valve first lifts, steam strikes the adjusting ring and is deflected downward. The reaction of the diverted steam flow causes the valve to pop open. This construction results in high lift and high capacity.

Escape pipes should be used if the discharge is located where workers might be scalded. A proper escape pipe is as essential to the safety of plant personnel as the safety valve is to the boiler. Too often a worker has been opening a stop valve when a safety valve, having no escape pipe and pointing directly at the person, pops. To be standing in the path of a high-pressure 3- or 4-in. jet of steam is usually fatal.

Every escape pipe should be at least 6 ft high. If headroom makes it impossible to terminate the escape pipe within a reasonable distance from the ceiling, it should extend out through the building wall or roof. If it is a flat roof where workers may be, the escape pipe should extend at least 6 ft above it. If a horizontal escape pipe is more practical, it should discharge at a safe location.

It is essential that the escape pipe diameter be at least equal to the size of the safety valve. If a length of over 12 ft is necessary, it is better to use a diameter ½ in. larger for each 12 ft in length. A long line with no increase in diameter will cause a back pressure because of flow friction and may cause serious chattering of the safety valve. All 90° bends should be avoided if possible.

The escape pipe should be supported independently of the safety valve. Serious stresses may be set up in the safety-valve body, connection, or boiler nozzle by the weight of a heavy, unsupported escape pipe.

After a safety valve has blown many times, it is not uncommon for slight leakage to develop. Condensation of this leakage may gradually fill an undrained escape pipe with water. This condition alone prevents the safety valve from blowing at its set pressure. The popping point will be increased 1 lb for every 2.3-ft elevation of water in the escape pipe. Also, in an outdoor escape pipe exposed to severe winters, ice may form and seriously interfere with proper safety-valve operation. Every escape pipe should have a ⅜- or ½-in. open drain at its lowest point. This drain should be conducted off the boiler top in order to prevent external corrosion induced by dampness. Figure 10-3 shows a correctly installed safety valve.

The number and capacity of safety valves required on boilers are governed by Code rules. The following rules on boilers must be followed:

1. The safety-valve capacity on a boiler must be such that the safety valve (or valves) will discharge all the steam that can be generated by the boiler (this is assumed to be the maximum firing rate) without allowing the pressure to rise more than 6 percent above the highest pressure at which any valve is set, and in no case more than 6 percent above the maximum allowable pressure.

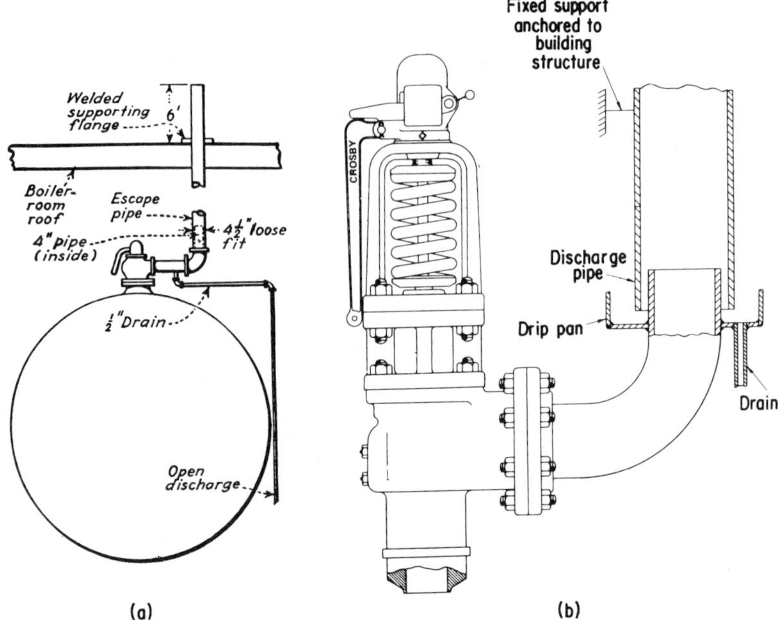

Fig. 10-3 Escape or discharge pipe from the safety valve should have a drain to remove condensate. *(a)* Low-pressure pipe; *(b)* high-pressure pipe.

2. The minimum safety-valve relieving capacity, for other than electric boilers, must be determined on the basis of pounds of steam generated per hour per square foot of boiler heating surface and waterwell heating surface, as given in Fig. 10-4. For electric boilers, the relieving capacity is determined by multiplying the kilowatt input by 3½ to obtain the pounds per hour of steam-relieving capacity.
3. For HTHW boilers, the required steam-relieving capacity in pounds per hour is determined by dividing the maximum Btu output (for the fuel being fired) of the boiler by 1000.

The popping-point tolerance that a valve must meet is the following on a plus or minus basis:

1. Two psi for pressures up to and including 70 psi
2. Three percent for pressures for 71 to 300 psi
3. Ten psi for pressures over 301 to 1000 psi
4. One percent for pressures over 1000 psi

One or more safety valves must be set at or below the maximum allowable pressure. The highest pressure setting of any safety valve cannot exceed the maximum allowable working pressure by more than 3 percent. The range

Minimum Pounds of Steam per Hr per Sq Ft of Surface on HP Boilers

Surface	Fire-tube boilers	Water-tube boilers
Boiler heating surface:		
Hand-fired	5	6
Stoker-fired	7	8
Oil-, gas-, or pulverized-fuel-fired	8	10
Waterwall heating surface:		
Hand-fired	8	8
Stoker-fired	10	12
Oil-, gas-, and pulverized-fuel-fired	14	16

NOTE: When a boiler is fired only by a gas having a heat value not in excess of 200 Btu per cu ft, the minimum safety-valve relieving capacity may be based on the values given for hand-fired boilers above.

Fig. 10-4 The capacity of safety valves is determined, per ASME rules, by the amount of heating surface, the method of firing, and the type of boiler.

of pressure settings of all the saturated steam safety valves on the boiler cannot exceed 10 percent of the highest pressure setting to which any valve is set.

Each boiler requires at least one safety valve, but if the heating surface exceeds 500 ft^2 or the boiler is electric with a power input over 500 kW, the boiler must have two or more safety valves. When not more than two valves of different sizes are mounted singly on the boiler, the smaller valve must be not less than 50 percent in relieving capacity of the larger valve.

Every superheater attached to a boiler with no intervening valves between the superheater and boiler requires one or more safety valves on the superheater outlet header. With no intervening stop valves between the superheater and the boiler, the capacity of the safety valves on the superheater may be included in the total required for the boiler, provided the safety-valve capacity in the boiler is at least 75 percent of the aggregate safety-valve capacity required for the boiler.

The superheater safety valves should always be set at a lower pressure than the drum safety valves so as to ensure steam flow through the superheater at all times. If the drum safety valves blow first, the superheater could be starved of cooling steam, leading to possible superheater tube overheating and rupture.

Corrosion and deposits on valve and valve seat are caused by the safety valve not having been lifted for a long period. To avoid this most dangerous condition on automatic-fired (especially low-pressure) boilers, the safety valve should be periodically raised by using the hand lever, or preferably by raising the steam pressure to the popping point. The latter practice should be done only with constant attendance at the boiler, and then only under the supervision of trained personnel who will carefully watch boiler pressure and immediately shut down the boiler if the pressure starts exceeding the maximum allowable. The lever testing of safety valves should be done with at least 75 percent boiler pressure on the safety valve.

WATER COLUMNS, GAUGE GLASSES, AND GAUGE COCKS

The gauge glass and gauge cocks are essential appliances for indicating the level of the boiler water. The water column (often omitted in railroad and marine practice) is installed between the gauge glass and the boiler. It serves to eliminate excessive fluctuations of water-level indication in the glass due to rapid boiler circulation or ebullition and thus acts as a steadying medium. See Fig. 10-5*a* and *b*.

Mirrors are sometimes used where, because of extremely high elevation, gauge glasses are not visible from the operating floor. Other styles of glass make use of remote gauge-glass indicators which may be located at the operating floor level.

The gauge-glass connections between the glass and the water column (or boiler if the column is omitted) should be at least ½ in., and a drain valve or cock should be provided in the bottom of the lower connection chamber. A shutoff valve is installed in both connections to permit replacement of a defective glass while the boiler is under pressure.

Gauge cocks are installed to serve as a check on the accuracy of the gauge glass and to determine the water level in the event of glass failure. If two independent gauge glasses are installed at the same elevation and at least 2 ft apart, the gauge cocks are usually dispensed with. Most codes require but two gauge cocks for locomotive boilers within 36-in. diameter or on firebox boilers within 5 HP. Larger boilers are fitted with three gauge cocks. These cocks may be mounted on the water column or directly on the boiler and are equally spaced, within the visible range of the gauge glass.

Water-column material is usually cast iron up to 250 psi. The ASME Code specifies that malleable iron may be used up to 350 psi, but steel is required for higher pressures. The piping connecting a water column to the boiler should be at least 1-in. diameter. Brass pipe for the lower connection is preferred up to about 150 psi. Where any bend or turn is necessary in the lower connection, crosses should be used instead of 90° elbows. The two unused openings should be fitted with pipe plugs, and these should be removed at each annual internal inspection (more often if required) to see that the connecting pipes are clear and to clean them if necessary.

Water-column drains are required at the bottom and in the lowest part of any pocket in the lower connection. A ¾-in. or larger gate-valve drain is used for sediment removal. A globe valve may be used on the nonpressure side of the gate valve for tightness if desired. Use of the gate valve permits a wire to be run through an obstructed valve and pipe without removing the boiler from service.

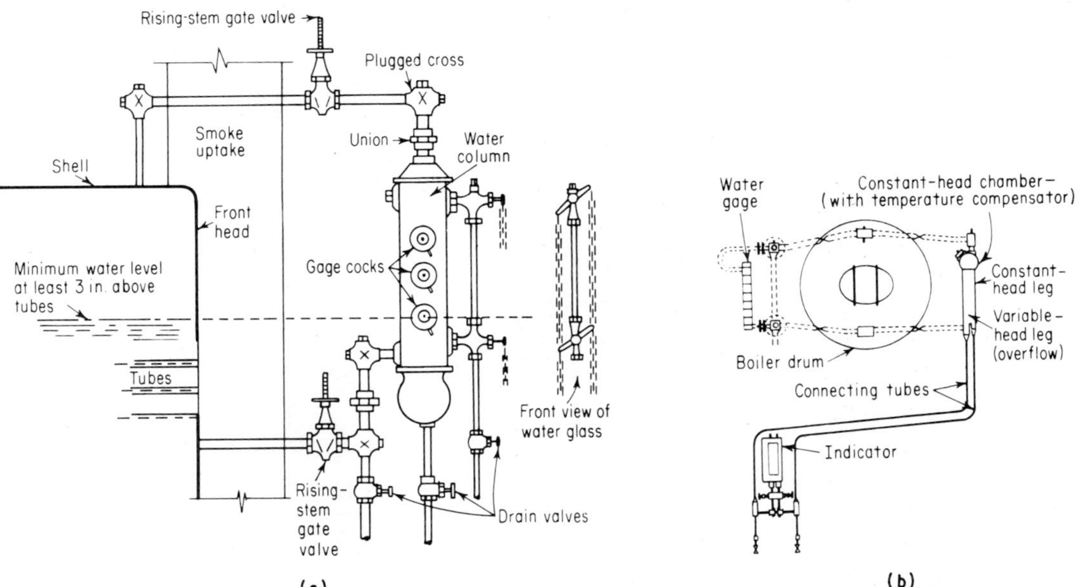

Fig. 10-5 *(a)* Water-column and gauge-glass connection for fire-tube boiler. *(b)* Remote water-level indicator permits operator to check water level at a floor remote from location of the drum.

The steam connection should pitch toward the column and the water connection toward the boiler so that a false indication of level will not be shown by water trapped in the column when the boiler water level is declining.

Leveling a water column or a gauge glass is important. The lowest visible point in the gauge glass should be at least 2 in. above the elevation of the lowest safe water level.

In fire-tube boilers, the lowest safe water level is at least 1 in. above the top row of tubes. The lowest safe water level in locomotive or scotch marine boilers is 1 in. above the highest part of the crown sheet; in VT boilers, one-third the height of the tubes; and in vertical submerged-head, or submerged-tube, fire-tube boilers, 1 in. above the upper tube sheet. In most standard types of watertube boilers, it is 6 in. above the bottom of the steam-and-water drum.

The ASME rule on gauge-glass connections for high-pressure boilers also requires that each boiler have at least one water-gauge glass, except boilers operated over 400 psi, which must have two water-gauge glasses, connected to a single water column or directly to the drum. For power boilers with all drum safety valves set at or above 900 psi, two independent remote level indicators may be used instead of one of the two gauge glasses for boiler drum water-level indication. When both remote level indicators are in reliable operation, the gauge glass may be shut off but must be maintained in serviceable condition. When the direct reading of the gauge-glass water level is not readily visible to the operator in her or his working area, two dependable indirect indications must be provided, either by transmission of the gauge glass or by remote level indicators.

Figure 10-5*b* shows a typical high-pressure remote level indicator. It is mounted on the boiler room instrument panel or installed at any other eye-level location convenient for the operator. The indicator is connected to fittings on the boiler drum by two small tubes. Changes in the boilerwater level cause corresponding change in static head in one of these tubes; static head in the other tube remains constant. Variations in the differential pressure at the indicator cause movement of the pointer which accurately indicates water level. The indicator is operated by the boiler water itself, using the pressure differential between a constant head of water and the varying head of water in the boiler drum. By means of a diaphragm-operated mechanism, with the diaphragm sides connected to high and low levels by tubing connection, the water level is shown by a graduated scale on the instrument.

VALVES AND PIPING

Valves on boilers include steam valves on the main headers; feed valves on the water feed to a boiler; drain valves on water columns, gauge glass, and drain connections; blowdown valves for both surface blowoff and bottom sediment blowoff; check valves on feed lines; and nonreturn valves on steam mains. Materials for power plant piping are mostly carbon steels or austenitic stainless steel. In nuclear applications, changes in metal properties as a result of radiation must also be considered in piping design. Factors affecting piping in power plants include thermal transients due to rapid changes of temperature from operating problems, water hammer due to entrapped water in piping, and vibration from connected machinery requiring close attention to balancing of machinery. Large-amplitude and low-frequency seismic vibration must be designed for in nuclear plants. Figure 10-6 shows some pipe support systems.

Primary caution to be observed when installing any check valve, whether swing or lift check type, is to see that flow enters at the proper end, i.e., that the disk opens with flow. If you follow the marking on the body of the valve, you can be sure the valve is properly installed and that the disk will be properly seated by backflow or by gravity when there is no flow.

A *nonreturn valve* (Fig. 10-7*e*) is used sometimes as a stop valve on the main steam line next to a steam boiler. The function of stop-checks, or nonreturn valves as they are frequently called, is as pop safety valves on boilers where two or more units are connected to the same header. They automatically prevent backflow from the header should a boiler fail. They simplify the work of cutting out a boiler or bringing a cold boiler into operation. They protect

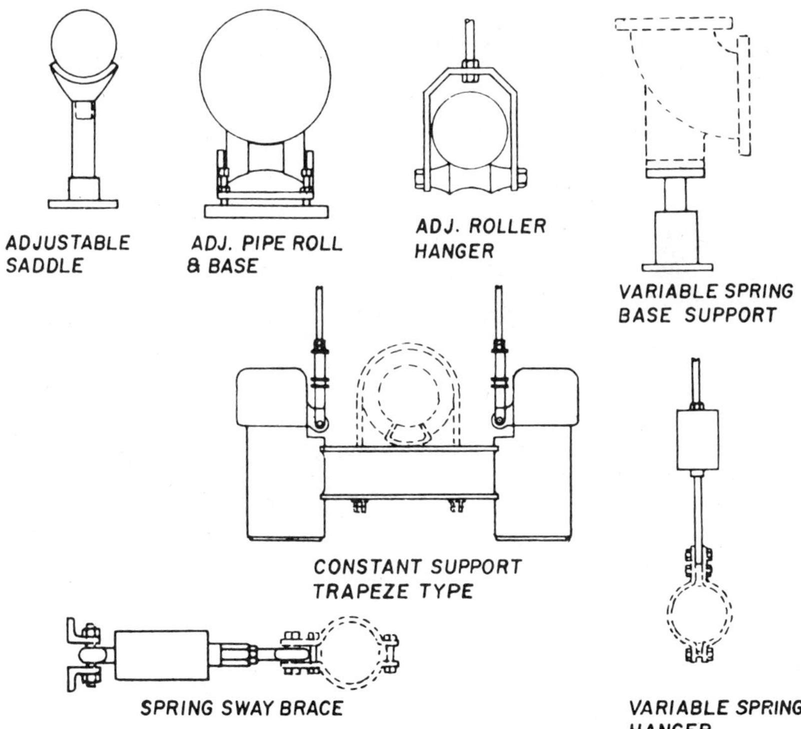

Fig. 10-6 Methods of supporting power plant piping.

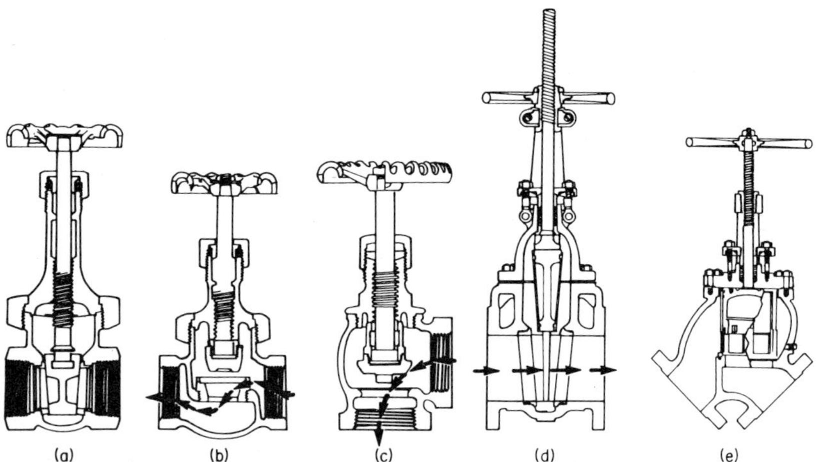

Fig. 10-7 Types of valves: *(a)* non-rising-stem type of gate valve; *(b)* globe valves, used to regulate flow; *(c)* angle valve; *(d)* outside screw-and-yoke type of gate valve; *(e)* nonreturn valve for steam line. *(Courtesy Crane Co.)*

boiler repair or inspection crews against steam backflow should the header valve be accidentally opened. No multiboiler plant should be without stop-checks. They are available in straight-way Y patterns, angle Y patterns, globe, and angle patterns, depending on working pressure.

Each main or auxiliary discharge steam outlet, except the safety valve and superheater connections, must have a stop valve placed as close to the boiler as possible. When the outlet size is over 2 in. (pipe size), the valves must be the outside-screw-and-yoke (O S & Y) type to indicate by the position of the spindle whether the valve is open or closed. When two or more boilers are connected to a common steam main, the steam connection from each boiler having a manhole must have two stop valves in series, with an ample free-blowing drain between them. The discharge of the drain must be in full view of the operator when opening or closing the valves. Both valves may be of the O S & Y type, but one should be an automatic nonreturn valve. This should be placed next to the boiler so that it can be examined and adjusted or repaired when the boiler is off the line. Steam mains going into a plant from the boiler should be adequately supported.

Blowdown Valves and Fittings A blowdown connection is required at the lowest waterspace of a boiler to serve three purposes:

1. To remove precipitated sludge or loose scale
2. To permit rapid lowering of the boiler water level if it has become too high accidentally
3. As a means of removing water from the boiler system so that fresh water may be added to keep concentration of solids in the boiler water below the point where difficulties may be experienced

Since at least one of these functions may be performed under emergency conditions, it is essential that the blowdown valves conform to rigid specifications. Globe valves are not permitted under any condition, for inherent in their design is a tendency for a dam or pocket to form, where sediment may build up.

Blowdown valves and fittings should be designed for at least 25 percent greater pressure than the maximum allowable pressure on the boiler.

Cast-iron fittings may be used between the blowdown valve and the boiler for pressures of 100 psi or less, but steel is preferred. Steel fittings are an ASME Code requirement for pressures over 100 psi, and steel valve construction for over 250 psi.

At least two blowdown valves (Fig. 10-9*a*) should be used if the pressure exceeds 100 psi. At least one of these valves should be a slow-opening valve. The ASME Code defines a slow-opening valve as one requiring at least five full 360° turns between full-open and closed position. Where there is more than one blowdown connection from a boiler, joining a common blowdown header, one valve on each independent line and a master valve on the header meet ASME Code regulations. The Code specifies further that a double blowdown valve in one casing is permissible provided that failure of one valve does not affect the other.

Valves of various types have been mentioned. As a rule, the globe valve (Fig. 10-7*b*) is used where positive tightness against leakage is desired and where the fluid controlled is practically free from suspended solids. It should be noted in the figure that both the valve disk and the seat are renewable. If they become worn, all that is necessary is to remove pressure from the pipe line, unscrew the valve bonnet, and renew these parts. The valve body does not have to be disturbed from its piping connections.

When a globe valve is used in the feedwater piping to a boiler, it is important for the flow to enter under the valve disk. If it entered from above and the disk became detached from the stem, the valve would automatically close, the feed to the boiler being thus prevented. As shown in Fig. 10-7*b*, fluids change direction when flowing through a globe valve. This seating construction increases resistance to—and permits close regulation of—fluid flow.

The disk and seat can be quickly and conveniently reseated or replaced. This feature makes them ideal for services that require frequent valve maintenance. Shorter disk travel saves operators time when valves must be operated frequently.

Angle valves (Fig. 10-7*c*) have the same operating characteristics as globe valves. Used when making a 90° turn in a line, an angle valve reduces the number of joints and saves makeup time. It also gives less restriction to flow than the elbow and globe valve it displaces.

The gate valve may be one of two types: the nonrising-stem inside-screw and the rising-stem outside-screw-and-yoke types (Fig. 10-7*a* and *d*). Gate valves are best for services that require infrequent valve operation and where the disk is kept either fully opened or closed. *They are not practical for throttling.* With the usual type of gate valve, close regulation is impossible. Velocity of flow against a partly opened disk may cause vibration and chattering and result in damage to the seating surfaces. Also, when throttled, the disk is subjected to severe wire-drawing erosive effects.

The plug gate valve, however, is excellently suited for throttling service. The gate valve operates on the wedge principle, with considerable seat contact. Also, since there is straight-through flow, no dam is offered to trap sediment or pieces of scale. Hence, the gate valve should be used for water-column drains and for similar services.

The rising-stem type gate valve is used where it is urgent that there be a visible indication that the valve is in the open position. This type is required when shutoff valves are used in the connecting pipes between a water column and a boiler.

Many other modifications of the globe and gate valves are available for special services with various fluids.

Check valves are used where unidirectional flow is essential, as when feedwater flows into a boiler. The swinging disk in the valve (Fig. 10-8*a*) closes against its seat if the flow tends to reverse. By bleeding pressure from the piping and removing the bonnet and side plug, the valve and seat may be ground to a new face when worn. Also, all these parts are renewable. Being nonreturn valves, check valves are used to prevent backflow in lines. In operating principle, all check valves conform to one of two basic patterns. Shown is the swing check type (Fig. 10-8*a*). Flow moves through these valves in approximately a straight line comparable to that in gate valves. In lift check valves (Fig. 10-8), flow moves through the body in a changing course, as in globe and

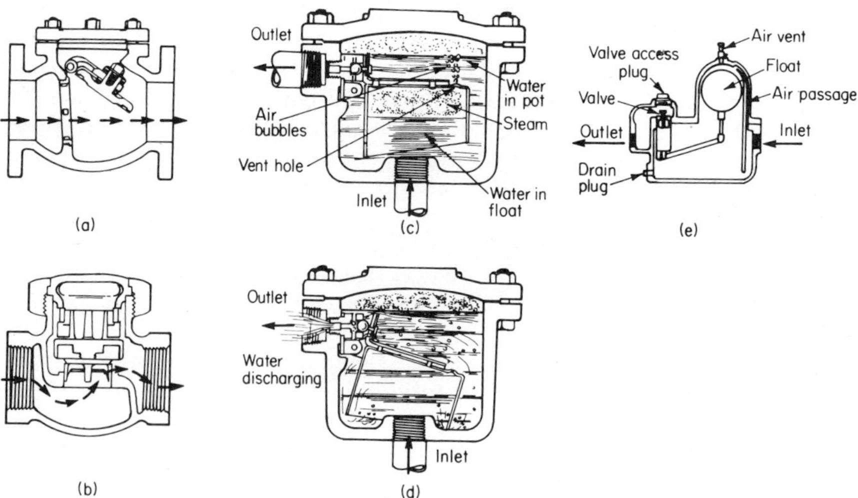

Fig. 10-8 Check valves and traps used on steam lines: *(a)* swing check valve; *(b)* lift check valve; *(c)* trap closed, incoming steam under the float buoys the float up, closing the valve; *(d)* trap open, the incoming condensate fills the float, sinks it, and opens valve; *(e)* ball float trap. *(Courtesy Crane Co.)*

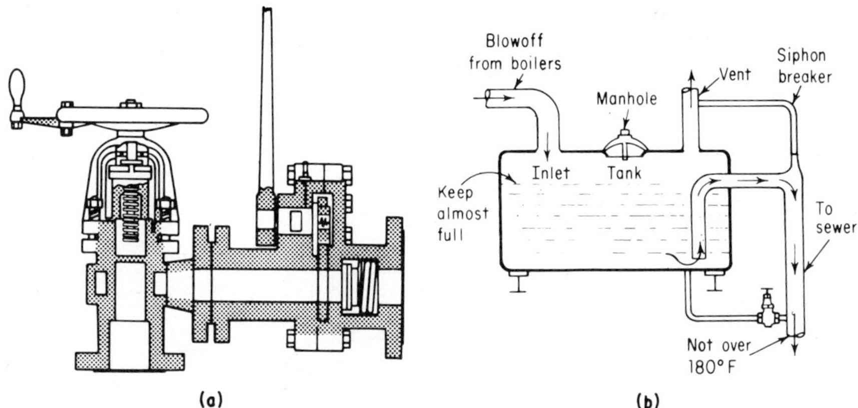

Fig. 10-9 Blowdown valves and blowdown tanks are used to remove sediment from boilers. *(a)* Angle and quick-opening blowdown valve. *(b)* Blowoff tanks are used to prevent discharging hot steam into sewers or open spaces to avoid injury of personnel.

angle valves. In both swing and lift types, flow keeps the valve open while gravity and reversal of flow close it automatically.

A blowoff tank is necessary when no open space is available into which blowoff from the boilers can discharge without danger of accident or damage to property. For example, discharging to a sewer would probably damage the sewer by blowing hot water under high pressure directly into it. A good blowoff tank installation is always nearly full of water (Fig. 10-9*b*). The National Board (NB) of Boiler and Pressure Vessels Inspectors has published recommended rules for boiler blowoff equipment, including methods of calculating the size of blowoff tank needed for the pressure and capacity of the boiler under consideration. Figure 10-10 shows the NB-recommended pipe-connection sizes for blowoff tanks, and minimum-size tank based on boiler capacity. The tank must be designed in accordance with Section VIII of the ASME Boiler Construction Code, for a working pressure of at least one-fourth the maximum working pressure of the boiler to which it is connected. In no case, however, can the plate thickness be less than ⅜ in.

STEAM TRAPS

Steam traps should be installed in lines wherever condensate must be drained as rapidly as it accumulates and wherever condensate must be recovered for heating, for hot-water needs, or for return to boilers. They are a "must" for steam piping, separators, and all steam-heated or steam-operated equipment.

Inverted open-float steam traps can be used to return condensate to regions where the pressure is less than that in the trap. They can lift condensate against a head not exceeding the pressure in the trap. Steam traps should

Fig. 10-10 ***(a)*** **Pipe sizes required for blowoff tanks.**

Boiler blowoff size, in.	Outlet size, in.	Vent size, in.
Minimum ¾	¾	2
1	1	2½
1¼	1¼	3
1½	1½	4
2	2	5
2½	2½	6

Fig. 10-10 ***(b)*** **Tank sizes required per boiler horsepower rating.**

	Minimum tank size		
Boiler rating, HP	Diameter, in.	×	Height, in.
2–20	18		24
21–50	24		30
51–100	30		36
101–200	36		36
201–400	36		42
401–800	42		48
801–1000	48		60

Note: Use 34.5 lb/(hr/HP) to convert to horsepower.

be selected on the basis of their discharge capacity—and not by pipe size. Capacity is determined by the design and construction of the trap, the size of the orifice, and the effective seat pressure.

Nonpumping traps are used to drain condensate from steam lines, separators, and steam chambers of a wide variety of machinery. An efficient trap should expel condensate but prevent the wasteful blowing through of steam. Many types are available. See Fig. 10-8*c*, *d*, and *e*.

PRESSURE GAUGES

The two main types of pressure gauges are the Bourdon tube and the diaphragm type. Figure 10-11*a* shows the interior mechanism of the single-tube Bourdon gauge with the dial removed. The bent tube of oval cross section is closed at one end and connected at the other to boiler pressure. The closed end is attached by links and pins to a toothed quadrant, which in turn meshes with a small pinion on the central spindle. As pressure builds up inside the oval tube, it attempts to assume a circular cross section, thus tending to straighten out lengthwise. This action turns the spindle by the links and gear-

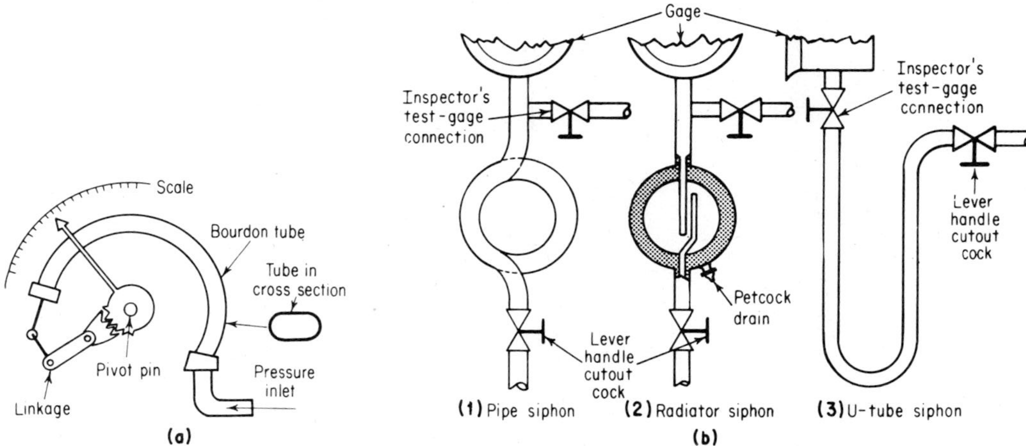

Fig. 10-11 *(a)* Bourdon steam pressure gauge movement. *(b)* Siphons are used on pressure gauges to protect the gauge from exposure to high-temperature steam.

ing, causing the needle to move and register the pressure on a graduated dial.

The boiler must have at least one pressure gauge so located and of such size that it is easily readable and which at all times indicates the boiler pressure. A valve or cock must be placed in the gauge connection (Fig. 10-11*b*) adjacent to the gauge so it can be removed for repairs. The gauge must be connected to the steam space or to the water column or its steam connection. For a steam boiler the gauge or its connection must have a siphon for maintaining a water seal to prevent steam from entering the gauge tube. The connection of a pressure gauge must be a minimum of ¼-in. ID.

For temperatures over 406°F, no brass or copper tubing should be used. The pressure gauge dial should be graduated to twice the safety-valve setting, but in no case less than 1½ times this setting. A valve connection of at least ¼-in. pipe size must be installed on the boiler for the exclusive purpose of attaching a test gauge and, when the boiler is operating, for checking the accuracy of the boiler pressure gauge. This connection is known as the inspector's connection. The pressure gauge must be illuminated, free from objectionable glare or reflection that can in any way obstruct an operator's view while noting the setting on the gauge. The pointer on the gauge must be in a near-vertical position when indicating the normal operating pressure. This is also true of other pressure gauges in the boiler room that are used on auxiliaries. Pressure gauges must not be tilted forward more than 30° from vertical and then only when it is necessary for proper viewing of the dial graduations.

The siphon is simply a pigtail or drop leg in the tubing to the gauge for condensing steam, thus protecting the spring and other delicate parts from high temperatures. Three forms are shown in Fig. 10-11*b*. If there is danger of freezing during long periods of shutdown, the siphon should be removed or drained.

FEED AND FEEDWATER HEATERS

Boilers having more than 500 ft² of heating surface should have two means of feed. However, the Code allows one means of feed under these conditions: Boilers fired by gaseous, liquid, or solid fuel in suspension may be equipped with a single feedwater system, provided means are furnished for the immediate shutoff of heat input if the water feed is interrupted. If the boiler furnace and fuel systems retain sufficient stored heat to cause damage to the boiler if the feedwater supply is interrupted, two means of feed are still needed.

For boilers firing solid fuel not in suspension, one means of feed must be steam-operated. The source of feed must be such as to supply water to the boiler at a pressure at least 6 percent higher than any safety-valve setting.

Feedwater heaters are used to bring feedwater nearer to the temperature of the boiler water. Each 10°F rise in feedwater temperature increases the overall boiler efficiency about 1 percent, owing to savings in fuel that would have been required to heat the boiler water an equal amount. An added advantage is that temperature stresses in the boiler may be avoided by feeding water at higher temperatures.

Two general classes of feedwater heater are used: open and closed. The open heater is sometimes classed as a *direct-contact* heater in that the water and steam mix, and the closed heater is sometimes termed an *indirect* heater because the steam and water are separated by tubes and the water is heated by conduction.

Under these classifications, the direct-contact heater has two definite subdivisions, namely, the standard open heater and the deaerating heater. The open heater was originally designed to utilize exhaust steam for feedwater heating and is essentially a low-pressure heater. It is always located on the suction side of the feed pump, and the heater must be at a sufficient elevation above the pump suction to prevent steam binding. (When hot water is subjected to vacuum, it flashes into steam. Thus, a pump handling hot water must have its suction fed under positive pressure, or no water will flow to the pump. See Fig. 10-12. A steam-bound pump will race and so may be damaged.) The required elevation depends on the maximum water temperature.

The principle of the open heater is to pass cold makeup water from the top down over a series of metal trays. Low-pressure steam enters between these trays, condensing and mixing with the water.

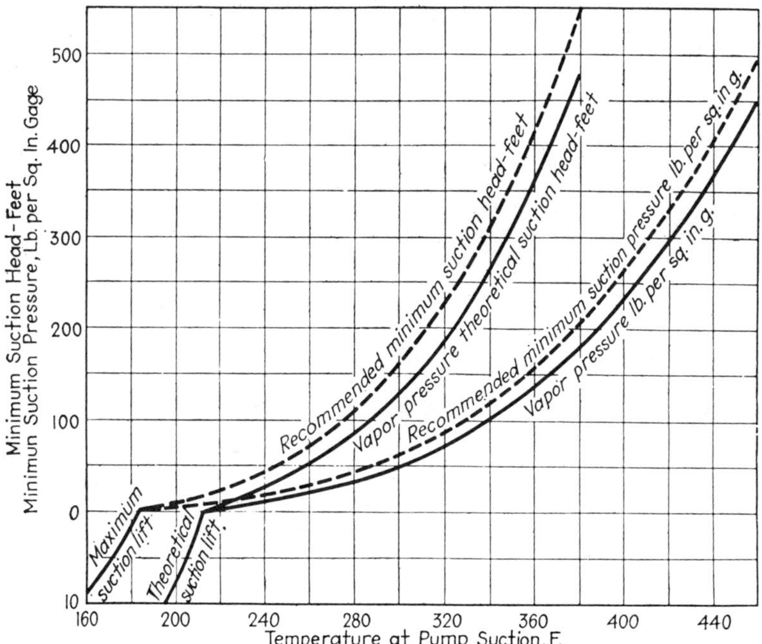

Fig. 10-12 Minimum suction head at eye of boiler feed pumps, as recommenced by Hydraulic Institute.

Important functions performed by the open heater in addition to raising the water temperature are:

1. Depositing solids causing "temporary" hardness in the water
2. Removing a considerable proportion of free oxygen by bringing the water to the boiling point and venting the gases to the atmosphere

Step 1 may reduce scale formation in the boiler; step 2 helps to reduce corrosion and pitting, which are accelerated by free oxygen.

The steam supply to open heaters is often exhaust from reciprocating engines or pumps. The pressure is seldom over 3 to 5 psi, and the heater shell is usually vented to the atmosphere through a small line. Thus, the maximum temperature attainable is slightly over 212°F. The shell should be protected

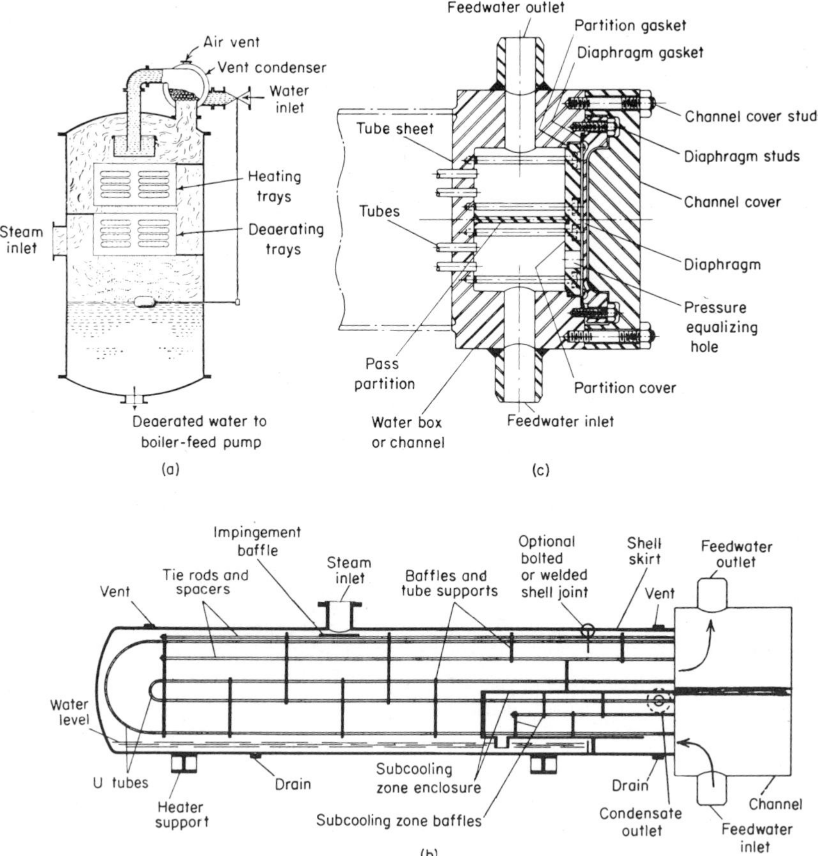

Fig. 10-13 *(a)* Deaerating heater; *(b)* U-tube closed feedwater heater; *(c)* channel-cover details of closed feedwater heater. *(Courtesy Foster Wheeler Corp.)*

against excessive pressure by an atmospheric relief valve (large-diameter safety valve) set at not over the maximum pressure for which the heater was constructed. This is often 15 psi.

The *deaerating heater* (Fig. 10-13*a*) is a development of the open heater and increases its oxygen-removal function by operating at temperatures corresponding to pressures above atmospheric. Although for this reason it is no longer an "open" heater, it is, nevertheless, still a direct-contact heater. It is used with excellent results in moderate- to large-sized plants where a sufficient volume of low-pressure steam (5 to 50 psi) is available for the heating process.

Oxygen and noncondensable gases are vented with steam through a vent condenser on top of the heater. Here the steam condenses and the condensate returns to the system, with the oxygen and other noncondensable gases being vented through a vacuum pump to the atmosphere.

The closed feedwater heater (Fig. 10-13*b*) was developed originally to operate where, because of oil contamination, the steam conditions were not satisfactory for mixing with feedwater. Since it is an indirect heater, the condensed steam was usually wasted, a condition seldom found in modern plants using this type of heater.

In large steam plants, closed feedwater heaters are frequently operated in series or cascade (Fig. 10-14). In this manner, the ultimate temperature is limited only by the temperature of available steam and the efficiency of heat transfer. The temperatures shown in Fig. 10-14 are attained in practice in moderate-pressure plants. Needless to say, clean steam is used in these installations, when extracted from a turbine. All condensate is reclaimed by trapping it back into the system. This extraction practice is common in high-pressure plants, for it permits the use of a smaller condenser for the turbine. Closed feedwater heaters used in this manner are often referred to as *bleeder* or *extraction* heaters.

The *evaporator* shown in Fig. 10-14 is a still in which raw (impure) water is evaporated into steam. This steam is condensed into pure condensate for feedwater. The condensate is often high in oxygen content, and it is customary to include with the installation a deaerating heater for oxygen removal.

The evaporator consists of a shell into which the raw water is fed to maintain a constant level. Tube coils through which steam at 10- to 150-psi pressure passes are submerged in the water. The condensate in the steam coils is trapped back into the feedwater system. The vapor or evaporated steam from the raw water passes through an evaporator condenser where the steam is condensed for use as makeup in the feedwater supply.

Most of the scale-forming impurities are left in the water in the evaporator shell. As the concentration of the raw water builds up, it should be reduced by blowing down the evaporator and refilling.

With a number of units, evaporators may be installed in series (multiple effect). Usually, four effects in series (quadruple effect) are sufficient to produce pure water from raw water as impure as is possible to use and yet be productive of maximum practical efficiency.

Steam supply to evaporators usually is extracted from steam turbines. The

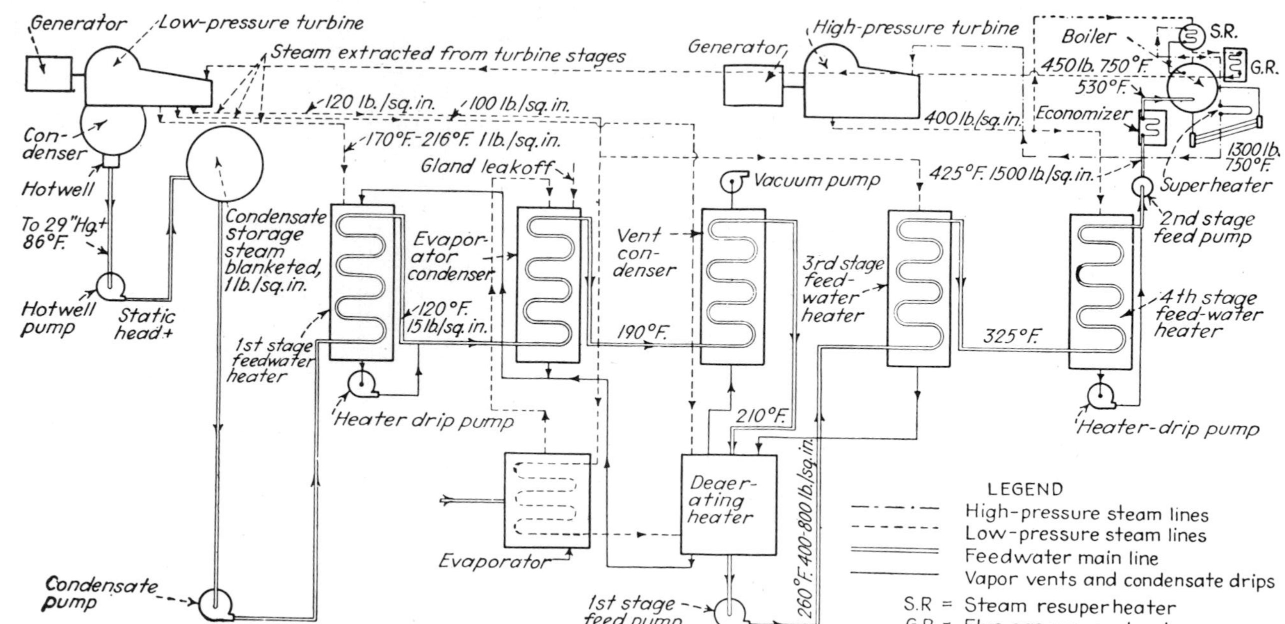

Fig. 10-14 Pressure temperatures for moderate-pressure power plant using feedwater heaters in series.

evaporator is more practical in moderate- to large-sized plants using a small percentage of makeup.

Evaporators are classified by the method of vaporization used as:

1. *Flash type.* Hot water is pumped or injected into a chamber under vacuum, where the water flashes into steam.
2. *Film type.* Water in a thin film is passed over steam-filled tubes.
3. *Submerged type.* Steam-filled tubes are submerged in the water to be evaporated.

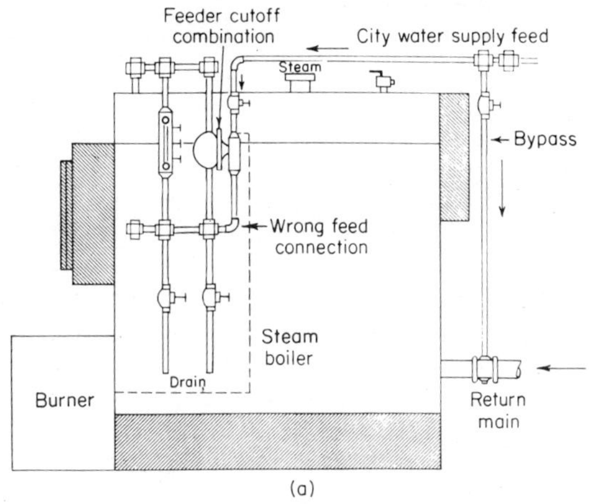

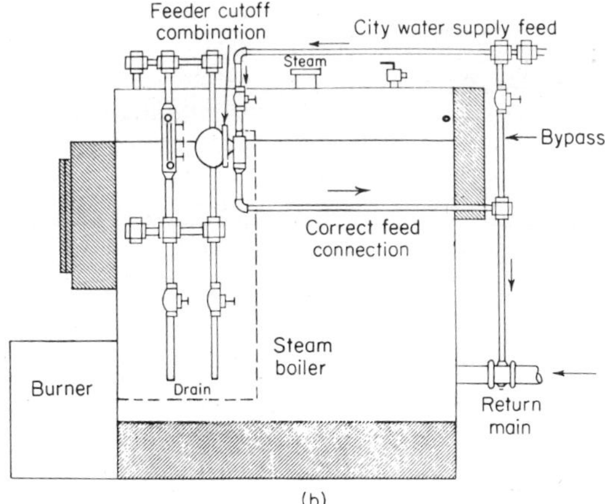

Fig. 10-15 Feeder–low-water fuel cutoff combinations can be connected incorrectly. *(a)* Makeup is at the wrong connection. *(b)* The preferred method of feed connection is through return lines.

Deaerators are also used. Air, oxygen, carbon dioxide, or other such entrained gases are carried by water into a boiler. These may come from raw water, from leakages within a system, or by chemical reactions of water and metals in a boiler loop system. The deaerator's main function is to remove these gases from the boiler water so as to prevent corrosion of metal parts in the boiler loop. (See Chap. 12.)

A common mistake on heating boiler installations utilizing a combination feed and low-water cutoff is to connect the makeup water line to the water-column cross-pipe connection as shown in Fig. 10-15*a*. This horizontal pipe connection can become obstructed, and both the gauge glass and the combination feed and low-water cutoff will not sense the true level of water in the boiler. Figure 10-15*b* shows the proper connection to avoid this pipe-obstruction possibility with the cross-tee connections having accessible plugs for cleaning out the horizontal pipe connections to the boiler.

Feedwater pumps in general use may be divided into general classes, *reciprocating, rotary,* and *centrifugal* types. The reciprocating type makes use of a water cylinder and a plunger directly mounted on a common rod from a direct-connected steam cylinder. One or two water (and steam) cylinders in parallel, known as *simplex* and *duplex* pumps, respectively, are the most common types of reciprocating feed pump (Fig. 10-16*a*). Triplex and quadruplex feed pumps often have each plunger rod connected by cranks to a mechanically driven crankshaft.

All centrifugal pumps are designed to operate on liquids. Whenever they are used on mixtures of liquid and vapor or air, shortened rotating-element life can be expected. If the liquid is high temperature or boiler feedwater with vapor (steam) present, rapid destruction of the casing can also occur. This casing damage is commonly called *wire drawing* and is identified by wormlike holes in the casing at the parting which allow liquid to bypass behind the diaphragms or casing wearing rings.

Whenever wire drawing is detected, an immediate check of the entire suction system must be made to eliminate the source of vapor. Vapor may be present in high-temperature water for several reasons. The net positive suction head (NPSH) available may be inadequate, resulting in partial or serious cavitation at the first-stage impeller and formation of some free vapor. The pump may be required to operate with no flow, resulting in a rapid temperature rise within the pump above the flash point of the liquid, unless a proper bypass line with orifice is connected and is open. (This can also cause seizure of the rotating element.) The submergence over the entrance into the suction line may be inadequate, resulting in vortex formation and entrainment of vapor or air.

When a pump becomes vapor-bound or loses its prime, a multistage pump becomes unbalanced and exerts a maximum thrust load on the thrust bearing. This frequently results in bearing failure; if it is not detected immediately, it may ruin the entire rotating element because of the metal-to-metal contact when the rotor shifts and probable seizure in at least one place of the pump.

Injectors, or *inspirators,* are used commonly for feeding water to small boilers or as an auxiliary means of mechanical feed for medium-sized boilers. They

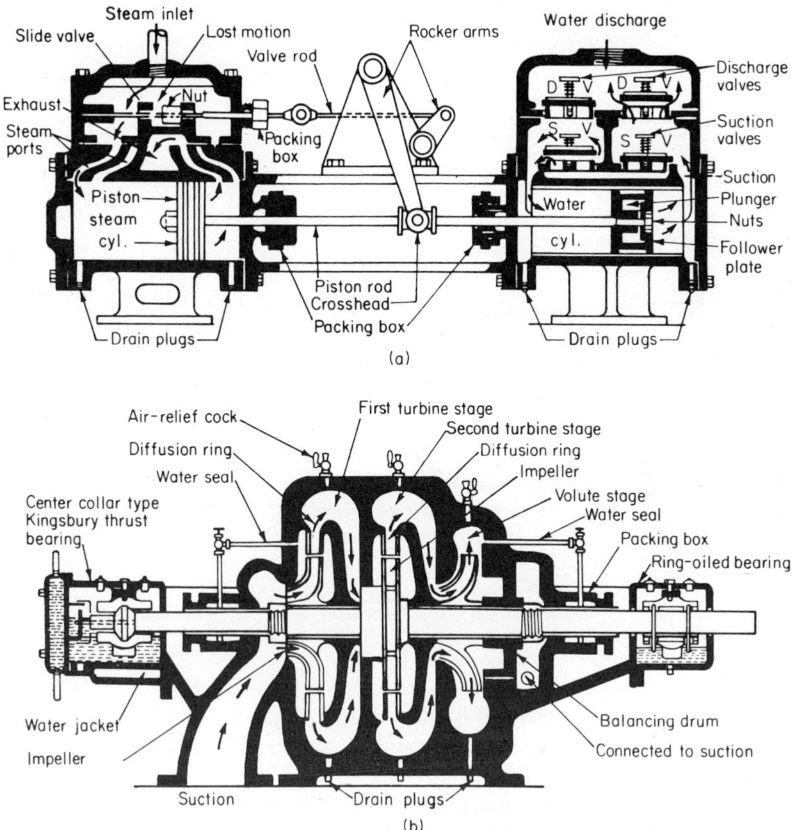

Fig. 10-16 Boiler feed-pump types most used: *(a)* duplex direct-acting steam type; *(b)* multistage centrifugal.

are quite common in railroad locomotive practice. They make use of an elongated nozzle, or *Venturi tube,* so that steam may feed water back against its own pressure.

Steam enters one end of the Venturi tube in a jet. The vacuum produced around this entering jet draws the feedwater fed to the jet chamber into the steam flow. As the steam-and-water mixture passes through the reduced area of the throat of the tube, a very high velocity of flow is produced. The weight of the water content in this steam-and-water mixture attains sufficient momentum to open the feed-pipe check valve against boiler pressure, with water being thus fed to the boiler.

Pressure-reducing valves, sometimes known as *pressure regulators,* are used to supply steam at a desired constant pressure lower than that of the supply. Their applications include supply for manufacturing processes, low-pressure feedwater or fuel-oil heaters, and other auxiliaries. See Fig. 10-17*a*.

When high-pressure boilers are used for heating low-pressure apparatus

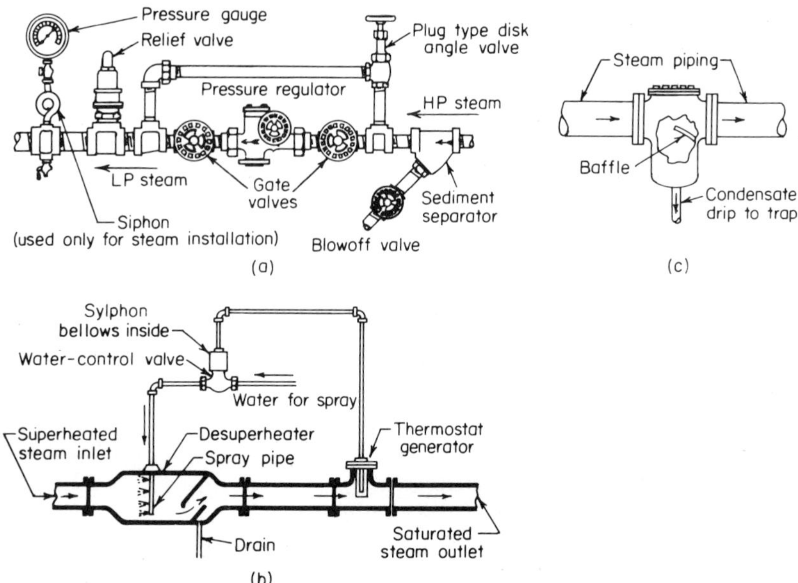

Fig. 10-17 Steam piping auxiliaries: *(a)* pressure-reducing valve arrangement; *(b)* desuperheating steam with water spray; *(c)* steam separator to eliminate slugs of water in steam line.

through a reducing valve, it is necessary to guard the low-pressure system against overpressure. To guard against overpressure, it is essential to install a safety relief valve on the low-pressure side of the reducing valve, with adequate capacity in pounds per hour to match the reducing-valve capacity. The pressure setting of the safety valve should always be based on the maximum allowable pressure of the weakest equipment on the low-pressure side of the reducing valve. The safety valve should be as near the reducing valve as possible. In the event of multiple vessels on the low-pressure side of the reducing valve, the total safety valves on each vessel should not be combined. It should be assumed that only one vessel may be operating, and, therefore, the total relieving capacity of all safety valves may be no protection.

The National Board has drawn up the following method to calculate the required relief-valve capacity in pounds per hour on the low-pressure side of a reducing station. These rules are drawn from thermodynamic principles involving flow through a nozzle. A coefficient of discharge had to be assumed (around 70 percent). Check with the reducing-valve manufacturer for the maximum valve flow and then install a safety valve with the recommended capacity. In lieu of data from the manufacturer, use the following formula:

$$RVC = \tfrac{1}{3} \times OC \times VSPA \qquad (10\text{–}1)$$

where

RVC = relief-valve capacity required, lb steam/hr

OC = orifice capacity, lb steam/(hr/in.^{2}) (Fig. 10-18*a*)

$VSPA$ = valve-size pipe areas, in.^{2} (Fig. 10-18*b*)

Orifice Relieving Capacities, lb per hr per sq in. for Determining the Proper Size of Relief Valves Used on Low-pressure Side of Reducing Valves

Outlet pressure, psi	Pressure-reducing valve inlet pressure, psi								
	125	100	85	75	60	50	40	30	25
110	4,550								
100	5,630								
85	6,640	4,070							
75	7,050	4,980	3,150						
60	7,200	5,750	4,540	3,520					
50	7,200	5,920	5,000	4,230	2,680				
40	7,200	5,920	5,140	4,630	3,480	2,470			
30	7,200	5,920	5,140	4,630	3,860	3,140	2,210		
25	7,200	5,920	5,140	4,630	3,860	3,340	2,580	1,485	
15	7,200	5,920	5,140	4,630	3,860	3,340	2,830	2,320	1,800
10	7,200	5,920	5,140	4,630	3,860	3,340	2,830	2,320	2,060
5	7,200	5,920	5,140	4,630	3,860	3,340	2,830	2,320	2,060

(a)

Nominal pipe size, in.	Standard		
	Actual ext diam, in.	Approx int diam, in.	Approx int area, sq in.
3/8	0.675	0.49	0.19
1/2	0.840	0.62	0.30
3/4	1.050	0.82	0.53
1	1.315	1.05	0.86
1 1/4	1.660	1.38	1.50
1 1/2	1.900	1.61	2.04
2	2.375	2.07	3.36
2 1/2	2.875	2.47	4.78
3	3.5	3.07	7.39
3 1/2	4.0	3.55	9.89
4	4.5	4.03	12.73
5	5.563	5.05	19.99
6	6.625	6.07	28.89
8	8.625	8.07	51.15
10	10.750	10.19	81.55
12	12.750	12.09	114.80

(b)

Fig. 10-18 Charts for determining relieving capacities of reducing valves. *(a)* Relieving capacity in pounds per hour per square inch of reducing-valve area and inlet and outlet pressure. *(b)* Standard nominal pipe sizes, internal diameters and areas in square inches.

Where a pressure-reducing valve is supplied by steam from the boiler, the capacity of the safety valve or valves on the low-pressure side of the system need not exceed the capacity of the boiler for obvious reasons.

Most pressure-reducing valves are arranged with a valved bypass, which also acts as a potential steam-source hazard in case the bypass is left open.

Where such a valved bypass is used, the following formula should be used to determine the steam flow rate through the bypass:

$$RVC = \tfrac{1}{2} \times OC \times BPA \qquad (10\text{–}2)$$

where

RVC = relief-valve capacity, lb steam/hr
OC = orifice capacity, lb steam/(hr/in.^{2}) (Fig. 10-18*a*)
BPA = bypass pipe area, in.^{2}

The larger of the relief-valve capacities calculated by Eqs. (10–1) and (10–2) should be used for selecting the relief value for the vessel.

DESUPERHEATERS

Large power-generating units are designed to operate more efficiently with a high degree of superheat. But small steam auxiliary units are often designed to operate with saturated-steam temperatures, for the use of superheated steam necessitates higher costs of construction with respect to close clearance and rotating expansion control than would be warranted, even when compared with the possible operating-expense reduction.

Rather than run a separate steam line from the boiler, independent of the superheater, it is often more practical (especially for temperature control) to pass all steam through the superheater and tap off a small line, from the superheater steam header, for auxiliary use. This small line passes steam through the desuperheater (Fig. 10-17*b*) which sprays a carefully proportioned amount of water into the flow. This proportion is regulated so that the amount of superheat to be removed will equal (or not quite equal) the amount of heat necessary to evaporate all water added to saturated steam.

Coal-burning and other solid-fuel-burning boilers require auxiliary equipment to remove fly ash and other particulates being emitted to the surrounding atmosphere. The equipment commonly used includes the following:

1. Bag house employing fabric filters now usually made of fiber glass that can withstand flue-gas temperatures of 275 to 550°F
2. Scrubbers that wash particulate emissions out of the flue gas and form a sludge that is disposed of in landfills
3. Electrostatic precipitators (Fig. 10-19*a*)

These precipitators produce an electric charge between two electrodes through which flue gas is passed. The particles in the flue gas become charged and are attracted to the positively charged and grounded collecting electrode. The particles so collected are discharged into hoppers by rapping the collecting electrode (Fig. 10-19*b*).

Induced-draft fans and forced-draft fans for steam generators require reliable service and an availability at all times in order to keep steam generators operating. The forced-draft fan supplies air for combustion of fuel as well as draft, while the induced-draft fan pulls the flue gas out of the boiler and into a

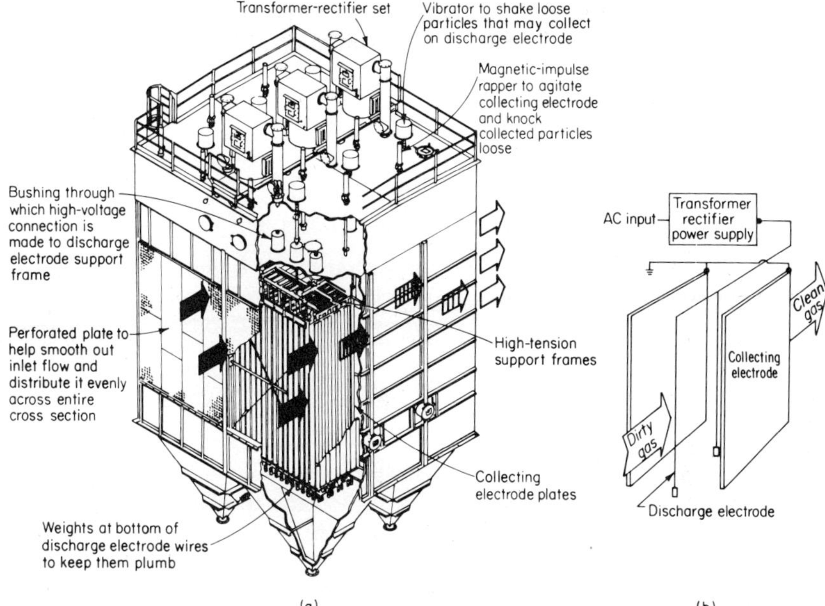

Fig. 10-19 *(a)* Electrostatic precipitator showing major components. *(b)* The basics of the electrostatic-precipitator operation: Particles in the dirty gas entering a precipitator are charged by the discharge electrode. The particles then migrate to the collecting electrodes, where they adhere and lose their charge. The particles are dislodged from the collecting plates by vibrator or rappers, and they fall into hoppers below the plates.

stack. Centrifugal and axial fans are used for forced-draft, induced-draft, gas-recirculation, and primary air fans. Interlocks are needed between fans and the combustion or burner equipment in order to avoid boiler combustion problems. Large induced-draft fans on large watertube boilers have caused high negative furnace pressures resulting from failure of interlock systems. The large negative pressures have caused *implosion* to occur in furnaces that included the cave-in of waterwall tubes. Blade erosion on induced-draft fans from particulates in the flue gas requires designs to prevent rapid blade material deterioration. Vibration monitoring is used to detect unbalance due to abrasive wear or abnormal deposits being lodged on the fan. Periodic inspection, cleaning, and balancing must be employed on induced-draft fans to keep them operating reliably.

Questions and Answers

10-1 What are the minimum appliances or appurtenances necessary for safe operation of a boiler?

Ans. Pressure gauge and test connection, safety valve, blowdown valve, gauge glass, gauge cocks, stop valve in steam line, and stop and check valves in the feed line.

10-2 What is the most important boiler appliance?
Ans. The safety valve.

10-3 What type of safety valve should you install?
Ans. ASME standard direct spring-loaded pop type.

10-4 How should a safety valve be connected to the boiler?
Ans. Directly to an independent nozzle with no intervening valve of any description. If over 3-in. diameter and over 15 psi, the valve should have a flanged connection.

10-5 What should be the maximum diameter for safety valves?
Ans. Heating Boiler Code stipulates 4½ in.

10-6 How do requirements of superheater safety valves discharging steam at over 450°F differ from requirements of those on the boiler drum?
Ans. They should have a flanged connection for all sizes. They should be constructed of steel or alloy steel suitable for the maximum temperature. The spring should be exposed so that it will not come in contact with high-temperature steam.

10-7 With a 4-in. safety valve, what is the minimum-size direct-connected nozzle and escape pipe which should be used?
Ans. 4 in. in each case.

10-8 What testing attachment is required on safety valves?
Ans. A lifting lever.

10-9 What is meant by blowback?
Ans. It is the number pounds per square inch of steam pressure drop from the point at which a safety valve pops to the pressure at which it reseats.

10-10 What controls the amount of blowback?
Ans. The blowback adjusting ring.

10-11 What is a huddling chamber?
Ans. It is a chamber exposing the underside of the valve disk in a safety valve to increased pressure area on its primary lift. Pressure acting on the increased area results in the pop, or secondary, lift.

10-12 What principle do some safety valves use in place of the increased area exposed by the huddling chamber?
Ans. The reaction principle.

10-13 Why go to the expense of installing huddling chambers or reaction flow to pop a safety valve?
Ans. Without such installation, a gradual lifting and seating of the valve

would rapidly ruin the valve seat by cutting or wire-drawing action of the steam.

10-14 When is more than one safety valve required on a boiler?

Ans. If the boiler has over 500-ft^2 heating surface or over 500-kW input for an electric boiler.

10-15 What is the purpose of a water column?

Ans. To steady the turbulence of boiler water between the drum and the gauge glass so that its level may be determined more accurately.

10-16 What attachments are permitted to pipe connections of a water column? Why limit the number of attachments?

Ans. Pressure gauge, damper regulator, feedwater regulator, drains, level indicators, or such connections as take negligible flow. Any appreciable flow would cause a false water-level indication.

10-17 Where should a gauge glass be located?

Ans. In an easily seen location with its lowest visible point at least 2 in. above the lowest safe water level in the boiler.

10-18 (1) What are gauge cocks, or try cocks? (2) How many are there and where should they be?

Ans. (1) They are valve cocks used to show the water level in a boiler as a check on the gauge glass. (2) Two are required on locomotive boilers up to 36-in. diameter or on firebox boilers up to 5 HP. Three are required on all other boilers over 15-psi pressure. They should be located equidistant within the visible range of the gauge glass. Try cocks are not required if two gauge glasses are installed at the same level at least 2 ft apart on a boiler.

10-19 Why is a drain required on a water column, and why should it be at least ¾-in. diameter?

Ans. To permit removal of sediment which might block the lower connection and cause a false water-level indication. Smaller sizes might become obstructed easily.

10-20 Why is a globe valve not desirable for water-column drain control?

Ans. Because the dam or pocket in this type of valve forms a natural trap for sediment and scale.

10-21 What three purposes does a blowdown valve serve?

Ans. Removal of sludge and loose scale, control of boiler-water concentration, and emergency control of abnormally high water levels.

10-22 Would you select cast iron or steel for elbows in a blowdown line between the boiler and the valve?

Ans. Cast iron is permitted up to 100 psi. Steel is required for higher pressures. Steel is preferred for all pressures over 15 psi.

10-23 On what principle do the majority of pressure gauges operate?

Ans. On the Bourdon-tube principle; that is, a curved tube tends to straighten when subjected to internal pressure.

10-24 Describe a fusible plug.

Ans. It is a threaded bronze or brass casing having a tapered core of nearly pure tin. A fusible plug is installed at the lowest safe water level in some low- and moderate-pressure boilers with the small end of the tapered core exposed to gases in the primary pass. Should the water level approach a dangerously low level, the core is designed to melt and escaping steam will sound the alarm.

10-25 Where and why are self-locking door latches required on firing doors?

Ans. On watertube boilers, in order to prevent the door from being blown open from positive furnace pressure resulting from tube ruptures, gas explosions, etc.

10-26 Where and why is a siphon required in pressure-gauge lines?

Ans. It is a pigtail or drop leg in the piping to the gauge, designed to trap condensate and to prevent live steam from entering the Bourdon tube. It prevents the tube, springs, and other delicate parts from being subjected to high temperatures.

10-27 What is the difference in principle between an open and a closed feedwater heater?

Ans. The open heater brings low-pressure steam in direct contact with the water and operates at or slightly above atmospheric pressure. The closed feedwater heater consists of shell and tubes with indirect contact between steam and water and may operate at high pressure.

10-28 What is the main difference in the purpose and function of a deaerating and an open feedwater heater?

Ans. The open heater reduces oxygen content by heating the feedwater to about 212°F, venting the contents at atmospheric pressure. The deaerating heater removes practically all oxygen by heating the feedwater with steam of about 30 psi or higher. Its shell is vented at pressure, through a vent condenser and vacuum pump.

10-29 What appliances would you recommend on an economizer?

Ans. Pressure gage, relief valve, blowdown valve, inlet and outlet water thermometers, inlet and outlet gas temperature indicators or recorders.

10-30 How is normal scale formation removed from evaporator tubes?

Ans. By shedding, resulting from forced expansion and contraction of the tubes.

10-31 What are the advantages and the disadvantages of motor- and steam-driven feedwater pumps?

Ans. A motor reduces the amount of steam piping and is preferred usually when the feedwater heating system secures steam bled from the main generating turbine. The steam drive is not affected by loss of electric circuits as would be the motor. At least one steam-driven feed pump is almost always installed for this emergency. Steam units may exhaust to a feedwater heater.

10-32 How does an injector force water against boiler pressure when it uses steam at the same pressure?

Ans. Because of the restricted area of the nozzle, high velocity carries the drops of water so that their momentum causes flow into the boiler.

10-33 What is a steam trap and where should you expect to find one in a boiler installation?

Ans. It is a device designed to remove condensate from steam space with minimum loss of steam. A steam trap is very often used on pockets or separators of steam lines.

10-34 What is the purpose of a pressure-reducing or pressure-regulating valve? Where might you find one?

Ans. It serves to reduce an available pressure to a lower, constant, desired pressure. For example, a steam line operating at 400 psi for power generation has to supply a branch line to a heater shell designed for 75 psi. A reducing valve would be installed on such a line and set to maintain heater pressure within the prescribed limits.

10-35 What is the purpose of a forced-draft and an induced-draft fan?

Ans. The forced-draft fan supplies air to the combustion space. The induced-draft fan draws gases and products of combustion and delivers them to the stack breeching.

10-36 What arrangement is desired for safety between the induced- and the forced-draft fans?

Ans. An interlock so that the forced-draft fan cannot be operated with the induced-draft fan shut down; otherwise, fire might be blown out of the doors and observation ports.

10-37 Under what conditions are shutoff valves permitted in connecting pipes between boiler and water columns?

Ans. If they are of the rising-stem outside-screw-and-yoke type or are straightaway valves or cocks marked plainly for their open and closed positions. They should be locked or sealed open.

10-38 Why are check valves required in the feedwater line?

Ans. So that boiler pressure will not force the boiler water back in case of piping failure; also, to assist the functioning of the feed pump or injector.

10-39 What are the minimum and maximum sizes permitted for blowdown valves or connections?

Ans. 1 in. minimum, 2½ in. maximum except for 100-ft^2 or less heating surface boilers where ¾-in. minimum size is permitted.

10-40 What are deflectors and when are they required on explosion doors?

Ans. They are sheet-metal plates placed in front of explosion doors in a boiler setting to divert any blast from operating floors, stairways, platforms, or anywhere anyone might be passing.

10-41 What force operates bucket and tilt traps?
Ans. Gravity.

10-42 Of what material are trap buckets or floats made?
Ans. Corrosion-resistant metal.

10-43 Why must a return trap be vented to the atmosphere after each discharge?
Ans. Because the body is at a pressure higher than that of the condensate returns, and the return check will be held closed. By venting the trap to atmosphere, the returns may flow freely into the trap until it tilts and closes these valves.

10-44 *(a)* In what units is a vacuum gauge graduated?
(b) What is the relation between these units and pounds per square inch?
Ans. *(a)* In inches of mercury (Hg) vacuum.
(b) Each inch of mercury equals 0.49 psi below atmospheric pressure (14.7 psi at sea level).

10-45 What is a compound gauge?
Ans. A gauge reading positive pressure in pounds per square inch on one side of the zero reading and negative pressure (vacuum) in inches of mercury on the other side.

10-46 Provision for what four details of installation must be made in a steam line transmitting steam a considerable distance from the boiler?
Ans. Provision for insulation, support, expansion, and drainage.

10-47 What are two essential appliances used in conjunction with the reduced-pressure side of the regulator? Explain your answer.
Ans. A pressure gauge and a safety valve. The former is required to check the operation of the regulator; the latter, to protect the low-pressure equipment against excessive pressure should the regulator fail.

10-48 An ASME boiler has 650 ft^2 of heating surface. In a boiler built for 150 psi, the safety valve is set to pop at 100 psi. Two means of feeding are used, city pressure at 110 psi and a pump. It is desired to operate the boiler at 150 psi. What would you recommend to do and why?
Ans. Recommend new springs for present safety valves. Recommend an additional means of feeding the boiler (pump or injector) since city pressure at 110 psi is not adequate. The fittings should be satisfactory for 150 psi. Two blowoff valves are required when the pressure is over 100 psi. The steam gauge should be graduated to approximately 250 psi.

10-49 In a boiler operating at a pressure of 65 lb, what is the smallest size of feed and blowoff connections if the safety valve is set at 90 psi?

Ans. ½-in. feed, ¾-in. blowoff.

10-50 Under what conditions does the Code permit stop valves or cocks in the connections to a water column?

Ans. They must be either the outside-screw-and-yoke type of gate valve or stop cocks with levers permanently fastened thereto and marked in line with their passage.

10-51 *(a)* What are the minimum and maximum amounts of blowdown permitted on a safety valve?

(b) How is this adjusted when needed?

(c) What tolerance, either plus or minus, would you allow on the opening pressure on a valve set for 150 psi?

Ans. *(a)* Minimum blowdown shall be not less than 2 lb and the maximum not lower than 96 percent of the safety valve set pressure.

(b) By the manufacturer or the manufacturer's representative only.

(c) Plus or minus 3 percent.

10-52 What is the allowable adjustment on the spring of an ASME safety valve set at *(a)* 290-psi pressure, *(b)* 190-psi pressure?

Ans. *(a)* 5 percent either way. *(b)* 10 percent either way.

10-53 Why should the safety valve not be placed on the same nozzle with the main steam line?

Ans. There would be a difference in pressure (because of the flow of steam in the pipe) between the boiler pressure and the pressure directly under the seat of the valve. The flow of steam in the pipe would cause the valve to chatter and damage the disk and seat when it blows. If the boiler has a dry pipe, it is liable to become obstructed, and then the valve could not blow to relieve overpressure. A stop valve is liable to be installed in steam line between the safety valve and boiler.

10-54 What are pump slip and volumetric efficiency?

Ans. Pump slip is the difference between the displacement or theoretical discharge of a pump and the actual discharge; it is usually expressed as a percentage of the displacement. Values of slip range from 3 to 15 percent. The volumetric efficiency of a piston pump is the ratio of the volume of water actually delivered to the displacement of the pump.

10-55 What causes safety valves to leak below the popping pressure?

Ans. Leakage is usually caused by damaged seats, lodged scale, or an operating pressure too close to the popping pressure. Strains set up in the valve body, either by pipe expansion or by weight of unsupported discharge piping, may also cause leakage. Such leakage can be stopped by relieving piping strain on the safety valve. Valves also leak or fail to pop at the set pressure because scale or dirt becomes wedged between the disk holder and the guide. Hardened boiler compounds deposited on the nozzle under the seat may also cause leakage or valve sticking.

10-56 What type of boilers require no gauge glass or gauge cocks?

Ans. Forced-flow steam generators with no fixed steam line and waterline (once-through boilers) and high-temperature water boilers of the forced-circulation type. The same applies to once-through hot-water-heating and hot-water-supply boilers having no fixed steam line and waterline.

10-57 What area of the boiler should be computed as heating surface?

Ans. That side of the boiler surface exposed to the products of combustion, exclusive of superheating surface. The areas to be considered for this purpose are tubes, fireboxes, shells, tube sheets, and the projected area of headers. For vertical fire-tube steam boilers, compute only the portion of the tube surface up to the middle gauge cock.

10-58 Calculate the heating surface required for an oil-fired, fire-tube boiler of 100 tubes, each 2½ in. in diameter, no. 20 gauge in thickness, each 15 ft long. The remaining heating surface of fire sheet and tube sheet totals 130 ft^2 at a working pressure of 125 psi. How many and what size valves should be installed?

Ans. Use the ID of the tubes as heating surfaces. Heat transfer is from the inside of the tube through the tube thickness to the waterside. Number 12 gauge tube has a wall thickness of 0.105 in.; thus the ID of the tube equals $2.5 - (2 \times 0.105) = 2.29$ in.

The area in square feet of all tubes equals the circumference times the length times the number of tubes $= \pi\ 2.29/12 \times 15 \times 100 = 897\ ft^2$.

$$\text{Total heating surface} = 897 + 130 = 1027\ ft^2$$

Because the boiler has over 500 ft^2 of heating surface, two or more safety valves are required.

10-59 A boiler feed pump has a discharge pressure of 200 psi gauge and delivers water weighing 62.3 lb/ft^3 at 1250 gal/min at rated load. If the mechanical efficiency of the pump is 85 percent and suction head is neglected, calculate the hydraulic or water horsepower required and the motor horsepower needed to drive the pump.

Ans. $$\text{Head developed} = \frac{200 \times 144}{62.3} = 462.3\ \text{ft}$$

$$\text{Weight of 1 gal of water} = \frac{62.3 \times 231}{1728} = 8.216\ \text{lb/gal}$$

$$\text{Weight of water handled per minute} = 1250 \times 8.216 = 10{,}270\ \text{lb/min}$$

$$\text{Hydraulic or water horsepower} = \frac{10{,}270 \times 462.3}{33{,}000} = 143.9\ \text{HP}$$

$$\text{Motor horsepower} = \frac{143.9}{0.85} = 169.3\ \text{HP}$$

11
Fuels, Firing, and Controls

The combustion process is a special form of oxidation in which oxygen from the air combines with fuel elements, which generally are carbon, hydrogen, and, though detrimental, sulfur. Important to combustion studies are the chemical thermodynamics and kinetics of flame travel and velocity of reactions. A proper mixture of fuel and air as well as an ignition temperature is required for the combustion process to continue. Fuel must be prepared so that thorough mixing of fuel and air is possible. The term *flammability* is used to describe a fuel's ability to be burned, or really its ability to be converted to a gas so that combustion can take place.

Three conditions must be satisfied for proper chemical reactions to take place in the combustion process:

1. Proper proportioning of fuel and oxygen (or air) with the fuel elements, as shown by chemical equations, is necessary.
2. The mixing of fuel and oxygen (or air) must be thorough, so a uniform mixture is present in the combustion zone and so every fuel particle has air around it to support the combustion. Solid fuels generally will be converted to gas first by the heat and presence of air. Liquid fuels will vaporize into gases and then burn. Atomization of liquids increases the mixing with air and increases the vaporization into a gas. Pulverization of coal will have the same effect.
3. The ignition temperature must be established and monitored so that the fuel will continue to ignite itself without external heat when combustion starts.

The chief heat-producing elements in fuels (except for atomic reaction and electricity) are carbon, hydrogen, and their compounds. Sulfur, when rap-

Fig. 11-1 Combustion reactions of fuel elements with combining weights and volumes and heating value of the fuel.

					Weights per pound of combustible						
Combustible element	Symbol	Chemical reaction	Combustion product	Volumes	Oxygen, lb	Nitrogen, lb	Air, lb	Gaseous products, lb	ft³ O_2/lb fuel	ft³ air/lb fuel	Heating value, Btu/lb
Carbon	C	$C + O_2 \rightarrow CO_2$	Carbon dioxide	1 vol C + 1 vol O_2 = 1 vol CO_2	2.67	8.85	11.52	12.52	31.65	151.3	14,600
Carbon	C	$2C + O_2 \rightarrow 2CO$	Carbon monoxide	2 vol C + 1 vol O_2 = 2 vol CO	1.33	4.43	5.76	6.76	—	—	4,440
Carbon monoxide	CO	$2CO + O_2 \rightarrow 2CO_2$	Carbon dioxide	2 vol CO + 1 vol O_2 = 2 CO_2	0.57	1.90	2.47	3.47	6.79	32.5	10,160
Hydrogen	H	$2H_2 + O_2 \rightarrow 2H_2O$	Water	2 vol H_2 + 1 vol O_2 = 2 vol H_2O	8	26.56	34.56	35.56	94.8	453	62,000
Methane	CH_4	$CH_4 + 2O_2 \rightarrow CO_2 + 2H_2O$	Carbon dioxide and water	1 vol CH_4 & 2 vol CO_2 = 1 vol CO_2 + 2 vol H_2O	4	13.28	17.28	18.28	47.4	226.5	23,850
Ethylene	C_2H_4	$C_2H_4 + 3O_2 \rightarrow 2CO_2 + 2H_2O$	Carbon dioxide and water	1 vol C_2H_4 + 3 vol O_2 = 2 vol CO_2 + 2 vol H_2O	3.43	11.38	14.81	15.81	—	—	21,600
Ethane	C_2H_8	$2C_2H_8 + 7O_2 \rightarrow 4CO_2 + 6H_2O$	Carbon dioxide and water	2 vol C_2H_8 + 7 vol O_2 = 4 vol CO_2 + 6 vol H_2O	3.73	12.40	16.13	17.13	44.5	212	22,230
Sulfur	S	$S + O_2 \rightarrow SO_2$	Sulfur dioxide	1 vol S + 1 vol O_2 = 1 vol SO_2	1	3.32	4.32	5.32	11.87	56.7	4,050

idly oxidized, is also a source of some heat energy, but its presence in a fuel has bad effects. The burning of coal, oil, or gas is a chemical reaction involving the fuel and oxygen from the air. Air is 23 percent oxygen by weight and 21 percent by volume. The remainder of air is mostly nitrogen, which takes no actual chemical part in combustion but does affect the volume of air required. The table in Fig. 11-1 represents some typical combustion reactions for various fuel constituents. It is always the carbon, hydrogen, or sulfur that produces the chemical reaction for heat by combining with oxygen.

Since oxygen in the air is known to be 23.15 percent by weight and 21 percent by volume (from combustion equations), the amount of air required can be calculated. For example, in the complete combustion of carbon, it can be determined that 2⅔ lb of oxygen is required to burn 1 lb of carbon. The amount of air required to burn 1 lb of carbon would then be

$$\frac{\text{Amount of oxygen}}{\text{\% oxygen in air by weight}} = \frac{2.67}{0.2315} = 11.52 \text{ lb}$$

This is shown in Fig. 11-1.

Incomplete combustion results in smoke and lowered operating efficiency. In order to obtain complete combustion, the furnace volume must be adequate to permit complete burning of fuel particles before they enter heating surfaces and are cooled below their ignition temperature.

In order to thoroughly mix oxygen with burning fuel gases and particles, the flame action must produce turbulence. Flexibility of flame control may be affected by control of the primary air supply. Primary air is that which conveys fuel to burners or mixes with fuel at burners or through the fuel bed. See Fig. 11-2*a*. Secondary air is supplied to the burning fuel so that oxygen may unite in combustion at advantageous points.

If not enough oxygen or air is supplied, the mixture is rich in the fuel; thus the fire is reduced, with a resultant flame that tends to be longer and smoky. The combustion also is not complete, and the flue gas (products of combustion) will have unburned fuel such as carbon particles or carbon monoxide instead of carbon dioxide. Less heat will be given off by the combustion process. If too much oxygen or air is supplied, the mixture and burning are lean, resulting in a shorter flame and cleaner fire. Excess air takes some of the released heat away from the furnace and carries it up the stack. Burning should always be with excess air to ensure that all the fuel is properly burned and thus attain better efficiency in heat release. This also reduces smoke formation and soot deposits, which today, with stricter pollution laws, is important.

When flue gas comes out of a stack as black smoke, it is an indication of insufficient air. Too much air usually causes a dense, white smoke. A faint, light-brown haze coming from the stack is a sign of a reasonably good air/fuel ratio. Of course, a more exact analysis is made with a flue-gas analyzer, such as an Orsat apparatus. From this analysis, the percentage of either excess or insufficient air can be determined. See Fig. 11-3.

A flue-gas analyzer measures the percentages of volume of carbon dioxide, carbon monoxide, and oxygen. Because air contains 21 percent oxygen and

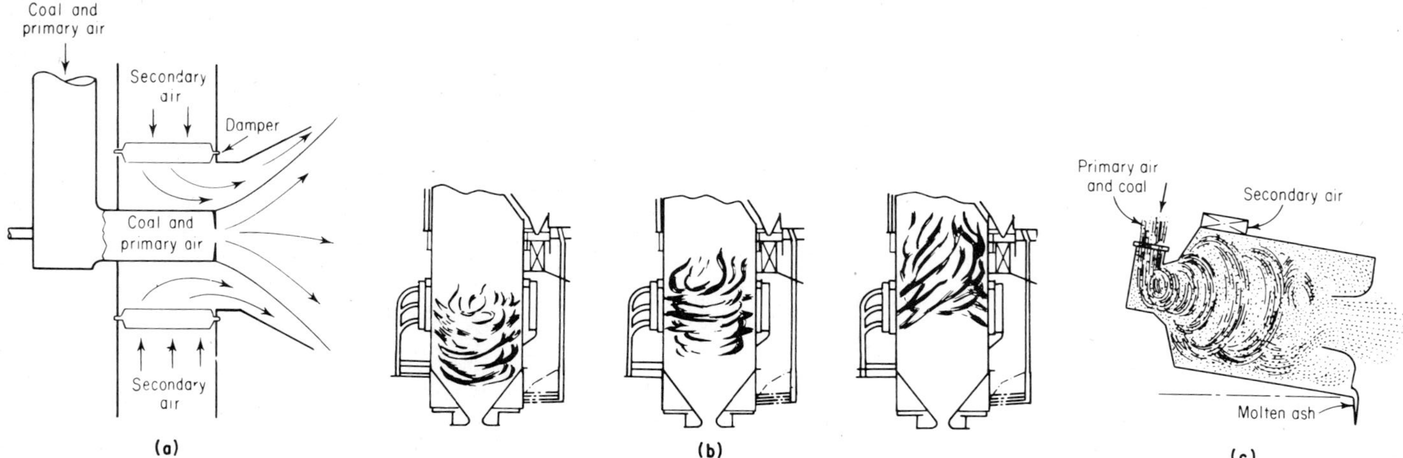

Fig. 11-2 Burner systems for pulverized-coal burning. *(a)* Primary air in turbulent burner aids fuel burning at burner, and secondary air completes combustion in furnace (secondary air is two-thirds of total). *(b)* Tilting burner in tangential firing is used for load control. *(c)* Cyclone burner whirls coal and air against the walls of the burner.

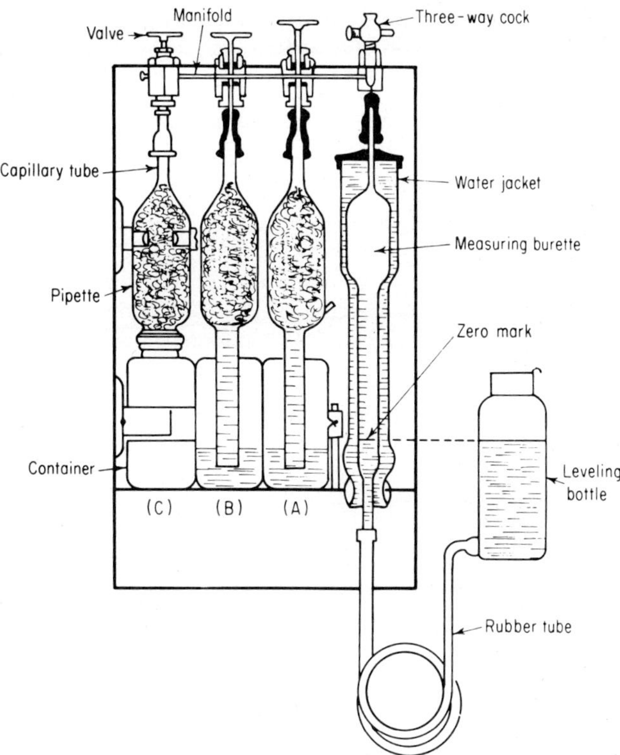

Fig. 11-3 Orsat apparatus used for analyzing flue gases.

79 percent nitrogen by volume and because nitrogen goes through a combustion process unchanged, the maximum percentage of CO_2 in flue gases that is possible (discounting nitrogen) is 21 percent. And since one volume of carbon combining with one volume of oxygen produces one volume of carbon dioxide, any unburned oxygen (excess) will reduce the percentage of carbon dioxide in the flue gas.

Example:

1. If there is no excess air, the flue-gas analysis will be

Carbon dioxide	21%
Oxygen	0
Nitrogen	79
Total	100%

2. If there is excess air of 100 percent,

Carbon dioxide	10.5%
Oxygen	10.5
Nitrogen	79.0
Total	100.0%

Note that the 21 percent maximum possible is now split evenly between oxygen and carbon dioxide at 10.5 percent each (100 percent excess air).

3. If there is excess air of 50 percent,

Carbon dioxide	14%
Oxygen	7
Nitrogen	79
Total	100%

In this case, carbon dioxide and oxygen percentages still add to 21 percent, but the oxygen is 50 percent of the carbon dioxide (50 percent excess air). It is also possible to calculate the amount of air used per pound of fuel from the flue-gas analysis.

Flue-gas analysis measurements can be used to calculate the weight of air used per pound of fuel burned by the following equation:

$$W_A = \frac{28N_2}{12(CO_2 + CO)\,(0.769)} \left(\frac{W_f C_f - W_r C_r}{W_f \times 100} \right)$$

where

CO_2 = percentage of carbon dioxide in flue gas by volume
CO = percentage of carbon monoxide in flue gas by volume
N = percentage of nitrogen in flue gas by volume
W_f = weight of fuel fired, lb
C_f = carbon content of fuel, percentage from ultimate analysis
C_r = carbon content of ash and refuse, percentage
W_r = weight of ash and refuse from W_f pounds of fuel, lb
W_A = actual weight of air per pound of fuel burned, lb

Example: To illustrate the use of this equation, assume that 700 lb of coal is fired in a boiler. The carbon content of this coal is 68 percent. Ash and refuse after burning amount to 60 lb, with the carbon in the refuse being 7.8 percent. Flue-gas analysis showed the following percentages by volume:

CO_2 (carbon dioxide) = 11.9%
N_2 (nitrogen) = 80.9%
CO (carbon monoxide) = 1%
O_2 (oxygen) = 7.3%

What is the actual weight of air used to burn this coal? Substituting values, we have

$CO_2 = 11.9$	$W_f = 700$
$CO = 1.0$	$C = 68$
$N_2 = 80.9$	$C_r = 7.8$
	$W_r = 60$

Substituting in the equation gives

$$W_A = \frac{28(80.9)}{12(11.9 + 1.0)(0.769)} \left[\frac{700(68) - 60(7.8)}{700 \times 100}\right]$$
$$= 19.0(0.68)$$
$$= 12.9 \text{ lb of air per pound of fuel fired}$$

or

$$700 \times 12.9 = 9030 \text{ lb of air for 700 lb of coal}$$

Another useful equation is used to determine the weight of flue gas W_{fg} formed by burning a pound of fuel (as determined from a flue-gas analysis):

$$W_{fg} = \frac{4CO_2 + O_2 + 700}{3(CO_2 + CO)} \left(\frac{W_f C_f - W_r C_r}{W_f \times 100}\right)$$

Example: The same coal is being burned with the same analysis as shown in the previous example.

Substituting, we get

$$W_{fg} = \frac{4(11.9) + 7.3 + 700}{3(11.9 + 1.0)} \left[\frac{700(68) - 60(7.8)}{700 \times 100}\right]$$
$$= 20.06(0.68)$$
$$= 13.6 \text{ lb flue gas per pound of fuel fired}$$

or

$$13.6 \times 700 = 9520 \text{ lb of flue gas for 700 lb of coal}$$

By using the specific heat (Btu per pound per degree temperature rise for a gas), it is possible to calculate the heat lost up the stack. Use the equation

$$H_L = WC_p(T_2 - T_1)$$

where

H_L = heat lost by flue gas
W = weight of flue gas going up the stack, usually lb/hr
C_p = mean specific heat of flue gas; can be taken as approximately 0.25 Btu (lb/°F)
T_2 = stack temperature, °F
T_1 = temperature of air entering furnace, °F

Example: In the previous problem, 9520 lb/hr of flue gas went up the stack. Assume the stack temperature is 650°F and the air inlet temperature is 80°F. Then the heat lost up the stack is

$$H_L = 9520(0.25)(650 - 80)$$
$$= 1{,}356{,}600 \text{ Btu/hr}$$

Since 700 lb of coal was fired, with an assumed average of 14,500 Btu/lb, the total energy input is

$$700 \times 14{,}500 = 10{,}150{,}000 \text{ Btu/hr}$$

The percentage of heat input going up the stack is then

$$\frac{1,356,600}{10,150,000} = 13.4\%$$

This indicates a boiler efficiency of $100 - 13.4 = 86.6$ percent.

Draft provides the differential pressure in a furnace to ensure the flow of gases. Without draft, stagnation in the burning process would result, and the fire or process of combustion would die from lack of air. Draft pushes or pulls air and the resultant flue gas through a boiler and up into the stack. The draft overcomes the resistance to flow of the tubes, furnace walls, baffles, dampers, and chimney lining (also slag).

Natural draft is produced by a chimney into which the boiler exhausts. The cool air admitted to a furnace (by means of damper openings) rushes in to displace the lighter hot gases in the furnace. Thus the hot gases rise (chimney effect), causing a natural draft.

Mechanical draft is produced artificially by means of forced- or induced-draft fans. The chimney is still necessary on mechanical-draft installations for venting the products of combustion high enough not to be offensive to the surroundings. Most modern boilers, including the domestic type, use some form of mechanical draft. Domestic burners may have a fan built into the burner unit.

The heat liberated by the complete and rapid burning of a fuel per unit weight or volume of the fuel is the *heating*, or *calorific, value* of the fuel. For solid and liquid fuels, this is usually expressed in Btu per pound. For gaseous fuels, it is expressed in Btu per cubic foot at a standard temperature and pressure, usually atmospheric pressure at 68°F.

Fuels which contain hydrogen have two heating values, higher and lower. The reason is that the burning of hydrogen produces superheated water vapor, which escapes at the temperature of the chimney gases. The lower heating value is the net heat liberated per pound of fuel after the heat necessary to vaporize and superheat the steam formed from the hydrogen (and from the fuel) has been deducted. The higher heating value is the one indicated by a fuel calorimeter and is usually used in engineering work.

A fuel calorimeter is a meter (also called *oxygen bomb*) to determine the heating value of 1 lb of fuel by burning a sample of the fuel under controlled conditions.

The main combustible elements in a fuel are carbon, hydrogen, and sulfur, which combine with oxygen. When each is burned separately, the following heating values are obtained for 1 lb of the element: carbon, 14,600 Btu; hydrogen, 62,000 Btu; sulfur, 4050 Btu.

The heating value of 1 lb of fuel can be calculated by Dulong's formula:

$$HV = 14{,}000C + 62{,}000\left(H - \frac{O}{8}\right) + 4050S$$

where

HV = heating value, Btu/lb of fuel
C = weight of carbon per pound of fuel
H = weight of hydrogen per pound of fuel
O = weight of oxygen per pound of fuel
S = weight of sulfur per pound of fuel

Example: An ultimate analysis of a coal shows the following percentages:

Carbon 82% =	0.82lb/lb of coal
Hydrogen 4.5% =	0.045lb/lb of coal
Oxygen 2.2% =	0.022 lb/lb of coal
Sulfur 1.8% =	0.018 lb/lb of coal
Ash 9.5% =	0.095 lb/lb of coal

Substituting gives

$$HV = 14{,}600(0.82) + 62{,}000\left(0.045 - \frac{0.022}{8}\right) + 4050(0.018)$$

$$= 11{,}972 + 2635 + 729$$
$$= 15{,}336 \text{ Btu/lb}$$

The ASME Power Boiler Code permits the values shown in Fig. 11-4*a* to be used in calculating safety-valve capacities where the exact heating value of the fuel is not known. The heating value is *H*.

COAL BURNING

Coal is a major source of energy in the United States and most probably will continue to be so for many years. However, one of the problems with coal as a source of energy is its sulfur content, present mostly as iron sulfide (pyritic sulfur) and in coal-borne organic systems, some containing the thiophene ring. Upon combustion of the coal as mined, harmful sulfur-containing gases and ashes are emitted into the atmosphere with deleterious effects on animal and plant life. In order to make coal burning environmentally safe and acceptable, some method of desulfurization must be applied. Stack-gas scrubbing is most often used, but it is economical only for large industrial and power plants.

Coal cleaning, and especially desulfurization prior to combustion, appears to be an attractive alternative. Many pollution controls and much monitoring can be omitted when desulfurized coal is available to plants carrying out relatively small operations.

Anthracite coal is very hard, is noncoking, and has a high percentage of fixed carbon. It ignites slowly, unless the furnace temperature is high, and requires a strong draft. The heating value is around 14,000 Btu/lb. Bituminous coal is soft, has a high percentage of volatile matter, burns with a yellow, smoky flame, and has a heating value of 11,000 to 14,000 Btu/lb. Semibituminous coal is the highest grade of bituminous. It burns with little smoke, is softer than anthracite, and has a tendency to break into small pieces when

Fig. 11-4 ***(a)*** **ASME heating values to be used in calculating relieving capacities of safety valves.**

Fuel	H = Btu/lb
Semibituminous coal	14,500
Anthracite	13,700
Screenings	12,500
Coke	13,500
Wood, hard or soft, kiln dried	7,700
Wood, hard or soft, air dried	6,200
Wood shavings	6,400
Peat, air dried, 25% moisture	7,500
Lignite	10,000
Kerosene	20,000
Petroleum, crude oil, Pennsylvania	20,700
Petroleum, crude oil, Texas	18,500
	H = Btu/ft³
Natural gas	960
Blast-furnace gas	100
Producer gas	150
Water gas, uncarbureted	290

Fig. 11-4 ***(b)*** **Typical coal properties used in analyzing coal-burning equipment needed.**

			Proximate analysis, as received				Ultimate analysis, dry and ash-free, percent					
Type of coal	Source	Heating value, Btu/lb	Moisture	Percent volatiles	Fixed C	Ash	S	C	H_2	O_2	N_2	Grindability ASTM
Anthracite	Pennsylvania	13,000	2	6.3	79.7	12	0.6	93.5	2.6	2.3	0.9	25
Bituminous	Pennsylvania	13,600	3	23.1	63.9	10	2.17	87.6	5.2	3.3	1.4	95
Bituminous	Ohio	12,450	6	34.8	49.2	10	2.44	82.2	3.5	7.7	1.7	66
Subbituminous	Colorado	9,200	24	30.2	40.8	5	0.36	75	5.1	17.9	1.5	58
Lignite	North Dakota	6,330	40	27.6	23.4	9	1.42	72.4	4.7	18.6	1.5	—

handled. The heating value is 13,000 to 14,500 Btu/lb. Subbituminous (black lignite) is a low grade of bituminous coal with a heating value between 9000 and 11,000 Btu/lb. Lignite is between peat and subbituminous coals, with a wood structure and claylike appearance. The heating value is 7000 to 11,000 Btu/lb.

Two methods of analyzing coal are *ultimate analysis* and *proximate analysis.* Ultimate analysis gives the percentages of the various chemical elements of which the coal is composed. Proximate analysis determines the percentage of moisture, volatile matter, fixed carbon, and ash with a fair degree of accuracy.

This analysis requires a laboratory and a skilled chemist. If a sample of coal is separated into its elements, certain proportions of oxygen, hydrogen, carbon, etc., will be found. These proportions are generally expressed as percentages of the weight of the original sample, the unit weight being 100 percent. The heating value of coal is estimated from the ultimate analysis by getting the percentages of carbon, oxygen, hydrogen, and sulfur in the coal and by measuring the heat of combustion available in 1 lb of coal.

Other conditions reported in a coal analysis are: (1) as-received; (2) air-dried; (3) moisture-free; (4) moisture- and ash-free; and (5) moisture- and mineral-free.

Because coals range over a broad spectrum of properties compared to gas and oil, burner and furnace designs for coal burning will vary a great deal. In addition, coal handling to the burners, the crushing and pulverizing equipment to be employed, the ash disposal methods to be used, and the type of environmental control that is to be exercised must all be considered.

Figure 11-4*b* shows some typical coal properties needed in burning coal. The grindability factor determines the ease with which it can be pulverized, as indicated by ASTM index numbers, also called the Hardgrove index. The base is taken as 100. A coal is difficult to grind if its index is below 100 and easier if it is above. Coal firing has progressed from the simple hand shoveling to stokers and pulverized firing. Fuel not burned in suspension is burned on various stokers. Two broad classes of stokers are overfeed, in which the fuel is carried into the furnace above the stoker, and underfeed, where the fuel is carried by the stoker underneath. Overfeed stokers are further classified into spreader and chain-grate stokers. See Fig. 11-5.

In the overfeed spreader stoker, raw coal is blown or thrown by air or steam or rotating paddles in suspension above the burning bed. The dust coal particles tend to burn in suspension. In the traveling grate stoker the fuel is added above the grate by a coal hopper through a gate, which regulates fuel-bed thickness. Coke is formed and burned as the grate moves the fuel to the back of the furnace, so that by the time the end is reached only ash remains, which is dumped off the grates. These grates travel to the front of the furnace by means of a sprocket drive for a fresh load of fuel to keep the cycle going. The underfeed type of stokers shown in Fig. 11-5*c* has the coal reach the fuel bed from below. The fuel is pushed along a feed trough, or retort, into the furnace and spills over onto the fuel bed at each side. Several names are used for the underfeed stoker types. The name of each is determined

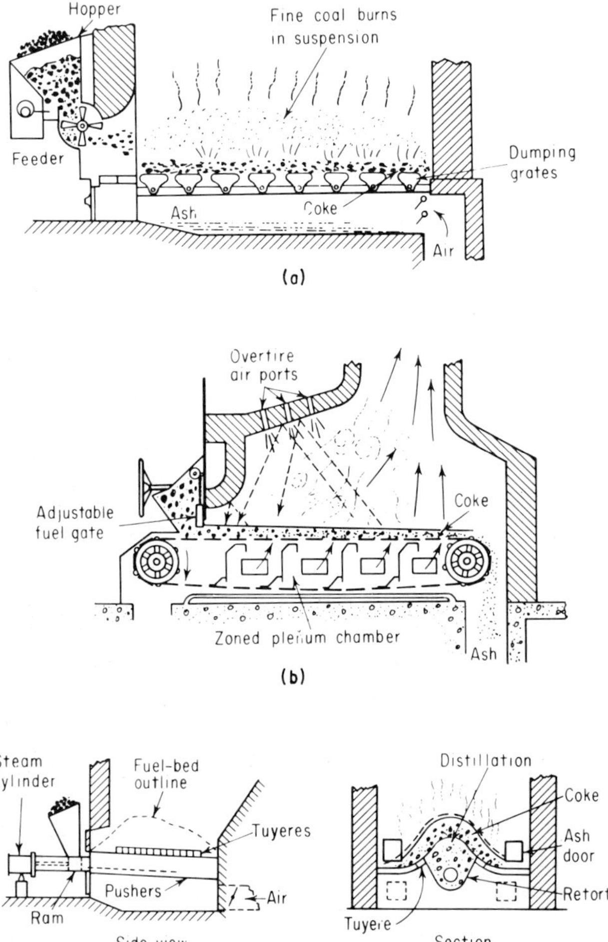

Fig. 11-5 Stokers are classified by *(a)* overfeed spreader, *(b)* travelling grate, *(c)* underfeed retort.

by the mechanism used to move the coal, such as single retort, multiple retort, screw feed, or ram feed. Single-retort units handle up to 50,000 lb/hr; multiple-retort designs handle up to 500,000 lb/hr.

Pulverized Coal Firing Pulverized coal firing is the most widely used method for burning coal in large boilers. The system requires coal to pass from feed bunkers through scales or feeders to the pulverizer. The grinding of the coal exposes the fuel elements in the coal to rapid oxidation (burning) as the ignition temperature is reached. More complete burning is thus possible than with fuel-bed burning.

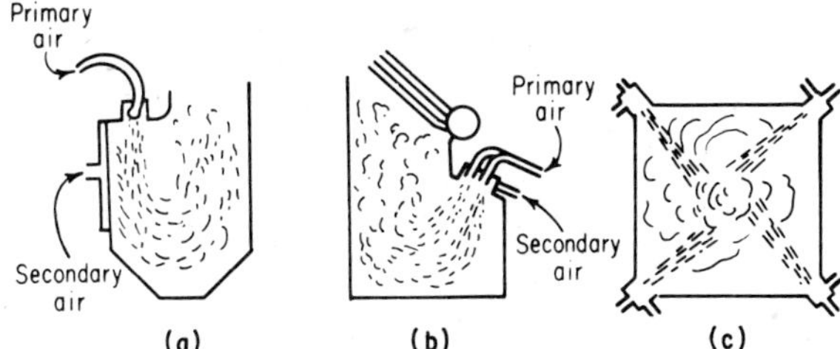

Fig. 11-6 Pulverized coal is fired in *(a)* long flame system, *(b)* shelf system, *(c)* corner or tangential system.

As these fine particles enter the furnace and become exposed to radiant heat, the temperature rises, and the volatile matter of the coal is distilled off in the form of a gas. Enough primary air is introduced at the burner to intimately mix with the stream of coal particles, which thus support combustion. The volatile matter burns first and then heats the remaining carbon to incandescence. Secondary air is introduced around the burner, which supplies the oxygen to complete the combustion of carbon particles in flames several feet long. Figure 11-6 shows pulverized-coal furnaces arranged for long-flame-system firing, shelf-system firing, and corner- (tangential) system firing.

In general, pulverizers (sometimes called *mills*) may be classified as attrition or impact types. To these might be added the shearing type, which is a form of the attrition type, impact type, or both. The impact mills generally have some attrition action. And conversely, while attrition may be the primary action of a mill, impact is usually present as a secondary action. Thus we have impact mills, including ball mills and hammer mills, and attrition mills, including bowl mills and ball-and-race mills.

Figure 11-7 shows a bowl-mill type of pulverizer. The grinding elements consist of three equally spaced, hollow toroidal rolls which run in a concave grinding ring. Force is applied to the rolls from above by a uniformly loaded thrust ring which fits smoothly into the necks of the rolls. The main drive shaft turns the table supporting the grinding ring, which transmits the motion to the rollers, driving them at about half the speed of the grinding ring. The rolls revolve about their own axes and simultaneously, in planetary fashion, they revolve about the axis of the mill, separated and spaced equally around the grinding ring by the satellite spacer.

Burners for pulverized coal must supply air and fuel to the furnace in a manner that permits stable ignition, effective control of flame shape and travel, and thorough and complete mixing of fuel and air. The air used to transport the coal to the burner forms the primary air; secondary air may be introduced in the burner (turbulent burners) or around or near the burner (nozzle burners). The turbulent burner (Fig. 11-2*a*) imparts a rotary motion to the coal-air mixture in a central nozzle and the secondary air issuing from a chamber around that nozzle, all within the burner. This gives some premixing for coal and air and considerable turbulence. In some burners the coal-air mixture issues

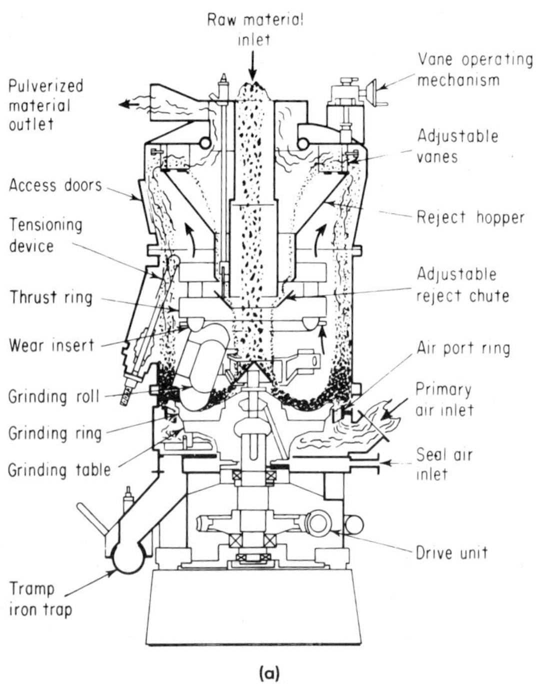

(a)

(b)

Fig. 11-7 Bowl-mill pulverizer. *(a)* Internal detail. *(Courtesy Foster Wheeler Corp.)* *(b)* External view of a unit. *(Courtesy Combustion Engineering Inc.)*

from a series of nozzles to mix within the furnace with secondary air admitted through separate openings. Tangential firing has burners in furnace corners that direct their flames tangentially to an imaginary circle in the furnace space (Fig. 11-6*c*). Turbulence can be set up in this fashion, and there is a tendency for unburned combustible material in the tail of the flame to be caught up in the second. A special form of tangential firing appears in Fig. 11-2*b*. These burners are adjustable to shift the flame zone vertically and so regulate the temperature of the furnace exit gas according to load. This, in turn, controls superheat over a wide load range. Newer installations use fully automatic control of burner inclination.

The cyclone burner illustrated in Fig. 11-2*c* receives crushed (not pulverized) coal in a stream of high-velocity air tangent to the circular burner housing, which forms a primary water-cooled furnace. Coal thrown to the rim of the furnace by centrifugal force and held by a coating of molten ash is scrubbed by fast-moving air. Secondary air enters at high velocity also and parallel to the path of the primary coal-air mixture. The coal in the sticky slag film burns as if it were in a fuel bed. Volatiles are distilled off, and carbon is burned out to leave ash. Combustion of volatile matter begins in the burner chamber and is completed in the secondary furnace into which the burner chamber discharges. Molten ash, under centrifugal force, clings to the burner-chamber walls, and the slight inclination causes slag to discharge continuously. The nature of this burning tends to reduce greatly the amount of ash carried in suspension, and hence fly-ash emission is negligible.

FLUIDIZED-BED COMBUSTION

Fluidized-bed boilers are being developed in order to take advantage of the large coal deposits that exist as well as overcome the pollution problem that exists in burning coal. The merits in using fluidized-bed combustion are the following: (1) High-sulfur fuels can be burned without resorting to flue-gas treatment. This is accomplished by injecting limestone into the bed which absorbs the sulfur dioxide. (2) Higher combustion efficiencies are obtainable in fluidized-bed burning. (3) Lower combustion temperatures are possible which minimize nitrogen oxide and furnace slag formation. (4) The waste product formed at a lower bed temperature is easier to handle and dispose.

A fluidized bed consists of granular particles lying on a nonsifting grid through which air is blown at a velocity sufficient to lift and float the particles. Bubbles are formed to the extent that the mass now acts like a boiling liquid. See Fig. 11-8. The particles held in suspension in the bed are usually limestone. Crushed coal is injected into the limestone and burned. To capture the sulfur in the fuel, the bed temperature is controlled at approximately 1550°F, at which optimum sulfur capture on lime occurs to form calcium sulfate. Raw limestone is continually injected into the bed while a gravity drain system withdraws the spent material including large coal-ash particles. The low operating temperature in the fluidized bed creates low nitrogen oxide formations. Particles being carried by the flue gas can be captured by convential electro-

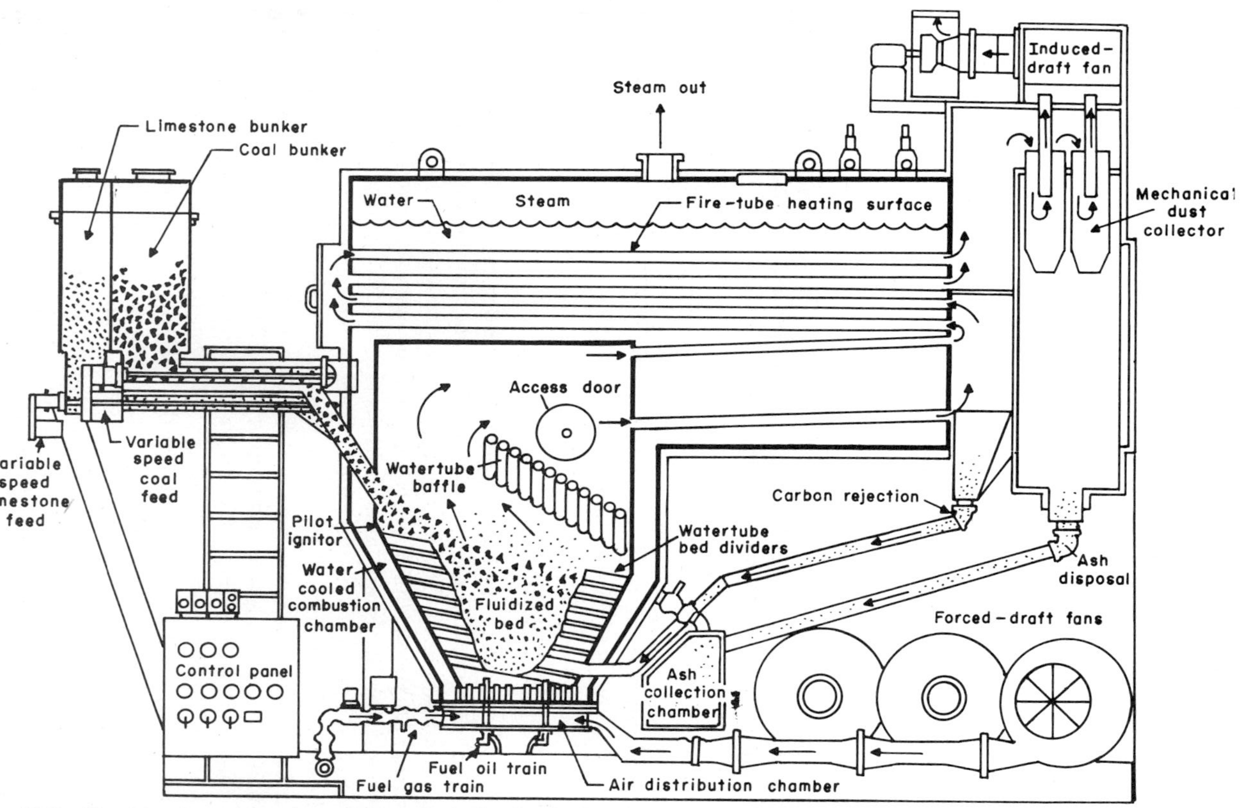

Fig. 11-8 Combination watertube and fire-tube boiler for fluidized-bed combustion and for burning oil or gas. *(Courtesy Johnston Boiler Co.)*

static precipitators or bag-house filters. Figure 11-8 shows a combination water-tube and fire-tube boiler designed for fluidized-bed burning. Fluidized-bed-burning boilers approach the oil-burning boiler in size because the former have heat-release rates of 100,000 Btu/(hr/ft^3) whereas a conventional coal-fired steam generator has a heat-release rate of around 20,000 Btu/(hr/ft^3).

FUEL OIL

Petroleum products are listed in Fig. 11-9*a* and *b*. Refining of petroleum involves separating and recombining the carbon and hydrogen molecules into

Fig. 11-9 ***(a)*** **Typical fractions of petroleum fuels.**

Type of fuel	Specific gravity	API gravity	Weight, lb/gal	Heating value	
				Btu/lb	Million Btu per barrel
Residual fuel	1.0	10	8.337	18,540	6.5
No. 4 fuel oil	0.966	15	8.053	18,840	6.35
Heavy distillate	0.910	24	7.587	19,190	6.1
Light distillate	0.865	32	7.215	19,490	5.95
Kerosene	0.825	40	6.879	19,750	5.7

Fig. 11-9 ***(b)*** **Fuel-oil properties arranged by API gravity scale at standard 60°F base.**

API gravity	Specific gravity	Weight, lb/gal	Btu/lb	Btu/gal	lb/42-gal barrel	lb/ft^3
3	1.0520	8.76	18,190	159,340	368.00	65.54
5	1.0366	8.63	18,290	157,840	362.62	64.59
7	1.0217	8.50	18,390	156,320	357.37	63.65
9	1.0071	8.39	18,490	155,130	352.46	62.78
11	0.9930	8.27	18,590	153,740	347.71	61.93
13	0.9792	8.16	18,690	152,510	342.88	61.07
15	0.9659	8.05	18,790	151,260	338.22	60.24
17	0.9529	7.94	18,890	149,980	333.64	59.42
19	0.9402	7.83	18,980	148,610	329.23	58.64
21	0.9279	7.73	19,060	147,330	324.91	57.87
23	0.9159	7.63	19,150	146,110	320.71	57.12
25	0.9042	7.53	19,230	144,800	316.59	56.39
27	0.8927	7.44	19,310	143,670	312.60	55.68
29	0.8816	7.35	19,380	142,440	308.70	54.98
31	0.8708	7.26	19,450	141,210	304.92	54.31
33	0.8602	7.17	19,520	139,960	301.18	53.64
35	0.8498	7.08	19,590	138,690	297.57	53.00
37	0.8398	7.00	19,650	137,550	294.04	52.37
39	0.8299	6.92	19,720	136,400	290.64	51.76
41	0.8203	6.83	19,780	135,090	287.23	51.16

fractions having the same range of boiling points. Typical fractions from light to heavy are naphtha, gasoline, kerosene, and gas-oil.

The API scale is the American Petroleum Institute scale for showing specific gravity. The API scale fixes a reading of 10°F as equal to a specific gravity of 1.00. Readings greater than 10°F indicate a specific gravity of less than 1.0 or an oil which is lighter. To obtain the actual specific gravity in relation to water from the API reading, use the following equation:

$$\text{Actual specific gravity} = \frac{141.5}{131.5 + \text{API deg}}$$

Specific gravity in degrees Baumé (°Bé) is found in the same way except that the numbers are 140 and 130, respectively. For practical purposes, the two specific-gravity scales may be considered the same.

Viscosity is the relative ease, or difficulty, with which an oil flows. It is measured by the time in seconds a standard amount of oil takes to flow through a standard orifice in a device called a *viscosimeter.* The usual standard in this country is the Saybolt Universal, or the Saybolt Furol, for oils of high viscosity. Since viscosity changes with temperature, tests must be made at a standard temperature, usually 100°F for Saybolt Universal and 122°F for Saybolt Furol. Viscosity indicates how oil behaves when pumped and, more particularly, shows when preheating is required and what temperature must be held.

Flash point represents the temperature at which an oil gives off enough vapor to make an inflammable mixture with air. The results of a flash-point test depend on the apparatus, so this is specified as well as temperature. Flash point measures an oil's volatility and indicates the maximum temperature for safe handling.

Pour point represents the lowest temperature at which an oil flows, under standard conditions. Including pour point as a specification ensures that an oil will not give handling trouble at expected low temperatures.

By centrifuging a sample of oil, the amounts of water and sediment present can be determined. These are impurities, and while it is not economical to eliminate them, they should not occur in excessive quantities (not more than 2 percent). Incombustible impurities in oil, from natural salts, from chemicals in refining operations, or from rust and scale picked up in transit, show up as ash. Some ash-producing impurities cause rapid wear of refractories, and some are abrasive to pumps, valves, and burner parts. In the furnace, they may form slag coatings.

Fuel oils are sold in six standardized grades, under the numbers or grades of 1, 2, 3, 4, 5, and 6. Grades 1, 2, and 3 are light, medium, and heavy domestic fuel oils. These usually do not require heating prior to burning in a furnace. Grades 4, 5, and 6 correspond to federal specifications for Bunkers *A, B,* and *C,* respectively. These oils are heavy and viscous; thus they require heating prior to being sprayed into a furnace.

OIL BURNERS

In addition to proportioning fuel and air and mixing them, oil burners must prepare the fuel for combustion. Two ways (with many variations) are: (1) oil may be vaporized or gasified by heating within the burner or (2) oil may be atomized by the burner so vaporization can occur in the combustion space. Vaporizing burners (first group) are limited in range to fuels they can handle and find little use in power plants. If oil is to be vaporized in the combustion space in the instant of time available, it must be broken up into many small particles to expose as much surface as possible to the heat. Atomization is effected in three basic ways: (1) by using steam or air under pressure to break the oil into droplets; (2) by forcing oil under pressure through a nozzle; and (3) by tearing an oil film into drops by centrifugal force. All three methods are used. In addition, a burner must provide good mixing of fuel and air so complete combustion of the oil droplets may ensue.

The steam-atomizing burners possess the ability to burn almost any fuel oil, of any viscosity, at almost any temperature. Air is less extensively used as an atomizing medium because its operating cost is apt to be high. These burners can be divided into two types:

1. Internal-mixing or premixing oil-and-steam (or air) (Fig. 11-10*a*) mix inside the body or tip of the burner (Fig. 11-10*b*) before being sprayed into the furnace
2. External-mixing, where oil emerging from the burner is caught by a jet of steam or air (Fig. 11-10*c*)

Steam consumption for atomizing runs from 1 to 5 percent of the steam produced, with the average around 2 percent. The pressure required varies from about 75 to 150 psi.

In the burner of Fig. 11-10*c*, oil reaches the tip through a central passage, with the flow being regulated by the screw spindle. Oil whirls out against a sprayer plate to break up at right angles to the stream of steam, or air, coming out behind it. The atomizing stream surrounds the oil chamber and receives a whirling motion from vanes in its path. When air is used for atomizing, it should be at 10 psi for lighter oils and 20 psi for heavier. Combustion air enters through a register. Vanes or shutters are adjustable to give control of excess air. Mechanical-atomizing oil burners depend on pressures forcing oil through nozzles to obtain a fine oil mist for combustion.

Good atomization results when oil under a pressure of 75 to 200 psi is discharged through a small orifice, often aided by a slotted disk. The disk gives the oil a whirling motion before it passes on through a hole drilled in the nozzle, where atomization occurs. For a given nozzle opening, atomization depends on pressure, and since pressure and flow are related, the best atomization occurs over a fairly narrow range of burner capacities. To follow the boiler load as steam demand goes up or down, a number of burners may be installed and turned on or off, or burner tips with different nozzle openings

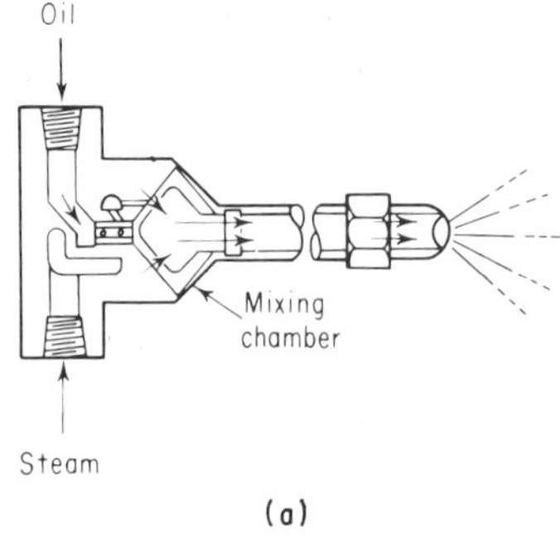

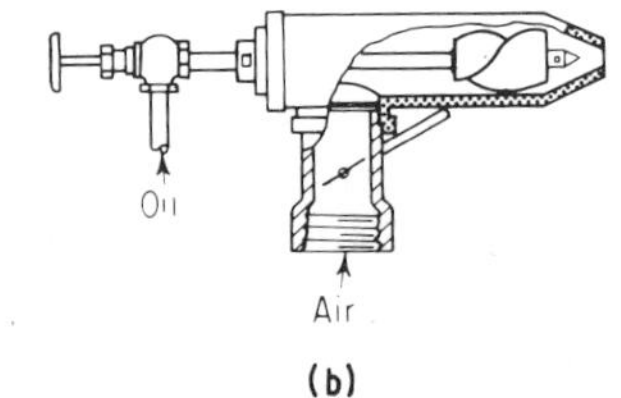

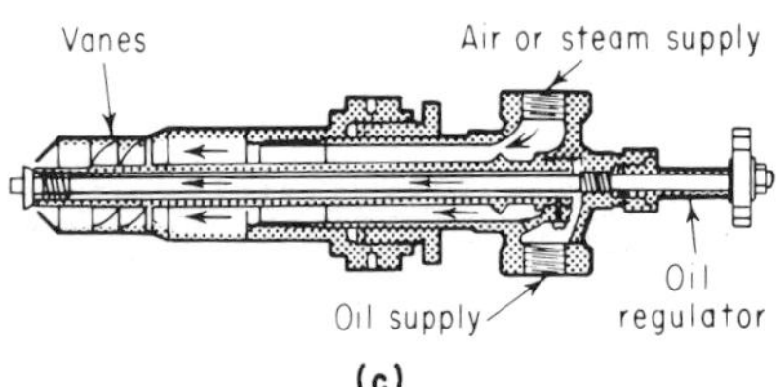

Fig. 11-10 *(a)* Internal mixing of oil and steam burner; *(b)* internal mixing of oil and air; *(c)* external mixing of oil and air.

may be used. All nozzle openings must be changed to the same size in a given system, never fired with mixed sizes.

There are many burner designs to extend the usual 1.4:1 capacity range of the mechanical-atomizing nozzle. One has a plunger that opens additional tangential holes in the nozzle as oil pressure increases. This gives a 4:1 range. The burner in Fig. 11-11*a* uses a movable control rod which, through a regulating pin, varies the area of tangential slots in the sprayer plate and the volume of the oil passing the orifice.

The wide-range mechanical atomizer (Fig. 11-11*b*) gives a capacity range of about 15:1 and much higher if needed. By use of either a constant-differential valve or pump, as shown, the difference in pressure between supply and return is held constant. This pump system offers advantages in many plants:

1. No hot oil is returned to the storage tank or pump suction.
2. Fuel enters the closed circuit at the same rate it is burned, thus simplifying fuel metering and combustion control.
3. The pump may be used to boost pressure on existing oil burner systems.

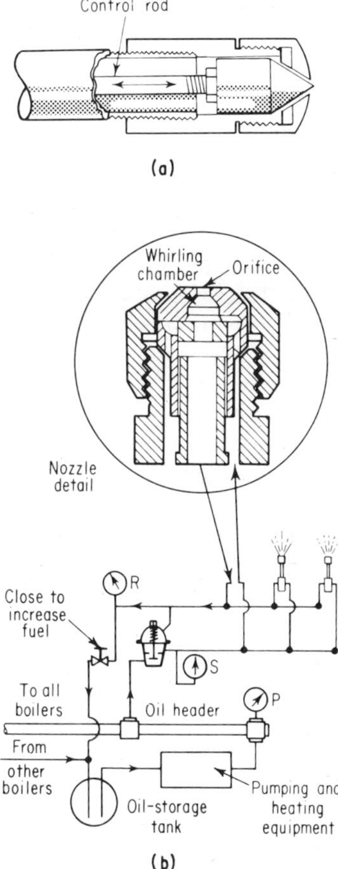

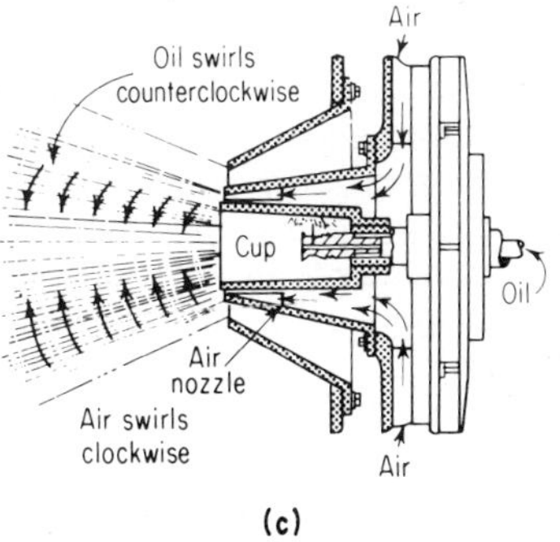

Fig. 11-11 Mechanical-atomizing burner types. *(a)* Movable control rod varies spray. *(b)* Constant pressure-differential pump or valve controls spray. *(c)* Rotary-cup burner atomizes fuel by centrifugal force.

The horizontal rotary cup atomizes fuel oil by literally tearing it into tiny droplets. A conical or cylindrical cup rotates at high speed (usually about 3500 r/min) if motor-driven. Oil moving along this cup reaches the rim where centrifugal force flings it into an air stream (Fig. 11-11*c*). This system of atomization requires no oil pressure beyond that needed to bring oil to the cup. But high oil preheat temperatures must be avoided since gasification may develop. The rotary cup can satisfactorily atomize oils of high viscosity (300 seconds Saybolt Universal, or SSU) and has a wide range of about 16:1.

Fuel-oil heaters are required when heavy, viscous oil is burned in order to facilitate flow and assist atomization. In these heaters, oil is pumped through tubes that are surrounded by steam and enclosed in a shell.

The maximum safe pressure of the shell should be ascertained definitely, and a safety valve set at not over this pressure should be installed on the shell or on the steam supply system.

The manufacturer's specifications of maximum oil pressure should be learned, and an oil-pressure relief valve installed between the oil pump and the first shutoff valve in discharge line to the heater. The discharge of this relief valve may be piped back to the oil storage tank for cleanliness. A second oil-pressure relief valve should be installed at the outlet from the heater, for, when valves are closed, expansion may create excessive pressures.

The condensate from the heater should be drained to waste unless a well-lighted and frequently observed gauge glass is on the trap body or other suitable means of oil detection are employed. One cannot be too careful with this installation, for a split tube in the oil heater might allow the fuel oil to pass into the feedwater system, flooding the inside of the boilers with oil, an extremely hazardous condition.

GAS FUELS

Natural gas is the main fuel used in steam generation because manufactured gases run too high in cost. By-product gases usually have low heating values and are produced in relatively minor quantities, so they are ordinarily used at the production point and not distributed. Natural gas is colorless and odorless. Composition varies with source, but methane (CH_4) is always the major constituent. Most natural gas contains some ethane (C_2H_6) and a small amount of nitrogen. Gas from some areas, often called *sour gas* contains hydrogen sulfide and organic sulfur vapors. The heating value averages about 1000 Btu/ft^3 (20,000 Btu/lb), but may run much higher. Natural gas is usually sold by the cubic foot, but may be sold by the therm (100,000 Btu).

Coal gas and coke-oven gas (manufactured gases) are produced by carbonizing high-volatile bituminous coal in retorts that exclude air and are heated externally by producer gas. Usually a number of by-products result. Cleaned of impurities, these gases are roughly one-half hydrogen and one-third methane, plus small amounts of carbon monoxide, carbon dioxide, nitrogen, oxygen, and illuminants (C_2H_4 and C_6H_6). The heating value runs around 550 Btu/ft³.

The gas served in a given area may be a mixture of two or more gases or a mixture of natural and manufactured gas. The heating value, usually held to 525 to 550 Btu/ft³, is often fixed by state or local ordinance.

Commercial butane and propane are essentially by-products from the manufacture of natural gasoline and from certain refinery operations. As supplied, propane (C_3H_8) is essentially pure, while butane (C_4H_{10}) usually contains a small amount of propane. Both have high heating values, are easily liquefied at low pressure, and are widely used as bottled fuels.

Blast-furnace gas, a by-product of iron making, has the lowest heating value of any commercial gas, about 90 Btu/ft³. It is close to three-quarters nitrogen and carbon dioxide, the only important combustible constituent being carbon monoxide. Raw gas, which usually contains a high concentration of solid impurities, is normally washed before use. But unwashed gas has been successfully burned in boiler furnaces.

Sewage-sludge gas runs about two-thirds methane and one-third carbon dioxide, with small amounts of hydrogen, nitrogen, and usually some hydrogen sulfide. The heating value is about 650 Btu/ft³. Although used mostly in internal-combustion engines, this gas is also burner-fired. See Fig. 11-12*a* for properties of fuel gases.

Burning gas requires no preparation of the fuel, as do other fuels. But proportioning with air, mixing, and burning can be handled in several ways. Also, the fuel's characteristics need to be known for sound selection of equipment and successful operation. Atmospheric burners are used for gas burning and differ mainly in the way air and fuel mix. The atmospheric burner is popular, as in home gas ranges. The momentum of the incoming low-pressure gas stream is used to draw in, or aspirate, part of the air needed for combustion. A shutter or similar device regulates the amount of air so induced. Gas and air together pass through a tube leading to the burner ports, mixing in the process. The mixture burns at the ports or openings in the burner head (with a blue, nonluminous flame). Secondary air is drawn into the flame from the surrounding atmosphere.

A single-port atmospheric burner is shown in Fig. 11-13*a*. A needle valve controls the gas flow through the spud; air is drawn in around the shutter at the end. With burner-port size and shape fixed, the nature of burning depends largely on the amount of primary air, or premix. With premix low, the flame is long and pale blue. It may have a yellow tip, indicating cracking and presence of free carbon.

Operation is usually satisfactory with 30 to 70 percent premix; in some special designs, 100 percent primary air is used. This premix range gives a turndown, or capacity, range of about 4:1. Usually premix and capacity ranges are somewhat narrower. Secondary air may be drawn in around the burner, the amount depending on the area of the opening and the draft. The high-pressure burner uses gas at about 20 to 30 psig and air at atmospheric pressure. Another type uses compressed air, with gas at atmospheric pressure.

Fig. 11-12 ***(a)*** **Properties of fuel gases**

Fuel	Source	Average composition	High heat value, Btu/ft³	Remarks
Blast-furnace gas	By-product of iron making	58% N_2, 27% CO, 12% CO_2, 2% H_2, some CH_4	90–100	Good fuel when cleaned—used mainly at source
Butane	By-product of gasoline making, also in casing-head gas	C_4H_{10} (usually has some butylene C_4H_8 and propane C_3H_8)	3200–3260	Liquefies under slight pressure, sold as liquid (bottled gas)
Casing-head gas	Oil wells	Varies, mostly butane, propane	1200–2000	Used mostly in oil fields
Carbureted water gas	Manufactured from coal, enriched with oil vapor	34% H_2, 32% CO, 16% CH_4, 7% N_2, 5% C_2H_4, 4% CO_2, 2% C_6H_6	500–600	Good fuel, but usually costly Part of most city gas
Coke-oven gas	By-product coke ovens	48% H_2, 32% CH_4, 8% N_2, 6% CO, 3% C_2H_4, 2% CO_2, 1% O_2	500–600	Good fuel when cleaned, often used at source
Natural gas	Gas wells	Varies, mostly CH_4, C_2H_6, C_3H_8	950–1150	Ideal fuel, piped to point of use
Oil gas	Manufactured from petroleum	54% H_2, 27% CH_4, 10% CO, 3% N_2, 3% CO_2, 3% C_2H_4	500–550	Used on West Coast, often mixed with coke-oven gas
Producer gas	Manufactured from coal, coke, wood, etc.	51% N_2, 25% CO, 16% H_2, 6% CO_2, 2% CH_4	135–165	Requires cleaning
Propane	By-product of gasoline	C_3H_8	2500	Similar to butane
Refinery gas	By-product of petroleum processing	Varies, mostly butane, and propane	1200–2000	Used mainly at refineries
Sewage gas	Sewage-disposal plants	65% CH_4, 30% CO_2, 2% H_2, 3% N_2, traces of O_2, CO, H_2S	600–700	Many disposal plants meet all power needs with this fuel

Fig. 11-12 ***(b)*** **Heating value of waste fuels.**

Waste	Average heating value (as fired), Btu/lb
Gases:	
Coke-oven	19,700
Blast-furnace	1,139
Carbon monoxide	575
Refinery	21,800
Liquids:	
Industrial sludge	3,700–4,200
Black liquor	4,400
Sulfite liquor	4,200
Dirty solvents	10,000–16,000
Spent lubricants	10,000–14,000
Paints and resins	6,000–10,000
Oily waste and residue	18,000
Solids:	
Bagasse	3,600–6,500
Bark	4,500–5,200
General wood wastes	4,500–6,500
Sawdust and shavings	4,500–7,500
Coffee grounds	4,900–6,500
Nut hulls	7,700
Rice hulls	5,200–6,500
Corn cobs	8,000–8,300

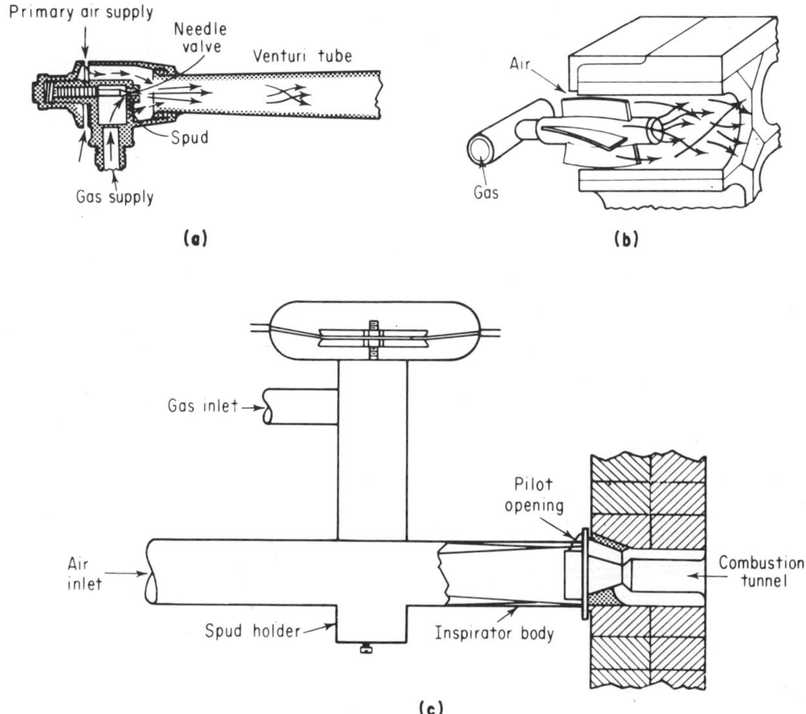

Fig. 11-13 Gas burner types. *(a)* Atmospheric gas burner. *(b)* Tunnel gas burner. *(c)* Inspirator burner premixes air and gas for burning.

A refractory gas burner is shown in Fig. 11-13*b*. It depends on natural or fan draft to draw in all the air required for combustion; hence draft conditions are important. One design uses multiple gas jets, which discharge into the airstream to cause violent agitation in a short mixing tube or tunnel of refractory. In the burner of Fig. 11-13*b* turbulence vanes impart a swirling motion to the air entering the tunnel.

Large steam-generating units often use a high-pressure (2 to 25 psi) gas burner of the gas-ring, center-diffusion-tube, or turbulent, design. The gas ring has an annular manifold located between the air register and the furnace wall surrounding the burner opening. Orifices drilled in this ring spray gas angularly across an incoming airstream, controlled in quantity, velocity, and rotation by the resistor.

Figure 11-13*c* shows a low-pressure-type burner in which gas is at atmospheric pressure and air is at 1 to 2 psi. This unit, through the inspirator governor, prepares a burning mixture to several burners. It can provide a higher burner-head pressure, to overcome variable draft conditions in the furnace, and good overload capacity with uniform air-gas mix at all loads.

WASTE FUELS

Combustion of industrial wastes and municipal refuse is gaining increased attention as fossil fuel becomes scarcer and more expensive. Among the solid fuels are the following: Wood from lumber and woodworking industries in the form of sawdust, slabs and shavings, and hog wood. Hog wood is wood refuse cut to uniform size before burning. Bark from pine, oak, and hemlock trees is burned in special furnaces. The heating value of wood varies from 2500 to 3000 Btu/lb. Bagasse is the crushed stalks of sugarcane from which the sap has been extracted. The heating value is from 3500 to 4500 Btu/lb. Coke is the solid remains after the destructive distillation of either petroleum oils or certain bituminous coals. The heating value of petroleum coke is from 11,500 to 15,000 Btu/lb. See Fig. 11-12*b*.

The boiler manufacturer should be consulted on any conversion of a fossil-fuel-burning boiler to a waste-fuel-burning one. There are many variables to consider in order to prevent future operating difficulties. In general, solid wastes are burned on a fuel bed, or by shredding, in suspension. Liquids are burned by atomizing-type burners.

The continuing emphasis on environmental control adds another design consideration: the problem of emissions from burning of waste fuel. Mechanical dust collectors are used as well as wet or dry scrubbers and electrostatic precipitators in order to comply with emission standards.

SO_2 Removal The new Clean Air Act of the federal government makes it mandatory in all fuel-firing systems to limit not only particulate emission normally handled by bag-house filters, or precipitators, but also sulfur dioxide emissions, which are a gaseous product. Present emphasis in stack-gas cleanup is on flue-gas desulfurization, but nitrogen oxides will also become an emission subject to control. Gas-scrubbing systems for cleaning boiler and incinerator flue gases are being used. The three major types of scrubbing systems designed are the lime or limestone slurry type, the magnesium oxide slurry, and the double alkali system. Scrubbing systems are an additional expense, and in power generation they absorb 2 to 7 percent of total power plant output because of the additional pumps, fans, and similar equipment that are needed. Material of construction for scrubber systems must be acid- and abrasion-resistant and must also withstand high temperatures during plant-upset conditions. Stainless and nickel-alloy steels are used to avoid stress corrosion cracking in the presence of chlorides. Rubber-lined mild steel is also being used, as are flakeglass-reinforced polyester linings and similar acid-proof coatings or linings. Figure 11-14 shows a typical limestone scrubbing system that is added to the flue-gas flow system after the boiler precipitator. The flue gas, after being cleaned in the precipitator, enters the bottom of the scrubber tower and passes up through the limestone slurry, which is being sprayed into the tower from above. The chemical reaction between the SO_2 and limestone results in removal of about 80 to 85 percent of the SO_2 from the flue gas. The sludge formed must be disposed of in approved landfill sites.

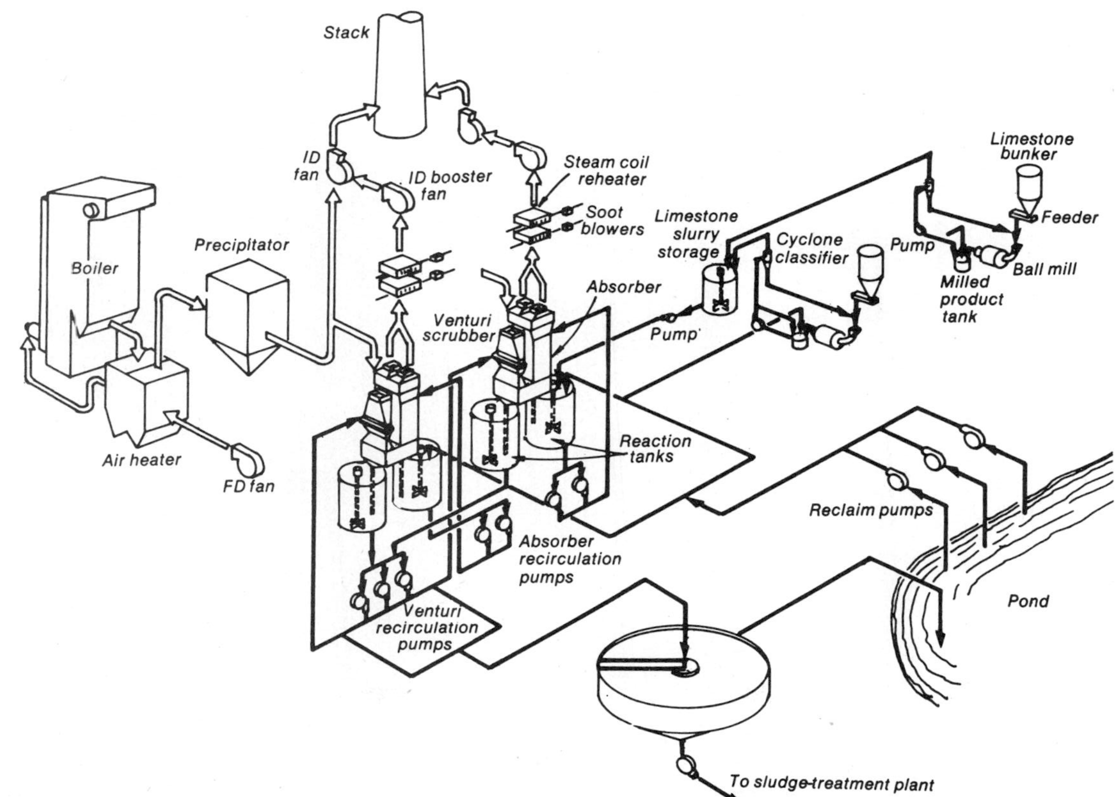

Fig. 11-14 Limestone scrubber system to remove SO_2 from flue gas. *(Courtesy Power Magazine.)*

CONTROLS

Controls on boilers are used extensively to regulate steam pressure and load within 1 to 2 percent of design points—steamdrum water levels within 1 in. of set points, fuel/air ratios within 5 percent of excess air requirements, and many more variables that can affect the operation of a boiler or its safety. A multitude of new sensors have been developed such as sodium analyzers, oxygen analyzers, and complete stack-gas analysis equipment in order to assist an operator or attendant in operating a boiler efficiently and safely. This section covers briefly general, feedwater, combustion, and pressure or load controls as well as safety controls and interlocks.

Sensors are devices that have the inherent ability to monitor changes in the medium being measured, such as temperature, pressure, flow, level, draft, percentage of CO_2, and similar quantities. Transducers are used to convert the changes noted by a sensor to an electric or a pneumatic signal. These signals are sent to controls, which are set to regulate a quantity within set points. The signals are thus compared to these set points, and if needed, a signal is then relayed to an actuator so that control of the medium is maintained within the established set points. Some controls incorporate the sensor, transducer, and actuator into a self-contained control device. Packaged boilers of the fire-tube type generally provide fully automatic operation because of the use of controls. The standard package will include a compact control cabinet containing motor starters, relays, switches, fuses, control transformers, and electronic flame safeguard controls. Indicating lights portray the sequence

Fig. 11-15 Scale and low water caused tube melting from lack of testing and maintenance of basic boiler controls. *(Courtesy Factory Mutual Engineering.)*

of operation and can be tied in with a remote station or even to a commercial surveillance system. Packaged watertube boilers are designed for gas or oil burning and are instrumented for basic automatic operation through control technology. However, complete reliance on automatic controls without periodic testing and maintenance can be dangerous. Figure 11-15 shows the result of neglectful operation. The tubes in the watertube boiler were melted from overheating resulting from a combination of scale in the tubes and failure of the low-water cutoff to stop the fuel input as the water level dropped to dangerous levels.

Safety controls generally are those that limit energy input and thus shut down the equipment when unsafe conditions develop. They are: (1) pressure-limit or temperature-limit switches; (2) low-water fuel cutoffs; (3) flame-failure safeguard systems; (4) automatic ignition controls; (5) oil and gas fuel-shutoff-valve controls; (6) air and fuel pressure intelock controls; and (7) feedwater regulating controls.

The safety valve (or relief valve) is the most important safety device. While not considered a control in the usual sense, it is the last measure against a serious explosion.

Safety controls guard against the following hazards:

1. Overpressure, leading to explosions from the waterside or steam side
2. Overheating of metal parts, possibly also leading to explosion in a fired boiler (mainly because of low water or poor circulation)
3. Fireside explosions (furnace explosions) due to uncontrolled combustible mixtures on the firing side

These types of accidents are considered major and may lead to loss of life and serious property damage. Other potential sources of accidents are cracking; bulging from local overheating because of scale; deforming, such as tubes bowing; thinning of vital pressure parts, which can lead to cracking or localized rupture; and expansion and contraction failures, causing cracking or rupturing of metal parts.

Manufacturers and state laws are trying to prevent, with safety control equipment, the three major types of accidents of overpressure, dry firing, and furnace explosions. While the other types of failures are controlled somewhat by automatic controls, prevention is mostly by legal inspection requirements and by proper operation and maintenance practices expected from the owner-user of a boiler. Included are good feedwater treatment and testing of controls at periodic intervals, including safety relief valves.

Any installation will be improved by the use of instruments when trained operating personnel are in attendance and make intelligent use of the data provided. Instrumentation of larger packaged boilers should include an Orsat apparatus for obtaining flue-gas analysis and determining combustion efficiency.

As a minimum, the ASME Boiler Code recommends the following instruments: (1) Steam pressure gauge; (2) feedwater pressure gauge; (3) furnace draft guage; (4) an outlet pressure gauge on the forced-draft fan and an inlet pressure gauge on the induced-draft fan; (5) steam-flow recorder for checking boiler output; (6) CO_2 recorder to check on combustion; (7) superheater inlet and outlet temperature recorder; (8) inlet and outlet temperature recorders for air heaters; (9) thermometers indicating inlet and outlet steam temperatures for boiler reheaters; (10) feedwater-temperature recorders for checking degree of deaeration and economizer operation; (11) pressure gauges on pulverizers to check differential pressure for fuel-air mixtures to burners; (12) pressure gauges for oil-fired boilers on oil lines to burners and temperature gauges before and after any oil preheaters; and (13) pressure gauges for gas-fired boilers on the main gas line to burners and on individual burners.

Load control on a boiler is generally geared to maintaining a constant pressure or temperature. Pressure variation is caused by:

1. *Load on the boiler.* An increase in load without additional fuel input causes a pressure drop. A decrease in load without an accompanying decrease in fuel input causes a pressure rise.
2. *Fuel input to the boiler.* Too high an input will cause a pressure rise, while too low an input will cause a pressure drop.

Thus pressure regulation and fuel regulation, or combustion controls, are directly related. For this reason combustion controls are geared for modulation by pressure variations within close limits. While airflow and exhaust flow usually follow fuel flow, the latter is determined by the pressure-set limits within which a boiler is to operate.

Although manufacturers differ in approach, the following factors must be considered in any control used on a boiler: (1) steam pressure and flow; (2) furnace pressure and draft; (3) air pressure and flow; (4) feedwater pressure and flow (including low water); (5) flue-gas flow and composition; and (6) proper ignition and burner-flame control.

Modes or the manner in which a control acts and reacts to restore a variable on a boiler may be classified into on-off or two-position control, positioning control, and metering control.

On-off controls are limited to fire-tube and small watertube boilers. As the name implies, a drop in pressure actuates a pressurestat or mercury switch to start the stoker or burner and open the air damper, or to reverse the process when pressure rises again. Since control is limited to varying the lengths of on and off periods, combustion efficiency is low. A typical on-off combustion-control circuit for a low-pressure steam-heating boiler has the low-water cutoff control and pressure control switches hooked up in electric series. Thus if either control opens, the current to the burner motor is interrupted. The primary control consists of an electromagnetic relay that is energized by the thermostat.

The combustion control generally used on packaged boilers is a *positioning control system,* because it is more flexible. (See Fig. 11-16 for combustion controls used for different fuels.) Steam pressure is the measured variable, and a master pressure controller responds to changes in header pressure and (by means of power units or actuators) positions the forced-draft damper to control airflow and the fuel valve to regulate fuel supply. An independent controller, positioning the uptake damper, maintains furnace draft within the desired limits.

Although positioning-type control systems are an improvement over the on-off type, airflow and fuel supply are at their theoretically correct ratio at only one setting. This is usually the point at which they are calibrated on installation. Positioning control also assumes that a given output signal from the master controller always produces the same changes in the flow of combustion air, in stoker speed, or fuel-valve setting. But stoker speed might be affected by line-voltage variations, and airflow by boiler slagging or barometric conditions. Thus manual adjustment is still necessary, not only on load changes but also to counteract these longer-term effects.

A *metering control* measures the fuel flow and airflow, then modifies the valve and damper positions to maintain these measured flows rather than implied ones. Thus it holds an optimum air/fuel ratio over a wide load range without manual intervention. Especially valuable is its inherent compensation for such variables as boiler cleanliness, voltage swings in electric actuators, lost motion in mechanical or pneumatic devices, and changes in fuel quality.

Large high-pressure boilers generate superheat and reheat steam. A steam temperature of 1000°F is common, and units have been installed for 1050 and 1100°F. Because these high temperatures are limited only by metallurgy, steam temperatures must be held to close limits for safety as well as economy. Six basic methods are used for controlling the temperatures of superheated steam leaving the boiler:

1. Bypass damper control with a single bypass damper or series-and-shunt damper arrangement for bypassing flue gas around the superheater as required
2. Spray-type desuperheater control where water is sprayed directly into the steam with a spray-water control valve for temperature regulation
3. Attemperator control where a controlled portion of the steam passes through a submerged tubular desuperheater and a control valve in the steam line to the desuperheater or attemperator is used
4. Condenser control with desuperheating condenser-tube bundles located in the superheater inlet header and water-control valve or valves to regulate a portion of the feedwater flow through the condenser as required
5. Tilting-burner control where the tilt angle of the burners is adjusted to change the furnace heat absorption and resultant steam temperature
6. Flue-gas-recirculation control where a portion of the flue gas is recirculated into the furnace by means of an auxiliary fan with a damper control to change the mass flow through the superheater and the heat absorption in the furnace, as required to maintain steam temperature.

The principle underlying flow measurement is shown in Fig. 11-17*a*. A pressure drop across an orifice, in this case in the steam line, can be measured by tapping the pipe at each side of the restriction. The resulting pressure differential is proportional to the square of the fluid velocity. But correct location of the tapping points is important. By using nozzle restriction (Fig. 11-

Fig. 11-16 Types of combustion controls used for different fuels and boiler capacities.

Type of fuel	Type of control systems	Boiler capacity range (Mbh)	Precision of pressure or temperature control—tolerance limits (in percent of set point values)	Precision of fuel/air ratio control (% O_2)
Coal	On/off	400–3,000	±6	Not applicable
	High/low/off	2,000–5,000	±5	Not applicable
	Modulating positioning	3,000–62,000	±3	±1.0 for firing rates of or above 33% of maximum; ±2.0 for firing rates below 33% of maximum
	Semimetering	35,000–72,000	±3	±0.9 for firing rates of or above 33% of maximum; ±1.8 for firing rates below 33% of maximum
	Full metering (steam flow/air flow basis)	38,000–80,000	±3	±0.8 for firing rates of or above 33% of maximum; ±1.6 for firing rates below 33% of maximum
	Full metering (steam flow/air flow basis) with O_2 compensating	68,000 and above	±3	±0.6 for firing rates of or above 33% of maximum; ±1.2 for firing rates below 33% of maximum
Oil	On/off	400–3,000	±6	Not applicable
	High/low/off	2,000–5,000	±5	Not applicable
	Modulating positioning	3,000–66,000	±3	±0.5 for firing rates of or above 33% of maximum; ±1.0 for firing rates below 33% of maximum

	Semimetering	36,000–66,000	±3	±0.4 for firing rates of or above 33% of maximum; ±0.8 for firing rates below 33% of maximum
	Full metering (fuel flow/air flow basis)	36,000 and above	±3	±0.3 for firing rates of or above 33% of maximum; ±0.6 for firing rates below 33% of maximum
	Full metering (fuel flow/air basis) with O_2 compensation	69,000 and above	±3	±0.2 for firing rates of or above 33% of maximum; ±0.4 for firing rate below 33% of maximum
Gas	On/off	400–3,000	±6	Not applicable
	High/low/off	2,000–5,000	±5	Not applicable
	Modulating positioning	3,000–66,000	±3	±0.4 for firing rates of or above 33% of maximum; ±0.8 for firing rates below 33% of maximum
	Semimetering	36,000–66,000	±3	±0.3 for firing rates of or above 33% of maximum; ±0.6 for firing rates below 33% of maximum
	Full-metering (fuel flow/air flow basis)	36,000 and above	±3	±0.2 for firing rates of or above 33% of maximum; ±0.4 for firing rates below 33% of maximum

Static pressure
inches of
water
Flow
Primary element
(orifice plate)
Pressure differential
Mercury
Secondary element

(a)

Upstream (h-p) tap
Downstream (h-p) tap
Low-pressure
connection
Flow
Mercury level
Flow-straightening vane
Nozzle throat
High-pressure
connection
Float
Check valves
U tube

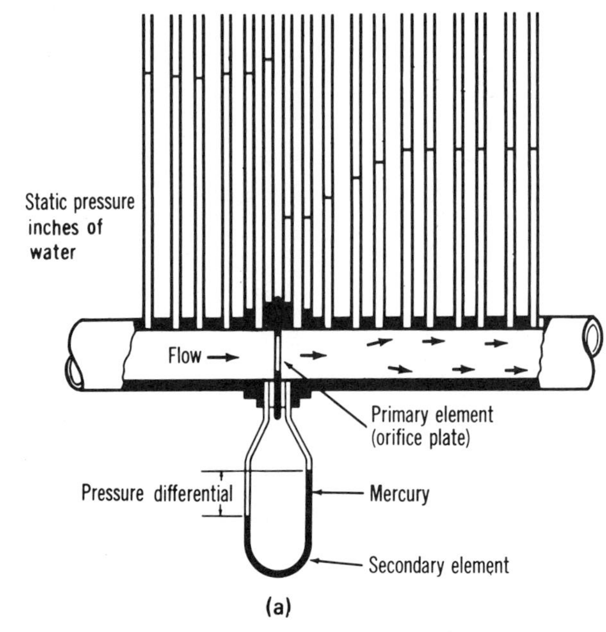

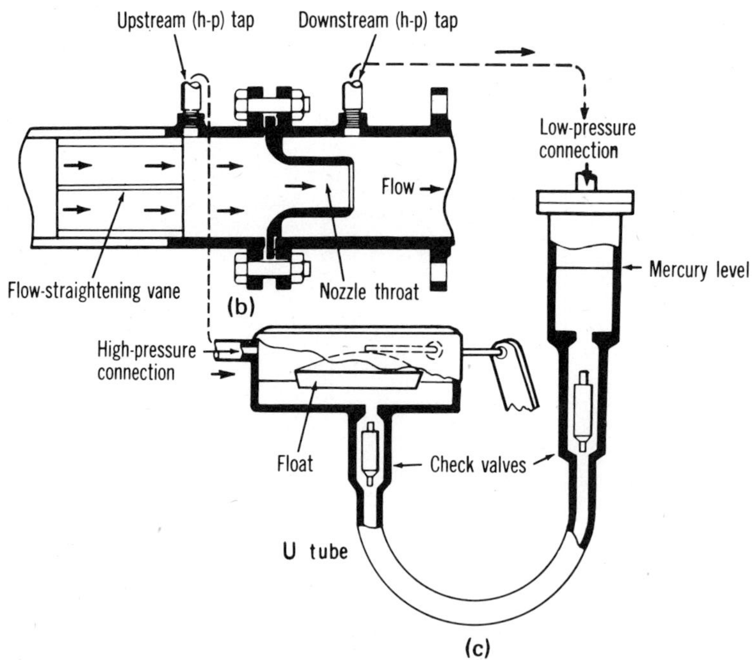

(b)

(c)

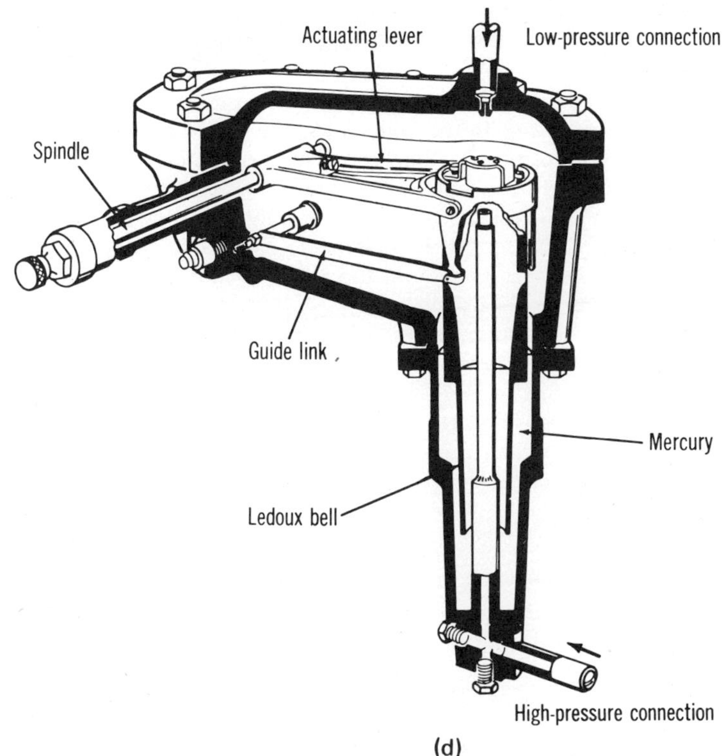

Fig. 11-17 *(a)* and *(b)* Flow measurements depend on the pressure differential across an orifice. *(c)* A mercury float manometer and *(d)* a Ledoux bell mechanism convert the differential pressure to an output signal.

17*b*), the best result (largest pressure differential for a given flow) is obtained when connections are located about one pipe diameter upstream and one-half diameter downstream from the nozzle's inlet face.

Converting differential pressure to a usable output signal may be done in various ways. Two widely used secondary elements are a mercury float manometer (Fig. 11-17*c*) or a Ledoux bell mechanism (Fig. 11-17*d*). In Fig. 11-17*c* the two pressure tappings from the primary element are connected to two mercury chambers, joined by a U tube. Pressure variations raise and lower the float. A pressuretight shaft conveys this movement to mechanical linkage within the controller. Check valves in the two legs of the U tube prevent damage to the mechanism resulting from sudden changes or reversals in pressure differential.

Feedwater Regulators There are three general classes of feedwater regulator in use for all sizes of power boilers: (1) the thermostatic-mechanical regulator; (2) the thermostatic-fluid type; and (3) the float-operated style.

The Copes feedwater regulator (Fig. 11-18*a*) is a thermostatic-mechanical type. Each end of the inclined tube is connected to the boiler; thus, when the boiler water level is normal, the water level will be about midway in the inclined tube. Owing to the inclined position of the thermostatic tube, a 1-in. vertical change in the boiler water level will cause several-inches change in the water position in the tube. This fact makes the regulator extremely sensitive in operation. One end of the tube is connected through a bell crank and linkage to the feedwater control valve. As the boiler water level falls, more of the tube length is exposed to steam; the tube expands and opens the feedwater control valve. As the water level rises, the tube cools, contracts, and closes the feedwater valve.

The Bailey regulator represents the thermostatic-hydraulic type (Fig. 11-18*b*). An inclined thermostatic tube is used. This tube is surrounded by a jacket having fins to dissipate heat to the atmosphere and make it rapidly responsive to temperature changes within the thermostatic tube. The jacket contains water which through a closed piping system connects with a metal bellows on top of the feedwater controlling valve. A spring tends to balance

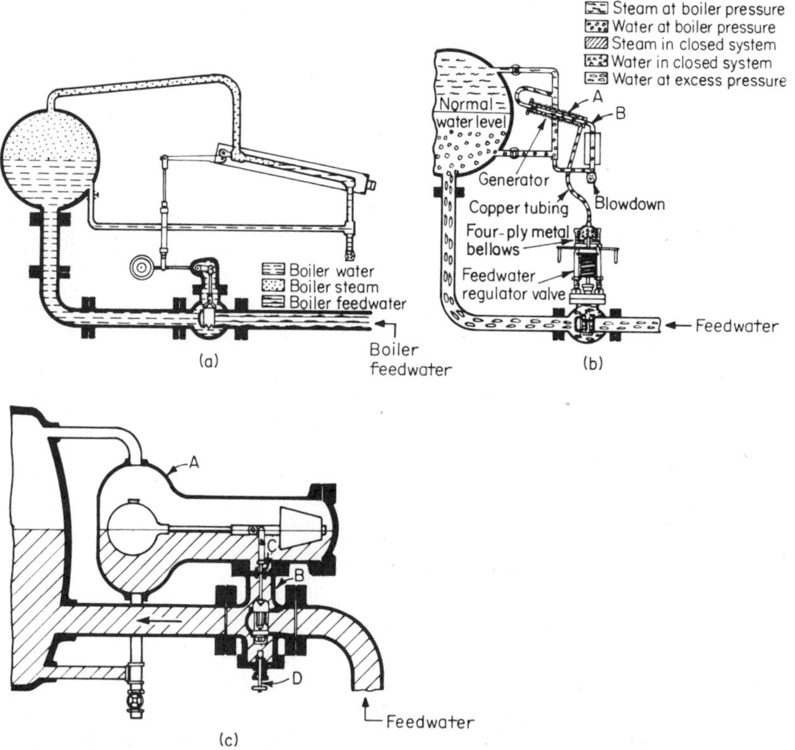

Fig. 11-18 Feedwater regulators: *(a)* Copes type, *(b)* Bailey type, and *(c)* Stets type. *(Courtesy Bailey Meter Co.)*

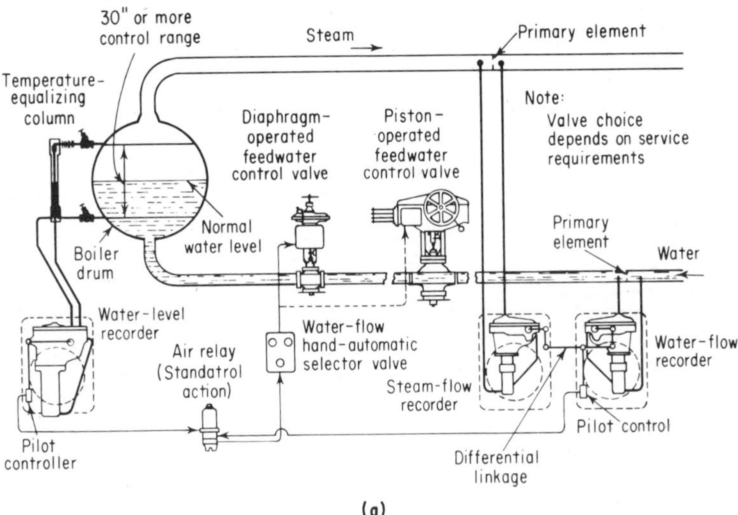

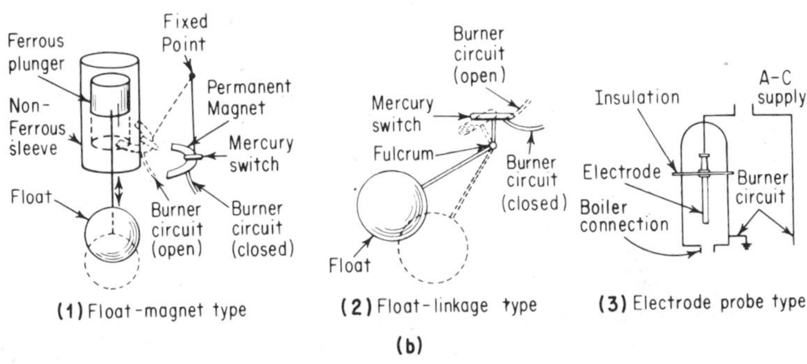

Fig. 11-19 *(a)* Three-element feedwater control regulates steam-and-water flow and also water level. *(b)* Types of low-water fuel cutoffs.

the valve against the fluid pressure on the syphon or bellows. The feedwater valve position is controlled by the pressure of the fluid as caused by temperature changes of water-level (and steam-space) variation in the thermostatic tube.

The Stets regulator represents the third type (Fig. 11-18*d*). Here a float chamber is installed and connected to the boiler at the same elevation as the normal water level in the boiler. A ball float rises and falls with the boiler water level and, through a linkage system, controls the position of the feedwater valve.

In a three-element feedwater control, as shown in Fig. 11-19*a*, steam flow, feedwater flow, and water level are measured and recorded by mechanically operated meters. Measurements of steam flow and water flow are balanced against each other with differential linkage. A pilot control is connected to the linkage so that any difference between the amounts of steam flow and of water flow causes a change in the pneumatic output signal. This signal is transmitted to an air relay where it is combined with the pneumatic signal from the water-level recorder.

A change in boiler load unbalances the differential linkage, thus producing a change in the output of the pilot control. That in turn changes the output of the air relay. This new signal repositions the feedwater control valve, admitting required water into the boiler equal to the steam flow out of the boiler. The resulting change in feedwater flow rebalances the differential linkage and brings the pilot-control signal back to its neutral point. As a final check, and to ensure having the proper drum level, the signal from the pilot control in the water-level recorder readjusts the feedwater control valve, if required. The selector valve in the system provides automatic or remote manual control. Under normal operating conditions, the control pressure gauge on the selector valve is an indication of valve position.

A low-water cutoff (Fig. 11-19*b*), separate from the programming-sequence control, immediately shuts down the boiler if the water drops to a dangerously low level. Three types are as follows:

1. The float-magnet type has a ferrous plunger on one end of a float rod. The plunger slides within a nonferrous sleeve. A permanent magnet, with a mercury switch affixed, is supported by a pivot adjacent to the nonferrous sleeve. Under normal water conditions, the ferrous plunger is above and out of reach of the magnetic field. In this position, the mercury switch is in a horizontal plane, keeping the burner circuit closed. But if the boiler water level drops, the float also drops, bringing the ferrous plunger within the magnetic field. Then the magnet swings through a small arc toward the plunger; the mercury switch tilts opening the burner circuit.
2. The float-linkage type has a float connected through linkage to a plate supporting a mercury switch. Because the plate is horizontal in the normal water-level position, the switch holds the burner circuit closed. If the water level drops, the float drops, tilting the plate so the switch opens the circuit.
3. The submerged-electrode type uses boiler water to complete the burner circuit. If the water level drops below the electrode tip, current flow is interrupted, shutting down the burner. On fire-tube boilers, the low-water cutoff generally includes an intermediate switch that controls the feed pump.

BURNER FLAME SAFEGUARD SYSTEMS

A flame safeguard system is an arrangement of flame detection systems, interlocks, and relays which will sense the presence of a proper flame in a furnace and cause fuel to be shut off to the furnace if a hazardous (improper flame or combustion) condition develops. Modern combustion controls are closely interlocked with flame safeguard systems and also pressure-limit switches, low-water fuel cutoffs, and other safety controls that will stop the energy input to a boiler when a dangerous condition develops. Thus it becomes obvious that a modern flame safeguard system performs actually two functions: It senses the presence of a good flame or proper combustion and programs the operation of a burner system so that motors, blowers, ignition, and fuel valves are energized only when they are needed, and then in proper sequence.

A furnace explosion is the ignition and almost instantaneous combustion of explosive or highly inflammable gas, vapor, or dust accumulated in a boiler setting. Often it is of greater expansive force than the boiler setting can withstand. In minor explosions, called puffs, flarebacks, or blowbacks, flames may blow suddenly for a distance of many feet from all firing and observation doors. Thus anyone in the flame path may be seriously or fatally burned. Such minor explosions indicate dangerous conditions, even if no real damage is done. Heavier explosions may shatter gas baffles, bulge setting walls, loosen refractory, blow brick tops of boiler settings through roofs, blow the sidewalls out from under the boiler, break connecting piping, and even demolish boiler housings. The main causes of firebox explosions are:

1. Flame failure resulting from liquids or inert gases entering the boiler fuel system
2. Insufficient purge before the first burner is lit
3. Human error
4. Faulty automatic fuel-regulating controls
5. Fuel-shutoff-valve leakage
6. Unbalanced fuel/air ratio
7. Faulty fuel supply systems
8. Loss of furnace draft
9. Faulty pilot igniters

There have been several incidents of *furnace implosions*, resulting in considerable damage to utility boilers. These furnace implosions have occurred in balanced-draft, fossil-fired boilers, where sufficient force was developed to exceed the structural strength of the boiler furnace. The implosions have occurred on boilers fired with oil or coal with the reported incidents divided approximately equally between the two fuels. No incidents have been reported on natural-gas-fired units, because few gas-fired boilers are balanced-draft. All the reported incidents occurred on balanced-draft boilers with induced-draft fans with high-head capability, operating, at least momentarily, under low-flow conditions. All the implosions have occurred without main fuel firing, some have occurred prior to light-off of main fuel, and others have occurred following a main fuel trip.

Two conditions have caused furnace implosions, both involving operation of high-head induced-draft fans under low-flow conditions:

1. A malfunction of the equipment regulating the boiler gas flow path, including air supply and gas removal, resulting in the furnace being exposed to the full induced-draft-fan head capability; usually occurs at the low-flow, high-head range of a centrifugal fan, and often combined with an interruption of the forced-draft airflow path.

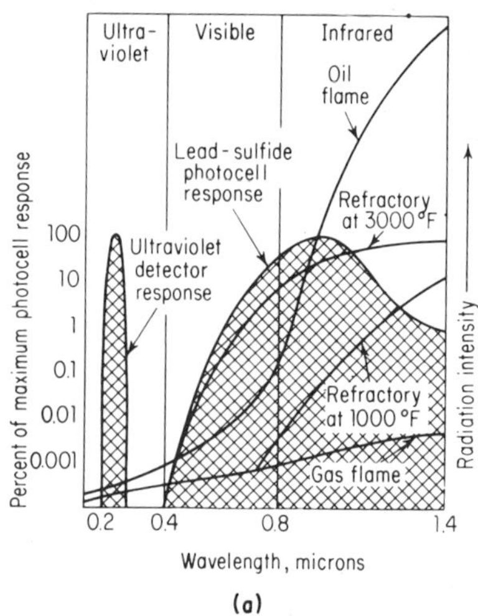

(a)

COMPARISON OF FLAME SAFEGUARDS

Principle of Flame Detection	Rectification		Infrared	Visible Light	Ultravision
Type of Detector	Rectifying Flame Rod	Rectifying Phototube	Lead Sulfide Photocell	Cadmium Sulfide Photocell	Ultravision Detector Tube
Advantages					
Same detector for gas or oil flame			▨		▨
Can pinpoint flame in three dimensions	▨				
Viewing angle can be orificed to pinpoint flame in two dimensions		▨	▨	▨	▨
Not affected by hot refractory	▨			▨	▨
Checks own components prior to each start	▨	▨	▨	▨	▨
Can use ordinary TW plastic-covered wire for general application, no shielding needed	▨	▨		▨	▨
No installation problem because of size			▨	▨	
Disadvantages					
Difficult to sight at best ignition point			▨		
Exposure to hot refractory may reduce sensitivity to flame flicker and require orificing			▨		
Flame rod subject to rapid deterioration and warpage under high tenperatures	▨				
Not sensitive to extremely hard premixed gas flow			▨	▨	
Temperature limit too low for some applications	▨	▨	▨	▨	▨
Shimmering of hot gases in front of hot refractory may simulate flame			▨		
Hot refractory background may cause flame simulation		▨			
Electric ignition spark may simulate flame					▨

(b)

Fig. 11-20 *(a)* The properties of a flame are used by detectors to scan the flame. *(b)* Comparison of flame detectors.

2. Rapid decay of furnace gas temperature and pressure following a rapid reduction in fuel input or a main fuel trip, similar to a flame-out condition.

Many flame-monitoring devices are based on the following physical principles of a flame (see Fig. 11-20*a*).

1. A flame produces an ionized zone, meaning that it can conduct a current through it. Conductivity-flame rod detectors use the principle of a conducting flame for flame detection monitoring.
2. A flame can rectify an alternating current. This is done by making one electrode across a flame larger than the other, thus making electrons flow through a flame much more readily in one direction than in the opposite direction.
3. Radiation of light is a known phenomenon of any fire. A flame radiates energy in the form of waves which produce heat and light. Three types of radiation from a flame are:
   - *(a)* Visible light that can be seen by the human eye. The wavelengths of visible radiation extend only from 0.4 to 0.8 micrometer (μm) (formerly microns). When cadmium is exposed to visible light, it emits electrons with the strength of the visible light. Thus, if a cadmium phototube is designed in an appropriate electronic circuit, electricity will flow through the circuit when the cadmium is exposed to sufficient light. This electricity can be used to trigger relay circuits for flame detection.
   - *(b)* Infrared radiation covers most of the useful band of wavelengths and also covers most of the radiation strength. Infrared detectors are suitable for gas and oil flames. Because hot refractories also radiate infrared, scanners must avoid hot refractories. Lead sulfide cells are used in photocells to sense infrared radiation. Unlike the cadmium phototube, it does not emit electrons, but has the property of having its electric resistance reduced while exposed to infrared radiation. The greater the strength of the radiation, the lower the resistance of the lead sulfide. If it is connected to a designed electronic circuit, this principle is used for flame detection purposes.
   - *(c)* Ultraviolet radiation is the latest flame detector based on the phenomenon of sensing the strength of ultraviolet radiation in a flame. It is insensitive to visual and infrared radiation and is not affected by hot refractories, since these usually do not give off appreciable ultraviolet radiation. When radiation from a flame passes through the typical quartz viewing window of one of these detectors into the flame-sensing tube, the tube becomes electrically conductive. The strength of the detector signal, or current passed through the sensing tube, depends on the kind of fuel, size and temperature of the flame, and distance between the flame detector and the flame. Figure 11-20*a* shows some typical wavelengths in a flame and response percentage of total wavelengths of typical flame pickup devices.

The type of flame detector depends on the fuel used, the type of burner, and the size and arrangement of the boiler. Flame detectors vary from those used on small domestic boilers to those on large boilers. Types of detectors and the task each does are:

1. Stack switch, heat sensing
2. Rectifying flame rod, heat sensing
3. Rectifying phototube, visible-light sensing
4. Lead sulfide photocell, infrared-light sensing
5. Cadmium sulfide photocell, visible-light sensing
6. Ultraviolet flame-detector tube, ultraviolet-light sensing

While each of the flame-sensing devices can offer a substantial amount of protection if properly installed, they are all subject to certain limitations that must be allowed for. For example, the lead sulfide cell and the photoelectric cell are subject to the following limitations (there is some variation between each type):

1. *Discrimination between burners.* With more than four burners in one firebox, it becomes difficult to locate the sensing cell where it will not be actuated by the flame of an adjacent burner when the burner on which it is mounted has been extinguished.

2. *Ambient temperature of sensing cell.* High ambient temperatures of the sensing cells (which are easily obtained at the locations where they must be mounted) can result in erratic signals, false signals, and a short life. Air and water cooling have been used to prevent or minimize this limitation.

The flame rod is subject to: (1) short service life at high flame temperatures; (2) fouling of insulators, causing short circuits and shutdowns; (3) difficulty in providing sufficient rod area to ground the flame; and (4) limited generally to pilots and small burners. Figure 11-20*b* makes some comparisons of the different flame safeguard systems.

The rectified impedance system (Fig. 11-21) operates on the principle that either a flame or a photocell sighted at a flame is capable of conducting, as well as rectifying, an alternating current. This alternating current is applied to either a flame electrode inserted in the flame or a photocell sighted at the flame. The resultant rectified current, which can be produced only when a flame is present, is in turn detected by the relay.

The prevention of furnace explosion is a requirement detailed in the National Fire Protection Association (NFPA) codes as well as by fire insurance engineering groups, such as the Industrial Risk Insurers and the Factory Mutual groups. Code 85 of the NFPA is useful in detailing requirements to prevent furnace explosions. For example, NFPA no. 85-B for gas firing and no. 85-T for oil firing stipulate that register-starting procedures must be programmed. The open-register start-up procedure for gas firing is outlined briefly thus:

1. Set all, or most, burner-air registers in the normal firing position. Then purge the furnace and boiler setting, using not less than 25 percent of full-load airflow for 5 min.

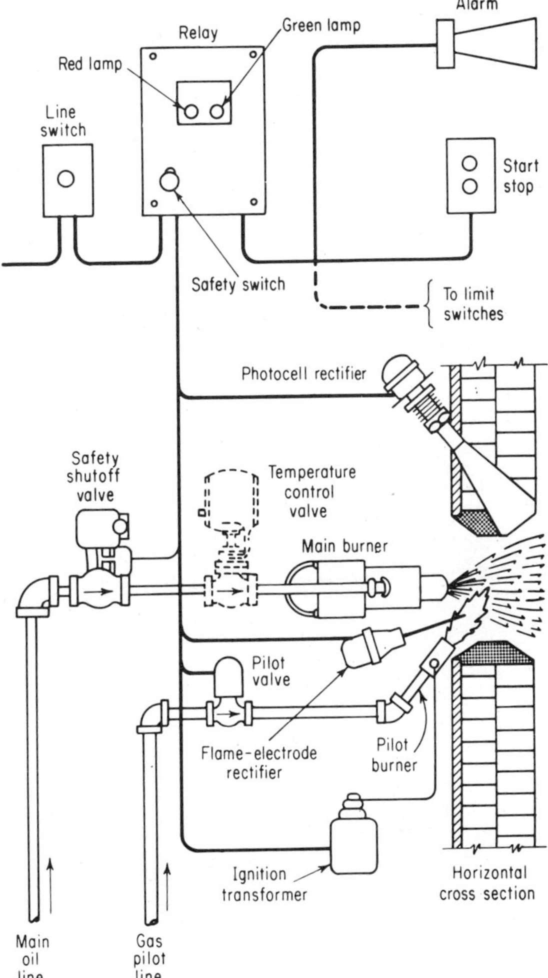

Fig. 11-21 Rectified impedance flame safeguard system operates on the principle that a flame is conducting.

2. Throughout the start-up period, maintain the same register settings and the same total airflow used for purge.

3. Set fuel header pressure at a value which will provide a burner fuel flow compatible with the burner airflow.

4. Light burners one at a time as increased heat input is required, keeping the burner fuel-header pressure and register settings at their initial settings. As each new burner is lit off, close the burner register to light-off position. Since the furnace is air-rich, additional burners may be cut in with no increase

in airflow until the fuel flow approaches 25 percent (or whatever airflow rate was used during the purge).

Another requirement found in fire codes, as well as in the ASME Safety Code entitled "Controls and Safety Devices for Automatically Fired Boilers," is the double shutoff-valve systems on fuel trains. This generally requires that the main gas-burner train have two manual-reset safety shutoff valves. This arrangement ensures burner cutoff even if one valve stays in an open stuck position.

State laws are being expanded to include the hazard of furnace explosion in state boiler inspection programs. For example, one state has the following requirements for automatic-heating boilers in its inspection code:

1. *Gas-fired boilers.*
   (a) Pilot has to be proved, whether manual or automatic, before permitting the main gas valve to open, either manually or automatically, by completing an electric circuit.
   (b) A timed trial for the ignition period is established based on the input rating of the burner. For instance, for input rating of 400,000 to 5,000,000 Btu/hr per combustion chamber, the trial for the ignition period for the pilot of automatically fired boilers cannot exeed 15 sec. And the main-burner trial for ignition also cannot exceed 15 sec.
   (c) The burner flame-failure controls must shut off the fuel within a stipulated time, again depending on the fuel input of the burner. For a burner rated with an input of 400,000 Btu/hr or more, the electric circuit to the main fuel valve must be automatically deenergized within 4 sec after flame failure. And the deenergized valve must automatically close within the next 5 sec.
2. *Oil-fired boilers.* Similar provisions have been adopted, with requirements on response time for controls to shut off the burner based on fuel input in gallons per hour, instead of Btu per hour. The flame must be continuously supervised by the controls.

Questions and Answers

11-1 What three conditions determine whether the chemical reactions required for burning will take place?

Ans. The three conditions generally needed are: (1) proper fuel-air mixture; (2) intimate mixing of fuel and air in the combustion zone so that all fuel particles have oxygen; and (3) maintenance of an ignition temperature.

11-2 A flue-gas analysis showed the following percentages by volume: CO_2, 16 percent; oxygen, 5 percent. What was the excess air based on these readings?

Ans.
$$\text{Excess air} = \frac{\text{Oxygen by volume}}{CO_2 \text{ by volume}} = \frac{5}{16} = 31.25\%$$

11-3 What air temperature must be provided a burner for the following conditions? Stack temperature, 625°F; coal burned at 12,000 lb/hr with a heating value of 15,000 Btu/lb; flue gas formed, 13.8 lb/lb fuel burned; boiler efficiency, 80 percent?

Ans. Heat lost up the stack is 12,000 (1 − 0.8)(15,000) = 36 million Btu/hr. The weight of the flue gas going up the stack is 12,000(13.8) = 165,000 lb/hr. The heat lost equals $WCp(T_2 - T_1)$. Substituting and solving for T_1 with $Cp = 0.25$, we get

$$36{,}000{,}000 = 165{,}000(0.25)(625 - T_1) \qquad T_1 = 245°\text{F}$$

Thus preheated air must be used.

11-4 What fuel is considered to have two heating values and why?

Ans. Fuels containing hydrogen have a higher and lower heating value. The higher heating value is determined in a calorimeter. From this, the heat lost by the water vapor that is formed in burning hydrogen must be deducted to obtain the lower heating value. In most engineering calculations, the higher heating value is used.

11-5 A boiler burning fuel oil with a heating value of 18,500 Btu/lb has an 82 percent efficiency as determined by test. Fuel-oil consumption averages 1150 lb/hr. How much in Btu's is lost up the stack?

Ans.
$$\text{Heat released} = 1150(18{,}500) = 21{,}275{,}000 \text{ Btu/hr}$$
$$\text{Heat loss} = 21{,}275{,}000(1 - 0.82) = 3{,}829{,}500 \text{ Btu/hr}$$

So about 3829 lb/hr of steam is lost.

11-6 What are primary and secondary air in burning fuels in a boiler?

Ans. Primary air is the air mixed with the fuel at or in the burner; it is about one-third of the total needed for complete combustion. Secondary air is air brought in around the burner or through the openings in a furnace floor or wall in order to provide additional air for complete combustion in the furnace. Secondary air is about two-thirds of the total needed.

11-7 What is the harmful effect of sulfur in a fuel?

Ans. Sulfur burns to sulfur dioxide which, when mixed with water or water vaopr, forms sulfurous acid that is corrosive to tubes, breechings, and economizer sections on larger boilers. The dew point (temperature at which water vapor condenses) of the flue gas has to be watched (especially with high-sulfur-content fuel) because the gas becomes cooler and cooler while going through a furnace, so as to prevent the combination of water vapor with sulfur dioxide. Recent air pollution laws are tough on high-sulfur-content fuels, because sulfur dioxide is known to pollute the air. Percentages of permissible sulfur content in fuels are slowly being lowered from the previous 4 percent maximum to 1 percent in the near future.

Sulfur in oil-fired boilers leads to fireside corrosion of tubes, especially in those boilers cycling on and off. In cooling on the off cycle, the sulfur dioxide combines with water vapor and with water from leaks to attack the tubes by means of the resultant acid formed.

11-8 Convert 3 in. of draft into pounds per square inch or into ounces per square inch.

Ans. Since water at normal temperature weighs 62.4 lb/ft³, dividing this by the number of cubic inches in a cubic foot gives

$$\frac{62.4}{1728} = 0.036 \text{ psi/in.}^3 \text{ water}$$

Thus for 3 in., pressure = 3 × 0.036 = 0.108 psi. And since there is 16 oz in a pound, the ounces per square inch = 0.108 × 16 = 1.73-oz pressure per square inch.

11-9 What is meant by balanced draft?

Ans. A boiler using both forced-draft and induced-draft fans can be regulated and balanced in the amount of air and flue gas handled, so that the furnace pressure is almost atmospheric. This results in better control of air leakage from the furnace and thus control of the fuel/air ratio in the furnace.

11-10 What are the usual percentages of excess air in burning the various common fuels in boilers?

Ans. For coal, usually 50 percent excess air is used. With oil, gas, or pulverized coal, excess air is 10 to 30 percent.

11-11 To what capacity of boiler do federal clean-air laws apply? What are some of the requirements on permissible limits?

Ans. The Environmental Protection Agency (EPA) has established minimum standards for fossil-fuel-burning boilers with capacities of 250 million Btu/hr of input or larger. However, in their general authority, severe violations below this capacity could also be cited for exceeding the following requirements: (1) no particulate emission in excess of 0.10 lb per 1 million Btu input; (2) no opacity exhibit greater than 20 percent, except up to 40 percent opacity for 2 min is permitted in any hour; and (3) sulfur dioxide emission limited to 1.2 lb for every 1 million Btu fired.

11-12 What factors determine the collection efficiency of a precipitator?

Ans. Time of flue-gas retention, amount of electric power supplied to the precipitator discharge system, the size of the fly-ash particles, and the resistance of the dust to flow.

11-13 Name three types of synthetic fibers used in fabric-type filters.

Ans. The limiting factor of fabric filters or collectors has been the rapid wear, or inability to withstand temperature and corrosiveness of flue gasses. The development of synthetic fibers such as polyesters, acrylics, and fiber glass has greatly expanded the application of fabric filters.

11-14 What is meant by low-temperature corrosion in a boiler?

Ans. Low-temperature corrosion occurs when the flue gas contacts surfaces that are at a temperature below the dew point of corrosive constituents in the gas. These low temperatures are found at the water inlet of economizers if the feedwater temperature is too low and at the cold end of an air heater. They may also occur in scotch marine boilers subject to on-off operation and furnace purging cycles. Low temperatures cause sulfur and other corrosive gases to form sulfurous and sulfuric acid that attacks boiler metal.

11-15 What are the chief combustibles in fuel oil?

Ans. Fuel oils come from petroleum, with the chief combustible ingredients being carbon and hydrogen. The specific gravities of fuel oils vary and are influenced by origin of the petroleum and by temperature. At 60°F, specific gravities can vary from 0.84 to 0.96.

Since hydrogen has a much higher heating value and lower atomic weight than the other principal elements in fuel oil, the proportions of carbon and hydrogen affect both specific gravity and heating value. Because of this, specific gravity forms a reliable guide to an oil's heating value. Specific gravity in degrees API is found by dividing specific gravity with respect to water (at 60°F) into 141.5 and subtracting 131.5 from the answer.

11-16 What is a potential source of oil getting into an oil-fired boiler?

Ans. A tube failure inside the oil heater can cause a boiler to be contaminated with oil. If the oil pressure is higher than the steam pressure, oil will be forced into the steam side of the oil heater and then travel to the waterside of the boiler. To avoid this, place a check valve on the steam line to the oil heater so reverse flow cannot take place. Where condensation is returned to the boiler, it should be piped to a condensate receiver equipped with a gauge glass. Then oil, if any, can be seen in the glass. Double-shell-type oil heaters are often used for extra safety.

11-17 What are the pour point, flash point, and fire point as applied to oil burning?

Ans. Liquid petroleum products can be increasingly viscous as their temperature declines. The pour point is the temperature at which an oil ceases to flow, or pour. The flash point is the temperature at which a sample of oil will flash, or catch on fire, when it comes in contact with an open flame. It generally indicates at what temperature the oil is safe to handle. Distillate fuels flash at about 150°F. The fire point is the lowest temperature at which an oil will burn continuously once it has been lighted.

11-18 Name four methods of atomizing oil for burning.

Ans. Oil burners atomize fuel oil into minute droplets which then readily vaporize by the following methods: (1) under high pressure using air or steam; (2) by discharging the oil through orifices or nozzles at high pressure; (3) by passing the oil over a spinning cylindrical cup in the burner mouth to give a centrifugal force to the oil droplets; and (4) by an acoustic-type burner that has a vibrating resonator located just ahead of the point at which the oil leaves the burner nozzle.

11-19 What difficulties can be expected with water in fuel oil?

Ans. Difficulties encountered with water in fuel oil are: (1) flame outage; (2) erratic or sputtering firing; (3) waste of fuel from slow or faulty ignition; (4) loss of heat by vaporizing the water in the oil; and (5) corrosion in lines and storage tanks from the water (this may lead to burner-tip plugging).

11-20 What are some reasons for incomplete combustion in oil burning?

Ans. Incomplete combustion may be caused by incorrect fuel/air ratio, improper atomization of fuel, oil too viscous or at wrong viscosity, nozzles partially plugged that interfere with maintaining correct spray pattern, water in the fuel, poor draft, and improper damper setting on forced-draft fan.

11-21 What is the effect of vanadium on boiler tubes in fuel oil?

Ans. Fuels high in vanadium have a high tendency to form a fluid plastic-type ash at relatively low furnace temperature of 1100°F. The deposits formed on the tubes act as an insulator, reduce heat transfer, and can cause tubes to blister from localized overheating. Vanadium also combines with iron in the molten state and removes it in layers, often called high-temperature corrosion. Removal of these hard vanadium deposits requires physical chipping of the fireside scale.

11-22 Determine the wattage needed to heat 40 gal/hr of fuel oil from 120 to 200°F.

Ans. Good approximation is obtained by using the equation

$$\text{Watts needed} = 1.25 \text{ gal/hr } (T_2 - T_1)$$

or

$$= 1.25(40)(200 - 120) = 4000 \text{ W or } 4 \text{ kW}$$

11-23 What maintenance is stressed on oil burners?

Ans. Make sure that the burner gets uniformly free-flowing oil, clear of sediment that clogs burner nozzles. This means avoiding sludge buildup in storage tanks and keeping strainers in good condition. The preheat temperature must be right for fuel and burner type and must be uniform. Watch for wear caused by abrasion of ash in fuel and for carbon buildup. In rotary-cup burners, worn rims cause poor atomization. If cups are not properly protected after being turned off, carbon forms on the rim. When the burner is shut down, always take out the cup and insert a flame shield. Worn or carbonized mechanical-atomizing nozzles give trouble. Always replace worn nozzles and keep them clean.

11-24 What are caking coal and free-burning coal?

Ans. A caking coal is one which fuses at the surface when burning to form a more or less heavy crust. The term *coking* is also used. A free-burning coal does not form a crust and is friable (easily crumbles) throughout the combustion process.

11-25 What are the mineral impurities of coal?

Ans. One is ash, which is the incombustible mineral matter left behind when coal burns completely. The amount and character of the ash constitute the biggest single factor in fuel-bed and furnace problems such as clinkering and slagging. An increase in ash content usually means an increase in the carbon carried to waste or imperfect combustion. Next are the incombustible gases such as carbon dioxide and nitrogen. When the volatile matter distills off, a solid fuel is left, consisting mainly of carbon but containing some hydrogen, oxygen, sulfur, and nitrogen that are not driven off with the gases. Sulfur in coal burns, but is undesirable. Besides causing clinkering and slagging, it corrodes air heaters, economizers, breachings, and stacks. It also causes spontaneous combustion in stored coal.

11-26 Why must soot blowers be provided on coal-burning boilers?

Ans. Soot blowers must be provided for cleaning ash from the furnace walls and from the convection-heating surfaces. Hoppers are provided at the bottom of the furnace and at other strategic points throughout the boiler to remove the ash collected at these points if the unit is large.

11-27 Why should coal be sampled and analyzed in a power plant?

Ans. Because coal is not homogeneous with a variable composition. Sampling a given shipment may require taking up to 100 increments in order to obtain a good cross section for laboratory analysis of the coal properties.

11-28 Which stoker system is the most widely used in industrial plants?

Ans. The two most widely used are the traveling-grate and spreader stokers.

11-29 Name three types of pulverizers.

Ans. The three most commonly used pulverizers are: (1) the pressurized, airswept table and roller mill; (2) ball mills using a rotating cylinder with steel balls in it to grind the coal; and (3) tube or rod mills using rods in a rotating cylinder.

11-30 What is meant by fluidization?

Ans. Fluidization is the condition that results when air is blown through a grid at a rate sufficient to overcome the static weight of the solids atop the grid to a point where the mass of these solids, referred to as a bed, behaves like a boiling liquid. Hence the term *fluidization*.

11-31 What are some of the advantages of fluidized-bed combustion?

Ans. Elimination of fine pulverizing, ability to inhibit molten-slag formation by holding combustion temperatures below the ash fusion point, and removal of sulfur by chemical interaction with limestone in the fluidized bed.

11-32 Name two methods of collecting fly ash.

Ans. Fly ash, depending on its size and characteristics, is collected in either an electrostatic precipitator or a bag house. Collection efficiency greater than 99.5 percent may be realized. The bag house is now becoming competitive with the electrostatic precipitator as a result of the development of fiber glass material.

11-33 What factors must be considered in modifying an oil- or gas-burning boiler to coal burning?

Ans. For a boiler that was not originally designed for coal burning, the following must be considered: (1) the furnace bottom may have to be modified so that furnace ash can be disposed; (2) pulverizers, stoker coal feeders, and bunkers and associated piping must be added; (3) soot blowers may have to be added; (4) superheaters and air heaters may have to be modified to prevent plugging by fly ash through the convection passes; and (5) wind boxes will have to be modified to accommodate new fuel burners for coal burning.

11-34 What is meant by mode of control?

Ans. Mode of control means the manner in which the automatic controller acts and reacts to restore a variable quantity on a boiler, such as pressure, flow, or temperature, to a designed control or desired value. The three controller systems used to control a boiler are pneumatic, electric, and electronic.

11-35 Why are manual-reset controls useful on pressure or temperature high-limit controls?

Ans. The manual-reset mechanism on the high-limit control calls attention if the operating control (not high-limit) has malfunctioned, thus prohibiting further boiler operation until corrected. At times, this malfunction may be due to fused contacts, a leaking gas valve, a shorted wire, etc. Thus the boiler should not be operated until this is corrected.

11-36 What conditions are necesssary to cause a furnace explosion?

Ans. Usually three: (1) accumulation of unburned fuel; (2) air and fuel in an explosive mixture; and (3) a source of ignition, such as hot furnace walls, improper ignition timing, faulty torch, and dangerous light-off procedures on manually started boiler combustion systems.

11-37 What is the purpose of a flame-sensing device?

Ans. Any flame-sensing device is an integral part of the combustion control system. The prime purpose of a flame-sensing device is to ensure that flame or burning conditions are safe and that during light-off proper sequences are followed in order to obtain a safe ignition. In the event of flame failure during operation, the flame-sensing device must be capable of sending a signal so that the fuel is cut off, preferably an alarm initiated, and the furnace and gas passes purged with air before any attempt is made to relight the burners.

11-38 What does a flame safeguard system prevent?

Ans. Furnace explosions (combustion explosions). These are caused by the sudden ignition of accumulated fuel and air in the fireside of the boiler. This can also lead to devastating property damage and loss of life. The flame safeguard system used on different boilers firing different types of fuel is designed to sense, and sometimes anticipate, this accumulation of unburned fuel and air in the fireside of the boiler. It safely shuts off the firing equipment, with purging of the furnace usually following in a time sequence so as to drive the unburned fuel-air mixture out of the furnace.

11-39 How is the efficiency of a boiler affected by the addition of automatic controls?

Ans. Through precise regulation of fuel, air, feedwater flow, and similar variables, tests have shown that the efficiency of a boiler is generally improved with the addition of automatic combustion controls. Uniform steam pressure is maintained, brickwork lasts longer, and better regulation is possible up to maximum load. The operator's skill is now directed to maintaining the controls in good condition and to detecting signs of malfunction before the control can become seriously impaired.

11-40 Why should low-water fuel cutoffs be inspected internally or dismantled at least once a year?

Ans. Dismantled inspections are made to check on the following possible places that could lead to this vital safety device not operating when needed: (1) binding of floats, rod, and associated pivot points; (2) linkage parts that may be broken, corroded, or worn; (3) excessive scale buildup in float chamber; (4) pipe connections to cutoff plugged with sediment; (5) float waterlogged; and (6) electric-type probe bridged with scale, giving false electric signal on low-water cutoff.

11-41 What type of feedwater control is especially vulnerable to possibly causing a low-water failure?

Ans. Units with combination pump return and/or feeder and low-water cutoff control, because a single failure of the float will make immediately inoperative the water feed to the boiler as well as nullify the low-water fuel cutoff, and because the nonfunctioning float controls both feed and the cutoff. An independent second low-water fuel cutoff should supplement the combination feeder and low-water fuel cutoff.

11-42 What limit devices are generally recommended on a combustion-control safety system?

Ans. The safety system should include the following limit devices incorporated into limit circuits so that the unit will shut down automatically if any indicates an unsafe condition: (1) high steam pressure; (2) forced-draft-fan interlock; (3) low fire on start; (4) high fuel pressure; (5) low oil temperature; (6) low water; (7) loss of flame; and (8) additional limit devices recommended by the following property groups—Factory Mutual, Industrial Risk Insurers, and the National Fire Protection codes.

11-43 What precautions must be observed in adjusting a burner on a two-boiler or more burner system?

Ans. When an operator makes any change to the fuel flow or airflow to one of the burners, he or she should also check the effect of these changes on the other burner's flame stability.

11-44 What precautions should an operator take to prevent furnace explosions?

Ans. Seven basic precautions are:

1. Check the operation of the boiler periodically.

2. If a burner goes out accidentally, shut off the igniter and fuel supply and thoroughly scavenge the furnaces and gas passes before again attempting ignition. Always determine and remedy the cause of the stoppage.

3. Keep burners and all allied equipment clean.

4. On boilers using both forced- and induced-draft fans, test the interlock periodically.

5. Do not attempt to secure excessively high CO_2 by using too rich a fuel/air ratio or by an inadequate secondary air supply.

6. Keep the temperatures and pressures for preheated air, drying air, fuel oil, etc. at the right levels.

7. Never allow an unstable flame condition to continue uncorrected.

12

Operating Problems, Cause of Failures, Water Treatment, and Repairs

OPERATING FUNDAMENTALS

Safe boiler operation requires management commitment to safety, efficiency, and continuity of dependable operation, and this requires management support of the operators and maintenance personnel in correcting plant problems as they appear. Many automatic devices have been developed to make boiler control easier, safer, more efficient. But all this automatic equipment demands more head work to replace the strong back required of yesterday's engineer. The automatic equipment must be understood and maintained. If it should fail, the operating engineer must be capable of picking up manual control of many operations on a split second's notice.

Boiler explosions (also fireside explosions) can occur. Here are some of the causes:

1. More unattended automatic operation of boilers, with complete reliance on automatic controls for overpressure and fireside explosion prevention. Although controls can malfunction in many ways, their installation can lead to a false sense of security.
2. Failure to test safety relief valves on a consistent, regular basis.
3. Failure to maintain boiler and auxiliaries properly. The latter includes reserve boiler feed and low-water fuel cutoff. Maintenance is often neglected on water treatment, cleaning, and checking of controls.
4. As automatic boilers become more complex in control arrangement, tampering with controls or blocking the safety controls may lead to a failure.

5. The higher firing rates with suspended fuel on today's more compact boilers can quickly lead to dry firing or to improper fuel/air ratios that trigger fireside explosions, again if safety controls do not work fast enough.

Many plants operate on a *breakdown maintenance* basis, which means equipment is allowed to operate to failure before it is repaired or replaced. This type of maintenance requires little planning, but it does produce inefficient use of workers, who are paid overtime pay in emergencies, and causes excessive unplanned downtime in a service or production. *Engineered preventive maintenance* combines predictive analysis and testing techniques to determine the frequency of overhaul or part inspection and repair or replacement in order to maximize operating time and to eliminate unnecessary overhaul work based only on frequency schedules. Some inspection work is dictated by jurisdictional requirements, and these are usually used as a starting base for preventive maintenance work. As pressures and temperatures rise and the effect of an accident or the expense of an outage becomes apparent, the legal inspections must be supplemented by plant-engineered preventive maintenance programs.

The results of a boiler explosion should not be minimized. Even a small boiler may cause terrific damage. See Fig. 12-1. Small boilers, indeed, must often be considered more hazardous than those in large plants, owing to their frequent lack of competent attendance. However, close adherence to ASME Code standards for construction and care of power boilers and a high grade of operation and inspection will practically eliminate the possibility of accidents of this nature.

Fig. 12-1 Boiler explosion due to low water, causing rupture of pressure parts. *(Courtesy Factory Mutual Engineering)*

The following paragraphs on the effects of improper water treatment and overheating all deal with conditions that may lead to an explosion if neglected.

SCALE FORMATION

Scale formation on the waterside of boiler heating surfaces is caused by the contact of certain impurities in boiler water with the hot surfaces. Most common among these impurities are calcium (Ca), magnesium (Mg), and silica (SiO_2). The calcium or magnesium may unite with sulfates (SO_4) or magnesium which are scale-forming.

Calcium is common in raw water because it is present in many forms, as marble, limestone, and chalk. Magnesium in various forms is found, too, in raw water from many sources. A well-known form is magnesium sulfate ($MgSO_4$) (Epsom salts). Silica, found in sand and glass, forms an exceedingly hard, dense scale, actually glasslike.

Scale-forming water is said to be "hard." This hardness is "temporary" or "permanent" or both. The temporary hardness may be eliminated by heating the feedwater to about 212°F in an open or deaerating heater where the salts causing temporary hardness are precipitated. The permanent hardness must be controlled by treatment in water softeners or by treatment in the boiler.

There are two definite objections to scale on boiler heating surfaces. (1) Scale is a very efficient *nonconductor* of heat, the degree of nonconduction varying somewhat with its density. Its presence in appreciable thickness means less heat absorption by the boiler water, with consequent loss of boiler efficiency. (2) Because of the fact that scale is a poor heat conductor, the heating surfaces thus insulated from boiler water on one side and exposed to hot gases on the other may soon reach a dangerously high temperature. Serious damage—rupture of tubes (Fig. 12-2) and even boiler shells—has resulted.

Scale formation often increases with the rate of evaporation. Thus, scale deposits will often be heavier where the gas temperatures are highest.

Scale is usually more serious in a watertube boiler than in a fire-tube boiler. A coating of scale ¹⁄₁₆ in. thick on watertubes exposed to radiant heat may cause tube failure, whereas much heavier deposits of scale on fire tubes cause loss in efficiency but may not be dangerous. The reason for this difference is that fire tubes absorb their heat by convection of gases, and not by radiant exposure.

Heavy scale deposits are usually an indication of neglect, for scale can be prevented in most cases by proper treatment of the water. Where scale has formed to an appreciable thickness, it should be removed; and once a clean boiler is attained, proper steps should be taken to prevent its recurrence. Scale formation can block or interfere with the proper operation of controls and even safety valves by plugging connections or causing the valve or control to stick. A primary cause of safety-valve failures is the buildup of deposits around the seat which cause the valve to stick in the closed position. A regular

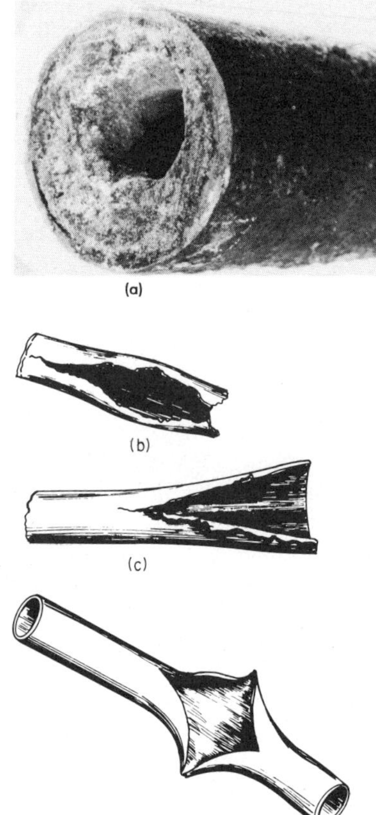

Fig. 12-2 Tube scale and ruptures: *(a)* excessive phosphate deposits; *(b)* deposits and chemical-attack-ruptured tube; *(c)* long-term overheating rupture, shown by thick fractured surface; *(d)* short-term overheating damage, shown by knife edges of fractured surface.

program of valve testing will prevent this type of failure as will improvement in water and steam purity.

Oil in boilers is a dangerous condition. Oil is an excellent heat insulator, and its presence on heating surfaces exposed to high temperatures may cause serious overheating and damage to the boiler.

A common cause of this condition is the use of reciprocating steam-equipment exhaust containing cylinder oil for condensate return to the boiler feed system. Also, fuel-oil heating equipment may leak oil into the steam system and cause this difficulty if the condensate is returned to the boiler. A minimum amount of high-grade properly compounded cylinder oil should be used for lubrication of steam engines and pumps where condensate is returned, and an efficient type of oil separator should be used in the exhaust system. Oil may also enter the feed through its presence in such raw-water supplies as rivers and streams contaminated by mill, marine, or trade wastes.

Oil may be removed from a boiler by boiling it out with soda ash and caustic soda, using 1 lb of each per 1000 lb of water in the boiler. Carry the boiler at about 5-psi pressure and continue boiling for two or three days, depending on the extent of oil contamination. Then empty the boiler, wash it thoroughly with freshwater, and again check the internal surfaces.

INTERNAL CORROSION

Internal corrosion is an electrochemical deterioration of the boiler surfaces, usually at or below the water line. See Fig. 12-3. The pH value of the water is a measure of its alkalinity or acidity and usually has a direct bearing on the corrosive properties. All water contains alkaline (hydroxyl, OH) ions and hydrogen (H) ions. The product of these concentrations is always approximately 10^{-14}. The pH value of the water is the log of the reciprocal of the H ion value.

If the water is neutral, the OH ion concentration is 10^{-7}; therefore, the H ion must also be 10^{-7}. Then the pH is 7. Waters with an H ion concentration greater than 10^{-7} are acid. Hence, a pH below 7 indicates acidity; over 7 designates an alkaline condition.

The detrimental effects of corrosion depend on its rate of penetration. Corrosion affecting large areas of boiler plate is not likely to penetrate as rapidly as localized corrosion on small areas. The former condition sometimes is difficult to see, and it may progress unnoticed to a dangerous extent.

Localized corrosion may be in the form of pitting or *grooving*. Pitting is caused by repeated breaks at the same spot in the protective H film. It is

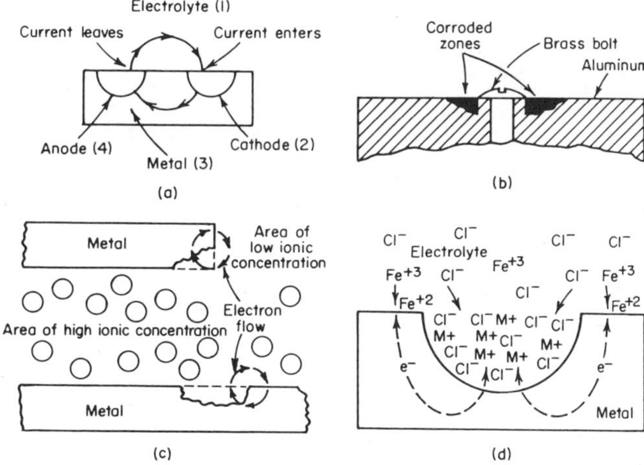

Fig. 12-3 Types of corrosion attack. *(a)* Electrochemical reaction theory favors current flowing in corrosion. *(b)* Galvanic corrosion attacks less noble of two metals. *(c)* Crevice corrosion occurs at localized structural fault. *(d)* Chemical pitting is caused by high chloride concentration.

affected by the type of surface, especially if mill scale or such surface irregularities are present. The pits may be as small as a pinhead or as large as a half dollar.

Weakening Effects of Corrosion The strength of the shell or drum is reduced as corrosion or the formation of closely spaced pits progresses. The ratio of actual thickness at any area so affected (for 4 in. or more in a longitudinal direction) to the original thickness should be considered as a percentage. If this percentage is below the efficiency of the longitudinal seam, the maximum allowable pressure should be reduced accordingly. The allowable pressure P in a corroded area can be calculated by the existing installation equation

$$P = \frac{TS \times t \times \text{efficiency}}{R \times FS}$$

where

P = allowable pressure, psi
TS = ultimate tensile strength of plate material, lb./in.^{2}
t = thickness of plate, in.
efficiency = efficiency of longitudinal joint, as a percentage
R = inside radius of drum or shell, in.
FS = factor of safety (for riveted boilers, a minimum of 5; for welded boilers, see Code at time of construction)

A comparison of the original joint efficiency to the deterioration in thickness percentage can be made. If the deterioration efficiency is less than the longitudinal efficiency, two methods of obtaining the allowable pressure can be used, with both giving the same answer.

Example A boiler drum has a 48-in. diameter, the longitudinal-seam efficiency is 90.0 percent, the plate is ½ in. thick, the steel has a tensile strength TS of 70,000 lb/in.^{2}, and the drum is designed for 262-psi maximum working pressure based on a factor of safety of 5. Corrosion and pitting have reduced the average thickness ⅛ in. in large areas. What is the safe pressure?

First,

$$\text{Deterioration efficiency} = \frac{0.5 - 0.125}{0.5} = 75\%$$

Using 75 percent as a new joint efficiency instead of 90 percent, with original thickness, we have

$$P = \frac{70{,}000 \times 0.5 \times 0.75}{24 \times 5} = 218.7 \text{ psi}$$

Using reduced thickness with a joint efficiency of 100 percent on the basis that deterioration is not in the joint area, we get

$$P = \frac{70{,}000 \times (0.5 - 0.125) \times 1}{24 \times 5} = 218.7 \text{ psi}$$

Grooving is a form of deterioration of boiler plate by a combination of localized corrosion and stress concentration. See Fig. 12-4*a*. It is found usually in areas adjacent and parallel to riveted seams or flanging, as in dished heads. The groove is usually ⅛ to ½ in. wide and may be several inches to several feet in length. Since the reduction in thickness occurs in a part that is subjected to stress concentration, grooving may be very serious. If it occurs to any extent in the seams of an unstayed boiler shell or drum, no repairs are possible.

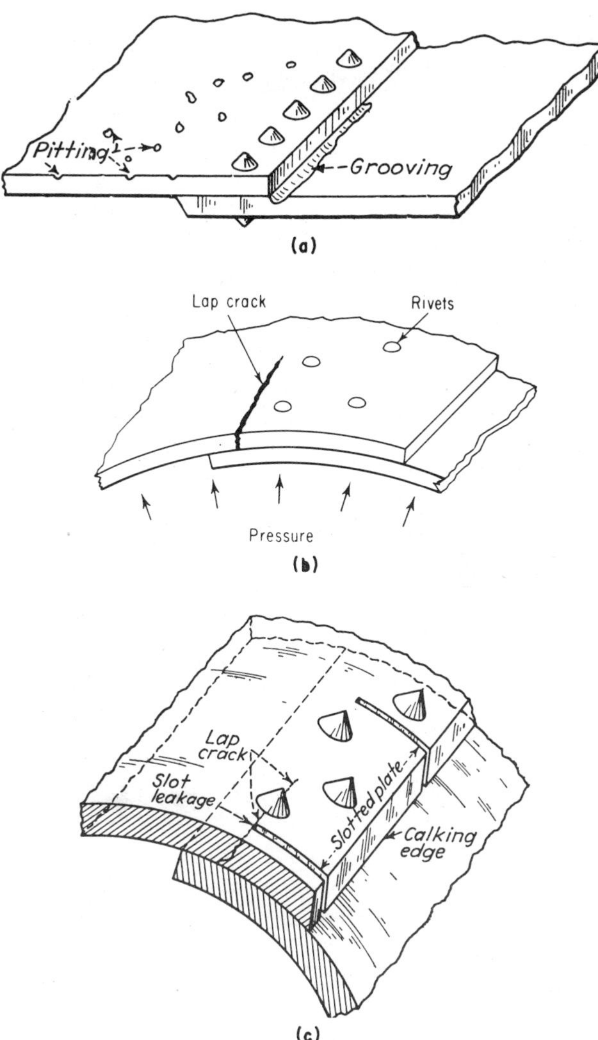

Fig. 12-4 Lap joints are susceptible to *(a)* grooving and *(b)* cracks at lap. *(c)* Lap-crack leakage is found by slotting plate.

The allowable pressure must be reduced considerably, or the boiler must be permanently removed from service. In all such cases, the advice of an authorized inspector should be followed.

Corrosion fatigue usually causes transgranular cracking of metals that are cycled under stress in a corrosive environment. *Stress corrosion* cracking occurs where a metal is in contact with a corrosive medium while under heavy stress. It requires the presence of highly stressed metal and a corrosive medium. Normal stresses are then magnified above normal.

Caustic cracking (caustic embrittlement) is a serious type of boiler metal failure characterized by continuous, mostly intergranular cracks. The following conditions appear to be necessary for this type of cracking to occur: (1) the metal must be stressed; (2) the boiler water must contain caustic sodium hydroxide; (3) at least a trace of silica must be present in the boiler water; and (4) some mechanisms, such as a slight leak, must be present to allow the boiler water to concentrate on the stressed metal. Caustic cracking is a particular problem in older boilers with riveted drums because of stresses and crevices in the areas of rivets and seams. While this type of cracking has become less frequent since the advent of welded drum boilers, rolled tube ends are still vulnerable areas of attack. The possibility of caustic cracking should be considered in establishing any water treatment program.

Modern failure analysis includes a metallurgical review of the reason for a pressure-part failure. Since most pressure vessels are fabricated by welding, the metallurgical review where weld failures occur may include such factors as hydrogen embrittlement from moisture in the electrodes used, stress corrosion cracking, effect of heat treatment or lack of same in the heat-affected zone (HAZ) of a weld, and sensitivity to notches that may have been in the weld ripples.

Hydrogen-induced cracking in welds has been experienced. Dissolved hydrogen can be introduced into a weld as a result of the welding process and the moisture that may be present. For example, absolutely dry electrodes should be used, and low hydrogen-coated electrodes are also most desirable. The concentration of hydrogen at which cracking can occur depends on the following factors: the existence of stress raisers, such as a notch, inclusion, HAZ microcrack; a stress level above which hydrogen embrittlement can occur (below this stress level it may not occur); and the effect of temperature on hydrogen embrittlement.

BOILER WATER TREATMENT

The main purpose of boiler water treatment is the elimination of the trouble that can be caused by scale, corrosion, embrittlement, and possibly carryover.

The treatment of feedwater is a problem requiring periodic testing of the water and proportioning the treatment according to the varying conditions. There are a number of reputable laboratories prepared to equip small or large plants with suitable test kits and to supply or advise the proper treatment indicated by the tests.

Extreme care should be taken in removing existing scale in a boiler by treatment of the water. If the scale is removed too quickly, it may drop down in large quantities, with serious damage to the boiler being the result because of restricted circulation and overheating. In a watertube boiler, ruptured tubes may be the consequence; in a fire-tube boiler, bulges and even rupture in the shell have followed.

There are five possible steps needed in water treatment depending on the supply, pressure, extent of makeup, and similar conditions: (1) pretreatment of raw-water supply; (2) treatment of makeup water going to the boiler; (3) internal treatment of the water in the boiler; (4) treatment of the condensate being returned to the boiler; and (5) blowdown control to remove precipitated sludge from the boiler.

Analyzing a water sample is the process of finding out how much of the various impurities and other chemical substances is present in the water. The results are usually expressed in parts per million (ppm) and tabulated as shown in Fig. 12-5a. Parts per million is a measure of proportion by weight, such as one pound in a million pounds. Grains per gallon is another way of expressing the amount of a substance present. One grain per gallon equals 17.1 ppm.

No matter what the chemical characteristics of the impurity may be, three states are possible:

1. If the impurity is a soluble solid, it appears in a dissolved state or in *solution* with the water.
2. If the solid is not soluble in water, it is not in solution, but in a state of *suspension*.
3. Those impurities of a gaseous nature that are partially soluble are in an *absorbed* state in the water.

Most impurities are found in a dissolved state or in solution with the water. The temperature of the water has a marked effect on solubility. Some impurities become less soluble as temperatures rise and start to precipitate, or form scale in the water. Calcium sulfate and calcium hydroxide are such impurities. The mere fact that some impurities stay soluble at higher temperatures and do not precipitate does not mean that they should be ignored. They may have a tendency to corrode metals, they may act as moisture carryovers, or they may cause embrittlement. *Pretreatment* of raw water may be necessary because water impurities vary with the source of supply. Surface waters are characteristically high in dissolved oxygen and organic and inorganic sediment, and low in total solids and hardness. Groundwaters are usually high in carbon dioxide (because of carbonates), total solids, and hardness but low in sediment.

Regardless of the treatment process used to reduce impurities, clarification is usually the first step. The presence of other troublesome water constituents such as hydrogen sulfide or iron may dictate further external treatment. *External* treatment of boiler water depends on many factors, including boiler pressure, amount of makeup, use of steam, and similar considerations. External-treatment systems may be installed to reduce one of or all the following feedwa-

REPORT OF WATER ANALYSIS			Parts per million	Equivalents per million
Date ______		Silica as SiO_2	5	
Source ______		Iron as Fe_2O_3	1.2	
Date analyzed ______		Calcium as Ca	62	
Total dissolved mineral solids	ppm	Magnesium as Mg	31	
Organic matter	none ppm	Sodium and potassium as Na	38	
Suspended solids	5 ppm	Bicarbonate as HCO_3	250	
Chloroform, extractable (oil, etc.)	none ppm	Carbonate as CO_3	0	
pH	7.7	Hydroxide as OH	0	
Phenolphthalein alkalinity as $CaCO_3$	0 ppm	Chloride as Cl	11	
Methyl orange alkalinity as $CaCO_3$	205 ppm	Sulfate as SO_4	138	
Hydroxide alkalinity as $CaCO_3$	0 ppm	Nitrate as NO_3	0	
Hardness as $CaCO_3$	282 ppm	Carbon dioxide as CO_2	10	
Specific conductance	micromhos	Turbidity	5	
		Physical characteristics of sample	Clear when drawn	

(a)

Method of treatment	Average analysis of treated water				
	Hardness ppm as $CaCO_3$	Alkalinity ppm as $CaCO_3$	CO_2 in steam (potential)	Dissolved solids	Silica
Cold lime-soda	30–85	40–100	Medium-high	Reduced	Reduced
Hot lime-soda	17–25	35–50	Medium-low	Reduced	Reduced
Hot lime-soda phosphate	1–3	35–50	Medium-low	Reduced	Reduced
Hot lime-zeolite	0–2	20–25	Low	Reduced	Reduced
Sodium-cation exchanger	0–2	Unchanged	Low to high	Unchanged	Unchanged
Anion dealkalizer	0–2	15–35	Low	Unchanged	Unchanged
Split-stream dealkalizer	0–2	10–30	Low	Reduced	Unchanged
Demineralizer	0–2	0–2	0–5 ppm	0–5 ppm	Below 0.15 ppm
Evaporator	0–2	0–2	0–5 ppm	0–5 ppm	Below 0.15 ppm

(b)

Fig. 12-5 *(a)* Water treatment starts with water analysis to record impurities. *(b)* External treatments are used to treat feedwater for listed analysis.

ter constituents: suspended solids, hardness, alkalinity, silica, and dissolved solids.

Feedwater can range from all makeup to all condensate. Many low-pressure boilers are able to use raw-water makeup with only internal treatment. On the other hand, high-pressure boilers may require almost complete removal of dissolved solids from the makeup. And there are a variety of feedwater treatments for boiler pressures that fall between these extremes.

Zeolite softeners are widely used for makeup water preparation for low- and medium-pressure boilers. These softeners usually remove the calcium and magnesium hardness to 0 to 2 milligrams per liter (mg/L) as $CaCO_3$. If inlet water hardness varies, automatic hardness analysis of the softener effluent is a good control. Hardness level determines the softener regeneration point.

The table in Fig. 12-5*b* shows what to expect in hardness reduction when the treatments shown are used. The quality of the makeup water needed depends mostly on boiler design, operating pressure, and temperature. The degree of treatment called for also depends on boiler capacity, quality of condensate returns available, and raw-water analysis. A large plant with a high percentage of makeup usually justifies more elaborate external treatment than does a small installation.

The cold, hot, and zeolite softening processes for water treatment removes four troublesome impurities in water—calcium, magnesium, silica, and oxygen—by chemical reaction with combinations of lime, soda ash, and caustic. When carried out at room temperature, these reactions are termed a *cold process.* When the water is heated well above room temperature, it is called a *hot process.* The hot process is used for treating boiler feedwater. The water is heated by spraying it into the upper steam space of the softener unit. The inlet flow actuates a proportioning device to control the amount of lime and soda ash fed to the heating and mixing zone. Chemical reactions take place almost instantly. Sedimentation proceeds at a rapid pace because of the elevated temperature. The sludge-collecting cone at the bottom of the unit receives the precipitates and then periodically discharges them to the sewer. Clarified water leaves the settling tank and is fed to the anthracite filters for polishing.

Hot-process equipment can treat long flows in relatively small units. Chemical dosages are much lower than required by cold-process units. Treatment jobs handled by the hot-lime soda softener include softening by removing calcium and magnesium, silica removal with magnesium salts, oxygen removal, and removing turbidity.

Ion exchange in water treatment is based on the principle that impurities which dissolve in water dissociate to form positively and negatively charged particles known as ions. These impurities, or compounds, are called electrolytes. The positive ions are named cations because they migrate to the negative electrode (cathode) in an electrolytic cell. Negative particles are then anions since they are attracted to the anode. These ions exist throughout the solution and act almost independently. For example, magnesium sulfate ($MgSO_4$) dissociates in solution to form positive magnesium ions and negative sulfate ions.

Ion-exchange material has the ability to exchange one ion for another, hold it temporarily in chemical combination, and give it up to a strong regenerating solution. The chart in Fig. 12-6 lists the capacity of the ion exchangers commonly used in water treatment.

Figure 12-7*a* shows a water softener of the cation-exchanger type. Water

Ion-exchange materials	Flow rate gpm/sq ft	Regeneration		Typical capacity kgr/cu ft
		Chemical	lb/cu ft	
Cation exchangers:				
Sodium cycle:				
Natural greensand	5.0	NaCl	1.25	2.8
Synthetic gel	6.0	NaCl	5.0	10.0
Sulfonated coal	6–8	NaCl	3.15	7.0
Styrene resin	8–10	NaCl	10.0	25.0
Hydrogen cycle:				
Sulfonated coal	6–8	H_2SO_4	2.0	8.0
Styrene resin	8–10	H_2SO_4	5.0	11.0
		H_2SO_4	11.0	25.0
		HCl	10.0	30.0
Anion exchangers:				
Weakly basic (aliphatic amine)	6.0	Na_2CO_3	4.2	18.0
Weakly basic (Phenolic)	6.0	Na_2CO_3	4.2	18.0
Weakly basic (Styrene)	6.0	NaOH	3.25	20.0
Strongly basic (Type I)	6.0	NaOH	4.0	11.0
Strongly basic (Type II)	6.0	NaOH	4.5	14.0

Fig. 12-6 Commonly used ion-exchanger materials with regenerative chemicals used to restore the properties of ion-exchange materials.

softening using the ion-exchange process is accomplished by means of passing hard water through a bed of synthetic resin. The hardness-forming calcium and magnesium ions in the water are removed by exchanging them for the nonhardness-forming sodium ions which are attached to the resin. When all the sodium in the resin has been used up, the resin bed no longer has the ability to soften the water and must be regenerated. This is done by passing

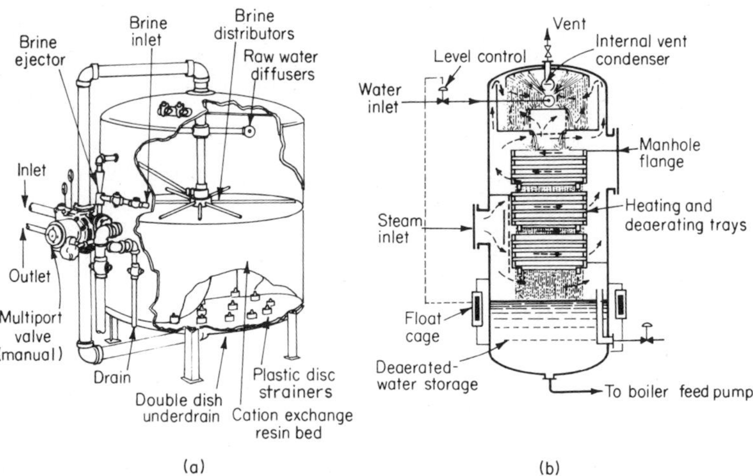

Fig. 12-7 *(a)* Water softener of the cation type. *(Courtesy The Permuttit Co.)* *(b)* Deaerator removes oxygen from water by heat.

an excess quantity of sodium chloride brine through the resin bed to drive off the calcium and magnesium and replace these elements with sodium. The brine is then rinsed out of the bed with water before being placed back into operation on the softening cycle.

Evaporators are used to remove solids from feedwater by the use of heat. Dissolved and suspended solids are removed from water effectively by the purely thermal process of vaporizing water with heat. The evaporator is simply a pressure tank filled with steam coils and a moisture separator on the outlet. Vapor is condensed in a separate heat exchanger to form pure water. Heating steam flows inside the tubes, and makeup water enters the shell. The makeup water is vaporized and then condensed to be used in the boiler. The impurities remain in the evaporator and must be cleaned out periodically.

Deaerators are used to remove oxygen in water also by the use of heat. Oxygen can be a dissolved gas in water that does not react chemically with the water and becomes less and less soluble as the water temperature increases. Thus it is easily removed by bringing the water to the boiling point corresponding to its operating pressure. Pressure and vacuum designs are used. Deaerators are used to heat water for boiler feed. But if water is used for cooling or other purposes where heating is not needed, vacuum units are used. Steam deaerators break up water into a spray or film and then sweep the steam across and through it to force out dissolved gases such as oxygen or carbon dioxide. The oxygen content can be reduced below 0.005 cubic centimeters per liter (cm^3/L), almost the limit of sample testing by chemical means. As carbon dioxide is removed, the increase in pH also gives an indication of deaeration efficiency. Fig. 12-7*b* shows a typical deaerator.

Internal treatment is used to adjust the water being supplied to a boiler in order to prevent the problems caused by scale, corrosion, and metal deterioration such as embrittlement. However, internal treatment requires care. The main disadvantage possible from internal treatment is that the treatment can develop precipitation within the boiler.

Scale-causing elements such as calcium and magnesium are usually controlled by a coordinated phosphate/pH program or by chelants or by a combination of both. When phosphate is added, hydroxide precipitate forms which can be easily removed by blowdown. Chelant compounds react with metal ions in the water to keep them soluble. In both treatment programs, the residuals of orthophosphate or chelant in the boiler water are an important control parameter and must be monitored.

To fight corrosion, a close pH control is important. Too low a pH will result in excessive iron corrosion while too high a pH causes corrosion of copper-containing alloys. Proper corrosion control usually includes some type of deaeration for removal of carbon dioxide and the addition of an oxygen scavenger (sulfite or hydrazine) and maintenance of alkaline pH in the boiler water. In addition, sometimes ammonia or neutralizing amines are applied to the boiler water to neutralize the carbon dioxide, or filming amines are applied to the steam to lay down a protective film on metal surfaces in the boiler to protect against dissolved oxygen.

In higher-pressure boilers it is necessary to control boiler-water silica concentration by automatic analysis. Silica is the firmest, toughest, and most difficult to remove of all the dissolved minerals. This removal is critical, for silica is apt to carry over with the steam. Silica's glasslike deposits inhibit heat transfer, resulting in tube burnout. When deposited on turbine blades, silica will reduce efficiency which often results in rotor unbalance, necessitating costly and premature shutdowns.

Figure 12-8 provides some boiler and feedwater limits that are recommended by various authorities. The ASME has published, through its Committee on Water in Thermal Power Systems, a guide on water quality for industrial boilers entitled "Consensus on Operating Practices for the Control of Feedwater and Boiler Water Quality in Modern Industrial Boilers." This paper should be used as a guide for establishing water-quality limits for the boiler size involved.

Condensate returns are a way of saving fuel; however, condensate also produces water problems. The condensate-return system may have corrosion products from steam and condensate piping which can form highly insulating deposits on boiler surfaces. Improved steam and return-line corrosion control must be used to limit this sludge. The most common attackers of the condensate systems are dissolved oxygen and carbon dioxide that find their way into the steam system.

The condensate-return system is a very revealing sample point to monitor total boiler water-treatment performance. The amount of contaminants found, and their nature, will often point out malfunctions and suggest corrective action in the rest of the system. Permissible contaminant levels would depend on the nature of the troublesome constituent, boiler design, and operating pressures. Dissolved oxygen and carbon dioxide, for example, can be directly or indirectly responsible for many system failures and are often monitored in condensate.

	BOILER WATER			FEEDWATER[3]			
Boiler Pressure	Total Solids ppm[1]	Total Alk ppm as $CaCO_3$[1]	Silica ppm as SiO_2[2]	Hardness ppm as $CaCO_3$	Iron ppm as Fe	Copper ppm as Cu	Oxygen ppm as O_2
0- 300	3500	700	75-50	0-1 Max.	0.10	0.05	0.007
301- 450	3000	600	50-40	0-1 Max.	0.10	0.05	0.007
451- 600	2500	500	45-35	0-1 Max.	0.10	0.05	0.007
601- 750	2000	400	35-25	0-1 Max.	0.05	0.03	0.007
751- 900	1500	300	20-8	0-1 Max.	0.05	0.03	0.007
901-1000	1250	250	10-5	0-1 Max.	0.05	0.03	0.007
1001-1500	1000	200	5-2	0	0.01	0.005	0.007
1501-2000	750	150	3-0.8	0	0.01	0.005	0.007
2001-2500	500[4]	100[4]	0.4-0.2	0	0.01	0.005	0.007
2501-3000	500[4]	100[4]	0.2-0.1	0	0.01	0.005	0.007

Feedwater organics should be zero and pH in the range of 8.0 to 9.5[3]

References and Notes:

1. American Boiler Manufacturers Assoc. 1958 Manual
2. Above 600 psig silica level selected to produce 0.02 ppm SiO_2 steam
3. Babcock & Wilcox Publications (a) Water Treatment for Industrial Boilers, BR-884, 8-68 and (b) J. A. Lux, Boiler Water Quality Control in High Pressure Steam Power Plants, 9/62
4. J. A. Lux, 3(b), recommends levels as low as 15 ppm TDS above 2000 psig.

Fig. 12-8 Recommended boiler water and feedwater limits for drum-type boilers.

Monitoring and control of hardness, conductivity, and specific contaminants (such as iron and copper) will also permit maximum condensate reuse.

Nuclear boiler systems deserve further considerations, for condensate return is usually considered a radioactive substance and requires special monitoring.

Condensate polishing is the term used for the removal of solid traces as well as dissolved solids that are found in condensate systems. In the utility field, the two condensate polishing systems used are the precoat filter and the mixed-bed demineralizer in order to eliminate the solids. Condensate systems can be chemically treated to control corrosion damage caused by water, carbon dioxide, and oxygen. The treatment chemicals include neutralizing amines, filming amines, hydrazine, and sometimes ammonia. Neutralizing amines control the condensate pH, as does ammonia. Filming amines provide a protective coating on metal surfaces. Hydrazine is used to minimize copper corrosion and is usually injected near copper-fabricated equipment, such as prior to a steam-turbine condenser equipped with copper tubes. Hydrazine also is an oxygen scavenger of the boiler water.

Blowdown is an integral part of the proper functioning of a boiler water treatment program and usually requires continuous monitoring for positive control. It is through blowdown that most of the dirt, mud, sludge, and other undesirable materials are removed from the boiler drum.

In most systems, surface blowdown is accomplished continuously, and the optimum blowdown interval is such that sludge or scale on heating surfaces is minimized. At the same time, the loss of heat and chemical additives is also kept to a minimum.

In the larger, more critical boilers, continuous surface blowdown is usually combined with a regular bottom blowdown. In many high-pressure boilers, it is desirable to minimize boiler blowdown to reduce heat and water losses.

Blowdown analysis is complicated by sample conditioning considerations (a sample cooler is usually required), but the control parameter is generally conductivity. Typically, a conductivity meter limit is maintained within a control range, and blowdown is activated by a certain deviation from that range. Other blowdown monitoring parameters include pH, silica, hydrazine, and phosphate.

Intermittent or bottom blowdown is taken from the bottom of the mud drum, waterwall headers, or lowest point in the circulation system. The blowoff valve is opened manually to remove accumulated sludge, about every 4 to 8 hr, or when the boiler is idle or on a low-steaming rate. But hot water is wasted, and control of concentrations is irregular.

Continuous *surface* blowdown automatically keeps the boiler water within desired limits. Continuously removing a small stream of boiler water keeps the concentration relatively constant. Savings by transferring heat in the blowdown to incoming makeup often pay for the investment.

Priming and foaming and *carryover* are factors usually controllable by the operating engineer. Priming is the lifting of boiler water by the steam flow. The water may be lifted as a spray or in a small body; as it enters the steam line, its weight and velocity may cause severe damage to equipment. Ruptured steam-line fittings or wrecked turbines or engines have resulted from "slugs" of water. Unless priming is induced by faulty boiler design (which is not common), it is caused by carrying too high a water level for the demands for steam flow. The water level in the drum should be kept several inches lower than normal if the steam flow fluctuates very much, for a sudden rush of steam sometimes tend to pick up water from the surface directly below the nozzle.

Foaming is more a chemical than a mechanical problem. High surface tension of the boiler water causes many of the steam bubbles to be encased by a water film. These film-encased bubbles rise and pass out in the steam flow. The cause of high surface tension is usually a high concentration of solids in the boiler water. Organic matter, too, may produce this trouble. Periodic checks on boiler-water concentration and control of blowdown to hold the concentration within allowable limits will prevent foaming. The density of the boiler water is a measure of its concentration. Specially calibrated hydrometers are available at low cost for direct reading of this condition.

Carryover from Steam Boilers Clean steam plays an important part in economical power plant operation. When the boiler is contaminated with water, mineral solids, or other impurities, numerous troubles develop and costs automatically increase. Foreign matter entrained in otherwise clean steam leaving a boiler drum is commonly termed *carryover.* It can be eliminated or minimized by determining its cause and then applying the right correction. The magnitude of the losses occasioned by carryover is not generally realized. Fuel consumption, equipment maintenance costs, and plant safety are all affected.

There are many cases where carryover still persists, even though all ordinary preventive measures are used. To reduce entrained solids to an absolute minimum, steam washers or steam separators are installed. Washers use the relatively pure incoming feedwater to wash outgoing steam. Separators remove entrained water and solids by impinging the steam against baffles or suddenly changing its direction of flow so that foreign particles are thrown out by centrifugal force. Separators or purifiers are also valuable as insurance against any unexpected slugs of water which might damage power plant equipment.

Waterside *scale removal* is accomplished by one of three methods: mechanical removal, water treatment, and acid cleaning. Mechanical removal of scale is effected while the boiler is idle and empty. The accessible parts of shells, drums, heads, and braces are chipped with a dull chisel or scaling hammer, care being taken not to score the metal. Scale may be ground off the internal surfaces of watertubes with a tube turbine. Water is generally used to wash out ground "scale sludge" while the tube turbine is in operation. Care should be exercised not to operate a tube turbine too long in one place or to force it unduly, for damage to the tube may thus result.

Scale deposits on external surfaces of fire tubes may be vibrated loose with a tube rattler or by shaking a long, heavy bar in each tube. Extreme care should be taken after such mechanical treatment to see that *all* loosened scale is removed from the boiler before closing it up for operation. Many cases of

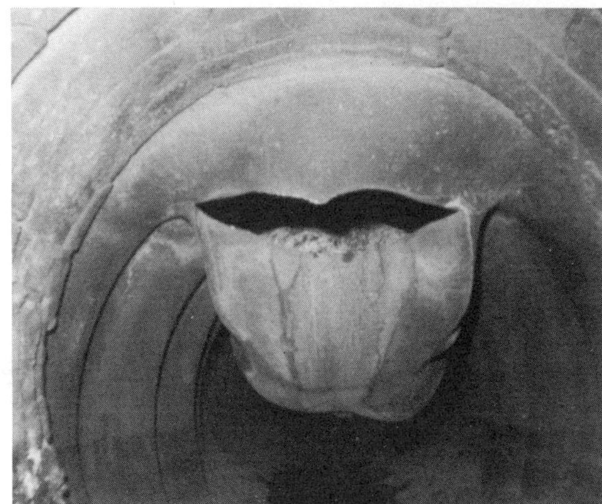

Fig. 12-9 Collapsed scotch marine corrugated furnace caused by scale and dry firing. *(Courtesy Royal Insurance Co.)*

serious damage have resulted from loose scale accumulations left in boilers. See Fig. 12-9.

Acid cleaning of boilers is often used to remove metallic oxide. Caustic soda and soda ash are the old standbys for cleaning oil from the waterside of boilers. One pound of each chemical is added for every 1000 lb of water required to fill the unit. After the boiler is filled with steam, drum vents open, and a light fire is started and maintained until the vents issue steam. After the vents are closed, pressure is built up to 25 psig and held while boiling for 24 hr. Some operators blow down to half a gauge glass after about 4 hr of boiling. But 24 hr after boiling, the solution is dumped, and the unit is refilled with fresh hot water with the vents open. After this flushing water is dumped, a thorough internal inspection is made. If necessary, a hose is used for spraying hot water for final cleaning.

The solvents used for acid cleaning are varied. Some use hydrochloric acid; others, phosphoric acid. The usual procedure is to fill the boiler until the solution overflows at the air vent (acid is added outside the boiler). The solution is *allowed* to soak the boiler from 4 to 6 hr, followed by refilling with a neutralizing agent. If hydrochloric acid is used for soaking, a weak solution of phosphoric acid is used. After draining, fresh warm water is used for flushing; then the boiler is immediately filled with an alkaline solution and boiled again for several hours. This solution is drained; the boiler is flushed again and then refilled with normal service water, with proper feedwater treatment started immediately.

A precaution to be observed in acid cleaning of boilers equipped with a superheater and other such bent tubes is to make sure all traces of acid are thoroughly cleaned out of U bends. This is critical in the neutralizing and flushing stage after the tubes are soaked with an acid solution. Compressed air may have to be used to force the solution out of dead pockets. If this is not done, the acid solution may not be completely cleared, and thinning of tubes will result. Acid cleaning of riveted boilers can be dangerous because the acid may settle under the butt straps or lapped plates and eat up these holding elements. Thus riveted boilers are usually not acid-cleaned.

Areas subject to high local stress and repetitive application of stress may be affected by acid treatment. Tubes that have been repeatedly rolled and acid-cleaned may develop tube-roll leakage. Then with further rolling impossible, new tubes may be required. But the prevention of scale in a boiler is still the best method of keeping a boiler clean.

To guard against the consequences of superheater chemical contamination, one of the most important practices prior to proceeding with a chemical cleaning operation is to carefully review the entire piping arrangement and procedures. All the cognizant personnel involved in the chemical cleaning must be familiar with the cleaning method and flow paths to be used in filling, draining, adding chemicals, backfilling the superheater, etc. All possible paths (such as drain lines from the superheater connected with lines or manifolds being used for filling or draining the boiler) should be examined closely to ensure that cleaning solutions being fed to or drained from the boiler do not have a possible flow path to the superheater. The lines used to fill the superheater with water during the boiler cleaning should preferably not be connected with boiler chemical fill or drain lines. If flow paths exist between the superheater and the boiler fill or drain lines or manifolds, positive means of isolation must be provided. Double valves with a telltale connection between them are the minimum recommendation.

External corrosion or deterioration of boiler surfaces on the fireside may be a continuous process. It is a chemical combination of the metal, known as oxidation or rust. Normally, this action would not progress appreciably in the life of a boiler. However, most boiler surfaces are coated with soot or ash on the fireside. The sulfur content of the soot combines with any moisture to form a sulfurous acid which is highly corrosive. Hence, a minor leak may cause a serious defect to develop within a few years; even though there is no leak, the boiler may "sweat" when idle in humid weather, and such moisture in combination with the soot will cause trouble.

Leaking tube-cap gaskets in box-header watertube boilers often cause moisture to saturate soot at the bottom of these headers, and rapid deterioration results. The blind heads of bent-tube watertube boiler drums sometimes extend into the brick wall and are practically inaccessible for inspection of their external surfaces. Here is a natural pocket to fill with soot, and trouble may be expected. To prevent serious explosions, the brickwork should be chipped back from the head seam so that the soot will not be held against the head, and sufficient clearance should be provided so that the head will be accessible for inspection. Ultrasonic tests for thickness from the waterside can detect external or fireside corrosion thinning.

Continued leaks from any source should not be tolerated whether they are from a roof, valve packing, gaskets, piping, or other sources. Water dripping onto a boiler will cause damage. Leaky soot blowers are a frequent source of external corrosion of watertubes. The soot-blower valves should be kept tight, and the piping drained of condensate before blowing soot.

Handhole- and manhole-gasket leakage frequently causes damage to the flange or surrounding plate by external corrosion. The dry sheet of HRT boilers should have a ¾-in. drain hole in the bottom so that any leakage from a manhole gasket, soot blower, or tube will drip through the hole and give an indication of the leak. Otherwise, water might collect on the bottom of the dry sheet and cause serious damage.

Mud-drum nipples of sinuous-header-type watertube boilers are ideal places for attack. Periodic boiler cleaning should include removal of soot accumulation from the mud drum and nipples. Care should be taken that all tube-cap gaskets are kept tight, since water dripping from these gaskets is the greatest enemy of mud-drum nipples.

Piping is often affected, particularly if it is buried, as are some blowdown pipes. It is best to have all piping accessible for inspection and general maintenance.

Erosion is closely allied with external corrosion in its effect, but it is purely a mechanical action, a wearing of external surfaces by abrasion. The gas-entrance ends of tubes in fire-tube boilers may become thin after 10 to 20 yr because of the scouring action of soot particles entering the tubes at high velocity. This effect may be due also to the fact that internal corrosion is more rapid where the high-temperature zones causes a higher evaporation rate.

Erosion by improperly adjusted soot blowers is not uncommon. In a few weeks of use, a hole may be worn through several tubes by one faulty jet of a soot blower. The action resembles sandblasting. Erosion as a result of flame scrubbing probably does not have the opportunity to become serious before the damage done by localized overheating makes the condition evident.

When large quantities of ash are produced in coal-fired boilers, slagging and fouling problems may occur in the furnace and convection sections of the boiler, especially if soot blowers are not operated sufficiently to remove the ash. Plugging of convection passes can cause the fly ash to become sticky, thus accelerating further slagging. Metal under slagging may be attacked in the presence of moisture, leading to fireside deterioration. Soot blowers and other boiler auxiliaries (Fig. 12-10*a* and *b*) require preventive maintenance to avoid boiler problems.

Flame impingement is a source of damage to boilers and refractory. If the flame impinges directly on the boiler shell (Fig. 12-10*c*), excessive evaporation will be caused on the water surface over that point. The high temperatures may cause damage through local scale formation or corrosion which otherwise would be dormant, or the temperature may be high enough to cause serious damage by overheating the plate.

Direct impingement of flame on tubes of watertube boilers may cause steam

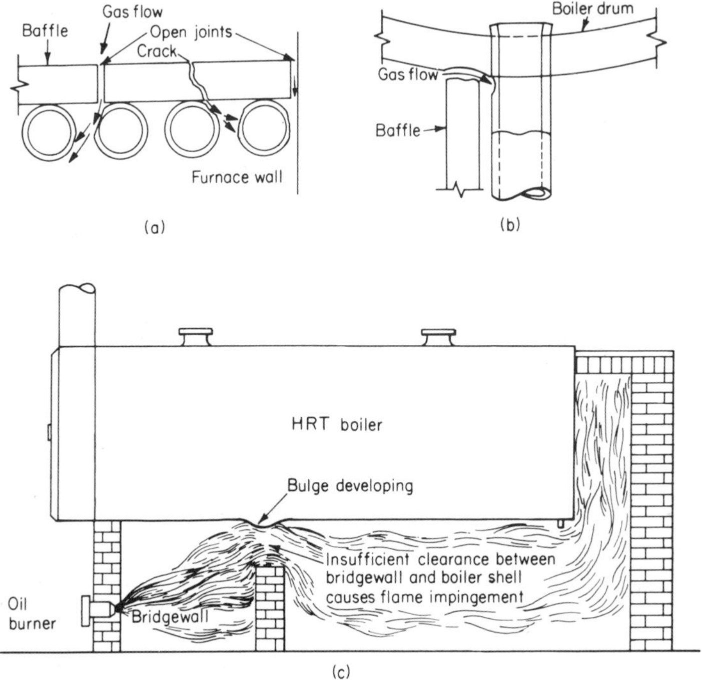

Fig. 12-10 Fireside problems. *(a)* and *(b)* Erosion of tubes from faulty baffles. *(c)* Flame impingement can cause local overheating.

pockets. That is, evaporation and resultant circulation upward in a tube may be more rapid than the rate at which cooler water can be supplied from its lower end. A steam pocket, serious overheating, and failure of the tube usually result. Watertubes are also susceptible to the same results of flame impingement as those mentioned with reference to fire-tube boilers.

A reduction in thickness, due to erosion by particles of burning carbon and of fly ash, will result if flame impingement continues. Flame scrubbing of refractory greatly shortens its life, owing to erosion and overheating.

Combustion processes and combustion problems are many and involve the burning of different fuels in proportioned furnaces so that proper air/fuel ratios and good ignition with a stable flame are maintained.

Oil fires sometimes pulsate or flutter to the extent that the entire boiler setting may vibrate. This effect may be traced usually to a pulsating oil-burner pressure resulting from a reciprocating oil pump. Use of an air-cushion chamber usually solves this problem.

Burners should never be lighted by the heat of the refractory or from the flame of a burner. Use the igniter. Furnace-gas explosions may result if these igniting precautions are neglected.

Ignition stability is important to safe burning of suspension fuels. Coal of low volatile (gas) content is sometimes unstable when pulverized and when operation is at low loads. Oil-burner instability can be traced usually to a clogged oil system or to improper oil temperature.

One of the most common causes of furnace explosion is the momentary failure of ignition during regular operation. During the pause, unburned fuel enters the furnace, and highly combustible gases, distilled by the heat of the firebox, fill the boiler setting. These gases may penetrate a crevice in the refractory or ash where red heat exists, and a blast results. Furnace explosions may be caused also by accumulations of unburned combustible igniting spontaneously. These explosions may cause serious damage.

Modern, properly installed combustion flame safeguard systems will help in eliminating many, if not most, furnace explosions resulting from poor burning. See Chap. 11. Main-flame and pilot scanning cells with electronic relays that can shut down a burner system within 2 to 4 sec after loss of flame are now commonly specified. Programmed purging prior to light-off and pilot proving are now required. Fuel trains must provide redundant safety shutoff valving.

The inherently limited combustion volume of internally fired boilers and the requirements of fuels such as wood-waste products sometimes demand additional volume. This is often produced by a Dutch oven, which is actually an external, primary furnace that leaves the boiler furnace as a secondary combustion chamber (Fig. 12-11).

In the paper industry, *recovery boilers* present an additional hazard besides fireside fuel explosions. This is the problem of smelt-water reactions taking place in the furnace. The best prevention is to maintain pressure-part integrity in order to avoid water entering the furnace in any form. The concentration of the black-liquor fuel must also be monitored so that too dilute a solution

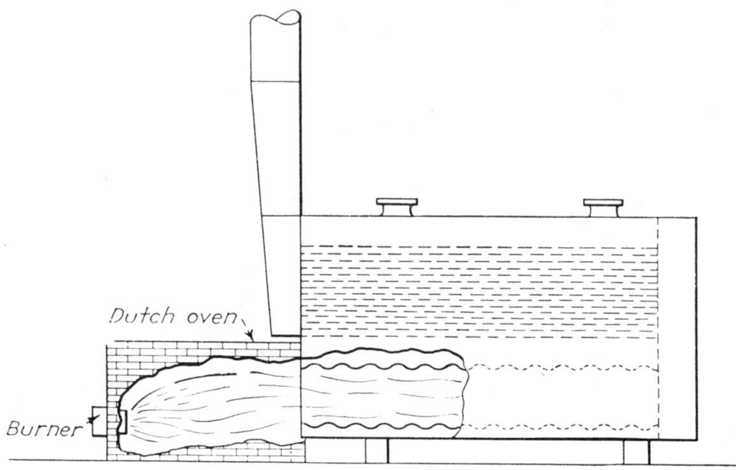

Fig. 12-11 Dutch oven adds to furnace length to permit more complete fuel combustion from the increase in furnace volume.

does not enter the furnace and cause a smelt-water reaction to take place. Use of water to wash down tubes or deposits in the presence of a smelt bed in the furnace must be avoided. The Black Liquor Recovery Boiler Advisory Committee, which consists of users, manufacturers, and insurance representatives, stresses the need to remove as rapidly as possible any water source that could enter the furnace. Emergency procedures are recommended under these conditions, one of which is to drain the boiler as rapidly as possible and in accordance with manufacturers' recommendations to a level 8 ft above the low point of the furnace floor.

To ensure proper draft-air opening for a packaged boiler installed in a closed machine room there should be a fixed opening for fresh air, an average area of 2 ft^2 for each 100 boiler HP. Opening windows is not the answer because they are often closed in cold weather. Then the boiler starves for air. For each boiler horsepower, about 10 ft^3/min of air is needed.

Some safety precautions to be observed in operating and maintaining an automatic burner system are the following:

1. Always close off all manual fuel valves before working on a burner or disconnect the wiring to automatic fuel valves or do both.
2. Never stand in *front* of a burner or boiler during start-up.
3. Never manually push in relays unless the manufacturer's instructions so advise.
4. Never permanently block in relays with rubber bands, sticks, or other devices.
5. Never change the safety-switch timing of a flame supervisory control. If the system is locking out, correct the cause, *not* the symptom.
6. Never install jumper wires or bypass any safety interlock switches.
7. Before starting a burner, visually inspect every combustion chamber to make sure there is no accumulation of combustible.
8. Regard every system lockout as a safety lockout until proved otherwise by competent personnel.

Low water in a boiler may lead to anything from leakage to an explosion, depending a great deal on the type of boiler, the rate of firing, and just how low the water gets. If the boiler is a type with a crown sheet over the firebox or combustion chamber, such as a locomotive or scotch marine type, a rupture of the crown is almost inevitable if the water drops below the level of the crown sheet, for the bared metal soon attains such a temperature (red heat) that its tensile strength drops to a dangerously low point. A rupturing crown sheet often is exceedingly violent, and many serious explosions have resulted.

In fire-tube boilers, the first result of the water level dropping below the safe level when a hot fire is carried may be leakage at the rear ends of the upper rows of tubes. As the water recedes from tubes exposed to high-temperature gases, the expansion of the tubes is so great that their rolled-in seat is broken. Leakage may appear from the rear ends of each succeeding row of tubes as the water level drops further, until distortion of the shell plates and heads, with leakage at the seams, usually occurs. An explosion due to low

Fig. 12-12 Melted-tube due to low-water in watertube boiler equipped with two LWCOs and a low-water alarm. *(Courtesy Factory Mutual Engineering.)*

water is uncommon in this type of boiler because of the many points of leakage to give warning. However, the tubes may collapse.

The effects of low water on watertubes are similar to those on fire tubes. The tubes expand as the water leaves them, and they break their expanded seats, leakage being caused. Excessively low water may result in tube rupture and tube melting. Figure 12-12 shows the tubes that are left after a severe dry firing. The boiler was equipped with two low-water fuel cutoffs, a feedwater regulator, and a low-water alarm. When these devices were periodically tested was not logged. Figure 12-13 shows a low-water fuel cutoff full of sludge deposits because of failure to periodically blow or flush the sludge out of the chamber. Persistent management insistence of logging actual testing of these vital safety devices would prevent a majority of low-water failures.

Improvements in the operation and care of a boiler to minimize the possibility of developing a low-water condition depend on frequent checking and testing of feed, condensate return, pumps, and similar components of a boiler water loop system that is supposed to keep the boiler supplied with water. Regulators and associated alarms with low-water cutoffs are the next and usually last defense against low water, unless an operator takes corrective actions. The best way to test the low-water cutout (LWCO) is by duplicating an actual low-water condition. Slowly drain the boiler (through the blowdown line) while under pressure. If a heating boiler does not have proper drains for doing so, be sure to correct this condition. Many operators drain only the float chamber of the cutout for this test. But the float-chamber drain is only for blowing

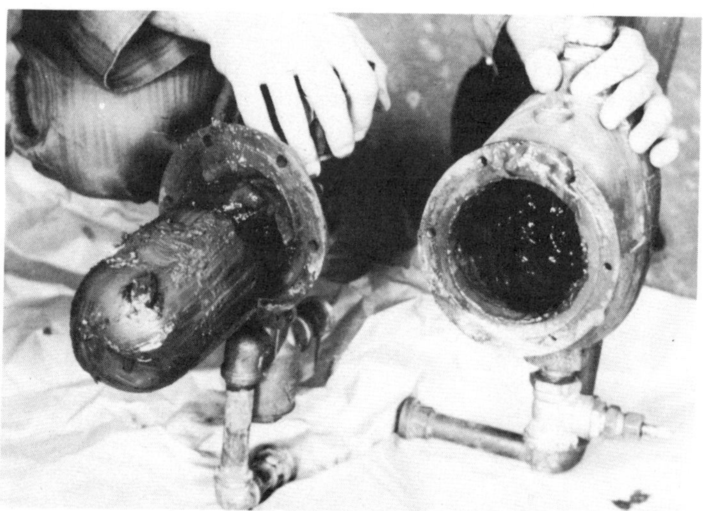

Fig. 12-13 Low-water fuel cutoff is plugged with sludge from lack of testing. *(Courtesy Factory Mutual Engineering.)*

sediment out of the float chamber. Usually the float will drop when this drain is opened because of the sudden rush of water from the float chamber.

Tests often show that draining the float chamber will indicate that the cutout performs satisfactorily, but when proper testing is done by draining the boiler, the cutout fails to work. Daily checking of the LWCO by draining the float chamber (or electrode chamber) is good practice. But at the beginning of each season, duplicate an actual low-water condition.

To test a boiler for whether the lower gauge glass is obstructed even though the gauge glass is half full of water, open the try cocks on the water column. If all show steam, it means the bottom connection is obstructed, permitting steam from the top connection to condense in the gauge glass. The boiler should be shut down immediately and inspected for possible dry-firing damage. Naturally, the bottom connection of the water column and gauge glass should be cleaned of all obstructions before the boiler is returned to service.

If an unusually high feedwater pressure is necessary to maintain the water in the boiler, check the feedwater valves and lines to make sure that a valve has not broken off its seat or that there is not some obstruction in the line itself. Some methods of feedwater treatment have been known to deposit chemicals inside the feedwater line, making it impossible to get water into the boiler. Also look for leaks due to cracked or corroded piping of the feedwater (or condensate line on heating boilers) especially if it is buried anywhere in the system.

If water is not visible in the gauge glass because of failure of the feedwater supply, immediately do the following:

1. Shut off the fuel to the burners and secure the burners.

2. Check the water level by trying the try cocks and water-column drain. If definite low water is indicated below the gauge-glass level, close the main steam valve and feedwater valve.

3. If the boiler is equipped with one, open the superheater drain.

4. Continue operating forced-draft and induced-draft fans until the boiler cools gradually.

5. Let the pressure reduce gradually and when the furnace area is sufficiently cooled, check for leaking tubes and other signs of overheating damage. On fire-tube boilers, look for cracked or warped tube sheets, broken and leaking stay bolts in the waterlegs. On scotch marine boilers, check for cracked or leaking furnace-to-tube sheet welds. On cast-iron boilers, look for cracked sections. On steel boilers, check for leaking joints on longitudinal or circumferential welds or riveted joints.

6. If no leakage is evident, give the boiler a hydrostatic test of 1½ times the allowable working pressure. Then again check for leakage at all critical parts of the boiler. If leakage is observed during the initial check or during the hydrostatic test, notify the authorized boiler inspector immediately so she or he can inspect the boiler and advise on permissible repairs.

BOILER LAY-UP

Controlling corrosion and eventual deterioration of boiler equipment is as important during shutdown as it is during operating periods. Corrosion can be initiated by oxygen, water, and low pH. If either water or oxygen is kept out of the system, the other will do no damage.

Cast-iron and steel heating boilers are susceptible to corrosion damage during summer outages. Soot, if not cleaned, will form sulfuric acid in damp or sweating basements. Water, if not treated, will corrode the inside of the boiler. Burning trash intermittently in a boiler can cause dry firing if the water level is not watched and can also cause acid-type attacks on the fireside when the boiler is idle. Heating boilers should be flushed and cleaned on the waterside. The fireside should be cleaned of all soot deposits. Then lay up the boiler either wet or dry. The big problem with the dry method is keeping the insides dry. Air-blast with independent outside hot air after draining. When it is dry, place shallow pans of quicklime inside; then close all openings tight. Place trays between tubes and one in each steam drum (watertube type) and bottom of shell (fire-tube type). Open the boiler every 30 days and if the quicklime is saturated with water, replace it (or whatever material is used).

The wet-storage method is best when freezing is not a problem and if the unit will not be needed for at least a month. After it is prepared for storage, fill the boiler to water level with deaerated water. If no deaerated water is used, open a top vent. Then build a light fire to boil the water for 8 hr so dissolved gases are driven to the atmosphere. Use 1½ lb of sodium sulfite for each 1000 gal of water stored in the boiler to protect against oxygen.

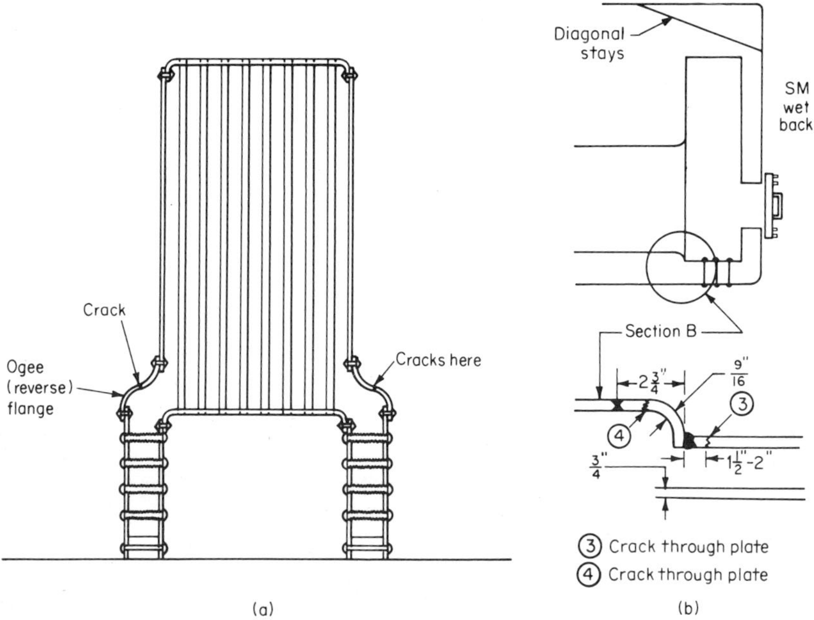

Fig. 12-14 Repetitive stresses can cause cracks at bends or sharp corners. *(a)* The VT boiler can crack at the ogee flange. *(b)* The SM boiler can crack on the furnace flange.

The concentration should be about 75 ppm. Use caustic soda to obtain alkalinity of 375 ppm. Keep the water temperature as low as possible and test the water weekly.

Cracking in boilers may be a result of faulty material, design, construction, or operating conditions.

In addition to flaws in design and construction already mentioned, the "notch" effect of gouges, tool marks, or grooves should be considered as a potential cause of cracking. Any such irregularity is a cause of stress concentration (see Fig. 12-4), and continued stress concentration may result in fatigue, one of the most common causes of cracking. Bending an iron wire back and forth eventually results in its fracture. Similarly, boiler plate may crack if it is continually bent back and forth. Rising and falling pressure and temperature in a boiler produce a "breathing" action, which also may cause fatigue of the metal after a period of years; cracking will then result in the sections subjected to greatest stressing.

Corrosion is a close partner of fatigue. See Fig. 12-14*b*. A fatigue crack may extend rapidly if affected by water of corrosive nature.

The reversed flange (ogee) of a Manning boiler is sometimes affected by fatigue cracking (Fig. 12-14*a*). This flange is subjected to fluctuating bending stress as the tubes expand and contract. Repairs to such a defect should not be attempted. If the crack extends more than about 3 in., the boiler should be removed from service permanently or until a new flange is installed. Since the latter operation practically entails dismantling and rebuilding the boiler and since these fatigue cracks do not appear usually until the boiler is quite old, it is seldom practical to install a new flange.

Sharp corners have caused a stress concentration and fatigue cracking in thick-walled drums for high pressures. Design procedure has overcome this tendency by fabricating such points in a sweeping contour to eliminate the corners.

Probably the most dangerous of all fatigue cracks is the "lap crack" developing unseen between rivet holes of the longitudinal lap-riveted seam of lap boiler shells. This defect is induced by the fact that a lap-seam boiler is not rolled into a true circle, and a bending stress is concentrated at the offset of the lap by the breathing action of the shell (see Fig. 12-4). Many serious explosions have resulted from such cracks. Usually, leakage from the seam appears as a warning of a lap crack. If any leakage exists or suspicion of this defect develops, the plate should be slotted (Fig. 12-4*c*). If any leakage shows through this slot, a crack is usually indicated. It is often necessary to remove several rivets in suspected regions so that the inside of the rivet holes may be examined. No repairs are permitted on lap cracks. The boiler should be condemned immediately.

Fatigue cracks sometimes appear on dished heads, though not commonly; usually they originate from the notch effect of toolmarks or from an incipient crack caused by the flanging process. Repairs may not be permissible, other than drilling and tapping each end of a small crack with a ¼-in. hole and fitting a plug to prevent extension of the crack. A new head is usually necessary if a crack extends more than 25 percent of the head thickness for more than several inches in length.

Embrittlement cracks may be treated as lap cracks insofar as their treatment goes. The cracks usually occur between rivet holes in a seam, and leakage may give warning of development of the defect. Removal of rivets and examination of the holes, and sometimes removal of the butt straps or heads, are necessary for final analysis of the problem. Such cracks may not be repairable, and boilers so affected may be discontinued from service. If an embrittling condition is encountered or suspected, it is essential that a reputable water treatment company be consulted for recommendations on feedwater treatment.

Fire cracks around rivet holes are not uncommon when the boiler plate is exposed to radiant heat. They are not dangerous as a rule. The cause of fire cracking is the comparatively great difference in expansion differential between the plate's water and fire surfaces. For this reason, fire cracking is more common in riveted seams of thick plates. The cracks usually extend from a rivet hole to the caulking edge of the plate, but sometimes extend across to the opposite side of the rivet hole. See Fig. 12-15.

Repairs are permitted if the crack does not extend over 3 in. beyond the lap of the plate. If the crack does extend into the solid plate within this limit,

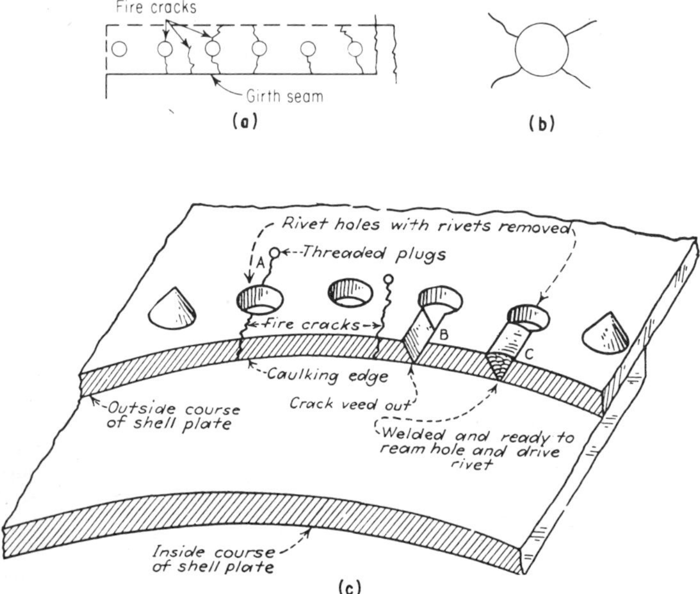

Fig. 12-15 *(a)* Fire cracks at thick section of riveted circumferential seam. *(b)* Star cracks around rivet holes which cannot be repaired. *(c)* Method of repairing fire cracks on circumferential joint.

the end of the crack should be drilled and plugged with a ¼-in. tapered or threaded pin to keep it from extending. If leakage occurs from fire cracks, the rivet should be removed from the holes affected and also two rivets on each side. The cracks should be veed out and electric-welded, the rivet holes reamed, and new rivets driven (Fig. 12-15*c*).

Common locations for fire cracks are the lower part of the front girth seam of HRT boilers and furnace seams of firebox boilers.

Weld Cracks Modern construction of boilers and pressure vessels has been possible with the advancement of welding knowledge. However, many variables must be considered in obtaining a proven, good welded joint. It is necessary to consider the following variable factors: weldability of the base metal, shape of the joint, welding process to be used, procedure to be followed in performing the weld, size and type of electrode, current (or temperature of weld) to be applied, preheat and postheat to be used, NDT to be applied to check the joint, and the use of properly qualified welders. Final acceptance considerations may also involve a hydrostatic test.

Defects in welding techniques or procedures can produce weld cracks (Fig. 12-16). Small cracks can usually be veed out and rewelded. Methods of NDT such as dye penetrant are used to determine if the end of a crack has been reached in the veeing-out process. Preheat and postheat treatment may be required on the weld repair following ASME Section IX rules. The repair must usually be approved and then witnessed by an authorized inspector.

Defects in welds considered unacceptable are cracks, areas with incomplete fusion or lack of penetration, and slag inclusions of ¼ in. for plate thickness up to ¾ in. with the maximum slag-inclusion length being ¾ in. for plate

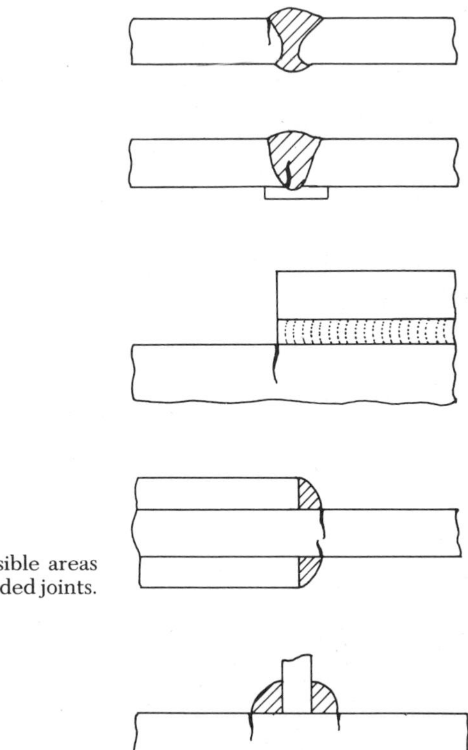

Fig. 12-16 Some possible areas of cracks to exist in welded joints.

thickness over 2¼ in. The ASME publishes porosity charts which detail the number and size of permitted porosity in a given length per weld thickness. Welds can fail from the application of repeated stress, especially where a discontinuity may be present. Figure 12-16 illustrates typical areas of welds that should be suspect during inspection for possible fatigue crack initiation. Heat-affected-zone cracking has been the subject of much research. Brittle fractures in the heat-affected zone may be initiated from the presence of hydrogen, reheat cracks, fatigue cracks, or lack of root or sidewall fusion. Postweld heat treatment will generally improve HAZ toughness of most grades of carbon steels and this will help in preventing brittle failures in the HAZ of a weld. As a quality control device in ensuring a sound weld, NDT inspections of welds have been materially beneficial in reducing weld crack failures.

BOILER REPAIRS

Most jurisdictions in the United States have reinspection laws on high-pressure boilers, as detailed in Chap. 1. The possible pressure-part failures that may result on a boiler will determine repairs that may be needed to correct the defect.

The National Board publishes a document entitled "The National Board Inspection Code," which is used as a guide by inspectors and others to help maintain the integrity of boilers and pressure vessels. It contains rules and guidelines for inspection after installation, repair, alteration, and rerating, thus helping to ensure that these objects may continue to be used safely. When a weld repair or alteration is completed, the organization that carried out the work is required to complete a record of weld-repair form which details the type of repair or alteration made by that organization. The record of weld-repair form is required to be signed by the organization carrying out the work, verifying that it was carried out in accordance with the National Board Inspection Code and the jurisdictional requirements. In addition, this form is to be signed by the authorized inspector who accepted the repair or alteration; the form verifies that the work was carried out in accordance with the National Board Inspection Code. The weld-repair form is forwarded to the enforcement authorities for boilers in a jurisdiction. In the event of a repair or an alteration to a boiler, the authorized inspector of the jurisdiction should be contacted for guidance in complying with the National Board rules, or those existing in a jurisdiction on repairs or alterations, usually similar to the nationally recognized NB rules.

Tube troubles in boilers can be caused by many factors, especially on large steam generators that are equipped with economizers, superheaters, and reheaters. Generating tubes fail mostly because of overheating due to the presence of scale or sludge, obstructed circulation, or heat concentration in an area so that the normal water and steam flow cannot remove the fireside input heat fast enough. Tubes are also attacked chemically. The problem of chemical attack is in some cases related to overheating, for high temperatures greatly speed up certain chemical reactions involving iron. Most chemical attack can be prevented, however, by proper boiler water treatment.

External thinning may be caused by the improper placement of soot blowers and by abrasive particles in the flue gas, such as fly-ash impingement. Failure of a fire tube by pitting or corrosion always requires renewal unless the defect is close to one end.

Superheater tubes rarely fail because of corrosion, unless it has occurred during periods of standby. Failures of these tubes are almost always attributed to overheating, which may be caused by the insulating effect of deposits carried over from the boiler water or by starvation.

Deposits having their origin in the boiler water are carried over because of foaming in the steam drum, resulting from excessive concentrations of dissolved solids or alkali or the presence of oil or other organic substances. Foaming caused by high concentration of solids and alkalinity may be minimized by increasing the boiler blowdown or by the addition of an antifoam material to the boiler water. The elimination of organic contaminants may require special pretreating equipment or the insertion of an oil separator in the feed-water line. It should be emphasized that pure hydrocarbon oils do not cause carryover, but additives in them frequently do.

Carryover in the form of slugs of water is another cause of superheater deposition, and this may be remedied by lowering the water level or by obtaining better water-level control. However, heavily swinging loads in excess of those for which the boiler was designed may also contribute to this condition.

Starvation of superheater tubes may result from poor start-up practice or by operating the boiler at undesirably low rating. If, during start-up, the furnace temperature is brought up too rapidly before full boiler pressure is attained, there will be insufficient steam to cool the tubes. In boilers with a radiant-type superheater, this same effect may be noted at low loads.

Creep cracking of tubes can be identified by the signs of extremely fine cracks or strain markings.

Repair of tubes involves considering the extent of the defects. Tube replacement is common. Localized repairs can be made by window patching or cutting out a section of the tube and circumferentially welding in a piece. Preheat and postheat treatment may be needed. An authorized inspector should approve the repair.

Removal of a tube is accomplished by cutting the beading or flaring off of each end. One end is slotted with a ripping chisel, with care being taken not to score the seat. The tube end is then reduced in diameter with a crimping tool (Fig. 12-17*c*) so that it may be drawn through the hole. An acetylene

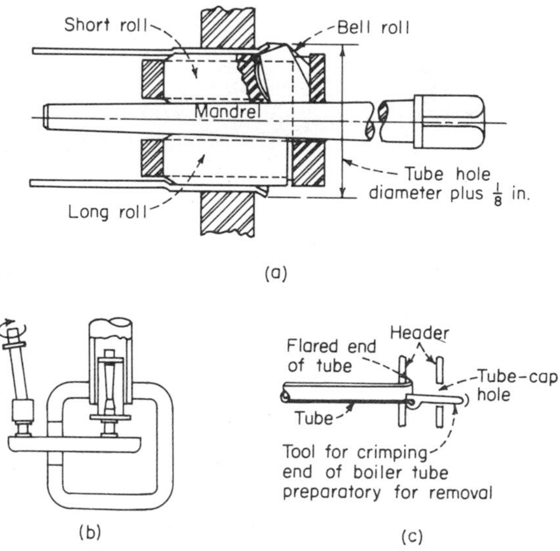

Fig. 12-17 Tools for tube work: *(a)* straight-run tube expander; *(b)* right-angle tube expander; *(c)* crimping tool for tube removal.

cutting torch may be used to advantage in some operations. The new tube is then put in place. If a fire tube, it is expanded and beaded. If a watertube, it may be expanded and flared (Fig. 12-17*a* and *b*).

If a tube hole should be damaged by scoring, it may be reamed out slightly oversize, and a copper or a soft-iron ferrule used between the tube and the tube hole. The tube is then expanded against this ferrule.

It is necessary sometimes to weld ends of a watertube, as with circulating tubes to a header which is being renewed. The edges of the tube should be ground to a V and a ⅛-in.-thick copper backing sleeve used inside the tube so that the weld metal will fuse into the bottom of the V, attaining full penetration without "icicles" hanging into the tube. The copper is used because the weld metal will not fuse to it; thus it may be removed easily. Steel backing rings as left in piping usually are not desirable in boiler tubes, for even a slight restriction in flow may not be tolerated.

Large-capacity steam generators have very long tube lengths that must be welded together during field erection of the boiler. While the ASME Code does not require circumferential tube welds to be x-rayed, as in shells or drum requirements, many users are specifying x-ray examination of all welds in the furnace-area waterwalls, some high-temperature superheater tube welds, and some tube-to-header welds. This is a form of quality control on the welding performed, but it is also an attempt to prevent an expensive outage from a relatively small weld failure, such as lack of fusion, porosity, or small HAZ crack. On recovery-type boilers, x-raying of welds around the furnace area, in order to prevent water getting into the furnace, is also becoming common practice.

A *bulge* is caused by overheating of shell plates and it affects the entire thickness of the plate, whereas a *blister* is caused by a slag inclusion forming a lamination at the time the plate was rolled in the steel mill. The entire thickness of the plate may not be affected. Usually, the area between the lamination and the fireside cracks and blisters. Bulges and blisters in the boiler shell often require repairs. If a bulge is not down more than 2 percent of its length, it is usually best to leave it alone unless the metal has been burned badly. If a blister has not reduced the thickness so that the percentage of new thickness to original thickness is less than the efficiency of the longitudinal seam, repairs are not usually necessary.

If a bulge is down more than 2 percent of its length, but not over about one-eighth of its length and the plate has not ruptured or burned, it is usually advisable for it to be driven back by experienced boilermakers. A form holding a charcoal fire may be used to heat the entire bulge simultaneously. A short-handled, heavy striking hammer is used around the circumference, and work is gradually toward the center.

Bulges or blisters of more serious extent or where the plate is burned badly are repaired only by patching. Usually the size of the area to be patched may be reduced by driving back the circumference of the affected section, provided that the plate is not burned. A blister is actually a separation of the metal from the shell plate, caused by impurities rolled into the shell plate when formed. But only the outside layer will blister from the heat because the remaining thickness is not affected. A blister cannot be driven back but must be cut out and the edges trimmed; if the remaining metal is sound, the pressure on the boiler must be reduced to correspond to the thickness of the remaining sound metal. If the pressure cannot be reduced, the entire blistered section, including sound metal, must be cut out, a flush-welded patch formed and butt-welded in, with localized stress relieving required after welding. If the length of the blister is over 8 in., the weld must be radiographed. After this is shown to be satisfactory, a hydrostatic test of 1½ times the maximum allowable pressure is required to check the repair. All welding must be done by a certified welder, and the repair must be approved by a commissioned boiler inspector.

Corroded areas of tube sheets may be built up by welding where tubes act as stays. But all tubes in such corroded areas must be removed before welding is applied. After welding, the tube hole should be reamed before new tubes are installed.

Stayed surfaces and tube sheets must also conform to the following requirements before welding. Corroded areas in stayed surfaces may be built up by fusion welding, provided the remaining plate has an average thickness of not

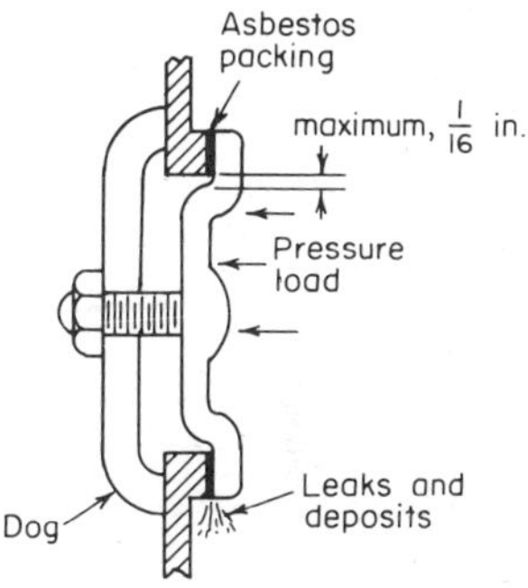

Fig. 12-18 Leaks around handholes and manholes can cause metal deterioration, requiring repairs if not corrected.

less than 50 percent of the original thickness and the areas so affected are not corroded sufficiently to impair the safety of the boiler per ASME Code. An authorized inspector must approve and witness this type of repair.

Handhole and manhole openings are commonly oval, with the cover fitted inside the boiler (Fig. 12-18). A woven asbestos gasket seals this joint. The gasket softens when heated so the nut against the dog must be tightened more as pressure is brought up on the boiler. Unless it is tightened gradually as pressure builds, the gasket may leak or be blown out, scalding personnel.

Persistent gasket leakage results in the boiler plate being thinned as a result of corrosion, which can only lead to expensive repairs. Always correct gasket leakage as soon as possible.

Handhole rings will be required to strengthen edges of handholes deteriorated badly by external corrosion. This corrosion is caused usually by leakage past the gasket. The elliptical ring must be of the same thickness as the sheet, and it should be flush-welded to the opening to sound metal. The thin edges of the opening should be cut back so that a bead may be applied by electric welding to seal the ring to the edge of the hole. Corrosion at the edges of handholes may be prevented by keeping the handhole gasket made up tight against leakage.

The subject of stay bolts and furnace-sheet repairs in firebox boilers is covered by NB repair rules. Threaded stay bolts sometimes develop leaks, which can be through telltale holes or around the boiler plate. Repairs are not permitted because the leaking telltale hole indicates the stay bolt is cracked inside the boiler. A new stay bolt should be installed.

Stay-bolt heads cannot be welded to stop leakage around the heads. The heads should be recaulked. If leakage persists, it may be an indication of a corroded sheet on the boiler. The old stay bolt must be removed, the sheet must be examined for corrosion and thinning, and if it is satisfactory, new stay bolts must be installed. If the sheets are corroded more than 50 percent of the original thickness, the defective section must be cut out and a flush-welded patch installed.

Threaded stays may be replaced by welded-in stays, provided that, in the judgement of the qualified boiler inspector, the plate adjacent to the stay bolt has not been materially weakened by deterioration or wasting away. Stress relieving other than thermal may be used as provided in NB rules for welding.

Violent furnace collapse may result, too, if a number of stay bolts in the same location are broken.

Bulged furnace sheets may be driven back by removing the stay bolts in that region and heating. Backing bars are sometimes valuable. If a handhole is not available in the shell opposite the bulge, it will be necessary to cut one.

Grooving at the bottom of the furnace sheet just over the mud ring may be repaired in one of two ways. Either the bottom of the waterleg may be cut off and the mud ring raised, or the defective plate should be cut off and a new circumferential strip riveted or butt-welded on. The permissible reduction in furnace volume and local regulations determine the method.

Braces or *stays,* unlike stay bolts, seldom break, for they are long enough to have considerable flexibility. If a brace or stay does break, repairs should not be attempted. A new brace or stay should be installed at once.

Through-to-head braces below the tubes in fire-tube boilers sometimes become bowed when the bottom of the shell is overheated owing to scale or oil. The shell may expand nearly 1 in. because of excessive temperatures; but the braces are subjected to the much lower temperature of the boiler water, and their heat expansion may be only half that amount. As the shell expands, it carries the lower part of the heads and brace attachments with it, stretching the braces far beyond their yield point. Thus, when the shell contracts, the braces have been permanently elongated and will bow. The direction of this bowing is often upward (but may be in any direction).

A slight bow need not be required. If the braces are bowed badly, they should be taken out and straightened. If they are elongated so much that a full number of threads is not in contact with the inside nut, new braces should be installed.

Piping problems constitute an important part of plant design and maintenance. Many industrial plants replace the boilers after 20 yr or so with new boilers designed and operated at higher pressure. Often, the original boilers operated at below 125 psi, and many of the pipe fittings and values throughout the mill were of cast iron, 125-lb standard. In installing new boilers for higher pressure, caution should be exercised to replace all low-pressure pipe fittings with fittings designed for the new operating pressure. Whenever a new system is started up, not only should the boiler be boiled out, but this process should be applied to the complete system. Every piece of pipe has foreign matter deposited inside. Unless it is cleaned out properly by personnel familiar with boiling out and flushing piping so as to avoid undue strains on piping, trouble may develop. Elbows, bends, and other dead pockets in piping are dangerous spots and must be thoroughly cleaned.

The size of pipe should be adequate for conservative flow velocities, or else excessive friction head will result. The maximum recommended velocity for steam flow is 5000 ft/min for heating service (up to 15 psi), 10,000 ft/min for high-pressure saturated steam, and 14,000 ft/min for high-pressure superheated steam. The Crane Co. has found velocities of up to 20,000 ft/min reasonable for high-pressure superheated steam in large pipes.

Example: In order to select the proper size of pipe, the following formula may be used:

$$D = 12\left(\sqrt{\frac{C}{0.7854F}}\right)$$

where

C = steam flow, ft³/min
D = required diameter of pipe, in.
F = permissible steam-flow velocity, ft/min

To select a proper pipe size for 12,000 lb/hr of saturated steam at 300 psi, first find the density for the pressure from the steam tables. For 300-lb pressure, it is 0.67 lb/ft³. Therefore,

$$\frac{12{,}000}{0.67} = 17{,}910 \text{ ft}^3/\text{hr or } 298 \text{ ft}^3/\text{min}$$

Substituting in the foregoing formula, we have

$$D = 12\left(\sqrt{\frac{298}{0.7854 \times 10{,}000}}\right) = 2.3 \text{ in.}$$

It would be customary to go to the nearest *larger* standard pipe size, and so 2½-in. pipe would probably be used in this case.

The density of steam rises with the pressure; thus, the *volume* for a given weight of steam is less at higher pressures, and smaller-sized piping may be used.

Pipe supports should be designed to carry the weight of the piping *full of water.* They should be spring-mounted for heavy service in order to provide proper support during vertical motion of the pipe resulting from expansion. Hanger-type supports should be adjustable by means of a turnbuckle or nut threaded to the top of the hanger rod so that compensation may be made for settling of the supporting structure.

Expansion of pipe anchored at each end may set up severe stresses. It is good practice to provide an expansion joint of the slip, bellows, or loop type for long horizontal runs of steam piping. The length of expansion is calculated by

$$(t_1 - t_2) \times L \times 0.0000065 = \text{expansion, in.}$$

where

t_1 = steam temperature
t_2 = room temperature
L = horizontal length of pipe section, in.

A Holly loop is a piping arrangement used to return condensate from steam-line separators to the boiler (Fig. 12-19*a*).

In operation, the flow of the condensate-and-steam mixture from the steam separator to the surge and separation tank is established by causing a slight pressure drop in the surge tank. This is accomplished by having the vent valve open slightly in a small line from the top of the surge tank to a heater or hot well. The surge tank is located at an elevation sufficient to give static head pressure so that the condensate will return against boiler pressure by gravity.

Use of a Hartford loop is confined usually to heating boilers and eliminates the need of a check valve to prevent condensate returns from backing up the return line in case the steam valve is closed (Fig. 12-19*b*).

Package boilers or economizers may be located *outdoors* between the boiler house and stack. In freezing temperatures, the economizer or boiler should be drained when out of service. When it is in use, a bypass around the blowdown valves should be provided, and this bypass should be opened slightly to prevent freezing of the blowdown line. Also, all exposed pipes out of the main circulatory system should be heavily insulated. Soot accumulations should be removed periodically, for they may impede heat transfer and draft. Large quantities of soot in the base of economizers may be washed out with a water hose.

Safety-valve care in the operating schedule should include frequent periodic tests of safety valves. In boilers of moderate pressure, the valve should be lifted by its lever at least once each week of operation, and the pressure should be raised to the popping point to test the safety valves at least once each year of operation. If the safety valve does not blow at its set pressure, the lever should be tried at that pressure, for the spindle may be stuck slightly

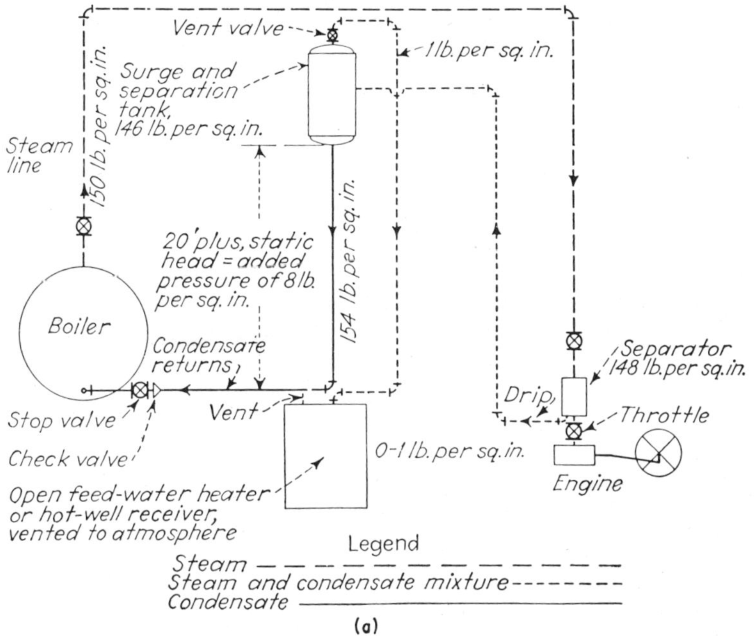

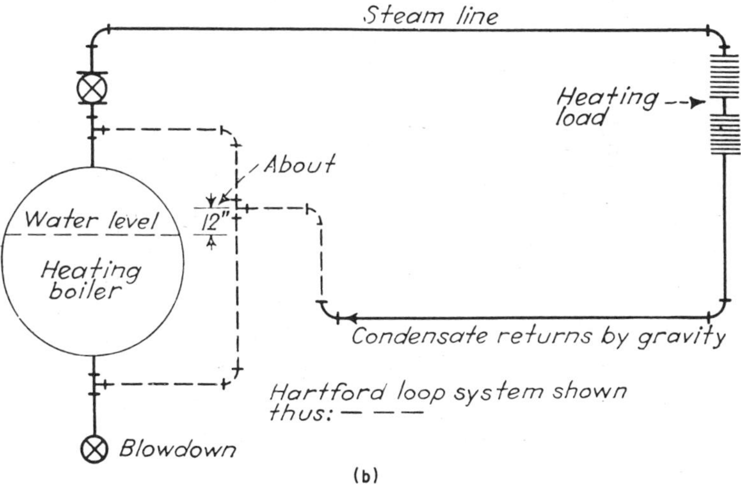

Fig. 12-19 *(a)* Holly loop returns condensate to the boiler from steam-line separators. *(b)* Hartford loop is used to return condensate in gravity return system.

in its bushing. If the valve is not freed or if it does not blow at its set pressure on several subsequent trials, the boiler should be shut down at once for safety-valve overhaul.

Owing to the cutting action of high-pressure steam, weekly tests of safety valves set at over 500 psi may not be advisable. The manufacturer and local boiler authorities should be consulted on this point, but it is seldom wise to allow more than a month to pass between tests except on superheater safety valves.

If the safety valve blows at a lower pressure than normal, according to the pressure gauge, the gauge should be tested. If the gauge is correct, the gauge piping should be blown out to be sure that it is clear. If the safety valve then blows at too low a pressure, it may be adjusted by a competent person, taking extreme care that plenty of clearance is left between the spring coils when the valve is wide open. It is always best to check the boiler pressure at two different points on the boiler to eliminate the possibility of incorrect pressure indication. Any considerable deviation from normal operation of safety valves should receive immediate competent attention.

Safety-valve escape pipes may become clogged with sediment owing to leaking safety valves. The periodic test should disclose this condition. Leaking safety valves should be reground, *never* tightened by adjustment to stop leakage.

Hydrostatic tests are made on boilers that have had any repairs, such as tube installation, or patching. They are applied also to determine the location, extent, or existence of defects. The boiler is filled to the top with water at room temperature, but not below 70°F so that if a defect should "let go," the slight expansion force of the water would not result in an explosion. A pressure of up to 1½ times the maximum allowable pressure is applied by either the feed pump or a special test pump.

The safety valves should be gagged or removed and the connections fitted with a blank flange. The safety-valve adjustment should not be screwed down for hydrostatic tests.

If other boilers are on the line, the drain valve between the stop valves should be left open. If water or steam leaks past the stop valves, showing from the open drip, the hydrostatic test should be delayed until the steam line is fitted with a blank flange. No chance should be taken of cold water pressure building up against a stop valve having steam temperature on its other side.

Any areas where defects are suspected, such as welded or riveted seams, should be exposed for the test by removing all brickwork necessary. The test pressure should be maintained long enough to examine all parts for leakage. A hammer test is often applied to suspected sections while under test pressure.

If any leakage or indication of distress develops, an authorized inspector should be consulted in regard to the cause and the procedure to follow. Wherever possible, it is advisable to have the test conducted by the inspector.

The first thing to do on completing the test is to remove the gags or blank flanges from the safety valves. This action should never be delayed, for if it is, it may be forgotten, and a serious boiler explosion would result from malfunctioning of controls or operator neglect.

Questions and Answers

12-1 Name three scale-forming elements in water.
Ans. Calcium (Ca), magnesium (Mg), and silica (SiO_2).

12-2 What are the two major objections to scale in a boiler?
Ans. (1) Scale is a heat insulator and thus may produce overheating of the parts affected. (2) Scale causes a considerable loss of efficiency.

12-3 Two identical boilers operate in the same plant with the same feedwater. Blowdown from each is the same. More scale forms in one boiler. Why?
Ans. Evaporation must be greater in the boiler having the most scale.

12-4 A watertube boiler is scaled heavily. What should you recommend?
Ans. Removing the scale mechanically by turbining the tubes and scaling (chipping) the drum surfaces if necessary. Acid cleaning caustic boiling may be beneficial. After the boiler is clean, treatment of the feedwater should be incorporated or corrected to prevent scale formation.

12-5 Can a tube turbine cause damage to a tube? How?
Ans. Yes, if it is forced or held in one position too long.

12-6 What care should be exercised in changing feedwater treatment or feedwater source of supply? Explain.
Ans. If scale of any quantity is present, it may be loosened suddenly and precipitated onto hot surfaces where it will cause overheating or damage. The change should be made as gradually as possible, and the boiler opened very often for inspection during and directly following change.

12-7 Is a tube turbine used to remove scale mechanically from fire tubes? If not, what is used?
Ans. No; a tube rattler or vibrator is used. A tube turbine is used on watertube boilers.

12-8 Of what benefit is a coating of oil on internal surfaces of a boiler? Explain your answer.
Ans. None. It is exceedingly dangerous. Oil is a heat insulator and thus may produce dangerous overheating of affected surfaces.

12-9 What is the most common source of oil in boilers?
Ans. Use of contaminated condensate returns from reciprocating steam equipment or process heat exchangers.

12-10 How might fuel oil get into a boiler when oil burners are used?

Ans. By failure of a tube or coil in a fuel-oil steam heating element when the condensate is returned to the feedwater system.

12-11 If the pH of water is 6.5, what may be said regarding alkalinity?
Ans. The water is acid. Neutral is 7.0.

12-12 What is pitting (briefly) and when is it serious?
Ans. It is a localized form of internal corrosion acting to deteriorate the boiler surfaces. It may be serious if active and is serious if the pits are extensive and closely spaced.

12-13 What is a protective coating used for in a boiler?
Ans. It is a very thin coating used to prevent direct contact of boiler water against the boiler surfaces. Internal corrosion and pitting may thus be reduced. Apexior is such a coating.

12-14 What is galvanic action?
Ans. It is an electrolytic flow between dissimilar metals in a boiler electrolyte, such as water, resulting in localized deterioration of the metal.

12-15 Is internal corrosion more serious in a drum having a quadruple-riveted longitudinal seam than in one having a lap seam? Explain your answer.
Ans. It is more serious with a quadruple-riveted boiler because a smaller percentage of deterioration would bring the resultant strength of the drum below the strength of the longitudinal seam of high efficiency, and the boiler pressure may have been calculated on this high joint efficiency. Lap joints have low efficiencies, thus the effect may not be as drastic.

12-16 What causes most external corrosion?
Ans. Sulfur in soot, coal, ash, and moisture.

12-17 What objection is there to burying piping under the boiler floor?
Ans. External corrosion may progress unnoticed to a dangerous point.

12-18 What is erosion and what is the most common cause of it in a boiler?
Ans. It is the wearing of surfaces by abrasion. It is usually caused by impingement of soot and ash particles.

12-19 What causes priming and foaming?
Ans. A high water level plus sudden demand for steam may cause priming. High concentrations of solids or organic matter may cause foaming.

12-20 What harm may priming and foaming do?
Ans. Priming may cause rupture of steam piping or wreck turbines or engines. Foaming may make it impossible to tell the correct water level in the boiler. It may cause severe wear of pipe fittings, valves, and steam equipment.

12-21 What should you recommend for priming and foaming?
Ans. Carrying a lower water level (but in safe range) and a higher pressure up to safe limits for the boiler, to reduce priming. Blowing the boiler down more frequently, to reduce concentration causing foaming.

12-22 What harm may be caused by flame impingement?

Ans. Localized overheating and damage to boiler parts if not protected by refractory.

12-23 What is the external indication of low water and overheating in a boiler other than possible distortion, discoloration, and leakage?

Ans. The soot would be burned off of affected surfaces.

12-24 Which is the more serious, a lap crack or a fire crack? Where may each be found?

Ans. A lap crack is more serious. Lap cracks are found in longitudinal lap-riveted seams. Fire cracks may be found in the lap of riveted seams exposed to radiant heat as in HRT girth seams and firebox-boiler furnace seams.

12-25 How may lap cracks be repaired?

Ans. No repairs are allowed other than an entire new course of boiler plate. Otherwise, the boiler is condemned.

12-26 Should you advise repairs to a fire crack extending from a rivet hole to the caulking edge? If so, how should the repairs be made?

Ans. Only if leakage occurs. Then remove the rivet and the two rivets on either side. Vee out the crack and electric-weld. Ream rivet holes and drive new rivets.

12-27 The ogee (reverse) flange of a Manning fire-tube boiler is cracked 17 in. circumferentially. What type of repairs would be in order?

Ans. A new ogee flange or a new boiler.

12-28 What is grooving? Where is it found?

Ans. It is a localized form of corrosion in highly stressed areas, such as those adjacent to a riveted seam or in the knuckle of a dished-head flange.

12-29 When is it dangerous to weld fire cracks?

Ans. When they extend over 3 in. beyond the lap into the solid plate or when they extend from rivet hole to rivet hole.

12-30 Name four causes of failure in tubes of watertube boilers.

Ans. (1) Scale; (2) oil; (3) low water; and (4) flame impingement.

12-31 What is safe-ending, and why is it used?

Ans. It is cutting off and replacing ends of fire tubes that have become eroded or corroded till they are too thin for safe use. A short length of new tube is welded to the cut end of the old tube.

12-32 What would be a likely cause of leakage at the upper ends of tubes in a vertical fire-tube boiler? Give repair and preventive recommendations.

Ans. Overheating the tubes by forcing fire in starting up the boiler is a common cause. The tube ends should be rerolled to make tight; or if the tubes are damaged by overheating, tube renewal may be necessary. A low fire should be carried until steam is raised to protect the upper ends of the tubes, or the boiler may be filled with water until hot, then blown down to operating level.

12-33 What is an approximate limit of extension for a bulge in boiler shell before repairs are recommended?

Ans. Depth not to exceed about 2 percent of length.

12-34 What is the limit of depth of a blister before repairs or a cut in allowable pressure is necessary?

Ans. When the ratio of sound metal remaining to the total original thickness, expressed as a percentage, is less than the percentage efficiency of the longitudinal seam, repairs and/or a cut in pressure is necessary.

12-35 Briefly, how is a bulged shell repaired?

Ans. If the depth of the bulge is not over about one-eighth its length and the plate is not ruptured or burned badly, the bulge may be heated and driven back. Otherwise, it is necessary to cut out the affected area and install a properly designed, flush-welded patch.

12-36 What is a common defect at edges of a handhole? Give cause, prevention, and repairs.

Ans. Leakage from a gasket may cause external corrosion to deteriorate the plate down to a knife edge, giving it insufficient strength to support the handhole plate. Leakage should be prevented. If deterioration does not exceed 40 percent of original thickness and the affected area is close to the handhole, the plate may be built up by electric welding. Otherwise, the hole should be trimmed back to sound metal and an elliptical ring seal welded inside the shell or head.

12-37 How is a broken stay bolt in an empty, idle boiler detected?

Ans. When the ends of solid-type stay bolts are struck with a hammer, a broken stay bolt may be detected by a dull sound. Flexible stay bolts are tested by removing the cap and using a heavy screwdriver in the ball slot.

12-38 What does leakage from the telltale hole in a stay bolt indicate?

Ans. The bolt is either broken or cracked halfway through.

12-39 What causes stay bolts to break, and what location is most susceptible in firebox boilers?

Ans. The expansion and contraction (breathing action) in the boilers bend the stay bolts. Fatigue may break them eventually. The upper rows are most susceptible to breakage, for the expansion is greatest there, the lower part of the firebox being stiffened by the riveted or welded connection to the outer sheet.

12-40 How should you repair a broken stay bolt?

Ans. Repairs are not allowed. A new stay bolt should be installed.

12-41 If a fire sheet is badly corroded externally or internally ⅛ in. around a number of stay bolts, what repairs would be advisable?

Ans. Removing the stay bolts, drill, and tap sheets and installing a larger stay bolt to contact sound metal.

12-42 Briefly, what causes through-to-head braces below the tubes in an HRT boiler to bow?

Ans. Overheating of the shell bottom causes excessive expansion and permanent elongation of the braces. When the shell cools and contracts, the braces bow.

12-43 What causes steam binding of a feed pump? How should you recognize it, and what should you do for it?

Ans. Excessive temperatures for the head pressure on the suction are the cause. A reciprocating pump would race, short-stroke, or hammer. A centrifugal pump would overheat, vibrate, and operate noisily. Temporary measures might include slowing the pump down, throttling the water discharge, and/or reducing the water temperature by introducing cold water into the suction if necessary. Playing a stream of cold water onto the suction pipe may aid in emergencies. Permanent correction requires greater suction-head pressure or lower temperature.

12-44 What is a common cause of excessive feed-pump suction temperature?

Ans. Steam blowing through into the feedwater storage owing to defective traps.

12-45 What operating precaution should be exercised with outdoor packaged boilers or economizers in cold weather?

Ans. Freezing of the blowdown should be guarded against either by use of a small line bypassing the blowdown valves or by leaving the blowdown valves open slightly.

12-46 How are superheater tubes protected against overheating in starting a boiler?

Ans. Flooding the tubes with water is recommended for a few types. More commonly, the superheater is vented to the atmosphere by a free blow valve until the boiler goes on the line, and the firing is kept at a low rate until steam appears from this vent, indicating that steam cooling of the tubes has started.

12-47 What happens if the lower connection to a gauge glass is obstructed?

Ans. The water will return to the glass very slowly after blowing it down. If the obstruction is complete, the glass will fill with water slowly, because of steam condensation. An incorrect water level will be shown by the glass.

12-48 With reference to question 12-47, how should the water level be checked? Outline the procedure.

Ans. By use of the try cocks. Try to remove the obstruction by attempting to blow down the glass with the steam connection closed. If this is not successful, close the connections and remove the glass. Try to run a stiff, bent wire through the lower connection after opening this valve partway, taking care to keep to one side so as not to get scalded. A careful check on the water level by the try cocks should be maintained. If the obstruction cannot be freed, the boiler should be shut down so that the connection may be dismantled for cleaning.

12-49 How should you know if the upper connection of a gauge glass were obstructed? What should you do if it were?

Ans. Water would rise rapidly to the top after blowing the glass down, although try cocks show a lower true level. Proceed as for lower obstruction.

12-50 What is the first thing to do on entering a boiler room?

Ans. Look at the gauge glass.

12-51 How would you know if a connection to the water column were obstructed?

Ans. The water in the glass would show action similar to that described for gauge-glass obstructions, but a check on the true level by try cocks would be impossible if they were mounted on the water column. If this condition is suspected, the boiler should be shut down immediately so that the connections may be cleaned.

12-52 If the distance between the bases of gauge-glass stuffing boxes is 12 in., how long should you cut a new gauge glass? Explain your answer.

Ans. About 11¾ in. to allow for expansion.

12-53 What tools and materials should you need to install a new gauge glass (tubular type)?

Ans. A gauge glass, a gauge-glass cutter, new packing rings, a knife, a packing hook, a wrench, a wiping rag, a stick, and a pail of cold water to cool the packing nuts.

12-54 What would you do if you saw a small amount of steam leaking through the brick covering on a boiler drum?

Ans. Shut down the boiler at once and have it inspected.

12-55 What is a hydrostatic test? How should you apply this test on a boiler installed in battery with others in operation?

Ans. It is a test on a boiler with cool water at a pressure of up to 1½ times the maximum allowable working pressure of the boiler. The test is performed on new boilers and on boilers in the field to check major repairs or suspected defects.

All areas to be examined should be exposed. The boiler should be filled to the top with water at room temperature but not under 70°F. The safety valves should be gagged or the connections blanked. Both stop valves on the steam line should be closed and the valved drip between them open. A test gauge should be used to check the pressure. If any leakage from the drip between the stop valves occurs, the test should be stopped and the steam line blanked off. The water pressure should be raised slowly to not over 1½ times the working pressure and held at that value long enough to make a complete examination of the boiler and a hammer test when advisable. Immediately after the test, the safety-valve gags or blank flanges should be removed.

12-56 What is the trouble if the try cocks on the water column all show steam but the glass is half full of water, and what should be your procedure?

Ans. The lower connection of the gauge glass is obstructed. Shut down the boiler at once, and have it inspected for possible damage.

12-57 Where on a boiler should you use gaskets cut from sheet rubber?

Ans. For gaskets below the waterline and only temporarily. Prepared gaskets of the asbestos-composition type should replace such rubber gaskets as soon as possible.

12-58 What would result from a nick *(a)* in the edge of a bowl of a mechanical-atomizing oil burner, *(b)* in the jet opening of a steam-atomizing oil burner?

Ans. *(a)* and *(b)* An unstable fire with poor combustion and excessive carbon deposits might result. *(b)* Also, flame shape and direction might be changed, flame impingement thus being caused.

12-59 If a safety valve sticks open, what should you do?

Ans. Reduce the firing rate. Tap the top of the safety-valve spindle sharply with a *light* hammer. After the valve seats, blow all dust from the external surfaces of the safety valve, and pour some kerosene around the spindle bushing. Operate the valve with the lifting lever until it works freely. If the valve will not seat, remove the boiler from service and overhaul the safety valve. Care should be taken to maintain proper water level when the safety valve is blowing.

12-60 How should you prepare surfaces for welding?

Ans. Wire-brush and wipe them clean of all paint, oil, rust, and scale deposits. Make sure that they are dry on both sides.

12-61 A horizontal tubular boiler is found to have three braces on one side of the top manhole and four braces on the other side. Should the odd brace be left in or cut out and why? How should braces be distributed?

Ans. If three braces will carry the load and they do not exceed Code requirements for maximum permissible spacing, then nothing need be done. If four braces are required to carry the load on one side, then an additional brace must be installed or the pressure reduced to Code requirements. If maximum spacing is exceeded, an additional brace would be required. Braces should be equally distributed and should not exceed Code requirements for maximum permissible spacing.

12-62 Where do you generally find the greatest number of broken stay bolts in a locomotive-type boiler and why?

Ans. In the upper corners of the side sheets toward the tube sheet and in the tube sheet opposite the throat, because of intense heat causing maximum expansion and contraction at these points.

12-63 A crack 14 in. long is discovered in a 30-in.-diameter corrugated furnace.

The condition of the furnace is otherwise satisfactory. Is it permissible to repair the crack by welding?

Ans. Yes, provided the welding is done by a qualified welding procedure and operator and the work is locally stress-relieved and approved by an authorized inspector.

12-64 A battery of several boilers is allowed 100-psi pressure, and all safety valves are set at that pressure. All steam gauges are accurate. One boiler shows 100 psi and its safety valve is blowing, while the other boiler gauges registered only 80 psi. What is wrong?

Ans. The stop valve on the boiler under 100 psi is closed, the connecting steam pipe is blanked off, or the dry pipe is stopped up to prevent passage of steam. This condition might also be created if the steam pipe connecting the 100-psi boiler were small compared to the size and steaming capacity of the boiler.

12-65 *(a)* Where is internal corrosion most likely to occur in a vertical tubular boiler?

(b) Where is internal corrosion most likely to occur in a locomotive boiler and why?

(c) Where is external corrosion most likely to occur on these boilers and why?

Ans. *(a)* Internal corrosion in vertical tubular boilers is most likely to occur in the water leg at the bottom ring or ogee flange because this is a point of poor circulation. Also it occurs along the waterline, of both shell and tubes, because this is where air or entrained oxygen is liberated in the ebullition of water while the boiler is steaming. Corrosion also occurs around the plate where the supporting stay bolts enter because of grooving.

(b) For the same reason, internal corrosion is most likely to occur in locomotive boilers around the mud ring, in the water leg, and along the waterline. This type of boiler is also found to corrode badly around the bottom of the barrel, especially toward the front or smokebox end.

(c) External corrosion of any boiler is most likely to occur at points where there may be leakage or moisture and especially where coupled with sulfur from soot and ashes. In vertical tubular boilers it occurs around the handholes, the lower part of the outside shell (where metal is in contact with ashes and with ogee construction at the bottom of the furnace sheet below the grate line), and the upper tube sheet and tube ends. In the locomotive type, it occurs around handholes, the lower part of furnace sheets at the grate line, the lower part of external water-leg sheets, the throat sheet, and the lower part of the front head at the tube sheet.

12-66 What are some common scale-forming substances found in boiler feedwater? What chemical is most frequently used as a solvent for scale deposits? Why is lime, caustic soda, or common soda ash used for removing scale in feedwaters?

Ans. Calcium, magnesium, and sodium compounds form scale. The ordinary solvent used is soda ash or trisodium phosphate. These chemicals are used to change the insoluble salts of calcium, magnesium, or sodium into soluble salts which can be removed through the blowoff pipe or by other means of cleaning. In some small plants, it is customary to place as much as ½ lb per boiler horsepower in the boiler when it is being taken off the line for cleaning. Other methods are to have some means between the feed pump and boiler to introduce chemicals while the boiler is in service with the feedwater. Also the water is treated with outside softeners before it enters the boiler. The latter method is preferable to internal treatment.

12-67 Describe the visible effects of the destructive action in boilers known as (1) pitting; (2) grooving; and (3) corrosion.

Ans. 1. Pitting takes the form of a local action by which small circular areas on shell plates, heads, or tubes are attacked. Pustules having a brown oxide covering form over the pits and when active are filled with black oxide of iron.

2. Grooving commonly occurs in the knuckles of flange or at rivet seams. It is generally caused by a breathing action and may result in cracks. Grooving is a combination of mechanical action and corrosion, each aiding the other.

3. Corrosion is oxidation or rusting of the metal, and it may be produced by acids in feedwater, exposure to air, or other oxidizing agents. Corrosion takes the form of rust and is usually general or local wasting away of the metal. Pitting attacks the inner surface of boiler from the waterline downward. Grooving occurs mainly in knuckles of flanges or near rivet sections on braced or stayed surfaces. Corrosion may occur anywhere on a boiler where the surface of metal is exposed to an oxidizing agent.

12-68 *(a)* In a watertube boiler, of longitudinal drum type, the lower tubes in one bank showed signs of overheating. Where would you look for the trouble?

(b) The nipples connecting the mud drum in this boiler are continually leaking. They have been renewed and rerolled, and they still leak. What would be the cause of this?

Ans. *(a)* The down flow pipe could be partially filled with scale, thus obstructing free circulation of water.

(b) The mud drum being bottomed or resting on brickwork would prevent movement caused by expansion and contraction.

12-69 *(a)* What appliances on a boiler must readily give the true water level? How can this be checked?

(b) What is the object of a water column? Why does the Code stipulate 1 in. as the minimum-size pipe connection?

Ans. *(a)* The water glass. It can be checked by using the try or gauge cocks.

(b) The water column is used to give enough volume of water to provide a steady level in the glass for a true reading when the glass cannot be connected directly to the boiler head. Smaller pipe could plug up too easily with scale and sediment.

12-70 On what basis are repairs allowed on boilers?

Ans. Repairs permitted are based on restoring the affected part or parts to as near the original strength as possible. They are governed by Code requirements for new construction, or by NB rules on permissible repairs, where the state has adopted NB rules for repairs.

12-71 Must all repairs be approved by an authorized inspector?

Ans. Yes, if the strength of the vessel has been impaired in any way requiring repairs involving Code enforcement and interpretation. Repairs not affecting the strength of the boiler or of a minor or routine nature may not require approval. But the inspector should be consulted on the problem if there is any doubt about the safety of the boiler. Crack repairs, welding, tube replacement, safety-valve replacement, and similar repairs or changes require approval. The best rule to follow on any structural repairs or changes on a boiler is to immediately contact an authorized boiler inspector.

12-72 Is a qualified welder permitted to make welded repairs on any part of a boiler?

Ans. Not necessarily. That a welder is qualified to make some welds may not mean he or she is qualified for welding: (1) the particular thickness of plate; (2) the type of material; (3) in the position of welding to be used; or (4) according to the method required.

12-73 List the impurities that produce hard and soft scale and corrosion in boilers.

Ans. Hard scale is caused by calcium sulfate, calcium silicate, magnesium silicate, and silica. Soft scale is caused by calcium bicarbonate, calcium carbonate, calcium hydroxide, magnesium bicarbonate, iron carbonate, and iron oxide. Corrosion is caused by oxygen, carbon dioxide, magnesium chloride, hydrogen sulfite, magnesium sulfate, calcium chloride, magnesium nitrate, calcium nitrate, sodium chloride, and certain oils and organic matter.

12-74 How can a dissolved impurity leave a solution and become a solid?

Ans. 1. By a temperature increase, reducing the solubility of the solid in the water.

2. By exceeding the saturation point of the dissolved impurity in the water. Water can hold only a limited amount of the dissolved impurity, so the concentration is important.

3. By chemical changes of the impurity with heat, causing it to break down and form insoluble substances.

12-75 What is the meaning of pH in water chemistry?

Ans. It is a number between 0 and 14 indicating the degree of acidity or alkalinity. The pH scale resembles a thermometer scale, but the pH scale indicates intensity of acidity or alkalinity. The midpoint of the pH scale is 7, and a solution with this pH is neutral. Numbers below 7 denote acidity; those above, alkalinity. Since pH is a logarithmic function, solutions having a pH of 6.0, 5.0, or 4.0 are 10, 100, or 1000 times more acid than one with a pH of 7.0

12-76 What is the purpose of an electrical conductivity test of boiler water?

Ans. To measure the extent to which dissolved substances are concentrated in the boiler water. This test then helps in controlling carryover of dissolved solids, which condense in lines or equipment, such as turbine blades, into the steam system.

12-77 Are there many water treatment methods available?

Ans. Yes, but they all come under three broad classifications: mechanical, heat, or chemical treatment. Mechanical treatment includes filtration and boiler blowdown. Heat treatment includes makeup distillation and steam purification. But distillation is limited to boilers with small amounts of makeup. Steam purifiers are used where the process demands very dry steam. Chemical treatment, both internal and external, is most widely used. External treatment adjusts raw-water analysis before the water enters the boiler. Internal treatment adjusts the boiler-water analysis by feeding chemicals directly to the boiler.

12-78 What impurity in water requires critical attention on very high pressure boilers?

Ans. Silica in high-pressure water volatilizes and passes over with steam to the turbine. As the steam expands and drops in pressure and temperature, the silica condenses and deposits on the turbine blades, cutting the machine's efficiency. Vaporous carryover involving the volatilization of matter (hot gases) increases most rapidly around 2000 psi. Above this range, mechanical carryover predominates, especially above the critical pressure (3206.2 psi), for there is no longer a liquid-to-steam phase. Steam quality on these units is no longer measured by a calorimeter in percentage of moisture but in parts per billion by the use of flame photometers.

12-79 What is clarification?

Ans. Clarification is the removal of suspended matter and/or color from water supplies. The suspended matter may consist of large particles which settle out readily. In these cases clarification equipment merely involves the use of settling basins and/or filters. Most often, however, suspended matter in water consists of particles so small that they do not settle out and even pass through filters. The removal of these finely divided or colloidal substances therefore requires the use of coagulants.

12-80 What is coagulation?

Ans. Coagulation is the clumping together of finely divided and colloidal impurities in water into masses which will settle rapidly and/or can be filtered out of the water. Colloidal particles have large surface areas which keep them in suspension; in addition, the particles have negative electric charges which cause them to repel each other and resist adhering together. Coagulation, therefore, involves neutralizing the negative charges and providing a nucleus for the suspended particles to adhere to.

12-81 What various types of coagulants are used?

Ans. The most common coagulants are iron and aluminum salts such as ferric sulfate, ferric chloride, aluminum sulfate (alum), and sodium aluminate. Ferric and alumina ions each have three positive charges; therefore their effectiveness is related to their ability to react with the negatively charged colloidal particles. With proper use these coagulants form a floc in the water which serves as a kind of net for collecting suspended matter. In recent years synthetic materials called polyelectrolytes have been developed for coagulation purposes. These consist of long, chainlike molecules with positive charges. In some cases organic polymers and special types of clay are used in the coagulation process to serve as coagulant aids. This aids coagulation in making the floc heavier, causing it to settle out more rapidly.

12-82 What is chemical precipitation?

Ans. In precipitation processes, the chemicals added react with dissolved minerals in the water to produce a relatively insoluble reaction product. Precipitation methods are used in reducing dissolved hardness, alkalinity, and in some cases silica. The most common example of chemical precipitation in water treatment is lime-soda softening.

12-83 What is ion exchange?

Ans. When minerals dissolve in water, they form electrically charged particles called ions. Calcium carbonate, for example, forms a calcium ion with plus charges (a cation) and a carbonate ion with negative charges (an anion).

Certain natural and synthetic materials have the ability to remove mineral ions from water in exchange for others. For example, in passing water through a simple cation-exchange softener, all the calcium and magnesium ions are removed and replaced with sodium ions. Ion-exchange materials usually are provided in the form of small beads or crystals which compose a bed several feet deep through which the water is passed.

12-84 What is demineralization?

Ans. This involves passing water through both cation- and anion-exchange materials. The cation-exchange process is operated on the hydrogen cycle. That is, hydrogen is substituted for all the cations. The anion exchanger operates on the hydroxide cycle which replaces hydroxide for all the anions. The final effluent from this process consists essentially of hydrogen ions and hydroxide ions, or water.

The demineralization process may take any of several forms. In the mixed-bed process, the anion- and cation-exchange materials are intimately mixed in one unit. Multibed arrangements may consist of various combinations of cation-exchange beds, weak- and strong-based anion-exchange beds, and degasifiers.

12-85 How does lime react in the softening process?

Ans. Hydrated lime (calcium hydroxide) reacts with soluble calcium and magnesium bicarbonates to form insoluble precipitates. This is shown by the following equations:

$$\underset{\text{Lime}}{Ca(OH)_2} + \underset{\text{Calcium bicarbonate}}{Ca(HCO_3)_2} \rightarrow \underset{\text{Calcium carbonate}}{\underline{2CaCO_3}} + \underset{\text{Water}}{2H_20}$$

$$\underset{\text{Lime}}{2Ca(OH)_2} + \underset{\text{Magnesium bicarbonate}}{Mg(HCO_3)_2} \rightarrow \underset{\text{Magnesium hydroxide}}{\underline{Mg(OH)_2}} + \underset{\text{Calcium carbonate}}{\underline{2CaCO_3}} + \underset{\text{Water}}{2H_2O}$$

Most of the calcium carbonate and magnesium hydroxide comes out of solution as a sludge and can be removed by settling and filtration. Lime, therefore, can be used to reduce hardness present in the bicarbonate form (temporary hardness) as well as decrease the amount of bicarbonate alkalinity in a water. Lime reacts with magnesium sulfate and chloride and precipitates magnesium hydroxide, but in this process soluble calcium sulfate and chloride are formed. Lime is not effective in removing calcium sulfates and chlorides.

12-86 How does soda ash react in the softening process?

Ans. Soda ash is used primarily to reduce nonbicarbonate hardness (also called sulfate hardness or permanent hardness). It reacts as follows:

$$\underset{\text{Soda ash}}{Na_2CO_3} + \underset{\text{Calcium sulfate}}{CaSO_4} \rightarrow \underset{\text{Calcium carbonate}}{\underline{CaCO_3}} + \underset{\text{Sodium sulfate}}{Na_2SO_4}$$

$$\underset{\text{Soda ash}}{Na_2CO_3} + \underset{\text{Calcium chloride}}{CaCl_2} \rightarrow \underset{\text{Calcium carbonate}}{\underline{CaCO_3}} + \underset{\text{Sodium chloride}}{2NaCl}$$

The calcium carbonate formed by the reaction tends to come out of solution as a sludge. The sodium sulfate and chloride formed are highly soluble and non-scale-forming.

12-87 On high-pressure boilers, in addition to conductivity of feedwater, what other measurement should be made of steam going to high-pressure turbines?

Ans. Stress corrosion cracking can occur if the recommended level of sodium values of the turbine manufacturer is exceeded. The maximum limit is about 25 parts per billion (ppb) for boilers operating over 600 psi. Ion electrode instruments are used to measure the sodium content of steam to the turbine. The possibility of exceeding permissible sodium levels occurs from upsets in demineralizers, condenser leakage, and leakage of sodium from condensate polishers. Turbine manufacturers are specifying even lower sodium levels as they try to find material that will perform satisfactorily under high stress in a possible hostile environment that can produce stress corrosion cracking with time exposure.

13 Inspection, Maintenance, and Boiler Plant Management

Boiler plant operation is generally rated by efficiency of operation, costs, reliability, and what is sometimes taken for granted, safe operation. As boiler size and capacity increase, the possibility of a forced outage, especially one resulting from a tube failure or explosion, takes on greater significance. The length of outage and the cost of repairs are proportionate to the size of the boiler. Thus every effort must be made to prevent failure of pressure parts by adequate inspection and maintenance. Visual inspections are still required to get as close to the parts of the boiler, both internal and external, as is practicable.

On older-type installations this was possible because a greater percentage of the surfaces of the pressure parts was accessible because of design. And more openings were provided in the setting. But with newer boilers this is not often possible. Thus inspection of large boilers should also include a review of instrumented readings so as to pinpoint areas of trouble. For example, an increase in pressure drop across a bank of tubes may indicate tube fouling in that bank, and is quicker than looking at the tubes.

The degree of availability of a steam generator and its many auxiliaries can be affected by errors in design, construction or erection, operation, and the preventive maintenance practiced in the boiler plant.

Various states and municipalities have laws pertaining to inspection of boilers. The main purpose of these laws is to protect the life and limb of employees and the public as well as to avoid property damage. The legislation sets up standards for design, installation, and inspection. These are usually ASME or NB rules, made legal requirements. Most of the laws provide for periodic inspection of boilers coming within their scope, by state, municipal, or insurance company inspectors. The owner or operator must arrange for these inspections at required, stipulated intervals. For power boilers, the requirement is usually a yearly internal inspection and an external inspection six months later. Low-pressure heating-boiler inspections may be on a yearly or biannual basis, depending on the jurisdiction.

Modern boilers and auxiliary equipment have been developed along such lines of automatic control that they seem to require little attention. Unfortunately, this development has had its effect on some operators with the consequence that they have acquired the idea that the equipment can think for itself. Regardless of how automatic a plant may be, the human element may still be the weakest link and may become responsible for a costly shutdown. Each individual connected with boiler plant operation and maintenance should be trained to understand that she or he has a highly responsible duty to carry out and that those who cannot assume this responsibility capably have no place in a boiler plant. This attitude should be assumed by each person in the organization, from the ash handler to the chief engineer. The success of the organization depends largely on the careful selection of each prospective employee—a real problem in personnel study.

High availability of power plant equipment is the result of correctly installing reliable hardware and associated systems, which are then properly operated and maintained by skilled people. Part of the effort to keep power plant equipment operating efficiently and safely requires setting up inspection programs of the boiler, auxiliaries, firing equipment, and controls.

One of the legal inspections generally required on high-pressure boilers is the internal inspection. The purpose of the internal inspection by plant personnel and an authorized inspector is to check on the structural soundness of the pressure-containing parts and to note any conditions that can affect its strength to confine the pressure. Wear, deterioration, corrosion, scale, oil, cracks, grooving, thinning, and other such weakening conditions require inspection. Most boilers develop their own areas of trouble spots, depending on design, operating conditions, and maintenance practices. Check all exposed metal surfaces inside the boiler for effectiveness of water treatment and scale solvents, also for oil or other substances that enter with feedwater. Oil or scale on heating surfaces weakens the metal, causing bagging or rupture. Corrosion areas next to a seam are more serious than in a solid plate away from seams. Thinning on a joint is dangerous because the strength of a joint is less than that of a solid sheet.

Check for evidence of grooving and cracks along longitudinal seams of shells and drums. Carefully look for internal grooving in fillets of unstayed heads. Inspect stays and stay bolts for even tension, fastened ends for cracks where stays or stay bolts are punched or drilled for rivets or bolts. Manholes and other openings are subject to corrosion thinning and cracks. See that openings to water column connections, dry pipes, and safety valves are free of obstructions such as mud and scale.

Ligaments between tube holes in heads (of all types of boilers) often crack, then leak and weaken the boiler. Also, on both watertube and fire-tube boilers, the beading and flaring on tube ends need checking for erosion and corrosion, cracks, and thinning. Welded nozzles and other such openings require inspections for weld washout, cracks, and evidence of deterioration of the joints.

Oil is usually hard to detect, especially if only a very small amount is present. Run the back of your finger along the waterline. If it is stained and the stain cannot be washed off with soap and water, oil is entering with the feedwater and is being distributed throughout the boiler by circulation. Oil, being lighter than water, rises gradually and forms a scum along the water level. The real danger comes from tiny solid particles sticking to the oil before it adheres to the drum sides. Then gradually this weighted oil settles to the heating surfaces, causing tubes to blister or completely fail.

Carefully inspect the plate and tube surfaces that are exposed to the fire. Look for places that might become deformed by bulging or blistering during operation. Solids in the waterside of lower generating tubes cause blisters when sludge settles in tubes and water cannot carry away heat. The boiler must be taken out of service until the defective part or parts have been properly repaired. Blistered tubes usually must be cut out and replaced with new ones.

Lap-joint boilers are apt to crack where plates lap in a longitudinal or straight seam. If there is evidence of leakage or trouble at this point, remove the rivets and examine the plate carefully if cracks exist in the seam. Cracks in shell plates are usually dangerous, except fire cracks that run from the edge of the plate into the rivet holes of girth seams. Usually, a limited number of such fire cracks are not very serious.

Test stay bolts by tapping one end of each bolt with a hammer. For best results, hold a hammer or heavy tool at the opposite end while tapping. A broken bolt is indicated by a hollow sound.

Tubes in fire-tube boilers deteriorate faster at the ends toward the fire. Tapping the outer surface with a light hammer shows if there is serious thinness. Tubes of vertical tubular boilers usually are thin at the upper ends when exposed to the products of combustion. Lack of water cooling is the cause. Tubes subject to strong draft often thin from erosion caused by impingement of fuel and ash particles. Improperly used soot blowers will also thin the tubes. A leaky tube spraying hot water on nearby sooty tubes will corrode them seriously from an acid condition. Short tubes or nipples joining drums or headers lodge fuel and ash and then cause corrosion if moisture is present. First clean and then thoroughly examine all such places.

Baffles in watertube boilers often move out of place. Then combustion gas, short-circuiting through baffles, raises the temperature on portions of the boiler, causing trouble. Heat localization from improper or defective burners, or operation causing a blowpipe effect, must be corrected to prevent overheating.

The inspections needed on the external fitting of boilers include the following: Safety valves are the most important attachments on a boiler. There should be no rust, scale, or foreign matter in casings to hinder free operation. The best way to test the setting and freedom of safety valves is by popping the valve with pressure. If this cannot be done, test by try levers. Inspect the discharge pipe to make sure it is secure. Operators have been killed because a valve discharging into the boiler room fills the space with steam in a few seconds. The opening in the discharge line must not be plugged.

Pressure gauges have to be removed to test by comparing with a standard test gauge. Blow out the pipe leading to the pressure gauge. Make sure water-column connections are free by removing plugs, or the tees. Examine the condition of the water column and gauge-glass attachments.

Examine the supports of the boiler structure. Make sure that ash and soot will not bind the boiler structure to produce excessive strains from expansion under operating conditions. Look also for evidence of corrosion from soot on structural supports. Check the blowdown valves to see that they work freely and are packed and that external piping and fittings are not corroded or damaged.

Nondestructive testing equipment is being used in boiler inspection to locate potential areas of failure. Five major nondestructive tests are used: ultrasonics, radiography, magnetic-particle, dye-penetrant, and eddy-current. Ultrasonic equipment is now portable for field use and is extensively used for plate- and tube-thickness checks. These instruments become useful as tracing instruments for determining causes of failure of a repetitive nature. For example, tube failures in waterwalls are a common problem. After one tube failure, adjacent tubes can be checked ultrasonically and thinned tubes replaced prior to failures.

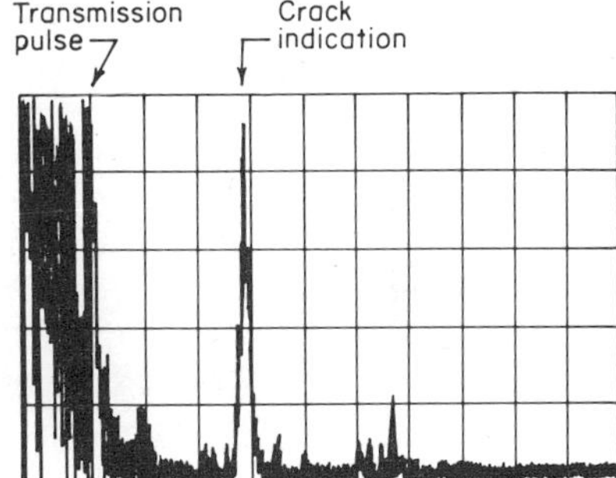

Fig. 13-1 Oscilloscope shows trace of a crack during pulse-echo flaw-detection tests.

A similar practice is followed on tubes subject to fly-ash erosion. The thickness of the tubes in a suspected area is checked by ultrasonic equipment, and those found thinned are replaced during normal outages. Plate thickness around manhole and handhole openings, water legs, shells, and heads is checked ultrasonically in order to determine allowable pressure.

Flaw detection, such as checking for laminations, cracks, porosity underneath plate surfaces, or welds on inaccessible visual parts, is playing a more important role. Pulse-echo instruments are now available for field testing to do flaw detection. See Fig. 13-1.

Radiography, so important in new construction, is extensively used in field testing. Welded repairs on high-pressure boilers are tested by x-ray or other radiographic equipment.

Magnetic-particle inspection finds its chief use in surface crack detection. Its main use is on piping and joints of boilers.

Eddy-current testing is finding its chief use in nonmagnetic-tube searching for defects, such as condensers and heat exchangers connected to a boiler.

Nondestructive testing of nuclear reactors in service of course supplements any traditional visual internal inspection which is limited because of the radiation hazard.

MAINTENANCE PROGRAMS

The problem of maintenance should be approached with the view of how much it will cost *not* to carry on an active maintenance program. The unexpected expenses will be sure to be far greater in frequency and extent when such a program is abandoned or deferred than when it is in effect.

Large plants use a card-index system. One card is made out for each piece of equipment and has on it identification information and a space for entering records of tests and remarks. A signaling system can be used sometimes to indicate the date on which equipment is due for tests, overhauls, and inspection. Even the smallest plant can create a system that will give this information in an alphabetical card-index file.

Preventive maintenance in an industrial boiler plant has been influenced by legal requirements and is performed to protect personnel and to prevent equipment damage that could lead to costly repairs and loss of productive capacity. In fact, preventive maintenance directed *specifically* to maintaining boiler efficiency has been the exception rather than the rule. But rising fuel costs have placed increasing emphasis on conscientious maintenance, which is necessary for preserving high efficiencies. These preventive maintenance practices are often justified easily on an economic basis.

Efficiency-related boiler maintenance is that directed at correcting any condition which increases the amount of fuel required to generate a given quantity of steam. Thus, at a specified boiler load, any condition which leads to an increase in: (1) flue-gas temperature; (2) flue-gas flow; (3) combustibles content of ash or flue gas; (4) convection or radiation losses from the boiler exterior, ductwork, or piping; or (5) blowdown rates is considered an efficiency-related maintenance item. Generally, attention to such items also can forestall more serious consequences, which could cause damage to equipment or injury to personnel.

The boiler tuneup is one of the most effective means for improving efficiency and for maintaining it at a high level. The primary objective of a tuneup is to achieve efficient combustion with a controlled amount of excess air, reducing the dry-gas loss and the power consumption of forced- and induced-draft fans.

A log in which each operator can keep hourly readings and remarks of any unusual occurrence is of value. The value is enhanced by recording notes on readings and observations from the more remote parts of the plant that may be visited but seldom during the normal course of duties. An operator, unless utterly incompetent, hesitates to walk past equipment to take a reading without observing its condition.

The central "clearinghouse" for daily study and writing up of these logs depends on the size of the plant. In any case, the reports should receive careful daily attention and then be systematically filed for reference at any time. Most important readings, such as fuel consumption, steam evaporation, and gas temperatures, should be broken down into efficiencies and equipment factors for recording in graphical form. A graph showing daily, weekly, or monthly progress in these results should be posted where everyone responsible for the operation may see them. A gradual depreciating operation is shown in this manner where otherwise a loss might mount unnoticed.

Individual boilers should have a log in order to record the operation and performance of controls and safety devices, as shown in Fig. 13-2. The mini-

To Test

WATER COLUMN AND GAGE GLASS–open drain valve quickly and flush water from glass and column. When drain is closed, water level should recover promptly.

LOW WATER FUEL CUT-OFF AND WATER LEVEL CONTROL–drain float chamber when firing equipment is operating. Proper operation of the control should shut off the firing equipment and start feed pump. If controls are of probe or other type that require lowering of water level in boiler to test, DO NOT lower water level to point below bottom of gage glass.

HIGH-PRESSURE POWER BOILER LOG
READINGS TAKEN EACH 8 HR. WATCH

BOILER NO. ____________

WEEK BEGINNING ____________

CHECK OR TEST OR RECORD EACH 8 HR WATCH	MON.			TUES.			WED.			THUR.			FRI.			SAT.			SUN.		
	8 to 4	4 to 12	12 to 8	8 to 4	4 to 12	12 to 8	8 to 4	4 to 12	12 to 8	8 to 4	4 to 12	12 to 8	8 to 4	4 to 12	12 to 8	8 to 4	4 to 12	12 to 8	8 to 4	4 to 12	12 to 8
WATER LEVEL-PROPER																					
STEAM PRESSURE, PSI																					
FEED PUMP PRESSURE, PSI																					
FEED WATER TEMPERATURE, °F																					
CONDENSATE TEMPERATURE, °F																					
FLUE GAS TEMPERATURE, °F																					
LOW WATER CUT-OFF-TESTED																					
WATER LEVEL CONTROL-TESTED																					
WATER GAGE GLASS-CLEAN																					
FEED PUMP IN GOOD REPAIR																					
CONDENSATE TANK & FLOAT TESTED																					
BURNER OPERATION-NORMAL																					
FUEL SUPPLY-ADEQUATE																					
FLAME FAILURE SAFEGUARD																					
WATER TREATMENT-TESTED																					
BOILER BLOWDOWN-PERFORMED																					
MONTHLY-SAFETY VALVE TESTED																					
OPERATOR'S INITIAL																					

REMARKS:

Fig. 13-2 High-pressure boiler log permits recording and checking of controls and safety devices. *(Courtesy Royal Insurance Co.)*

mum tests and checks shown will assist operators in determining whether the boiler and its controls are in good operating condition. The tests must be performed at established frequencies, and if any malfunctions are found, they should be corrected immediately. A boiler with a defective safety valve, improper water-level control, or nonoperating low-water fuel cutoff should never be operated unattended until these vital devices are in good working order.

Low water is a frequent cause of boiler outages, especially on automatically fired packaged boilers. Three major areas must be checked and tested periodically to prevent low-water accidents:

1. *Feedwater system.* It must have the capability to maintain normal water level in the boiler.
2. *Low-water safety controls.* They must sense when the water level in the boiler drops below prescribed limits.
3. *Fuel-cutoff actuators.* They must be able to interrupt fuel flow when receiving a signal from the low-water sensors.

It is necessary to test *float controls* periodically to make sure they move freely. On power boilers operating around the clock, this should be done on every shift; once a week is adequate for heating boilers. Note that float controls often become sluggish or stuck when sludge and sediment build up in the float bowls. Avoid this by blowing down the float bowl daily. Also, dismantle the entire component once each year and clean scale and sediment manually from the operating mechanisms. One further note: Blow down the water column after examining float controls; clean it thoroughly during the annual inspection.

Probe-electrode-type controls require periodic checking in the following areas:

Examine insulators where electrodes penetrate the boiler shell or water column to make sure they are not cracked or bridged with dirt or moisture. This could cause current to flow from the electrode to the shell, destroying the sensing value of the electrode.

Look for broken electric connections to the electrodes and for wiring touching the boiler. These conditions also can limit probe's sensing value.

Make sure electrodes of the proper length are installed on replacement jobs. Otherwise, it may be possible to insert a unit that senses at a point below the safe water level.

Minimize scale, mud, and sediment in a water-column-mounted electrode to stop these impurities from completing the circuit at a false water level.

Log sheets are a forced reminder to check certain components of a boiler to prevent trouble from developing later and to note if proper operation is taking place. A log sheet should record all important operating data, such as pressures and temperatures, and it should also record procedures such as the time at which the soot blowers and water columns are blown and when blowdown valves were operated. A continuous record of operating data and important procedures carried out is then at hand when needed. It is also important to log when testing safety appurtenances.

Any irregular operation or event should be recorded in a separate book, with a description of the irregularity and the corrective measures taken. In this book all orders should be written and initialed by the operator.

Firemen or shift engineers reporting for duty should read the notations made by the previous watches. Then they will know what the past operation has been, what orders have been issued for future operation, and what trouble spots to watch for. They should then initial those items for which they are responsible so as to indicate that they are familiar with the situation. Log sheets are supplied by some insurance companies for small low-pressure plants and industrial high-pressure boilers. Many plants design their own log sheets to record important data pertaining to their specific plant details and layout.

Cleanliness cannot be carried to an extreme in the boiler plant. However, increased operating costs and maintenance, accidents, and lost time due to illness may be expected in a dirty plant. But the operating personnel cannot be expected to maintain cleanliness if not all reasonable steps to prevent dirt have been taken. Nothing will break down the aims of those cleaning the plant quicker than to have clouds of soot and dust pour from cracks in a boiler setting just after they have cleaned the plant. Savings gained by long deferment of needed repairs are not economy. It is wasteful ignorance. All steps possible to prevent dirt should be taken, and then all energy necessary may be expected of those responsible for maintaining cleanliness.

Training for emergencies of the operating personnel is indispensable to the boiler plant. In the small plant, individual instruction may suffice, but in larger plants, as with a fire drill, the operating force should be schooled for every conceivable emergency. Whether the emergency is low water in a boiler, loss of auxiliary power, failure of an interlocking system, or a rupture in the pressure system, each worker should have a definite station and specific duties to fulfill. Everyone should be drilled to understand the reason for these duties and to carry them out quickly with no confusion or overlapping in their scope.

Certain alarm systems are invaluable for warning of emergencies, but there is no alarm system that can replace the almost instinctive anticipation of many emergencies by a well-trained, alert engineer.

Floods have created emergencies that normally require power plant equipment to be reconditioned after the waters recede. The following precautions should be taken before a boiler that has experienced flooding is returned to service:

1. *Safety valves.* Clean, lubricate, inspect, and test the valves. Inspect the discharge pipe for mud obstructions.
2. *Low-water cutoffs.* Clean, rewire if necessary, renew probes and mercoid switches if necessary, inspect, and test.
3. *Limit controls.* Clean, inspect, and test. Renew if necessary.
4. *Flame safeguard devices.* Clean, inspect, and test. Renew if necessary.

5. *Electric motors, transformers, relays, and wiring.* Electric motors should be cleaned, baked out, and tested before energizing. Conduit, BX, wiring, and relays should be checked by a qualified electrician. If any is defective, renew. Failure to follow safe procedures regarding electrical facilities could lead to fatal electrocution and/or electrical failures including burnouts and fires.

6. *Refractory, brickwork, and insulation.* After inspection and repair, all these should be dried out as much as possible before firing. Initial firing should be light and in short intervals, in order to allow moisture to escape gradually. Rapid firing could build up steam pockets with resultant pushing out of refractory.

7. *Internal inspection.* Boilers should be thoroughly inspected internally, and any accumulation of slime and mud should be removed.

8. *Start-up.* As soon as the boiler is started, the following should be done: Test flame safeguard devices, low-water cutoffs, safety valves, and limit controls.

9. *Surveillance.* Maintain the boiler log after the boiler becomes operational. Diligently monitor the operation of controls and safety devices. Keep the boiler under strict surveillance for leaks, dirty gauge glass, muddy blowdown, and similar evidence that the effect of the flooding may require further cleaning and repairs.

Forced outages must always be considered a possibility. Most plants have no spare capacity or not enough capacity when the largest unit suffers an unexpected failure. Emergency planning should include an adequate inventory of spare parts, such as tubes, plugs, gaskets, and similar components, that will help in expediting a repair. Extensive repairs of a key unit may require considering renting a boiler while the damaged unit is being repaired (Fig. 13-3). Much time can be gained if plant engineers preplan for such an emergency by having water, steam, and fuel piping installed for the rental boiler so that the rental unit can be quickly connected to the plant's system.

Study programs should be developed. The size of the plant and available facilities will determine the best approach. In large plants this may consist of periodic classes for the operating personnel under the supervision of a capable executive; in small plants it may outline home study for those interested. Some means of instilling ambition and promoting action to acquire further knowledge of the equipment with which they work will be to the benefit of all concerned.

There are several organizations of operating and power engineers which are represented in larger cities and whose aims are purely educational. The educational and social functions of such groups are recommended highly.

Management should annually review the training available for the power plant personnel to note if the relationship to the rest of the plant is clearly understood. Many plant engineers or powerhouse superintendents simulate plant upsets in frequent in-plant drills. This is an excellent method of preparing for all types of contingencies.

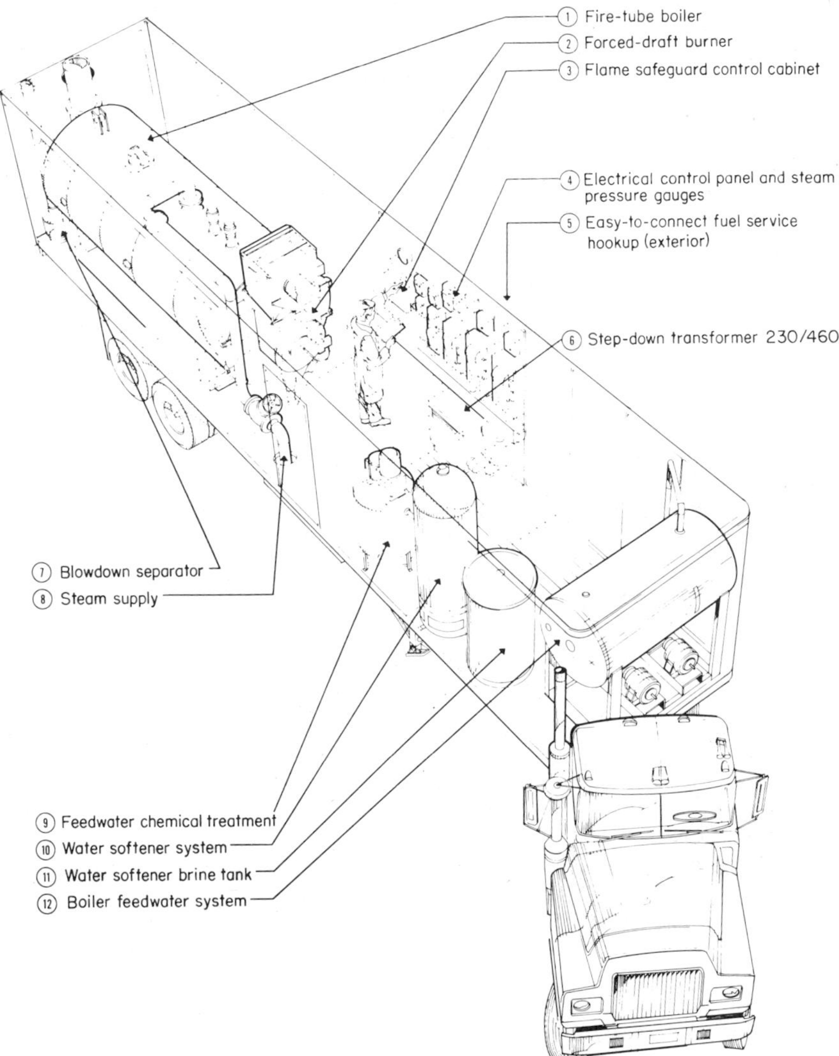

Fig. 13-3 An emergency boiler can be rented to maintain production in the event of a casualty to a plant unit. *(Courtesy Indeck Power Equipment Co.)*

An efficient way to become familiar with a new plant is to trace every important piping system and make a sketch of all the valves and equipment in each system. Do the same with electrical systems. An operator should know the plant well enough to be able to go to key valves and controls in the dark during an emergency. All new operators should trace and sketch each system. A file of literature pertaining to machinery, instrumentation, and equipment in the plant should be maintained for study. Also, a filing system for blueprints of important equipment should be kept in case they are needed in a hurry. Set up a spare-part inventory of critical components so they are always on hand. This includes boiler tubes and components of equipment that would shut down the plant if they were not in stock and needed because of failure.

Large boiler systems require special training programs because of the complexity of the controls, valving, instrumentation, and depth of hardware. Large boiler manufacturers sell training programs to utilities and large industrial customers. With government regulations on emissions standards affecting operations, training by personnel from the boiler manufacturer should help operators in taking over a new plant. Simulators are more and more used as the boiler plants become more complex.

Many state laws require full-time or part-time operator attendance on automatic-fired boilers. The meaning of *attendance* in many jurisdictions is as follows: Watertube boilers with capacities over 20,000 lb/hr and those fired with pulverized coal should have full-time operator attendance. High-pressure firetube boilers should have full-time or part-time operator attendance. *Full-time attendance* means the presence of a competent operator who never leaves the boiler room for more than 20 min. *Part-time attendance* means the presence of an operator who may leave the boiler room for more than 20 min, leaving the boiler operating unattended, but making periodic checks at least every 2 hr. A boiler is considered operated unattended when it is operated for more than 2 hr without a competent operator checking it for proper operation. State and city laws must be checked on operator-attendance requirements.

Where no legal requirements exist for operator attendance, it is good practice to train people to the following minimum requirements:

1. Trace and sketch a boiler-fuel system and associated valves, strainers, gauges, and controls.
2. Trace and sketch main and auxiliary steam systems and condensate-return and boiler-feed systems, including valves, gauges, controls, and interlocks.
3. Inspect and test boiler casings and settings.
4. Inspect firesides and watersides for leakage, corrosion, cracking, bulging, blistering, and other conditions that weaken the boiler.
5. Clean, inspect, and test oil and gas burners and registers.
6. Check fuel tanks and, if necessary, clean strainers, lines, and valves.
7. Line up, recirculate, check for leaks, and light off a fuel-oil system. Check combustion safeguards.
8. Observe and record fuel-system pressures and temperatures. Check and adjust control settings.
9. Inspect and operate soot blowers properly.
10. Light off, fire, and bank fires in a coal-burning boiler (if installed).

11. Inspect and regulate stoker and pulverizer operation in a coal-fired boiler. Check and adjust controls and interlocks.
12. Start, regulate, and secure forced-draft and induced-draft fans. Check for proper operation, controls, and interlocks.
13. Inspect breechings, uptakes, and stack.
14. Determine draft, wind-box, and furnace pressures and furnace and stack temperatures. Adjust the draft where needed.
15. Take and analyze flue gases for CO_2, CO, and O_2. Interpret the results and make the necessary corrections.
16. Blow down a boiler, both bottom and surface blows.
17. Blow down gauge glasses and water columns.
18. Test high- and low-water alarms and low-water cutoff. Dismantle and clean alarms and low-water fuel cutoff.
19. Regulate feed pumps and change-over pumps and adjust feedwater governors.
20. Regulate boiler water level.
21. Cut in, adjust, and secure feedwater-level regulator.
22. Test safety valves.
23. Put a boiler on line.
24. Warm up and cut in steam lines.
25. Take a boiler off line, pull fires (for coal-burning boilers), and secure the boiler.
26. Conduct boiler-water analysis, interpret the results, treat the feedwater, and adjust continuous blow according to a water-treatment specialist's advice, if required.
27. Shift combustion control from manual to automatic and back again. Check the safety controls in doing this.
28. Prepare boiler logs and operating records.
29. Cut in and out superheaters properly.
30. Adjust feedwater heater pressures and temperatures.
31. Renew and repack gauge glasses.
32. Remove, regasket, and replace manhole and handhole plates.
33. Inspect, repair, and set safety valves within Code limits.
34. Clean the fireside and waterside of the boiler.
35. Make refractory and other furnace repairs.
36. Conduct a hydrostatic test.
37. Lay up a boiler.
38. Know how to remove and replace tubes in the boiler, superheater, economizer, air heater, and feedwater heater.
39. Adjust soot blowers and lances.
40. Inspect, clean, and repair boiler gauges, instruments, and controls.

In addition to the above, an operator should continuously study: (1) boiler and steam system design; (2) construction; (3) operation; (4) maintenance; (5) state or city laws; (6) insurance company requirements and inspections; (7)

emergency measures; (8) safety in a boiler plant; and (9) ASME Boiler and Pressure Vessel Codes.

SAFETY

Large plants sometimes employ a "safety" engineer whose duty is to see that all hazards are reduced to a minimum, to educate the personnel to be safety-minded, and to select educational material on safety for posting. Obviously, the small plant cannot afford such a specialist. But the same ideas of safety can be developed among a smaller personnel at a proportionately lower cost. Such a program supplies another instance of how spending ten dollars may save hundreds.

No matter how small a plant may be, a bulletin board should be included in its equipment. If employee's compensation insurance is carried (as required in many states), the insurance companies usually furnish excellent safety bulletins free. These should be posted. A set of safety rules for the particular plant should be suitably framed and posted permanently in a conspicuous place. A set of these rules typical of those observed in many plants is shown in App. 4.

Safety laws involving power plant equipment and surroundings are covered by state statutes, Occupational Safety and Health Administration (OSHA) standards, and fire and casualty insurance company requirements. The modern approach to safety is to consider an entire system or loop and then determine the effect to the system if one of the components were to fail or if an operator did not respond to a situation properly. The extent of analysis needed will depend on the criticalness of the plant. Certainly in a nuclear plant, the malfunction of a valve may require more instrumentation, alarms, and redundancy than would be necessary in a small industrial plant. The term *plant interaction reviews,* in the analysis of the effect of a component failure, is gaining increased attention in reliability studies.

Large plants depend heavily on control-room-type operations. This practice has eliminated continuous area surveillance of large power plant sections, with heavy reliance placed on instrumentation and alarms to alert the control room to an abnormal situation. It is not possible to instrument an alarm for all possibilities of malfunction. Therefore, care is needed in avoiding too much reliance on control-room-type operation unless it is supplemented by area surveillance of operation at stipulated, safe intervals. As a result of several nuclear plant experiences, the following concepts in control room operation are being developed. Control rooms should display normal, marginal, and out-of-limit conditions on a system. Control room operators should be trained to identify when a safety system on the loop may be involved and what the implications to safe operation may be as a result. An upper-limit alarm system should force the operator to decide what preventive or corrective actions must be taken to avoid passing through a potentially dangerous condition. Critical information displayed in a control room should have means provided for rapid validation of readings. Alarms should ring on a priority system based on urgency for operator action whenever a plant upset jeopardizes equipment.

General safety rules that are applicable to boiler work include the following:

Use safe ladders with no defects as well as scaffolds, hoists, and cables in gaining access to boiler components.

When entering a boiler, make sure all valves, lines, and similar connections for steam, water, fuel, air, and flue gas are tightly closed or blanked off.

Use low-voltage lamps and extension cords with proper guards and insulation.

Make sure drums and furnaces are properly purged and vented before entering.

Wear hard hats in any area where head injury is possible from falling objects or from accidentally bumping into protruding parts, such as valves or elbows of piping.

Wear special clothing to avoid skin contact of potentially harmful contents in the boiler.

There are many other safety rules to be followed and detailed in company or legal statutes, and these should be observed around powerhouse equipment. Hydrostatic tests require consideration of support to handle the weight of water involved, test pressure and relief-valve requirements, accuracy of test gauges, venting of vessel prior to filling with water, and suitability of connected apparatus to withstand the test pressure.

Test programs in the boiler plant are essential if operating efficiencies are to be maintained at a high level. Even though a full complement of recording and indicating instruments is installed to permit most economical operation, these instruments should receive periodic calibration and check by actual test.

There are some plant functions for which instruments have not yet been developed to give direct readings and for which analytical methods of securing the desired information are necessary. Tests on fuel and water are examples of the methods employed in checking on this group of functions.

In the small plant, it is sufficient usually to test each lot of fuel received. But large plants storing coal outdoors find this system impractical, and for them a periodic test of the fuel as fired is necessary. Similarly, a small plant may find it sufficient to test the boiler water once a week, but a large plant operating under widely varying conditions finds it advisable to make this test at least once a day.

Testing and calibration of pressure gauges and meters should be carried out on routine schedule. The date of the test and the initials of the person making it should be printed on a slip and pasted on the instrument or otherwise recorded for future reference.

Computer logging and analysis of data are now being practiced on large boiler systems. This does not eliminate the need for periodic review of developed data in order to maintain the efficiency of the equipment, especially when a conflict may arise between production needs and reliable operation of the equipment. In large plants, purchasing of supplies is confined to a department for this purpose. However, in the small plant, the chief engineer often has to assume this duty.

A study should be made of all products available to fill a need, comparing their cost per unit of service (not first cost) with the results accomplished. For example, one grade of packing for a high-speed centrifugal pump cost $3.60 a set, compared with another brand costing $9.00. The first set required replacement every 30 days, but the latter set operated satisfactorily for 6 months (mo). Thus, the cost per day for the former was $0.12 versus $0.05 for the latter, or the more expensive first cost was less than one-half as expensive per unit of service.

In cases of the foregoing type, the apparent cost is not all that should be considered. One type of pump packing may cause excessive shaft-sleeve wear; a certain grade of lubricating oil may last longer but may be causing oxidation; a particular type of valve may be easier to operate but may not meet ASME Code specifications for safety; and so on down the line.

Replacement of equipment is an important problem of plant management. The chief engineer may be called on to investigate and to make up a comprehensive report on the advisability of replacing an older piece of equipment, such as a feedwater heater that continues to have tube problems, a manual-firing system with an automatic combustion control system, etc. In some organizations, the engineering department continually keeps track of operation and maintenance costs and compares them with possible new equipment for savings potential. The analysis for possible replacement must consider all factors of cost and return on investment. Even under favorable economic analysis, equipment may have to be installed or replaced, because of environmental or similar legal requirements, in order to avoid costly citations.

A typical cost-comparison analysis can be made to determine if a new feedwater heater should be bought or the old heater retubed. Tests showed that a new heater would result in a fuel savings of $250/mo, while retubing the heater would result in a fuel saving of $50/mo. A new heater costs $15,000 installed, and the older heater would have a scrap value of $500. Retubing would cost $5000 with no scrap value.

New heater:

$$\text{Fuel saving per year} = 250(12) = \$3000/\text{yr}$$
$$\text{Net cost of heater replacement} = \$15{,}000 - \$500 = \$14{,}500$$
$$\text{Investment returned in } \frac{14{,}500}{3000} = 4.83 \text{ yr}$$

Retubing old heater:

$$\text{Fuel saving per year} = 50(12) = \$600/\text{yr}$$
$$\text{Net cost of retubing} = \$500$$
$$\text{Investment returned in } \frac{5000}{600} = 8.3 \text{ yr}$$

If the expected life of the new heater is 15 yr, definite savings can be realized by installing a new heater. Additional charges to be considered are interest on the money that could be earned if it were not spent on the heater, depreciation, and taxes. Operating-cost savings generally favor replacing older, inefficient, and sometimes obsolete equipment with more modern, efficient equipment on which parts are readily available.

In the design for installation of new or replacement boiler plant equipment it is well to submit the blueprints, proposed plans, and specifications to the plant engineer for comments or approval. The engineer should know the "ins and outs" of the plant, its general setup, and its peculiarities, whereas the designer or consultant, lacking familiarity with the particular plant, may unintentionally slip on an important item. Many costly errors may be eliminated by close cooperation of the designer or consultant and the chief engineer.

Shift schedules are another important function of plant management. These must be arranged sometimes for a 24-hr day, and various schedules are followed in order to limit the work week to a certain number of hours.

Many plants that employ a number of operating engineers find it necessary to have at least one spare worker, or "floating" operator. This person should be capable and trained to fill in any shift vacancy due to illness, absence, or vacation periods. During normal conditions, the floating operator may be absorbed in a maintenance or test schedule.

It should be a strict rule in shift work that no operator should leave the station to prepare to go home until the relief operator has reported at the post ready to carry on the duties. It is often during the few minutes' interim of neglect when one is preparing to leave the instant relief shows up that an accident happens.

Insurance of boilers is an important item usually handled by the company executives, but one in which the engineering department quite often is involved, because boiler insurance is generally bought under a boiler and machinery contract that may cover other plant equipment besides boilers. Traditionally, the companies that sell boiler and machinery coverage also provide legal inspection service as part of their effort to prevent accidents. Thus, this service serves a public good by protecting the public from potential dangerous accidents, and this is, of course, the main reason why legal authorities accept insurance company inspections on boilers and pressure vessels.

Boiler and machinery coverage is twofold in scope. It is designed to provide financial reimbursement for any possible accidents, and it usually also provides a legal inspection service that will make it possible to reduce accidents to a minimum. This service helps point out any conditions in the boilers, overlooked by the plant personnel, that might be corrected to reduce operating or maintenance costs.

Two general types of boiler and machinery insurance coverage may be written. One covers damages caused by accidental breakdown of the boiler proper under pressure, and the other reimburses the user for loss of production due to outage of equipment operated by the boiler in case of such an accident.

The former is known as "breakdown or property coverage," and the latter is classed as "business-interruption coverage."

Many states require that boilers not under federal control be inspected by a state inspector (for whose services there is a charge) unless the boiler is insured by an authorized insurance company and inspected by one of its qualified inspectors. When a boiler is insured (usually by a company licensed to write boiler and machinery insurance in that state), the engineering department of the insurance company sends the state legal jurisdiction a notice that the boiler is insured. If the company has commissioned inspectors in their employ (and most companies writing boiler and machinery insurance do), the legal jurisdiction does not schedule an inspection on that boiler for which a notice was received. Why? Because it will be inspected and reported to them on a formal state report when the inspection comes due by the commissioned insurance company inspectors.

The owner must prepare the boiler for the internal inspection. On some low-pressure boilers, internals (inspections) may be scheduled only when it is deemed advisable, depending on the state law. In others, the boiler must be drained and opened for inspection whether it is of the high-pressure, low-pressure, steam, or hot-water-heating type.

The commissioned inspector makes the inspection, and if the conditions are satisfactory, files a report to the state, requesting the state to renew the certificate on the boiler. In some jurisdictions, the insurance company issues the certificate directly. If repairs are needed or if conditions need correction, certificates are withheld until the violation is removed. If this is not done, the inspector notifies the state of the violation and requests that no certificate be issued until the violation is removed. The state or legal jurisdiction can then use its police power to enforce the requirements.

When the insurance on a boiler is canceled or not renewed, the insurance company notifies the legal jurisdiction and gives the reasons for it. Dangerous conditions on a boiler can be reported to the state over the phone, if necessary, by the commissioned inspector. The owner will then have the full pressure of the legal jurisdiction to correct the condition or take the boiler out of service. Most insurance companies also have provisions for immediate suspension of insurance by the inspector under these adverse conditions.

A boiler owner can ensure that all possible steps to prevent boiler failure have been taken by doing the following: First, purchase the best equipment available for a given service. Second, see that the boiler is properly installed and equipped with all necessary Code-approved appurtenances and safety devices. Third, make sure the boiler is inspected regularly by a commissioned inspector. Only then will all the legal requirements be covered. Keep a logbook check system and a set of preventive maintenance and testing procedures. See that such checks and testing procedures are followed and that the results are always recorded. Immediately correct any malfunctions found during any check or test. And never operate the boiler unattended until the proper repairs or replacements have been made.

Questions and Answers

GENERAL

13-1 What precaution must be observed before closing a new boiler or a boiler that has been opened for cleaning or repair?

Ans. Make sure all tools, pipes, welding rods, rags, and other such items are removed from drums, headers, furnaces, and tubes. At times, a mirror and flashlight must be used to check headers that are otherwise not accessible for inspection of foreign material. Bent tubes that cannot be looked at from end to end (such as superheater tubes) should be thoroughly flushed, one tube at a time. On new boilers, or where work was done in a tube area, drop rubber balls, and even steel balls, to make sure the tubes are free of obstruction. Water and air can be used to push through the rubber balls.

13-2 What precaution must be observed in turbining the tubes?

Ans. Turbining tubes can cause local tube wear or nicking if the turbining tool is forced through a tube or held in one position too long.

13-3 Is it safe to use portable lamps and extension cords inside boiler drums, shells, or headers?

Ans. Only if low-voltage lamps, 32 V or less, supplied by transformers or batteries are used to avoid electric shocks in case a lamp or bulb breaks and creates a current flow through the boiler shell. Never use extension cords without proper waterproof fittings. Make all connections outside the boiler. And light bulbs should have explosionproof guards. Fittings, sockets, and lamp guards must be grounded.

13-4 On packaged boilers, list the causes of the burner continually going off and on.

Ans. (1) Fluctuating water level tripping low-water cutoff; (2) loose connections or defective control; (3) controls not connected properly or not properly adjusted; (4) partial electric ground; (5) intermittent low voltage; and (6) combustion control opening the circuit when the burner goes to low fire. The potential hazards of each are overpressure and dry firing.

13-5 What does an inspector do in following the construction of a boiler through the shop?

Ans. Initially the inspector checks the chemical and physical properties of the steel from the mill test reports to see if they meet Code requirements. He or she then checks the melt and slab numbers on these reports with the numbers stamped on the plates to identify them. The shop standing is checked to see if the shop is authorized to construct boilers for the state into which the boiler is to be carried and installed. The plates are examined for any visible defects, such as scars, gouges, or laminations. The thickness of the plates is gauged, a tolerance up to 0.010 in. under that specified being allowed. The design of the proposed boiler is checked to see if it meets Code specifications. Subsequent visits are made to check methods and procedures used for welding, preparing welding edges, rivet holes, tube holes, welder qualifications, and assembling the boiler.

A visit is made on completion to view a hydrostatic test and examine the general work; if the boiler has been completed satisfactorily in accordance with Code specifications, the boiler is stamped and the manufacturer's data sheets are signed by the inspector.

13-6 What precautions should be taken before entering a boiler shell?

Ans. If the blowdown enters a common line with other boilers in operation, it must be certain that all valves on the line to the open boiler are closed. If other boilers are operating on the same steam header, *both* stop valves must be closed and the drip valve between them must be open. Any other valves on lines under pressure leading to the boiler must be checked. The engineer in charge and the operator on duty must be told that someone is going inside. A responsible person should be stationed at the manhole or entrance doors while anyone is inside the boiler.

13-7 How are suspended solids removed from water by mechanical means?

Ans. Water is filtered by passing it through fine strainers or other porous media to remove suspended solids mechanically. The degree of filtration depends on the fineness of the filtering media and the type of filter aid used, such as cartridge filters. Since fine solids, in the size range from 1 to 75 μm, are readily removed, it is important to know the size of the particles in the water.

13-8 What assistance should the owner or operating engineer give the legal boiler inspector during inspections?

Ans. It is the responsibility of the owner to prepare the boiler for the required legal internal inspection. All openings must be removed. All scale and mud must be removed so metal surfaces are exposed for inspection. Firesides must be cleaned of soot so tubes can be inspected for corrosion, thinning, erosion, and evidence of overheating.

The inspector must be given all the help she or he needs. Point out any known defects. Station someone immediately outside the boiler during the internal inspection. If the boiler is in battery with others, make sure that all steam, water, and blowoff valves are locked and cannot be opened. Make provision for the hydrostatic test if the inspector deems it advisable. In general, assist in every way to make the examination thorough and complete.

13-9 Is it necessary to repeat the preheat treatment and postweld heat treatment on a power boiler that underwent repairs as a result of defects found during a hydrostatic test?

Ans. The preheating and postweld heat-treating rules of a welded joint must be repeated or reapplied to all weld repairs made to pressure parts.

13-10 In what kind of welding process would you expect to find tungsten inclusions?

Ans. Tungsten inclusions occur in welds made by the inert-gas tungsten arc-welding process that uses a nonconsumable tungsten electrode.

13-11 Name three causes of bagging or bulging of boiler shells and tubes.

Ans. Bagging or bulging is due primarily to overheating a plate or tube to a point where the tensile strength is lowered and plastic flow is induced by the boiler pressure. Contributing factors are deposits of scale or oil or both on the internal surfaces. Plastic flow may also be induced on clean boiler surfaces where severe flame impingement or faulty circulation causes steam blanketing. Bulging can also develop when a tube or plate has been reduced by corrosion or by erosion due to cutting of a soot-blower jet or fly ash until the wall is too light to withstand the boiler pressure.

13-12 What are the purposes of stacks or chimneys on a steam boiler?

Ans. To remove products of combustion; to create draft by the temperature differential between products of combustion in the stack and the air outside, the draft producing turbulence within the furnace to aid combustion of fuel.

13-13 What is meant by a shielded arc? Give reasons for using this type.

Ans. It is an electric arc produced by a coated electrode. The coating in melting back of the arc produces an inert gas which blankets the molten pool to prevent oxidation. As the arc moves on, a brittle crust is left to blanket the cooling metal.

13-14 Name three defects which would cause a boiler to be disconnected from service.

Ans. A longitudinal crack, excessive corrosion, and a badly deformed or burned plate as a result of low water.

13-15 What is a common defect at the edges of a handhole? Give cause, prevention, and repair.

Ans. Eroded and steam-cut bearing surface. The cause is surfaces not properly cleaned, or an improperly installed gasket. The bearing surfaces should be properly cleaned, and proper gaskets installed. Repairs can be made: Build up the corroded area by welding providing that better than 50 percent of the original metal thickness remains or a reinforcing ring may be used. If the area is built up by welding, the weld is then machined and made smooth for the gasket surface.

13-16 What information must an inspector furnish to the state inspection authorities after making the first inspection of a new high-pressure boiler?

Ans. Upon completion of the first inspection after installation, a data report must be furnished by the inspector to the chief inspector of the state in which the boiler is located. This report should show that the boiler complies in every way with the state requirements or submit a list of recommendations for the elimination of the state violations.

13-17 In fitting and heading up stay bolts, what defective material or work would cause your disapproval?

Ans. Insufficient number of threads in sheet; poor threads and fit; telltale hole not drilled straight and to proper depth; bolts too short to get proper heads on bolt; broken or split heads; cracked or split bolts.

13-18 What is the purpose of a quality control program in fabrication of boilers?

Ans. The manufacturer or assembler should have and maintain a quality control system which will establish that all Code requirements, including material, design, fabrication, examination (by the manufacturer), and inspection (by authorized inspector), will be met. Provided that Code requirements are suitably identified, the system may include provisions for satisfying any requirement by the manufacturer or user which exceeds minimum Code requirements and may include provisions for quality control of non-Code work. In such systems, the manufacturer may make changes in parts of the system which do not affect the Code requirements without securing acceptance by the authorized inspector.

13-19 What type of documentation must a boiler or pressure-vessel manufacturer maintain on file to be stamped by the ASME?

Ans. The manufacturer is required to keep certain documentation. The type of documentation needed to satisfy Code requirements varies in different sections of the code. In general, the manufacturer must keep all radiograph film and must also prepare a manufacturer's data report, which must be signed by both the manufacturer's representative and the authorized inspector. The authorized inspector should not sign the data report until he or she has carefully checked it to make certain it properly describes the boiler or vessel to which it applies, that the boiler or vessel complies with the Code, and that the data report has already been signed by the manufacturer's representative. These data reports may be registered with the National Board. For nuclear vessels, more extensive documentation is required.

13-20 What is a common reason for tube failures in a once-through steam generator?

Ans. A frequent cause of tube failure in this type of boiler is insufficient circulation of water to match the local heat input, leading to overheating of tubes. Poor circulation may be caused by operating problems or errors during low-load periods, sudden load upsets, and warm-up periods in placing a unit on the line. Strict observance of manufacturer's instructions is required to avoid tube failures.

13-21 What is the purpose of a welding procedure and welder qualification?

Ans. The purpose of having a welding procedure is to demonstrate that the proposed method of welding, if followed, will produce a satisfactory welded joint within acceptable standards and that the welded joint will satisfy the stipulated service requirements. The purpose of welder qualification is to confirm that the welder can produce a weld with the specified freedom from defects in the welding procedure being followed.

13-22 What is meant by external and internal feedwater treatment?

Ans. External treatment is the reduction or removal of impurities from water outside the boiler. In general, external treatment is used when the amount of one or more of the feedwater impurities is too high to be tolerated by the boiler system in question. There are many types of external treatment (softening, evaporation, deaeration, etc.) which can be used to "tailor-make" feedwater for a particular system. Internal treatment is the conditioning of impurities within the boiler system. The reactions occur either in the feed lines or in the boiler proper. Internal treatment may be used alone or in conjunction with external treatment. Its purpose is to properly react with feedwater hardness, condition sludge, scavenge oxygen, and prevent boiler-water foaming.

13-23 What areas of a scotch marine boiler experience the greatest number of failures?

Ans. 1. Tube leakage at tube to tube-sheet rolled joints
2. Tube leakage from metal thinning, wasting of tubes, or bowing of tubes
3. Weld leakage at furnace to tube-sheet welds
4. Weld leakage at tube sheet to shell welds
5. Tube-hole and ligament cracks and leakage
6. Furnace bags, bulges and cracking
7. Furnace bags, rupture and explosion

Item 7 can be a very serious type of failure causing not only extensive property damage but also personal injury and death.

13-24 On a packaged boiler, what would cause fire puffs when it is started up?

Ans. (1) Poor starting draft; (2) incorrect firebox; (3) wrong burner nozzle; (4) lean fire; (5) insufficient gas pilot or excessive gas pilot; and (6) water in the oil. The potential hazard of each is a furnace explosion.

13-25 Does the set pressure of the safety valves or safety relief valves (and not the maximum allowable pressure stamped on the boiler) determine the minimum required working pressure of the trim and valves installed on the boiler? This includes the steam valve, drain valves, feedwater valves, blowoff valves, gauge-glass valves, and all other valves and trim.

Ans. The set pressure of the safety valve or safety relief valve determines the minimum required working pressure of the trim and valves installed on the boiler at or below the maximum allowable working pressure for which the boiler was originally designed and constructed.

13-26 Is more than one blowdown valve required when the maximum allowable working pressure stamped on the boiler exceeds 100 psi, but the safety relief valve is set for a pressure of 100 psi or less?

Ans. When the allowable working pressure of a boiler exceeds 100 psi, the blowoff pipe must be fitted with two slow-opening valves or one slow-opening valve and one quick-opening valve. This requirement does not apply to high-temperature water, traction, and/or portable boilers. When the allowable working pressure of a boiler is 100 psi or less or the safety valve setting (trim) is less than 100 psi, only one blowoff valve is required.

Appendix 1 Glossary of Boiler Terms and Definitions

ABSOLUTE PRESSURE See *pressure.*

ACID CLEANING The process of cleaning the interior surfaces of steam-generating units by filling the unit with a dilute acid accompanied by an inhibitor to prevent corrosion and by subsequently draining, washing, and neutralizing the acid by a further wash of alkaline water.

ACIDITY Represents the amount of free carbon dioxide, mineral acids, and salts (especially sulfates or iron and aluminum) which hydrolize to give hydrogen ions in water; is reported as milliequivalents per liter of acid, or ppm acidity as calcium carbonate, or pH, the measure of hydrogen ion concentration.

AGGLOMERATION Groups of fine dust particles clinging together to form a larger particle.

AIR-ATOMIZING OIL BURNER A burner for firing oil in which the oil is atomized by compressed air which is forced into and through one or more streams of oil, breaking the oil into a fine spray.

AIR/FUEL RATIO The ratio of the weight, or volume, of air to fuel.

AIR HEATER OR AIR PREHEATER Heat-transfer apparatus through which air is passed and heated by a medium of higher temperature, such as the products of combustion or steam.

1. *Regenerative air preheater.* An air heater in which heat is first stored up in the structure itself by the passage of the products of combustion, and which then gives up the heat so stored to the subsequent passage of air.

2. *Recuperative air heater.* An air heater in which the heat from products of combustion passes through a partition which separates the products from the air.

   (a) Tubular air heater. An air heater containing a group of tubular elements through the walls of which heat is transferred from a flowing heating medium to an airstream.

   (b) Plate air heater. An air heater containing passages formed by spaced plates through which heat is transferred from a flowing heating medium to an airstream.

AIR PURGE The removal of undesired matter by replacement with air.

AIR-SWEPT PULVERIZERS A pulverizer through which air flows and from which pulverized fuel is removed by the stream of air.

AIR VENT A valved opening in the top of the highest drum of a boiler or pressure vessel for venting air.

ALKALINITY The amount of carbonates, bicarbonates, hydroxides, and silicates or phosphates in the water; reported as grains per gallon, or parts per million as calcium carbonate.

ALLOWABLE WORKING PRESSURE The maximum pressure for which the boiler was designed and constructed; the maximum gauge pressure on a complete boiler; and the basis for the setting on the pressure-relieving devices protecting the boiler.

ANALYSIS, PROXIMATE Analysis of a solid fuel determining moisture, volatile matter, fixed carbon, and ash; expressed as a percentage of the total weight of the sample.

ANALYSIS, ULTIMATE Chemical analysis of a solid fuel determining moisture, volatile matter, fixed carbon, and ash; expressed as a percentage of the total weight of the sample.

ANTHRACITE ASTM coal classification by rank: Dry fixed carbon 92 percent or more and less than 98 percent; and dry volatile matter 8 percent or less and more than 2 percent on a mineral-matter-free basis.

APPROVED The word *approved* as used in a Code means acceptable to the authority having jurisdiction.

ASH The incombustible inorganic matter in the fuel.

ASH SLUICE A trench or channel used for transporting refuse from ash pits to a disposal point by means of water.

ATOMIZING MEDIA A supplementary medium, such as steam or air, which assists in breaking the fuel oil into a fine spray.

ATTEMPERATOR Apparatus for reducing and controlling the temperature of a superheater vapor or of a fluid. See also *desuperheater*.

1. *Shell-and-tube type.* An attemperator consisting of a pressure vessel containing tubular elements through the walls of which heat is transferred.
2. *Spray type.* An attemperator in which a lower-temperature fluid is injected at relatively high velocity in an atomized state into the superheater vapor to reduce its temperature by direct contact with the atomized fluid.
3. *Submerged type.* An attemperator consisting of tubular elements located in the boiler circulation below the waterline.

AUTHORIZED INSPECTION AGENCY The inspection agency approved by the appropriate legal authority of a state or municipality of the United States or a province of Canada, which has adopted a section of the ASME Code.

AUTOMATIC LIGHTER OR IGNITER A means for starting ignition of fuel without manual intervention. Usually applied to liquid, gaseous, or pulverized fuel. See *igniter.*

AVAILABLE DRAFT The draft which may be utilized to cause the flow of air for combustion or the flow of products of combustion.

BACKING RING A strip of thin plate used on the inner surfaces of the abutting ends of pipe, tubes, or plates which are to be butt-welded. Its purpose is to prevent irregularities at the base of the weld and to permit penetration at its root.

BAG A deep bulge in the bottom of the shell or furnace of a boiler.

BAG FILTER A device containing one or more cloth bags for recovering particles from the dust-laden gas or air which is blown through it.

BALANCED DRAFT The maintenance of a fixed value of draft in a furnace at all combustion rates by control of incoming air and outgoing products of combustion.

BANKING Burning solid fuels on a grate at rates sufficient to maintain ignition only.

BARREL The cylindrical portion of a fire-tube-boiler shell that surrounds the tubes.

BITUMINOUS COAL ASTM coal classification by rank on a mineral-matter-free basis and with bed moisture only.

1. *Low volatile.* Dry fixed carbon 78 percent or more and less than 86 percent; dry volatile matter 22 percent or less and more than 14 percent.
2. *Medium volatile.* Dry fixed carbon 69 percent or more and less than 78 percent; dry volatile matter 22 percent or less and more than 31 percent.
3. *High volatile (A).* Dry fixed carbon less than 69 percent; dry volatile matter more than 31 percent. Btu value equal to or greater than 14,000 moist, mineral-matter-free basis.
4. *High volatile (B).* Btu value 13,000 or more and less than 14,000 moist, mineral-matter-free basis.
5. *High volatile (C).* Btu value 11,000 or more and less than 13,000 moist, mineral-free basis commonly agglomerating, or 8300 to 11,500 Btu agglomerating.

BLACK LIQUOR Liquid by-product fuel extracted from wood in the alkaline pulp-manufacturing process and containing the chemical used to accomplish the extraction.

BLOWBACK The number of pounds per square inch of pressure drop in a boiler from the point where the safety valve pops to the point where the safety valve reseats.

BLOWBACK RING An adjustable ring in a safety valve, used to control the amount of blowback.

BLOWDOWN The drain connection including the pipe and the valve at the lowest practical part of a boiler, or at the normal water level in the case of a surface blowdown. The amount of water that is blown down.

BOILER A closed vessel in which water is heated, steam is generated, steam is superheated, or any combination thereof, under pressure or vacuum by the application of heat from combustible fuels, electricity, or nuclear energy. The term does not include such facilities of an integral part of a continuous processing unit but does include fired units of heating or vaporizing liquids other than water where these units are separate from processing systems and are complete within themselves.

BOILER ASSEMBLER Means a corporation, company, partnership, or individual who assembles a boiler which has been delivered knocked down in multiple pieces by bolting, threading, welding, or other methods of fastening to produce a finished pressure vessel. A boiler assembler may also be a boiler installer.

BOILER, AUTOMATICALLY FIRED A boiler which cycles automatically in response to a control system.

BOILER HEADER (BOX) A pressure part of a boiler consisting of a flat tube sheet into which the ends of the water tubes are rolled. In a parallel plane is a tube cap or handhole sheet. The two sheets are spaced about 4 to 8 in. or more apart. The top and bottom and both ends are flanged together and riveted or may be closed by a narrow flanged strip of plate riveted to each sheet. Circulating nipples connect the top of the header and drum, or the header may be flanged and riveted directly to the drum.

BOILER, HIGH-PRESSURE, STEAM OR VAPOR A boiler in which steam or vapor is generated at a pressure exceeding 15 psig.

BOILER, HOT-WATER-HEATING A boiler in which no steam is generated and from which hot water is circulated for heating purposes and then returned to the boiler.

BOILER, HOT-WATER-SUPPLY A boiler functioning as a water heater.

BOILER, LOW-PRESSURE, STEAM OR VAPOR A boiler in which steam or vapor is generated at a pressure not exceeding 15 psig.

BOILER MANUFACTURER A corporation or company which manufacturers complete pressure parts for boilers or whose shop assembles parts into completed boilers.

BOILING OUT The boiling of a highly alkaline water in boiler pressure parts for the removal of oils, greases, etc. prior to normal operation or after major repairs.

BOURDON TUBE A hollow, metallic tube, bent semicircular, which forms the actuating medium of a pressure gauge.

BREECHING A duct for the transport of the products of combustion between parts of a steam-generating unit or to the stack.

BRIDGEWALL A wall in a furnace over which the products of combustion pass.

BRINELL TEST A hardness test performed by pressing a steel ball of standard hardness into a surface by a standard pressure.

BRITISH THERMAL UNIT The mean British thermal unit (Btu) is $\frac{1}{180}$ of the heat required to raise the temperature of 1 lb of water from 32 to 212°F at a constant atmospheric pressure. It is about equal to the quantity of heat required to raise 1 lb of water 1°F [251.9957 cal or 1054.35 joules (J)].

BUCKSTAY A structural member placed against a furnace or boiler wall to limit the motion of the wall against furnace pressure.

BULGE A local distortion or swelling outward caused by internal pressure on a tube wall or boiler shell due to overheating. Also applied to similar distortion of a cylindrical furnace due to external pressure when overheated provided the distortion is of a degree that can be driven back.

BUNKER C OIL Residual fuel oil (no. 6 fuel oil) of high viscosity commonly used in marine and stationary steam power plants.

BURNER A device for the introduction of fuel and air properly mixed in correct proportions to the combustion zone.

BURNER ASSEMBLY A burner that is factory-built as a single assembly or as two or more subassemblies which include all parts necessary for its normal function when installed as intended.

BURNER, ATMOSPHERIC A gas burner in which all air for combustion is supplied by natural draft, the inspirating force being created by gas velocity.

BURNER, AUTOMATICALLY LIGHTED A burner in which fuel to the main burner is normally turned on and ignited automatically.

BURNER, NATURAL-DRAFT TYPE A burner which depends primarily on the natural draft created in the flue to induce the air required for combustion into the burner.

BURNER, POWER A burner in which all air for combustion is supplied by a power-driven fan that overcomes the resistance through the burner to deliver the quantity of air required for combustion.

BURNER WIND BOX A plenum chamber around a burner in which an air pressure is maintained to ensure proper distribution and discharge of secondary air.

BYPASS TEMPERATURE CONTROL Control of vapor or air temperature by diverting part of or all the heating medium from passing over the heat-absorbing surfaces, usually by means of a bypass damper.

CAKING Property of certain coals to become plastic when heated and form large masses of coke.

CALCIUM A scale-forming element found in some boiler feedwaters.

CALORIE The mean calorie is $\frac{1}{100}$ of the heat required to raise the temperature of 1 g of water from 0 to 100°C at a constant atmospheric pressure. It is about equal to the quantity of heat required to raise 1 g of water 1°C (4.184 J).

CARRYOVER The moisture and entrained solids forming the film of steam bubbles; a result of foaming in a boiler. Carryover is caused by a faulty boiler-water condition. See also *foaming.*

CASING A covering of sheets of metal or other material such as fire-resistant composition board used to enclose all or a portion of a steam-generating unit.

CAULK To make the contacting surfaces of a seam tight against leakage by upsetting or forcing (by distortion) the edge or abutment of the plate into the surface of the adjoining plate. Also, to close any pinhole or fissure in a metal plate, by virtue of the ductility of boiler plate, by distorting its surface to close a slight opening. A blunt tool is used in caulking.

CHAIN GRATE STOKER A stoker which has a moving endless chain as a grate surface, onto which coal is fed directly from a hopper.

CHECKER WORK An arrangement of alternately spaced brick in a furnace with openings through which air or gas flows.

CHECK VALVE A valve designed to prevent reversal of flow. Flow in one direction only is permitted.

CINDER Particles of partially burned fuel from which volatile gases have been driven off, which are carried from the furnace by the products of combustion.

CIRCULATING TUBE A boiler tube used to connect the water spaces of two drums or the pressure parts of a boiler.

CLOSED FEEDWATER HEATER An indirect-contact feedwater heater; that is, one in which the steam and water are separated by tubes or coils.

CLOSING-IN-LINE The sealing by plastic refractory between a boiler shell or head and the firebrick wall; used to prevent hot gases from contacting the boiler above the lowest safe waterline.

COLLOID A finely divided organic substance which tends to inhibit the formation of dense scale and results in the deposition of sludge, or causes it to remain in suspension, so that it may be blown from the boiler.

COMBINED FEEDER CUTOFF A device that regulates makeup water to a boiler in combination with a low-water fuel cutoff.

COMBUSTION Chemical combination of the combustible (that part which will burn) in a fuel with oxygen in the air supplied for the process. Temperatures may range from 1850 to over 3000°F.

COMBUSTION (FLAME) SAFEGUARD A system for sensing the presence or absence of flame and indicating, alarming, or initiating control action.

CONDENSATE Condensed water resulting from the removal of latent heat from steam.

CONDUCTION The transmission of heat through and by means of matter unaccompanied by any obvious motion of the matter.

CONDUCTIVITY The amount of heat (Btu) transmitted in 1 hr through 1 ft^2 of a homogeneous material 1 in. thick for a difference in temperature of 1°F between the two surfaces of the material.

CONTROL A device designed to regulate the fuel, air, water, steam, or electrical supply to the controlled equipment. It may be automatic, semiautomatic, or manual.

CONTROL, LIMIT An automatic safety control responsive to changes in liquid level, pressure, or temperature; normally set beyond the operating range for limiting the operation of the controlled equipment.

CONTROL MANUFACTURER A corporation or company which manufactures operating and safety controls for use on boiler and furnace units.

CONTROL, OPERATING A control, other than a safety control or interlock, to start or regulate input according to demand and to stop or regulate input according to demand and to stop or regulate input on satisfaction of demand. Operating controls may also actuate auxiliary equipment.

CONTROL, PRIMARY SAFETY A control responsive directly to flame properties, sensing the presence of flame and, in event of ignition failure or unintentional flame extinguishment, causing safety shutdown.

CONTROL, SAFETY Automatic controls and interlocks (including relays, switches, and other auxiliary equipment used in conjunction to form a safety control system) which are intended to prevent unsafe operation of the controlled equipment.

CONSTANT IGNITION Usually a gas pilot that remains lighted at full volume whether the main burner is in operation or not.

CONVECTION The transmission of heat by the circulation of a liquid or a gas such as air. Convection may be natural or forced.

CORNER FIRING A method of firing liquid, gaseous, or pulverized fuel in which the burners are located in the corners of the furnace. See also *tangential firing*.

CORROSION The wasting away of metals as a result of chemical action. In a boiler, usually caused by the presence of O_2, CO_2, or an acid.

COURSE A circumferential section of a boiler shell or drum. With usual diameters, the number of courses will equal the number of plates forming the shell or drum.

CRIMPING TOOL A tool used to reduce the diameter of the end of a boiler tube preparatory to its removal from a boiler.

CRITICAL PRESSURE AND CRITICAL TEMPERATURE That point at which the difference between the liquid and vapor states for water completely disappears.

CROSS BOX A boxlike structure to the longitudinal drum of a sectional header boiler for connecting circulating tubes.

CROWFOOT The end of a brace in a boiler, split in two directions for riveting to the plate.

CROWN SHEET The plate forming the roof of an internally fired furnace or a combustion chamber.

DAMPER A device for introducing a variable resistance of regulating the volumetric flow of gas or air.

1. *Butterfly type.* A single-blade damper pivoted about its center.
2. *Curtain type.* A damper consisting of one or more blades, each pivoted about one edge.
3. *Flap type.* A damper consisting of one or more blades, each pivoted about one edge.
4. *Louvre type.* A damper consisting of several blades, each pivoted about its center and linked together for simultaneous operation.
5. *Slide type.* A damper consisting of a single blade which moves substantially normal to the flow.

DEAERATING HEATER A type of feedwater heater operating with water and steam in direct contact. It is designed to heat the water and to drive off oxygen.

DESIGN PRESSURE The pressure used in the design of a boiler for the purpose of determining the minimum permissible thickness or physical characteristics of the different parts of the boiler.

DESUPERHEATER Apparatus for reducing and controlling the temperature of a superheated vapor. See also *attemperator.*

1. *Shell-and-tube type.* A desuperheater consisting of a pressure vessel containing tubular elements through the walls of which heat is transferred.
2. *Spray type.* A desuperheater in which a lower-temperature fluid is injected at relatively high velocity in an atomized state into the superheater vapor to reduce its temperature by direct contact with the atomized fluid.
3. *Submerged type.* A desuperheater consisting of tubular elements located in the boiler circulation below the waterline.

DIAGONAL STAY A brace used in fire-tube boilers between a flathead or tube sheet and the shell.

DISTILLATE OIL Light fraction of oil which has been separated from crude oil by fractional distillation. See *fuel oil.*

DOLLY A riveting tool.

DOWNCOMER A tube or pipe in a boiler or waterwall circulating system through which fluid flows downward between headers.

DOWTHERM An organic chemical with an exceedingly high boiling point, sometimes used in special types of boilers for high-temperature service. It is composed of diphenyl and diphenyloxide.

DRAFT The difference between atmospheric pressure and some lower pressure existing in the furnace or gas passages of the steam-generating unit.

DRAFT CONTROL, BAROMETRIC A device that controls draft by means of a balanced damper which bleeds air into the breeching on changes of pressure to maintain a steady draft.

DRAFT DIFFERENTIAL The difference in static pressure between two points in a system.

DRIFT PIN A tapered steel bar used to drive into and align rivetholes or boltholes in plates or pipe flanges.

DRIP LEG The container placed at a low point in a system of piping to collect condensate and from which it may be removed.

DRUM A cylindrical shell closed at both ends, designed to withstand internal pressure.

DRY BOTTOM FURNACE A pulverized-fuel furnace in which the ash particles are deposited on the furnace bottom in a dry, nonadherent condition.

DRY PIPE A perforated or slotted pipe or box inside the drum and connected to the steam outlet.

DRY STEAM Steam containing no moisture. Commercially dry steam containing not more than 0.5 percent moisture.

DUCTILITY A plastic property of metal to withstand deformation without failure.

DUMP GRATE STOKER One equipped with movable ash trays, or grates, by means of which the ash can be discharged at any desirable interval.

DUTCHMAN A wedge or tapered plug used in butt-and-double-strap longitudinal seams of some boilers to fill the space between the abutting edges of the plate from the end of the inside butt strap to the edge of the adjoining course.

DUTCH OVEN An extended furnace, external to the main setting of a boiler, used to increase the volume of an existing furnace.

ECONOMIZER A series of tubes located in the path of flue gases. Feedwater is pumped through these tubes on its way to the boiler in order to absorb waste heat from the flue gas.

EFFICIENCY *Of boiler operation:* Output in heat units divided by input in heat units. The number of Btu's contained in all steam evaporated is the useful output. The number of Btu's contained in all fuel supplied to the boiler is the input. *Of a riveted seam:* Ratio of the strength of a unit length of a riveted seam to the same unit length of the seamless plate.

EJECTOR A device which utilizes the kinetic energy in a jet of water or other fluid to remove a fluid or fluent material from tanks or hoppers.

ELASTIC LIMIT The maximum tensile load to which a metal may be subjected without becoming permanently deformed upon cessation of the load.

ELECTRIC BOILER A boiler converting electric energy to heat energy.

ELECTRIC FURNACE A furnace used for the refinement of high-grade steel.

ELECTROSTATIC PRECIPITATOR A device for collecting dust, mist, or fume from a gas stream, by placing an electric charge on the particle and removing that particle onto a collecting electrode.

EMBRITTLEMENT An intercrystalline corrosion of boiler plate occurring in highly stressed zones. Cracking may result.

ENTHALPHY A thermal property of a fluid which is a function of state and is defined as the sum of stored mechanical potential energy and internal energy. It is generally expressed in Btu per pound of fluid (joules per kilogram).

ENTRAINMENT The conveying of particles of water or solids from the boiler water by the steam.

EQUALIZING TUBE A boiler tube used to connect the steam spaces of two steam drums, or pressure parts of a boiler.

EROSION The wearing away of refractory or of metal parts by the action of slag or fly ash.

EVAPORATION RATE The number of pounds of water evaporated in a unit of time.

EVAPORATOR A pressure vessel used to evaporate raw water by means of a steam coil. The steam is condensed by means of cooling water coils, and this distilled water is used as makeup boiler feed.

EVAPORATOR CONDENSER That section of an evaporator installation which condenses the vapor.

EXCESS AIR Air supplied for combustion in excess of that theoretically required for complete oxidation.

EXPANDED JOINT The pressuretight joint formed by enlarging a tube end in a tube seat.

EXPLOSION DOOR A door in a furnace or boiler setting designed to be opened by a predetermined gas pressure.

EXPLOSION FIRESIDE Combustion which proceeds so rapidly that a high pressure is generated suddenly.

EXTENDED SURFACE Heating surface in the form of fins, rings, or studs, added to heat-absorbing elements.

EXTERNAL CORROSION A chemical deterioration of the metal on the fireside of boiler heating surfaces.

EXTRACTION FEEDWATER HEATER A closed feedwater heater supplied with steam extracted or bled from a stage of a steam turbine. See also *feedwater heater*.

FACTOR OF SAFETY The ratio between that stress which will cause failure and the working stress. This ratio often applies to pressures instead of stresses.

FAN PERFORMANCE A measure of fan operation in terms of volume, total pressures, static pressures, speed, power input, and mechanical and static efficiency, at a stated air density.

FAN PERFORMANCE CURVES The graphical presentation of total pressure, static pressure, power input, and mechanical and static efficiency as ordinates and the range of volumes as abscissas, all at constant speed and air density.

FATIGUE LIMIT A measure of the ability of a material to withstand repeated stress reversals without fracture or damage to the crystalline structure. A piece of soft iron wire may be broken easily by hand when it is bent back and forth a few times. Its fatigue limit is low. Conversely, a piece of spring steel may be flexed many thousands of times without showing any indication of distress. In this case, the fatigue limit is high. This property is of special value in steam-boiler construction.

FEED THROUGH A trough or pan from which feedwater overflows in the drum.

FEEDWATER HEATER A device used to heat feedwater with steam. See also *extraction feedwater heater.*

FEEDWATER REGULATOR A device for admitting feedwater to a boiler automatically on demand. Practically a constant water level should result.

FERRULE A short, metallic ring rolled into a tube hole to decrease in diameter. Also a short, metallic ring rolled inside of a rolled tube end. Also, a short, metallic ring for making up handhole joints.

FIN Usually a strip of steel welded longitudinally or circumferentially to a tube.

FIN TUBE A tube with one or more fins.

FIRE CRACK A crack starting on the heated side of a tube, shell, or header resulting from excessive temperature stresses.

FIRE-TUBE A tube in a boiler having water on the outside and carrying the products of combustion on the inside.

FIRING RATE CONTROL A pressure or temperature flow controller which controls the firing rate of a burner according to the deviation from pressure or temperature set point. The system may be arranged to operate the burner on-off, high-low, or in proportion to load demand.

FIXED CARBON The carbonaceous residue less the ash remaining in the test container after the volatile matter has been driven off in making the proximate analysis of a solid fuel.

FLAME DETECTOR A device which indicates if fuel, such as liquid, gaseous, or pulverized, is burning or if ignition has been lost. The indication may be transmitted to a signal or to a control system.

FLANGE A circular metal plate threaded or otherwise fastened to an end of a pipe for connection with a companion flange on an adjoining pipe. Also that part of a boiler head (dished or flat) which is fabricated to a shape suitable for riveted or welded attachment to a drum or shell.

FLANGE To fabricate the flange in a head or similar plate.

FLAREBACK A burst of flame from a furnace in a direction opposed to the normal flow, usually caused by the ignition of an accumulation of combustible gases.

FLARED TUBE END The projecting end of a rolled tube which is expanded or rolled to a conical shape.

FLUE GAS The gaseous products of combustion in the flue to the stack.

FLY ASH Suspended ash particles carried in the flue gas.

FOAMING Formation of steam bubbles on the surface of boiler water due to high surface tension of the water. See also *carryover.*

FORCED-DRAFT FAN A fan suppling air under pressure to the fuel burning equipment.

FORGE-WELD The welding together of metals by raising the temperature to the plastic point and by applying pressure or blows.

FREE-BLOW A pipe open and free to blow to atmosphere.

FUEL OIL A petroleum product, requiring comparatively minor refinement, used as a combustible for steam boilers. The following terms are used to describe its properties:

1. *Flash point.* The flash point of a fuel oil is an indication of the maximum temperature at which it can be stored and handled without serious fire hazard.
2. *Pour point.* The pour point is an indication of the lowest temperature at which a fuel oil can be stored and still be capable of flowing under very low forces.
3. *Viscosity.* The viscosity of an oil is a measure of its resistance to flow. In fuel oil it is highly significant since it indicates both the relative ease with which the oil will flow or may be pumped and the ease of atomization. See also *viscosity.*

FUEL-OIL HEATER A tank and coil-type heater using steam as a heating medium to reduce heavy low-priced fuel oil to the proper viscosity for good atomization and combustion. Also an electric-coil heater making use of an electric-resistance coil because the heating medium is used sometimes where steam is not available for starting up a "cold" boiler plant.

FURNACE EXPLOSION A violent combustion of dust or gas accumulations in a furnace or combustion chamber of a boiler.

FURNACE RELEASE RATE Furnace release rate is the heat available per square foot of heat-absorbing surface in the furnace. That surface is the projected area of tubes and extended metallic surfaces on the furnace side including walls, floor, roof, partition walls, and platens and the area of the plane of the furnace exit which is defined as the entrance to the convection tube bank.

FUSIBLE PLUG A brass bushing having a tapered core composed of 99 percent pure tin and a melting temperature of 400 to 500°F and installed at the lowest safe water level of a boiler. The large end of the tapered core is exposed to boiler pressure; the other end is exposed to products of combustion. The core of fusible plug is designed to melt if the boiler water level approaches a dangerously low level. When the core melts, escaping steam will sound warning.

FUSION-WELD To weld the edges or surfaces of metal by raising the temperature to the fusion point and adding a "filler" metal (of the same characteristics as the metal being welded) at the same temperature.

GAG A clamp designed to prevent a safety valve form lifting. Used in applying a hydrostatic test at higher pressure than the safety-valve setting.

GAS RECIRCULATION The reintroduction of part of the combustion gas at a point upstream of the removal point, in the lower furnace for the purpose of controlling steam temperature.

GATE VALVE A stop valve using the wedge-and-double-seat principle. It may be used to control fluids containing some solids, for when wide open, it operates on a straight-through flow. There is little likelihood of its becoming obstructed.

GAUGE GLASS A glass-enclosed visible indicator of the water level in a boiler. Many gauge glasses are tubular, but modern high-pressure practice and railroad locomotives use two thick, flat strips of glass bolted between flanged plates, with the water and steam between the glass strips.

GAUGE PRESSURE The pressure above that of the atmosphere, 14.7 psi at sea level; absolute pressure minus 14.7 at sea level.

GENERATING TUBE A boiler tube used for evaporation.

GIRTH SEAM A roundabout, or circumferential, seam connecting two courses of a boiler shell or drum.

GLOBE VALVE A stop valve using the round-disk-and-seat principle. Used where the fluid controlled is comparatively clean.

GRAINS PER CUBIC FOOT The term for expressing dust loading in weight per unit of gas volume (7000 grains equals 1 lb).

GRATE The surface on which fuel is supported and burned and through which air is passed for combustion.

GRINDABILITY Grindability is the characteristic of coal representing its ease of pulverizing and is one of the factors used in determining the capacity of a pulverizer. The index is relative, with the large values, such as 100, representing coals easy to pulverize such as Pocahontas and smaller values such as 40 representing coals difficult to pulverize.

GROOVED TUBE SEAT A tube seat having one or more shallow grooves into which the tube may be forced by the expander.

GROUND A conducting connection, whether intentional or accidental, between an electric circuit or equipment and either the earth or a conducting body which serves in place of the earth.

GROUNDED Connected to earth or to some conducting body which serves in place of the earth.

GROUNDED CONDUCTOR A system or circuit conductor which is intentionally grounded.

GROUNDING CONDUCTOR, EQUIPMENT The conductor used to connect non-current-carrying metal parts of equipment, raceways, and other enclosures to the system-grounded conductor at the service and/or the grounding electrode conductor.

GUARDED Covered, shielded, fenced, enclosed, or otherwise protected by means of suitable covers, casings, barriers, rails, screens, mats, or platforms to remove the likelihood of contact by persons or objects.

GUN (1) A pneumatic riveter. (2) A gun-type oil burner, of the kind having a long-shaped flame. (3) An injector, in railroad terminology.

HANDHOLE An inspection, a sight, or a cleanout opening in a boiler; often elliptical and closed by a handhole plate.

HAND LANCE A manually manipulated length of pipe carrying air, steam, or water for blowing ash and slag accumulations from heat-absorbing surfaces.

HARDNESS A measure of the amount of calcium and magnesium salts in a boiler water. Usually expressed as grains per gallon or parts per million as Ca CO_2.

HARD PATCH A riveted patch made pressuretight by caulking.

HARD WATER Water which contains calcium or magnesium in an amount which requires an excessive amount of soap to form a lather.

HEADER A distribution pipe supplying a number of smaller lines tapped off of it. A main receiving pipe supplying one or more main pipe lines and receiving a number of supply lines tapped into it. Typical is the boiler *header and superheater header.*

HEATING SURFACE That surface which is exposed to the heating medium for absorption and transfer of heat to the heat medium per American Boiler Manufacturers Association (ABMA) rules as follows:

1. *Boiler and Waterwall Heating Surface.* This surface consists of all the apparatus in contact on one side with the water or wet steam being heated and on the other side with gas or refractory being cooled in which the fluid being heated forms part of the circulating system; this surface is measured on the side receiving heat.

Waterwall heating surface in the furnace, including walls, floor, roof, partition walls, and platens, consisting of bare or covered tubes, is measured as the sum of the projected areas of the tubes and the extended metallic surface on the furnace side.

Continuation of furnace tubes beyond the furnace gas outlet is included as boiler heating surface, and this surface is measured on that portion on the circumferential and the extended metallic surface receiving heat.

All other boiler surfaces, including furnace screen tubes, are measured on that portion of the circumferential and the extended metallic surface receiving heat. The surface is not included in more than one category.

2. *Superheater and Reheater Surface.* This heating surface consists of all the heat-transfer apparatus in contact on one side with steam being heated and on the other side with gas or refractory being cooled; this surface is measured on the side receiving heat.

The radiant superheating or radiant reheating surface in the furnace, including walls, floor, roof, partition walls, and platens, is measured as the sum of the projected areas of the tubes' extended metallic surface on the furnace side.

Continuation of superheater tubes beyond the furnace gas outlet is included as convection superheater surface, and this surface is measured on that portion of the circumferential and the extended metallic surface receiving heat.

All other superheater and reheater surface, including screen tubes, is measured on the basis of the circumferential and the extended metallic surface receiving heat.

HEAT RELEASE The total quantity of thermal energy above a fixed datum introduced into a furnace by the fuel, considered to be the product of the hourly fuel rate and its high heat value, expressed in Btu per hour per cubic foot of furnace volume or square foot of heating surface.

HIGH FIRE The input rate of a burner at or near maximum.

HIGH-HEAT VALUE or HIGHER HEATING VALUE The total heat obtained from the combustion of a specified amount of fuel which is at 60°F before the quantity of heat released is measured.

HOPPER BOTTOM FURNACE A furnace bottom with one or more inclined sides forming a hopper for the collection of ash and for the easy removal of same.

HOT-SHORT Brittle when hot.

HOT WELL A tank used to receive condensate from various sources on its passage back to a boiler through the feedwater system. It usually is vented to atmosphere.

HUDDLING CHAMBER A space provided under the valve disks of many safety valves, permitting the steam pressure in the boiler to act on an increased area when the valve disk lifts, to permit the valve to pop open rather than to rise gradually.

HYDROGEN DAMAGE Temporary reduction in ductility of steel without significant reduction in tensile strength as a result of absorption of hydrogen by the steel

HYDROSTATIC TEST A pressure test by water at room temperature applied to a boiler to determine its safety, as a check on repairs or to trace suspected leakage.

IGNITER A burner smaller than the main burner, which is ignited by a spark or other independent and stable ignition source and which provides proven ignition energy required to immediately light off the main burner.

IGNITION A system in which the fuel to a main burner or gas or oil pilot is ignited directly either by an automatically energized spark or glow coil or by a gas or oil pilot.

IGNITION TEMPERATURE Lowest temperature of a fuel at which combustion becomes self-sustaining.

IMPELLER The rotating wheel of a centrifugal pump.

IMPINGEMENT The striking of moving flame against boiler parts, causing local overheating.

INCOMPLETE COMBUSTION The partial oxidation of the combustible constituents of a fuel.

INHIBITOR A substance which selectively retards a chemical action. An example in boiler work is the use of an inhibitor, when using acid to remove scale, to prevent the acid from attacking the boiler metal.

INJECTOR A device for feeding water into a boiler, making use of the high-velocity-momentum principle to feed water back against boiler pressure by use of steam at the same pressure.

INPUT RATING The fuel-burning capacity of a burner at sea level in Btu per hour (watts) as specified by the manufacturer.

INTEGRAL ECONOMIZER A segregated portion of a watertube boiler in which the feedwater is preheated before its admixture with the circulating boiler water.

INTERBANK SUPERHEATER A superheater located in a space between the tube banks of a bent-tube boiler.

INTERDECK SUPERHEATER A superheater located in a space between tube banks of a straight-tube boiler.

INTERLOCK A device to prove the physical state of a required condition and to furnish that proof to the primary safety control circuit.

INTERMITTENT FIRING A method of firing by which fuel and air are introduced and burned in a furnace for a short period after which the flow is stopped, this succession occurring in a sequence of frequent cycles.

INTERMITTENT IGNITION An igniter which burns during light-off and while the main burner is firing and which is shut off with the main burner.

INTERNALLY FIRED BOILER A fire-tube boiler having an internal furnace such as a scotch, locomotive firebox, vertical tubular, or other type having a water-cooled plate-type furnace.

INTERTUBE ECONOMIZER An economizer, the elements of which are located between tubes of a boiler convection bank.

INTERTUBE SUPERHEATER A superheater, the elements of which are located between tubes of a boiler convection bank.

ION A charge atom or radical which may be positive or negative.

ION EXCHANGE A reversible process by which ions are interchanged between solids and a liquid. These ions exist throughout the solution and act almost independently.

LAGGING Blocks of asbestos or magnesia insulation wrapped on the outside of a boiler shell or steam piping.

LAMINATION As applied to boiler plate, a slag stratum or inclusion rolled into a piece of steel plate during rolling-mill operation.

LAZY BAR A bar fitting across the latches of firing doors of hand-fired boilers; used as a balance and rest for long, heavy firing tools.

LIGAMENT A series of holes in one or more rows.

LINING The material used on the furnace side of a furnace wall. It is usually high-grade refractory tile or brick or plastic refractory material.

LISTED Equipment or materials included in a list published by a nationally recognized testing laboratory that maintains periodic inspections of production of listed equipment or materials. Listing indicates compliance with nationally recognized standards.

LIVE STEAM Steam which has not performed any of the work for which it was generated.

LONGITUDINAL SEAM A riveted or welded seam along the longitudinal axis of a boiler shell or drum.

LOW-HEAT VALUE The high heating value minus the latent heat of vaporization of the water formed by burning the hydrogen in the fuel.

LOW-OIL-TEMPERATURE SWITCH A cold-oil switch; a control to prevent burner operation if the temperature of the oil is too low.

LOW-WATER CUTOFF A device to stop the burner on unsafe water conditions in the boiler.

LUG As applied to boiler suspension, a steel eyepiece fitted and riveted or welded to the curvature of a boiler shell or drum and connected by a steel U bolt or sling rod to overhead steel structure; used to support the weight of the boiler.

MAGNESIUM A scale-forming element found in some boiler feedwaters.

MAKEUP WATER The amount of raw water necessary to compensate for the amount of condensate that is not returned in the feedwater supply to the boiler.

MANHOLE An access opening to the interior of a boiler, elliptical and 11 in. by 15 in. or larger or circular 15-in. diameter or larger.

MANUAL-RESET DEVICE A component of a control which requires resetting by hand to restart the burner after safe operating conditions have been restored.

MECHANICAL-ATOMIZING OIL BURNER A burner which uses the pressure of the oil for atomizing.

MECHANICAL STOKER A device consisting of a mechanically operated fuel-feeding mechanism and a grate; is used for the purpose of feeding solid fuel into a furnace, distributing it over the grate, admitting air to the fuel for the purpose of combustion, and providing a means for removal or discharge of refuse.

MICROMETER One millionth of a meter, or 0.00039 in. (1/25400 in.); formerly called a micron. The diameter of dust particles is often expressed in micrometers.

MILL SCALE An iron oxide scale formed on the surface of a steel plate by cooling and exposing the plate to air just after it has been rolled at high temperatures.

MILL TEST REPORT An affidavit from a steel mill testifying as to the physical and chemical properties of the steel referred to by the report.

MINIATURE BOILER A boiler, the dimensions and working pressure of which do not exceed the following limits: diameter, 16 in.; working pressure, 100 psig; gross volume, 5 ft^3; or heating surface, 20 ft^2.

MUD OR LOWER DRUM A pressure chamber of a drum or header type located at the lower extremity of a watertube-boiler convection bank which is normally provided with a blowoff valve for periodic blowing off of sediment collecting in the bottom of the drum.

MULTIFUEL BURNER A burner by means of which more than one fuel can be burned either separately or simultaneously, such as pulverized fuel, oil, or gas.

MULTIPLE-RETORT STOKER An underfeed stoker consisting of two or more retorts, parallel and adjacent to one another, but separated by a line of tuyeres and arranged so that the refuse is discharged at the ends of the retorts.

NATURAL CIRCULATION The circulation of water in a boiler caused by differences in density; also referred to as thermal or thermally induced circulation.

NFPA National Fire Protection Association.

NIPPLE A short length of pipe or tubing.

NONRETURN TRAP A trap designed to discharge its condensate at atmospheric pressure or at considerably lower pressure than at its inlet.

NOZZLE A short flanged or welded neck connection on a drum or shell for the outlet or inlet of fluids; also a projecting spout for the outlet or inlet of fluids; also a projecting spout through which a fluid flows.

OGEE FLANGE A flange in the form of a reverse curve, used to connect the edges of two concentric shells.

OIL BURNER A burner that atomizes fuel oil and blows it into the combustion chamber in the form of a fine mist or vapor. Steam or mechanical motion plus air may be used as the operating medium.

ONCE-THROUGH BOILER A steam-generating unit usually operated above the critical pressure in which there is no recirculation of the working fluid in any part of the unit. In the case of a supercritical steam generator, there is a constant increase in temperature and enthalphy from inlet to outlet.

OPERATING CONTROL A control to start and stop the burner; it must be in addition to the high-limit control.

ORSAT An instrument for determining the chemical analysis of flue gas.

OXYGEN ATTACK Corrosion or pitting in a boiler caused by oxygen.

PACKAGED STEAM GENERATOR A boiler equipped and shipped complete with fuel-burning equipment, mechanical draft equipment, automatic controls and accessories; usually shipped in one or more major sections.

PALM The end of a brace in a boiler, forged flat or riveted to the shell plate; used to stay flat surfaces.

PATCH A piece of boiler plate used to replace a defective section cut out of a boiler.

PENDANT-TUBE SUPERHEATER An arrangement of heat-absorbing elements which are substantially vertical and suspended from above.

PERFECT OR STOICHOMETRIC COMBUSTION The complete oxidation of all the combustible constituents of a fuel, utilizing all the oxygen supplied.

pH The hydrogen ion concentration of a water to denote acidity or alkalinity. A pH of 7 is neutral. A pH above 7 denotes alkalinity while one below 7 denotes acidity. This pH number is the negative exponent of 10 representing hydrogen ion concentration in grams per liter. For instance, a pH of 7 represents 10^{-7} g/L.

PILOT A small burner which is used to light off the main burner.

PILOT, CONSTANT A pilot that burns without turndown throughout the entire time the boiler is in service.

PILOT FLAME ESTABLISHING PERIOD The length of time fuel is permitted to be delivered to a proved pilot before the flame-sensing device is required to detect pilot flame.

PILOT, PROVED A pilot flame which has been proved by flame-failure controls.

PIT Corrosion localized in a small spot.

PITCH The unit spacing of a series of holes, tube holes, or other holes in a plate.

PLATEN A plane surface receiving heat from both sides and constructed with a width of one tube and a depth of two or more tubes bare or with extended surface.

PLATEN SUPERHEATER A superheater made up of close back-spaced tubes forming plane elements located so as to absorb heat primarily by radiation.

PLENUM An enclosure through which gas or air passes at relatively low velocities.

PORCUPINE BOILER A boiler consisting of a vertical shell from which project a number of dead end tubes.

POSTPURGE A period after the fuel valves close during which the burner motor or fan continues to run, to supply air to the combustion chamber.

POWER-ACTUATED RELIEF VALVE A safety or relief valve, actuated by a separate power source, usually electrical or pneumatic; set to operate slightly below the spring-loaded pressure-actuated valve. This valve is for the express purpose of saving wear and tear on the Code valve and may be installed with an isolating valve to permit maintenance and repair without the necessity of shutting down the boiler. The relieving capacity of this valve is not to be included in the relieving capacity in calculating total Code required.

PREPURGE PERIOD A period on each start-up during which air is introduced into the combustion chamber and associated flue passages in volume and manner as to completely replace the air or fuel-air mixture contained therein prior to an attempt to initiate ignition.

PRESSURE Absolute pressure; the pressure above a perfect vacuum; gauge pressure plus 14.7, at sea level.

PRIMARY AIR Air introduced with the fuel at the burners. In direct-fired systems this may be the same as pulverizer air bypassed around the pulverizer or bled in at the exhauster suction.

PRIMING An induction of boiler water caused by the steam flow into the steam line. The water may be in the form of a spray or a solid body.

PROJECTED GRATE AREA The horizontal projected area of the stoker grate.

PROPORTIONAL CONTROL A mode of control in which there is a continuous linear relation between value of the controller variable and position of the final control element (modulating control).

PULVERIZED FUEL Solid fuel reduced to a fine size, such as 70 percent through a 200-mesh screen.

PULVERIZER A machine which reduces a solid fuel to a fineness suitable for burning in suspension. Types used are:

1. *High speed* (over 800 r/min)
   (a) *Impact pulverizer.* A machine in which the major portion of the reduction in particle size of the fuel to be pulverized is effected by fracture of larger sizes by sudden shock, impingement, or collision of the fuel with rotating members and casing.
   (b) *Attrition pulverizer.* A machine in which the major portion of the reduction on particle size is by abrasion, either by pulverizer parts on coal or by coal on coal.

2. *Medium speed* (between 70 and 300 r/min)
   (a) *Roller pulverizer.* A machine having grinding elements consisting of conical or cylindrical rolls and a bowl, bull-ring mating rings, or table, any of

which may be the rotating member, the fuel to be pulverized being reduced in size by crushing and attrition between the rolls and the rings.
(b) *Ball pulverizer.* A machine in which the grinding elements consist of one or more circular rows of metal balls arranged in suitable raceways, wherein the fuel to be pulverized is reduced in size by crushing and attrition between the balls and raceways.

3. *Low speed* (under 70 r/min)
   (a) *Ball or tube pulverizer.* A machine having a rotating cylindrical or conical casing charged with metal ball or slugs and the fuel to be pulverized, with reduction in particle size being effected by crushing and attrition resulting from continuous relative movement of the charge on rotation of the casing.

RADIANT As applied to heat, having the property that permits heat to be transmitted by rays similar to those of light. To absorb radiant heat, an object must be in the "light" of the fire.

RADIANT SUPERHEATER A superheater exposed to the direct radiant heat (light) of the fire.

RAM A form of plunger used in connection with underfeed stokers to introduce fuel into retorts; a form of pusher.

RATED CAPACITY The manufacturer's stated capacity rating for mechanical equipment, for instance, the maximum continuous capacity in pounds of steam per hour for which a boiler is designed.

RAW WATER Untreated feedwater.

RECYCLE The process of sequencing a normal burner start-up following shutdown.

RED LIQUOR An acid-water mixture of organic material (wood residue) and spent inorganic pulping chemicals, generally associated with a sulfite pulping process in the paper mill industry.

REFRACTORY A heat-insulating material, such as firebrick or plastic fire clay, used for such purposes as lining combustion chambers.

REHEATER A device using highly superheated steam or high-temperature flue gases as a medium serving to restore superheat to partly expanded steam; used often between high- and low-pressure turbines.

RELAY A device that is operative by a variation in the conditions of one electric circuit to start the operation of other devices in the same or another electric circuit (such as pressure or temperature relay).

RETARDER A straight or helical strip inserted in a fire tube primarily to increase the turbulence.

RETRACTABLE BLOWER A soot blower in which the blowing element can be mechanically extended into the boiler and retracted or pulled back.

RETURN TRAP A trap designed to discharge its condensate against boiler pressure and feed to the boiler without additional mechanical equipment.

RINGLEMANN CHART A series of four rectangular grids of black lines of varying widths printed on a white background, used as a criterion of blackness for determining smoke density from chimneys.

RISER TUBE A tube through which steam and water pass from an upper waterwall header to a drum.

ROLLED JOINT A joint made by expanding a tube into a hole by a roller expander.

ROTARY OIL BURNER A burner in which atomization is accomplished by feeding oil to the inside of a rapidly rotating cup.

SAFE-END To replace a deteriorated end of a fire tube by cutting off the end and welding on a short length of new tube.

SAFETY VALVE A valve that automatically opens when pressure attains the valve setting which is adjustable; used to prevent excessive pressure from building up in a boiler.

SAFETY-VALVE DRAIN A hole of at least ⅜-in. diameter required through the body below the valve-seat level in safety valves larger than 2-in. diameter; used to prevent condensate from collecting at this point.

SAFETY-VALVE ESCAPE A pipe conducting steam discharged from a safety valve to a safe location.

SAFETY-VALVE LIFTING LEVER A lever by which a safety valve may be lifted from its seat.

SAFETY-VALVE MUFFLER A silencer designed so that it will not cause appreciable restriction to steam flow.

SAFETY-VALVE NOZZLE A flanged nozzle by which a safety valve is connected to a boiler shell or drum.

SCALE A deposit of medium to extreme hardness occurring on water heating surfaces of a boiler because of an undesirable condition in the boiler water.

SCRUBBER An apparatus for the removal of solids from gases by entrainment in water.

SEAL WELD A weld used primarily to obtain tightness and prevent leakage.

SECONDARY COMBUSTION Combustion which occurs as a result of ignition at a point beyond the furnace. See also *delayed combustion.*

SECONDARY TREATMENT Treatment of boiler feedwater or internal treatment of boiler water after primary treatment.

SEPARATOR A tank-type pressure vessel installed in a steam pipe to collect condensate to be trapped off and thus providing comparatively dry steam to connected machinery.

SHORE SCLEROSCOPE A device to test the hardness of a material, performed by dropping a diamond-pointed hammer from a standard height.

SILICA A scale-forming element found in some boiler feedwaters.

SINGLE-RETORT STOKER An underfeed stoker using only one retort in the assembly of a complete stoker. A single furnace may contain one or more single-retort stokers.

SINUOUS HEADER A header of a sectional header-type boiler in which the sides are curved back and forth to suit the stagger of the boiler tubes connected to the header faces.

SIPHON A pigtail-shaped pipe or a drop leg in the pipe leading to a steam pressure gauge, serving to trap water in the gauge and prevent its overheating from direct contact with steam.

SLAG A residue deposited by ash particles that have attained their softening temperature (1900 to 2700°F) depending on their composition. Slag may be plastic and viscous when hot. It hardens and is rather porous and brittle when cool.

SLAG-TAP FURNACE A pulverized-fuel-fired furnace in which the ash particles are deposited and retained on the floor in molten condition, and from which molten ash is removed by tapping either continuously or intermittently.

SLICER A slicing bar; a long steel bar used for breaking up a fuel bed in coked or caked condition.

SLUG A solid body of boiler water passed into the steam flow by priming or picked up from a pocket of condensate in the steam line.

SMOKE Small gas-borne particles of carbon or soot, less than 1 μm in size, resulting from incomplete combustion of carbonaceous materials and of sufficient number to be observable.

SOFTENING The act of reducing scale-forming calcium and magnesium impurities from water.

SOFT PATCH A patch applied with tap bolts, with a gasket under the patch plate to prevent leakage.

SOOT BLOWER A tube from which jets of steam or compressed air are blown for cleaning the fireside of tubes or other parts of a boiler.

SPALLING The breaking off of the surface refractory material as a result of internal stresses.

SPECIFIC GRAVITY The ratio of the weight of a unit volume of a material to the weight of the same unit volume of water.

SPECIFIC HEAT The quantity of heat, expressed in Btu (joule) required to raise the temperature of 1 lb (kilogram) of a substance 1°F(C).

SPONTANEOUS COMBUSTION Ignition of combustible material following slow oxidation without the application of high temperature from an external source.

SPRAY NOZZLE A nozzle from which a liquid fuel is discharged in the form of a spray.

SPUD A flange nut wrench, open at one end and pointed at the other, as a drift pin. The pointed end is used for aligning boltholes of pipe flanges.

STACK A steel "chimney."

STAY BOLT A stay threaded and riveted over at each end, used to connect two flat or curved pressure parts of a boiler.

STEAM Water vapor produced by evaporation. Dry saturated steam contains no moisture and is at a specific temperature for every pressure; it is colorless. The white appearance of escaping steam is due to condensation at the lowered temperature; it is the water vapor that shows white.

STEAM-AND-WATER DRUM A pressure chamber located at the upper extremity of a boiler circulatory system in which the steam generated in the boiler is separated from the water and from which steam is discharged at a position above a water level maintained there.

STEAM-ATOMIZING OIL BURNER A burner for firing oil which is atmoized by steam. It may be of the inside or outside mixing type.

STEAM BINDING A restriction in circulation due to a steam pocket or a rapid steam formation.

STEAM-GENERATING UNIT A unit to which water, fuel, and air or waste heat are supplied and in which steam is generated. It can consist of a boiler furnace and fuel-burning equipment and may include as component parts waterwalls, superheater, reheater, economizer, air heater, or any combinations.

STEAM QUALITY The percentage by weight of vapor in a steam-and-water mixture.

STRESS The internal resistance of a material to an external force changing, or tending to change, the shape or position of the material. Also *total stress* or *unit stress.*

STRESS-RELIEVE To dissipate pent-up stresses caused by welding, by means of heat treatment.

STRONGBACK A heavy steel bar bolted to tube sheets of fire-tube boilers during construction, while braces are being installed, to prevent the tube sheet from buckling before installation of the tubes.

STUD A projecting pin serving as a support or means of attachment.

STUD TUBE A tube having short studs welded to it.

SUBPUNCH To drive a pilot hole through a plate preparatory to drilling a larger hole.

SUPERHEATED STEAM Steam heated to a temperature higher than that corresponding to the temperature equivalent to the pressure.

SUPERHEATER A series of tubes exposed to high-temperature gases or to radiant heat. Steam from the boiler passes through these tubes to attain a higher temperature than would be possible otherwise. This superheated steam ensures dryness. See also *radiant superheater.*

SUPERHEATER HEADER A large-diameter (about 4- to 8-ft) thick-walled shell or drum into which a row of superheater tubes is rolled.

SURFACE BLOWOFF Removal of water, foam, etc. from the surface at the water level in a boiler; the equipment for such removal.

SURGE The sudden displacement or movement of water in a closed vessel or drum.

SUSPENDED SOLIDS Undissolved solid in boiler water.

SWITCH, AIR-FLOW-PROVING A device installed in an airstream which senses air flow or loss thereof and electrically transmits the resulting impulses to the flame-failure circuit.

SWITCH, HIGH-PRESSURE A device to monitor liquid, steam, or gas pressure and arranged to open and/or close contacts when the pressure value is exceeded.

SWITCH, LOW-PRESSURE A device to monitor liquid, steam, or gas pressure and arranged to open and/or close contacts when pressure drops below the set value.

SWITCH, OIL-TEMPERATURE-LIMIT A device to monitor the temperature of oil between preset limits and arranged to open and/or close contacts should improper oil temperature be detected.

TACK To hold edges of plate in correct position for riveting by a few scattered bolts, known as "tack bolts," placed through rivet holes or by small, scattered spot welds known as "tack welds" or "stitch welds."

TANGENTIAL FIRING A method of firing by which a number of fuel nozzles are located in the furnace walls so that the centerlines of the nozzles are tangential to a horizontal circle. Corner firing is usually included in this type.

TANGENT TUBE WALL or TUBE-TO-TUBE WALL A waterwall in which the tubes are substantially tangent to one another with practically no space between the tubes.

TAP HOLE An opening for the removal of slag from a slag tap furnace.

TELLTALE HOLE A hole drilled into the ends of a stay bolt. The hole extends at least ½ in. inside the inner surface of the stayed sheets; or if the stay bolt is reduced in diameter at its middle portion, the hole extends ½ in. inside the point of diameter reduction. The purpose is to show leakage, through the telltale hole, if the stay bolt breaks or cracks.

TENSILE STRENGTH (ULTIMATE) That stress which causes breaking in tension.

TERTIARY AIR Air for combustion supplied to the furnace to supplement the primary and secondary air.

THEORETICAL AIR The quantity of air required for perfect combustion.

THERM A unit of heat applied especially to gas. One therm equals 100,000 Btu.

THERMAL SLEEVE A spaced internal sleeve lining of a connection for introducing a fluid of one temperature into a vessel containing fluid at a substantially different temperature; used to avoid abnormal stresses.

THERMOSTATIC TRAP A nonreturn trap using a thermostatic expansion and contraction principle as its actuating medium.

THROUGH-STAY A brace used in fire-tube boilers between the heads or tube sheets.

TIE BAR A structural member designed to maintain the spacing of furnace waterwall tubes.

TIE ROD A tension member between buckstays or tie plates.

TILE A preformed, burned refractory, usually applied to shapes other than standard brick.

TIME DELAY A deliberate delay of a predetermined time in the action of a safety device or control.

TITRATION A chemical process used in analyzing feedwater.

TITRATION POINT That point at which a solution changes color when an indicating chemical is introduced drop by drop.

TOTAL STRESS The total resistance of a material to an external force on its entire cross-sectional area in a plane perpendicular to the direction of the force. See also *stress.*

TRAP A device designed to remove condensate from steam automatically, with negligible loss of steam. See *nonreturn trap, return trap,* and *thermostatic trap.*

TRAVELING-GRATE STOKER A stoker similar to a chain-grate stoker with the exception that the grate is separate from but is supported on and driven by chains. Only enough chain strands are used as may be required to support and drive the grate.

TRIAL FOR IGNITION That period of time during which the programming flame-failure controls permit the burner fuel valves to be open before the flame-sensing device is required to detect the flame.

TRIAL FOR MAIN-FLAME IGNITION A timed interval when, with the ignition means proved, the main valve is permitted to remain open. If the main burner is not ignited during this period, the main valve and ignition means are cut off. A safety-switch lockout follows.

TRIAL FOR PILOT IGNITION A timed interval when the pilot valve is held open and an attempt is made to ignite and prove it. If the persence of the pilot is proved at the termination of the interval, the main valve is energized; if not, the pilot and ignition are cut off, followed by a safety lockout.

TRY COCK One of three valves mounted on a boiler or water column within the visible range of the gauge glass and used to check the water level.

TUBE CAP An elliptical or a circular handhole plate used opposite the end of a watertube in a header of a watertube boiler; used for inspection, cleaning, or tube removal.

TUBE RATTLER A vibrating tool designed to be passed through fire tubes to crack scale loose from the tube as a result of vibration.

TUBE SHEET A flat head of a boiler or that part of a boiler drum into which boiler tubes are rolled.

TUBE TURBINE A rotating tool used with water or compressed air pressure, designed to be passed through watertubes to remove scale.

TUBULAR-TYPE COLLECTOR A dust collector utilizing a number of essentially straight-walled cyclone tubes in parallel.

TURBIDITY The optical obstruction to the passing of a ray of light through a body of water, caused by finely divided suspended matter; used to check feedwater.

TURBULENT BURNER A burner in which fuel and air are mixed and discharged into the furnace in such a manner as to produce turbulent flow from the burner.

TUYERES Forms of grates, located adjacent to a retort, through which air is introduced.

UL Underwriters Laboratories, Inc.

UNIT STRESS A value expressed in pounds per square inch and found by dividing the total stress or force by the cross-sectional area stressed. See *stress, total stress.*

UPSET To enlarge or increase the cross-sectional area of any part of a metal by forging it back to a shorter length.

VALVE See *check valve, gate valve, globe valve, safety valve.*

VALVE, MANUAL-RESET SAFETY SHUTOFF A manually opened, electrically latched, electrically operated safety shutoff valve designed to automatically shutoff fuel when de-energized.

VALVE, SAFETY SHUTOFF A valve automatically closed by the safety control system or by an emergency device to completely shut off fuel supply to the burner.

VANE A fixed or adjustable plate inserted in a gas or airstream used to change the direction of flow.

VANE CONTROL A set of movable vanes in the inlet of a fan to provide regulation of airflow.

VANE GUIDE A set of stationary vanes to govern direction, velocity, and distribution of air or gas flow.

VAPOR GENERATOR A container of liquid, other than water, which is vaporized by the absorption of heat.

VENT An opening in a vessel or other enclosed space for the removal of gas or vapor.

VENT VALVE (GAS BURNER) A normally open, power closed valve piped between the two safety shutoff valves, vented to a safe location.

VISCOSITY Measure of the internal friction of a fluid or its resistance to flow.

VORTEX ELIMINATOR Baffles, screens, or plates at the entrance to a large downcomer designed to prevent the formation of a free vortex.

WASHOUT PLUG An inspection, sight, and cleanout opening, circular, threaded, and fitted with a threaded pipe plug, and not to be used for any pipe connection.

WASTE FUEL Any by-product fuel that is waste from a manufacturing process.

WATER COLUMN A vertical, hollow chamber located between a boiler and the gauge glass for the purpose of steadying the water level in the glass through the reservoir capacity of the column. Also, the column may eliminate the obstruction of small-diameter gauge-glass connections by serving as a sediment chamber.

WATER HAMMER A sudden increase in pressure of water due to an instantaneous conversion of momentum to pressure.

WATER LEG That space which is full of boiler water between two parallel plates. It usually forms one or more sides of internally fired furnaces.

WATER SCREEN A screen formed by one or more rows of water tubes spaced above the bottom of a pulverized-fuel furnace.

WATERTUBE A boiler tube through which the fluid under pressure flows. The products of combustion surround the tube.

WATERTUBE BOILER A boiler in which the water or other fluid flows through the tubes and the products of combustion surround the tubes.

WATERWALL A row of watertubes lining a furnace or combustion chamber, exposed to the radiant heat of the fire; used to protect refractory and to increase capacity of the boiler.

WELD To join two edges or surfaces of metal by the application of heat. Also *forged-weld, fusion-weld.*

WELDED WALL A furnace closure wall made up of closely spaced waterwall tubes welded together or to an intermediate fin to form a continuous airtight structure.

WINDBOX A chamber below the grate or surrounding a burner, through which air under pressure is supplied for combustion of the fuel.

WIRE DRAWING A cutting of surfaces caused by the abrasive action of high-velocity flow under restricted outlet.

WRAPPER SHEET The outside plate enclosing the firebox in a firebox or locomotive boiler. Also the thinner sheet in the shell of a two-thickness boiler drum.

YIELD POINT The point at which a metal, under a mounting tensile load, exceeds its elastic limit. At the yield point the metal becomes permanently deformed and will not return to its original shape or position upon cessation of the load.

Appendix 2
Boiler Scale: Data on Water Treatment

Control of the water and/or steam environment inside economizer, boiler, superheater, and reheater tubes is a prerequisite for trouble-free performance of a steam generator. When water-and-steam chemistry is not maintained within limits recommended by the boiler manufacturer or a qualified consultant, corrosion or corrosion-related damage may occur in waterwall and economizer tubes. And overheating damage may occur in these tubes, too, as well as in superheater and reheater tubes, if poor water treatment and improper boiler operation permit deposits to build up in them.

Physically, carbonate scale is usually of low density, having a granular appearance. Often, however, the carbonate crystals are bound with fine particles of other material, with a smooth, uniform appearance resulting. If a piece of carbonate scale is dropped into a solution of acid, carbon dioxide bubbles will come rapidly to the surface.

Silica scale is the hardest type found. It is most often light-colored, very brittle, and dense. It is not soluble in acid.

Sulfate scale is harder and denser than carbonate scale, but not as dense as the silica. It is also more brittle than carbonate scale. If dropped in concentrated sulfuric acid, it will not effervesce; but, if the acid is heated, the scale will dissolve.

Organic scale formed by oil, sewage, trade wastes, or vegetable matter is usually dark-colored, often brown. It is very light in weight and is of low density, and it will often burn if ignited. It is usually soluble in strong nitric acid.

Soda ash is of some value in changing sulfates into an insoluble carbonate sludge that may be removed by blowing down the boiler. The advisability of using soda ash in a boiler operating at over 150-lb pressure is questionable. The soda ash may break down and, with some waters, cause embrittlement. For precipitation of sulfate at higher pressures, it is best to use trisodium phosphate.

A simple method for determining the amount of soda ash necessary for such treatment is now described; it may be of value to those plant engineers who have decided that the use of soda ash is advisable. First, the relative hardness of the water in terms of calcium carbonate ($CaCO_3$) is determined by use of a standard soap solution and the hardness curve (Fig. A-1).

Soap solution is added, 1 cm^3 at a time, to 50 cm^3 of the water in a clear bottle. The bottle is shaken after each addition, until the lather covers the surface of the solution for 5 min, when the bottle is placed on its side. To find the equivalent hardness in parts per million calcium carbonate ($CaCO_3$), locate on the abscissa the amount of soap solution used. Project vertically to the hardness curve and then horizontally to the ordinates for the answer.

Turning to the soda ash curve, Fig. A-1, we locate the hardness point from the ordinates. Dropping down to the soda-ash scale in the abscissa, we may find the approximate number of ounces of soda ash necessary to soften 1000 gal of water. Thus, a water taking 8-cm^3 soap solution to produce sufficient lather is shown by the hardness curve as having a calcium carbonate hardness of 102.5 ppm. Following over to the soda-ash curve and checking on the abscissa, we see that 3.6 oz of soda ash is required for each 1000 gal of water

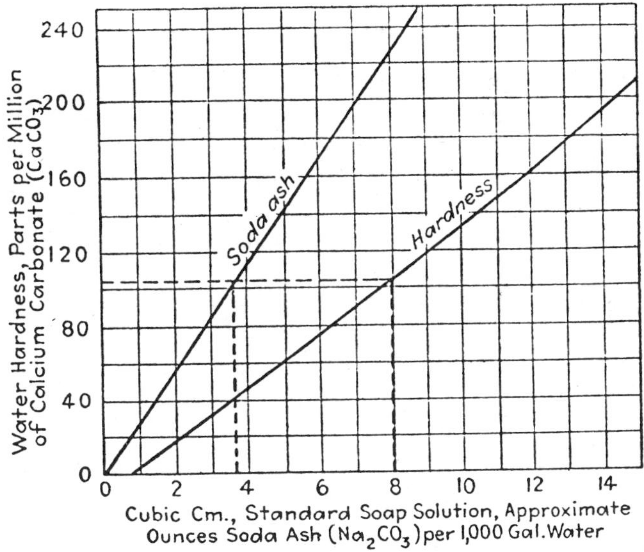

Fig. A-1 Curves for determining hardness and weight of soda ash required to soften water for boilers operating under 200 psi.

Cations	Ion Formula	Ionic Weight	Equivalent Weight
Aluminum	Al^{+++}	27.0	9.0
Ammonium	NH_4^+	18.0	18.0
Calcium	Ca^{++}	40.1	20.0
Hydrogen	H^+	1.0	1.0
Ferrous Iron	Fe^{++}	55.8	27.9
Ferric Iron	Fe^{+++}	55.8	18.6
Magnesium	Mg^{++}	24.3	12.2
Manganese	Mn^{++}	54.9	27.5
Potassium	K^+	39.1	39.1
Sodium	Na^+	23.0	23.0
Anions			
Bicarbonate	HCO_3^-	61.0	61.0
Carbonate	CO_3^{--}	60.0	30.0
Chloride	Cl^-	35.5	35.5
Fluoride	F^-	19.0	19.0
Nitrate	NO_3^-	62.0	62.0
Hydroxide	OH^-	17.0	17.0
Phosphate (tribasic)	PO_4^{---}	95.0	31.7
Phosphate (dibasic)	HPO_4^{--}	96.0	48.0
Phosphate (monobasic)	$H_2PO_4^-$	97.0	97.0
Sulfate	SO_4^{--}	96.1	48.0
Sulfite	SO_3^{--}	80.1	40.0

Compounds	Formula	Molecular Weight	Equivalent Weight
Aluminum hydroxide	$Al(OH)_3$	78.0	26.0
Aluminum sulfate	$Al_2(SO_4)_3$	342.1	57.0
Alumina	Al_2O_3	102.0	17.0
Sodium aluminate	$Na_2Al_2O_4$	164.0	27.3
Calcium bicarbonate	$Ca(HCO_3)_2$	162.1	81.1
Calcium carbonate	$CaCO_3$	100.1	50.1
Calcium chloride	$CaCl_2$	111.0	55.5
Calcium hydroxide (pure)	$Ca(OH)_2$	74.1	37.1
Calcium hydroxide (90%)	$Ca(OH)_2$	——	41.1
Calcium sulfate (anhydrous)	$CaSO_4$	136.2	68.1
Calcium sulfate (gypsum)	$CaSO_4 \cdot 2H_2O$	172.2	86.1
Calcium phosphate	$Ca_3(PO_4)_2$	310.3	51.7
Disodium phosphate	$Na_2HPO_4 \cdot 12H_2O$	358.2	119.4
Disodium phosphate (anhydrous)	Na_2HPO_4	142.0	47.3
Ferric oxide	Fe_2O_3	159.6	26.6
Iron oxide (magnetic)	Fe_3O_4	321.4	——
Ferrous sulfate (copperas)	$FeSO_4 \cdot 7H_2O$	278.0	139.0
Magnesium oxide	MgO	40.3	20.2
Magnesium bicarbonate	$Mg(HCO_3)_2$	146.3	73.2
Magnesium carbonate	$MgCO_3$	84.3	42.2
Magnesium chloride	$MgCl_2$	95.2	47.6
Magnesium hydroxide	$Mg(OH)_2$	58.3	29.2
Magnesium phosphate	$Mg_3(PO_4)_2$	263.0	43.8
Magnesium sulfate	$MgSO_4$	120.4	60.2
Monosodium phosphate	$NaH_2PO_4 \cdot H_2O$	138.1	46.0
Monosodium phosphate (anhydrous)	NaH_2PO_4	120.1	40.0
Metaphosphate	$NaPO_3$	102.0	34.0
Sodium bicarbonate	$NaHCO_3$	84.0	84.0
Sodium carbonate	Na_2CO_3	106.0	53.0

Fig. A-2 Chemical compounds used in water treatment. *(Courtesy Nalco Chemical Co.)*

Chemical	Purpose	Comment
Sodium hydroxide $NaOH$ (caustic soda)	Increase alkalinity, raise pH, precipitate magnesium	Contains no carbonate, so doesn't promote CO_2 formation in steam
Sodium carbonate Na_2CO_3 (soda ash)	Increase alkalinity, raise pH, precipitate calcium sulfate as the carbonate	Lower cost, more easily handled than caustic soda. But some carbonate breaks down to release CO_2 with steam
Sodium phosphates NaH_2PO_4, Na_2HPO_4, Na_3PO_4, $NaPO_3$	Precipitate calcium as hydroxy-apatite $[Ca_{10}(OH)_2(PO_4)_6]$	Alkalinity and resulting pH must be kept high enough for this reaction to take place (pH usually above 10.5)
Sodium aluminate $NaAl_2O_4$	Precipitate calcium, magnesium	Forms a flocculent sludge
Sodium sulfite Na_2SO_3	Prevent oxygen corrosion	Used to neutralize residual oxygen by forming sodium sulfate. At high temperatures and pressures, excess may form H_2S in steam
Hydrazine hydrate $N_2H_4 \cdot H_2O$ (35 % solution)	Prevent oxygen corrosion	Removes residual oxygen to form nitrogen and water. One part of oxygen reacts with three parts of hydrazine (35 % solution)
Filming amines Octadecylamine, etc	Control return-line corrosion by forming a protective film on the metal surfaces	Protects against both oxygen and carbon dioxide attack. Small amounts of continuous feed will maintain the film
Neutral amines Morpholine, cyclohexylamine, benzylamine	Control return-line corrosion by neutralizing CO_2 and adjusting pH of condensate	About 2 ppm of amine is needed for each ppm of carbon dioxide in steam. Keep pH in range of 7.0 to 7.4 or higher
Sodium nitrate $NaNO_3$	Inhibit caustic embrittlement	Used where the water may have embrittling characteristics
Tannins, starches, glucose and lignin derivatives	Prevent feed line deposits, coat scale crystals to produce fluid sludge that won't adhere as readily to boiler heating surfaces	These organics, often called protective colloids, are used with soda ash, phosphate. Also distort scale crystal growth, help inhibit caustic embrittlement
Seaweed derivatives (Sodium alginate, sodium mannuronate)	Provide a more fluid sludge and minimize carryover	Organics often classed as reactive colloids since they react with calcium, magnesium and absorb scale crystals
Antifoams (Polyamides, etc)	Reduce foaming tendency of highly concentrated boiler water	Usually added with other chemicals for scale control and sludge dispersion

Fig. A-3 Chemicals and purpose for using in boiler water treatment. *(Courtesy Power magazine.)*

evaporated. It should be emphasized that this amount is only for the water in condition as tested. The treatment should vary according to changes in condition of the water and load changes.

Figures A-2 and A-3 show some typical chemical compounds and their purpose and use in boiler water treatment—a highly specialized subject on large high-pressure boilers.

Appendix 3
Riveted Joints

Because there are still boilers in service with riveted joints, this appendix is to be referred to whenever a riveted-boiler problem may be encountered. Previous chapters have detailed the method of calculating allowable pressure on shells and drums. The appropriate joint efficiency must be used in the shell equations, and these efficiencies can vary over a broad range on riveted joints.

RIVETED SEAMS

The five most common types of riveted seam are shown in Fig. A-4*a* to *e*. The single-lap-riveted seam is used largely for circumferential seams or for straight seams in stayed plates. It has seldom been used in recent years for the longitudinal seam of a steam boiler, although it was not uncommon during the nineteenth century. Practically all codes in this country prohibit lap-riveted longitudinal seams (either single or double) for boiler drums or shells in excess of 36-in. diameter or for those to be operated at over 100 psi. The major disadvantage of a lap-riveted longitudinal seam is that the plate is "lapped"; thus, the drum or shell does not form a true circle. A certain degree of bending stress may occur along the lap of the seam when the shell is subjected to pressure. This causes a stress concentration that eventually may result in fatigue and cracking of the plate, thus producing a disastrous boiler explosion. About the only "advantages" of the lap-riveted seam are low cost, simplicity, minimum width of space, and minimum total thickness where the seam is exposed to the fires, as in girth seams of some fire-tube boilers.

In the butt-riveted seam, the edges of the shell plate butt together so that the shell or drum is formed into a true circle. A butt strap (strip of boiler plate) is riveted on the inside and outside of the shell or drum along the abutting edges to form the seam. This is called a *butt-and-double-strap joint.* The outer butt strap is usually narrower than the inside strap (Fig. A-4*c* to *e*), and with good reason, for caulking of the edges of the outside butt strap is much more satisfactory when the row of rivets nearest to the caulking edge has a comparatively close pitch. Caulking a plate or strap edge adjacent to a long rivet pitch tends to spring the plate, and it may be difficult to make it tight against leakage. The outer row of rivets passes through the shell and the inside strap only, a slightly stronger joint thus resulting because the greater pitch in the outer rows brings up the overall strength, owing to the greater

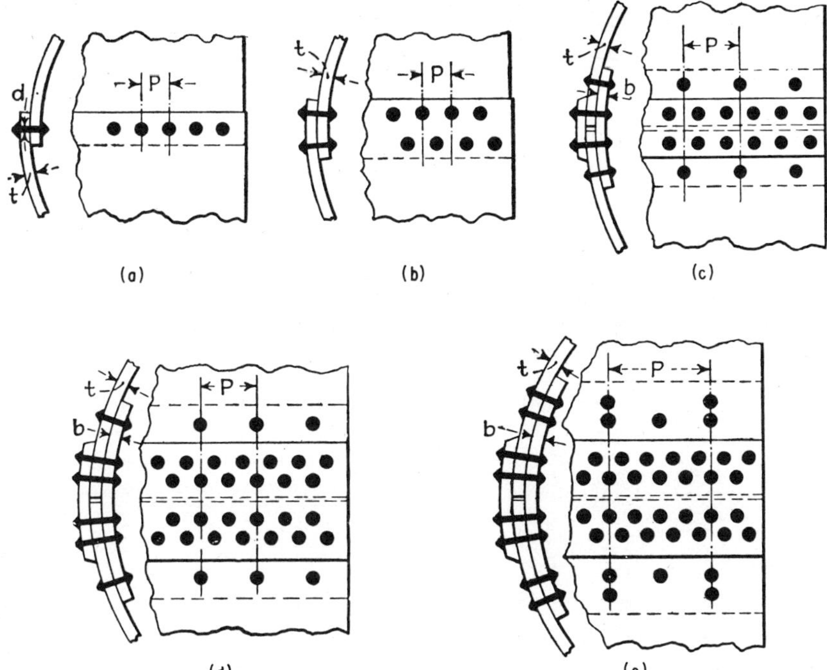

Fig. A-4 Types of riveted joints: *(a)* single-lap, *(b)* double-lap, *(c)* double-strap, double-butt, *(d)* double-strap, triple-butt, *(e)* double-strap, quadruple-butt.

strength of the plate between more widely spaced rivet holes compared with the localized shearing and crushing stresses at each rivet hole in the unit of length selected. This unit is usually equal to the greatest pitch.

Efficiency of Riveted Seams The following terms are used in the formulas for the various types of riveted seam:

TS = tensile strength stamped on the plate
P = pitch of rivets in row having rivets spaced farthest
t = plate thickness, in.
d = diameter of rivet hole, in.
a = area of rivet hole, in.^{2}
s = strength of rivet in single shear (44,000 lb/in.^{2} for low-carbon steel rivets)
S = strength of rivet in double shear (88,000 lb/in.^{2} for low-carbon steel rivets)
n = number of rivets in single shear in section of joint used for calculation *(P)*
N = number of rivets in double shear in section of joint used for calculation *(P)*
c = crushing strength of boiler plate (95,000 lb/in.^{2} for flange and firebox steel, 120,000 lb/in.^{2} for alloy steel)

The net section is the plate between the P rivets, area of rivet holes being deducted.

SINGLE-LAP-RIVETED SEAMS

See Fig. A-4*a*.

1. $P \times t \times TS =$ strength of solid plate
2. $(P - d) \times t \times TS =$ strength of net section
3. $n \times s \times a =$ strength of one rivet in single shear $(n = 1)$
4. $n \times t \times d \times c =$ crushing strength of plate in front of one rivet $(n = 1)$

The strength of the solid plate, step 1, is considered 100 percent. Then, the weakest value as found by steps 2, 3, or 4 is used in the following ratio, with X being the efficiency of the seam in percentage, based on the weakest part: 100 is to X as the strength of solid plate is to the strength of weakest part, or

$$\frac{100}{X} = \frac{\text{strength of solid plate}}{\text{strength of weakest part}}$$

Transposing, we get

$$X = \frac{100 \times \text{strength of weakest part}}{\text{strength of solid plate}}$$

This equation is used for finding the efficiency of any type of riveted seam.

Example:

$P = 2\frac{1}{2}$ in.	$t = \frac{1}{2}$ in.	$d = \frac{15}{16}$ in.	$TS = 55{,}000$ lb/in.^{2}
$a = 0.690$	$s = 44{,}000$	$c = 95{,}000$ lb/in.^{2}	

Find the efficiency of a single-lap-riveted seam of the foregoing specifications.

1. $P \times t \times TS = 2.5 \times 0.5 \times 55{,}000 = 68{,}750$
2. $(P - d) \times t \times TS = (2.5 - 0.9375) \times 0.5 \times 55{,}000 = 42{,}968$
3. $n \times s \times a = 1 \times 44{,}000 \times 0.690 = 30{,}360$
4. $n \times d \times t \times c = 1 \times 0.9375 \times 0.5 \times 95{,}000 = 44{,}531$

Since step 3 is the weakest in this case,

$$X = \frac{100 \times 30{,}360}{68{,}750} = 44.2\%$$

DOUBLE-LAP-RIVETED SEAMS (CIRCUMFERENTIAL OR LONGITUDINAL)

The method of calculation and the sequence of steps in determining the efficiency of the double-lap-riveted seam (see Fig. A-4*b*) are exactly the same as for the single-lap-riveted seam. The only variation is in step 3, where n is 2 instead of 1, for the number of rivets in single shear; and in step 4, where n is 2 instead of 1, for the number of rivets tending to crush the plate.

Example: A longitudinal seam is double-lap-riveted and has 3-in. pitch, ⅜-in. plate having a TS of 55,000 lb/in.^{2} and $^{13}\!/_{16}$-in.-diameter rivet holes; $s = 44{,}000$, and $c = 95{,}000$. Find the efficiency of this seam.

$$TS = 55{,}000 \qquad P = 3 \qquad t = 3/8 \qquad d = 0.8125 \qquad a = 0.519$$

1. $P \times t \times TS = 3 \times 0.375 \times 55{,}000 = 61{,}875$
2. $(P - d) \times t \times TS = 2.1875 \times 0.375 \times 55{,}000 = 45{,}117$
3. $n \times s \times a = 2 \times 44{,}000 \times 0.519 = 45{,}672$
4. $n \times d \times t \times c = 2 \times 0.8125 \times 0.375 \times 95{,}000 = 57{,}891$

Since step 2 is the weakest,

$$X = \frac{45{,}117 \times 100}{61{,}875} = 72.9\%$$

BUTT JOINTS WITH STRAPS OF EQUAL WIDTH (CHAIN-RIVETED)

The calculation of this type of seam is carried out as for a lap joint, except that the rivets are in double shear and S is used as 88,000, instead of $s =$ 44,000. The efficiencies are often higher than for lap seams, and there are no restrictions for the use of these seams as long as they are not exposed to high-temperature gases.

BUTT AND DOUBLE STRAP (UNEQUAL WIDTH), DOUBLE-RIVETED

This joint (Fig. A-4*c*) usually has a somewhat higher efficiency than a lap-riveted seam. Here one finds several additional steps in calculating the seam efficiency, for there are more possible combinations of failure of the joint.

1. Strength of the section of solid plate equal to P in length:

$$P \times t \times TS$$

2. Strength of the net section of plate, rivet holes in the outer row being deducted:

$$(P - d) \times t \times TS$$

3. Strength of all rivets in shear, two double and one single shear:

$$N \times S \times a + (n \times s \times a)$$

4. Strength of the net section of plate in the inner row, plus strength of outer rivet in single shear:

$$(P - 2d) \times t \times TS + (n \times s \times a)$$

5. Strength of the net section of plate in the inner row, plus crushing strength of the strap in front of the outer rivet:

$$(P - 2d) \times t \times TS + (n \times b \times d \times c)$$

Note: The strap is considered in crushing, rather than the plate, for the strap is not so thick.

6. Crushing strength of the plate in front of the inner row rivets plus the crushing strength of the strap in front of the outer rivet. Note: The plate is considered in crushing at the inner row, rather than the straps, for the combined thickness of two straps exists at this point.

$$N \times d \times t \times c + (n \times b \times d \times c)$$

7. Crushing strength of the plate in front of the inner row rivets plus the shearing strength of the outer rivet:

$$n \times d \times t \times c + (n \times s \times a)$$

Example: A longitudinal seam is butt and double strap, double-riveted, and has 5 in. by 2½ in. pitch, 7⁄16-in.-thick shell plate, and 55,000-lb/in.² *TS;* butt straps are each 5⁄16 in. thick, and rivet holes are ¾ in. in diameter. Find the efficiency.

1. $5 \times 0.4375 \times 55{,}000 = 120{,}313$
2. $(5 - 0.75) \times 0.4375 \times 55{,}000 = 102{,}266$
3. $2 \times 88{,}000 \times 0.442 + (1 \times 44{,}000 \times 0.442) = 97{,}240$
4. $(5 - 1.5) \times 0.4375 \times 55{,}000 + (1 \times 44{,}000 \times 0.442) = 103{,}667$
5. $(5 - 1.5) \times 0.4375 \times 55{,}000 + (1 \times 0.75 \times 0.3125 \times 95{,}000) = 106{,}485$
6. $2 \times 0.75 \times 0.4375 \times 95{,}000 + (1 \times 0.75 \times 0.3125 \times 95{,}000) = 84{,}610$
7. $2 \times 0.75 \times 0.4375 \times 95{,}000 + (1 \times 44{,}000 \times 0.442) = 81{,}792$

$$X = \frac{81{,}792 \times 100}{120{,}313} = 68.0\%$$

BUTT AND DOUBLE STRAP (STRAPS OF UNEQUAL WIDTH) TRIPLE-RIVETED

The efficiency of this seam (see Fig. A-4*d*) is found exactly as is that of the double-riveted seam in the last example. The only difference in the calculations is that there are a larger number of rivet holes in the unit length *P*.

Example: A triple-riveted butt-and-double-strap seam has a pitch of 7 in. by 3½ in. The plate is ½ in. thick and has a *TS* of 55,000 lb/in.². The rivet-hole diameter is 15⁄16 in., and the strap thickness is ⅜ in. Find the efficiency.

1. Strength of unit section of solid plate equal in length to *P*:

$$7 \times 0.5 \times 55{,}000 = 192{,}500$$

2. Strength of net section of plate between rivet holes of outer row:

$$(7 - 0.9375) \times 0.5 \times 55{,}000 = 166{,}719$$

3. Strength of all rivets in shear, four double and one single:

$$4 \times 88{,}000 \times 0.690 + (1 \times 44{,}000 \times 0.690) = 273{,}240$$

4. Strength of net section of plate, between rivet holes of second row plus shearing strength of outer rivet (single shear):

$$(7 - 2 \times 0.9375) \times 0.5 \times 55{,}000 + (1 \times 44{,}000 \times 0.690) = 171{,}298$$

5. Strength of net section of plate between rivet holes of second row plus crushing strength of strap in front of outer rivet:

$$(7 - 2 \times 0.9375) \times 0.5 \times 55{,}000 + (1 \times 0.9375 \times 0.375 \times 95{,}000) = 174{,}336$$

6. Crushing strength of plate or strap in front of all rivets:

$$4 \times 0.9375 \times 0.5 \times 95{,}000 + (1 \times 0.9375 \times 0.375 \times 95{,}000) = 211{,}523$$

7. Crushing strength of plate in front of inner two rows of rivets plus shearing strength of outer rivet:

$$4 \times 0.9375 \times 0.5 \times 95{,}000 + (1 \times 44{,}000 \times 0.690) = 208{,}485$$

$$X = \frac{166{,}719 \times 100}{192{,}500} = 86.6\%$$

BUTT AND DOUBLE STRAP (STRAPS OF UNEQUAL WIDTH), QUADRUPLE-RIVETED

Two additional steps are necessary in calculating the efficiency of this type of joint (see Fig. A-4*e*), which, if properly designed, has an efficiency of over 90 percent.

Example: A quadruple-riveted butt-and-double-strap joint has a rivet pitch of 15 in. in the outer row. The plate is ⅝ in. thick and has a *TS* of 55,000 lb/in.^{2}. The rivet-hole diameter is 1 in., and the butt-strap thickness is 7⁄16 in. What is the efficiency?

1. Strength of unit section of solid plate equal in length to *P*:

$$15 \times 0.625 \times 55{,}000 = 515{,}625$$

2. Strength of net section of plate between rivet holes of outer row:

$$(15 - 1) \times 0.625 \times 55{,}000 = 481{,}250$$

3. Strength of all rivets in shear, eight double and three single:

$$8 \times 88{,}000 \times 0.7854 + (3 \times 44{,}000 \times 0.7854) = 656{,}595$$

4. Strength of net section of plate between rivet holes of second row, plus shearing strength of outer rivet (single shear):

$$(15 - 2 \times 1) \times 0.625 \times 55{,}000 + (1 \times 44{,}000 \times 0.7854) = 481{,}433$$

5. Strength of net section of plate between rivet holes of third row, plus shearing strength (single) of two rivets in second and one in outer row:

$$(15 - 4 \times 1) \times 0.625 \times 55{,}000 + (3 \times 44{,}000 \times 0.7854) = 481{,}798$$

6. Strength of net section of plate between rivet holes of second row, plus crushing strength of strap in front of outer rivet:

$$(15 - 2 \times 1) \times 0.625 \times 55{,}000 + (1 \times 1 \times 0.4375 \times 95{,}000) = 488{,}538$$

7. Strength of net section of plate between rivet holes in the third row, plus crushing strength of the strap in front of two rivets in the second and one in the outer row:

$$(15 - 4 \times 1) \times 0.625 \times 55{,}000 + (3 \times 1 \times 0.4375 \times 95{,}000) = 502{,}813$$

8. Crushing strength of the plate or strap in front of all rivets:

$$\underset{(8 \times 1 \times 0.625 \times 95{,}000)}{\text{Plate strength}} + \underset{(3 \times 1 \times 0.4375 \times 95{,}000)}{\text{strap strength}} = 599{,}688$$

9. Crushing strength of the plate in front of the third- and fourth-row rivets, plus the shearing (single) strength of the three rivets in the outer two rows:

$$8 \times 1 \times 0.625 \times 95{,}000 + (3 \times 44{,}000 \times 0.7854) = 578{,}673$$

$$X = \frac{481{,}250 \times 100}{515{,}625} = 93.3\%$$

PREPARATION OF PLATE FOR RIVETED SEAMS

Preparation of the boiler plate for a riveted seam is important. In order to facilitate proper caulking, the edges of the boiler plate, heads, and straps are beveled to an angle of not less than 70° to the plane of the plate. The edges are then planed, milled, or chipped back to a depth of not less than one-fourth the thickness, but in no case less than ⅛ in., the latter requirement ensuring that the edges of the plate which are distorted and stressed by the shearing process will be removed.

Before a riveted seam can be made tight against leakage, the exposed edges of the plate must be caulked. A blunt-nosed, chisellike caulking tool is used to upset the edge of the exposed plate slightly against its companion plate, making a tight metal-to-metal seam.

Rivet holes are often drilled, but they may be punched to within ⅛ in. of full diameter for plate not over 5⁄16 in. thick or to within ¼ in. of full diameter for plate exceeding 5⁄16-in. thickness. However, punching of holes was not allowed by the ASME Code if the plate thickness exceeds 5⁄8 in. This requirement is to ensure that residual stresses will not be set up in the plate surrounding the rivet holes. After the plates and straps have the rivet holes punched, they are held together in position by tack bolts, and the holes are drilled or reamed to full size.

Whenever two or more plates are tacked together for drilling or reaming holes, it is important that the plates be separated for removal of all chips or burrs before final riveting.

The location of the rivet holes was specified thoroughly by the ASME Power Boiler Code which used to read as follows: "On longitudinal joints of all types of boilers and on circumferential joints of drums having heads which are not supported by tubes or through stays, the distance from the centers of rivet holes to the edges of plates, except rivet holes in the ends of butt straps, shall not be less than 1½ and not more than 1¾ times the diameter of the rivet holes, this distance to be measured from the center of the rivet holes to the calking edge of the plate before calking."

The diameter of a rivet hole is usually 7⁄16 in. plus the thickness of the plate in *sixteenths* of an inch. Thus a 3⁄8-in. plate would be likely to have 13⁄16-in. rivet holes. The rivet diameter, before driving, is 1⁄16 in. less than that of the rivet holes, except where machined rivets are used. Thus, a difference of 1⁄32 in. is customary. The length of the rivet should be sufficient to allow the rivet body to fill the hole completely and to form a head at least as strong as the body.

The usual procedure in riveting is to use tapered pins (drift or barrel pins) to line up the holes. The holes should line up closely enough that undue sledging will not be necessary. Then tack bolts are placed in scattered locations along the seam to hold the plates together. A rivet is driven on each side of a tack bolt before the bolt is removed. Though pneumatic riveting machines must be resorted to where construction makes it impossible to use other means, the bull riveter (a hydraulically operated unit) is used in most cases and is preferable, for it can maintain full pressure on the rivet until cooled below red heat.

Appendix 4 Boiler Safety Rules

Each of these rules is based on past accidents.

NEVER

NEVER fail to anticipate emergencies. Do not wait until something happens to start thinking.

NEVER start work in a strange plant without tracing every pipe line and learning the location and purpose of every valve. Know your job.

NEVER leave an open blowdown valve unattended when a boiler is under pressure or has a fire in it. Play safe; memory can fail.

NEVER allow sediment to accumulate in gauge-glass or water-column connections. A false water level may fool you and make you sorry.

NEVER give verbal orders for important operations or report such operations verbally with no record. Have something to back you up when needed.

NEVER light a fire under a boiler without a double check on the water level. Many boilers have been ruined and many jobs lost this way.

NEVER light a fire under a boiler without checking all valves. Why take a chance?

NEVER open a valve under pressure quickly. The sudden change in pressure, or resulting water hammer, may cause piping failure.

NEVER cut a boiler in on the line unless its pressure is within a few pounds of header pressure. Sudden stressing of a boiler under pressure is dangerous.

ALWAYS

ALWAYS study every conceivable emergency and know exactly what moves to make.

ALWAYS proceed to proper valves or switches rapidly but without confusion in time of emergency. You can think better walking than running.

ALWAYS check the water level in the gauge glass with the gauge cocks at least daily and also at any other time you doubt the accuracy of the glass indication.

ALWAYS blow out each gauge-glass and water-column connection at least once each day. Forming good habits may mean longer life for you.

ALWAYS accompany orders for important operations with a written memorandum. Use a log book to record every important fact or unusual occurrence.

ALWAYS have at least one gauge of water before lighting off. The level should be checked by the gauge cocks. You will not be fired for being too careful.

ALWAYS be sure blowdown valves are closed and proper vents, water-column valves, and pressure-gauge cock are open.

ALWAYS use the bypass if one is provided. Crack the valve from its seat slightly, and await pressure equalization. Then open it slowly.

ALWAYS watch the steam gauge closely and be prepared to cut the boiler in, opening the stop valve only when the pressures are nearly equal.

NEVER

NEVER bring a boiler up to pressure without trying the safety valve. A boiler with its safety valve stuck is nearly as safe as playing with dynamite.

NEVER take it for granted that the safety valves are in proper condition. The power plant is no place for guesswork.

NEVER increase the setting of a safety valve without authority. Serious accidents have occurred from failure to observe this rule.

NEVER change adjustments of a safety valve more than 10 percent. Proper operation depends on the proper spring.

NEVER tighten a nut, bolt, or pipe thread under steam or air pressure. Many have died doing this.

NEVER strike any object under steam or air pressure. This is another sure path to the undertaker's.

NEVER allow unauthorized persons to tamper with any steam-plant equipment. If they do not injure themselves, they may cause injury to you.

NEVER allow anyone to enter a boiler without proper protecting signs. Do not remove signs until you have personally checked that everyone is clear.

NEVER allow major repairs to a boiler without authorization. If you do not break a law, you may break your neck.

NEVER light a burner without a torch. You cannot dodge a furnace blast.

NEVER attempt to light a burner without venting the furnace until clear. Burns are painful.

NEVER fail to report unusual behavior of a boiler or other equipment. It may be a warning of danger.

ALWAYS

ALWAYS lift the valve from its seat by the hand lever when the pressure reaches about three-quarters of popping pressure.

ALWAYS raise the valve from its seat with the lifting lever periodically while the boiler is under pressure. Test by raising to popping pressure at least once per year.

ALWAYS consult an authorized boiler inspector and accept his or her recommendations before increasing the safety-valve pressure setting.

ALWAYS have the valve fitted with a new spring and restamped by the manufacturer for changes over 10 percent.

ALWAYS play safe on this rule. The one that is going to break does not have a special warning sign.

ALWAYS play safe on this rule. You cannot tell which straw might break the camel's back.

ALWAYS keep out loiterers and place plant operation in the hands of proper persons. A boiler room is not a place for a club meeting.

ALWAYS put a sign "Worker Inside" on a boiler at the point where the person enters. Lock all valves closed that might endanger the person if they were opened accidentally. Station a person outside for emergencies.

ALWAYS consult an authorized boiler inspector before proceeding with boiler repairs.

ALWAYS assume delayed ignition is going to cause a furnace explosion. Use proper ignition methods.

ALWAYS allow draft to clear furnace of gas and dust for prescribed purge period. Change draft conditions slowly.

ALWAYS consult someone in authority. Two heads are better than one.

Selected Bibliography

ASME Boiler and Pressure Vessel Codes, Sections I through VI, IX, and XI, American Society of Mechanical Engineers, New York.

Elonka, S. M.: *Standard Plant Operator's Manual,* McGraw-Hill Book Company, New York, 1975.

———, and A. L. Kohan: *Standard Boiler Operators' Questions and Answers,* McGraw-Hill Book Company, New York, 1969.

Fundamentals of Welding, American Welding Society, Miami, Fla.

National Board Inspection Code, National Board of Boiler and Pressure Vessel Inspectors, Columbus, Ohio.

National Fire Protection Codes, National Fire Protection Association, Boston.

Power Piping Code, ANSI B31.1, American National Standards Institute, New York, 1977.

Recommended Practices for NDT Personnel Qualifications and Certification, American Society for Nondestructive Testing, Evanston, Ill.

State, County, and City Synopsis of Boiler and Pressure Vessel Laws on Design, Installation, and Reinspection Requirements, Uniform Boiler and Pressure Vessel Laws Society, Hartford, Conn.

Index

About the Authors

Harry M. Spring, Jr. (deceased) authored the first edition of this book, published in 1940. The present edition was revised, brought up-to-date, and expanded by Mr. Kohan.

Anthony Lawrence Kohan is Manager, Boiler and Machinery Technical Specialists, Boiler and Machinery Department, Royal Insurance Companies. He has more than 30 years' experience as a power plant technician, tester, and insurance company inspector and supervisor of inspectors for boilers, pressure vessels, and machinery. He has written articles for *Power* magazine, and he is the coauthor (with Steve Elonka) of *Standard Boiler Operators' Questions and Answers* (McGraw-Hill).